Donal .Elias

D1498079

To my friends of MMI

John W. ~

4-29-91

The Chemistry and Uses
of Fire Retardants

THE CHEMISTRY AND USES OF FIRE RETARDANTS

John W. Lyons

WILEY-INTERSCIENCE

A Division of John Wiley and Sons Inc.
New York · London · Sydney · Toronto

Library of Congress Catalogue Card Number: 71-112595

ISBN 0-471-55740-4

Printed in the United States of America

10 9 8 7 6 5 4 3 2 1

Dedicated to
GRACE and the CHILDREN

PREFACE

Research into what makes things burn and how combustion processes may be altered has been conducted for many years. In Chapter 5 the history of attempts to improve the fire safety of fabrics is traced back to the early seventeenth century. At the turn of the twentieth century research was under way in the laboratories of many renowned chemists, including William Henry Perkin. What is new at mid-century is the breadth and depth of the concern on the part of many nontechnical as well as technical persons about the condition of our environment—the impact on it of man and the subsequent deterioration of many of its more vital aspects. Thus we now find increasing opposition to continued pollution of the air, the water, and, indeed, the ground. Congress is passing legislation to improve the environment, to restrict the most harmful of our activities, and to attempt to reverse the trend of deterioration.

In this context safety has become more important than ever. Automobiles are now required to have safety devices, legislation designed to eliminate certain harmful residues from pesticides has been passed, and pharmaceuticals are coming under increasing regulation and scrutiny. The burden of proof of the safety of many products is being firmly placed on their manufacturers. The fire-safety attributes of materials for many different uses are now the subject of this same overall concern.

Fire hazards have always been with us, but for the most part people have become inured to them and have lived and worked in highly combustible surroundings with scarcely a thought for safety. However, as people crowd into the metropolitan areas city officials find that they must be more concerned with fire hazards. As our national forests dwindle in size foresters become more protective of the remainder. As more everyday articles are made from synthetic polymers, the first of which were often more flammable than the natural materials they displaced, means of controlling their flammability are sought. This trend culminated in the mid-1960s with new federal laws that define the national interest in fire safety, set forth some guidelines, and establish the means of moving ahead with better products. These laws were concerned principally with the flammability of certain fabrics, aircraft interiors, and, to a lesser extent, the interiors of other means of transportation. Funds for stepped-up research into fire-safety questions were also allocated. So efforts to alter and control the behavior of materials in combustion have been greatly increased in the 1960s. This book is an attempt to review where we have been and to set the stage for the new surge of technical activity just getting under way.

A word about what this book is and is not. The original search of the literature made for me by Michael Dub of Monsanto's Information Center, was directed primarily toward the applied aspect of fire safety. We sought out all papers and patents on fire-safe materials. All references to fire retardants and fire retardancy, flame retardants, and similar terms were retrieved from *Chemical Abstracts* from Volume 1 on. Various federal technical reports, patent search tapes, and bibliographies published elsewhere were reviewed with the same objective. We believe the final number of references in our files was close to six thousand: there were too many to count. Some of them were culled as being trivial, too early, and not of real historical interest; or untrustworthy, suspect, or otherwise in doubt. The final result was nonetheless close to or perhaps in excess of four thousand citations, many in multiple references or in tables with no reference numbers attached. The magnitude of the task appeared to us only after we were well along and found that the few other reviews in print—in bibliographical form—contained only four hundred to five hundred items each. We believe that this survey of the applied literature on fire retardants and their uses is as complete as it is humanly possible to make it.

The intent of the publisher of this book was, originally, to issue a companion volume on the mechanisms of thermal degradation of polymers and the effect of fire retardants on them. The present book was written on the assumption that such a volume would be issued nearly simultaneously. Such has not occurred as yet and its prospects are uncertain. Therefore the current void in critical reviews of basic studies will remain unfilled for a time.

There is a definite need for someone to amplify and update Madorsky's landmark volume on thermal degradation, now almost six years old. The present book refers in several places to Madorsky's work, and I have tried, with the introduction of each family of polymers, to touch at least lightly on what is known about their thermal degradation. Much more research must clearly be done in this area, especially with an oxidizing atmosphere. In any event, the mechanistic aspect of the problem was not covered in detail by design and deliberately. I leave it to others or to another day to complete that task.

I am indebted to many people at Monsanto for making this book possible: to Mike Dub for his painstaking search of the literature, to Leo Weaver, Clayton F. Callis, and Barton MacDonald for their encouragement, and to E. A. Matzner, H. L. Vandersall, M. M. Crutchfield, C. E. Miles, W. W. Morgenthaler, C. W. Heitsch, R. S. Mitchell, A. R. Handleman, G. W. Ingle, C. C. Kemp, G. H. Birum, and many others for helpful comments or for references and suggestions. I am indebted also to many workers outside Monsanto who read and criticized individual chapters. The book would never have gone to press without the yeoman service of Charolyn Manley and Kathleen Pierson who prepared the final form of most of the manuscript and who were most helpful in setting up the tables and figures and keeping track of innumerable changes in the references. Many others also assisted with the typing and contributed to the final result. Finally, I thank my family for tolerating this project for the last three years. They not only gave up weekend and evening activities so that I might write but they also pitched in with the sorting and writing of reference lists. They will be glad to see the McBee cards and stacks of articles gone from the house at last. To all of them and many more whom I have undoubtedly overlooked I extend my thanks for invaluable help and encouragement.

JOHN W. LYONS

Webster Groves, Missouri
May 1970

CONTENTS

The Chemistry and Uses
of Fire Retardants

Chapter 1

INTRODUCTION: AN OVERVIEW

This chapter reviews briefly certain aspects of the theory of how fire retardants work, fire testing, and new trends in fire-safe materials. It also anticipates the results of Chapters 4–8 and offers some broader insights than perhaps can be gained from reading later, more detailed sections. Comparisons, contrasts, and rough generalizations are offered in the hope of providing some guidance to the reader for his particular problem. The serious student should pass on to the subsequent chapter or section in which his subject is amplified and then, of course, he should delve into the original literature.

The very broad kind of review article that might place this whole question in perspective is lacking in the literature. Babcock has summarized certain aspects of fire hazards and the like in an encyclopedia article [1]. A review of where information may be found referring to various code writing groups, underwriters' organizations, testing groups, and the like was published in 1966 and is useful in terms of addresses and functions [2]. The Federal Register is the official organ in which all actions of the government appear. It requires some experience in reading to interpret correctly the items reported. Some examples of useful papers are a review of research in England (1946–1964 [3]), a lengthy report of a worldwide series of personal contacts with fire research groups (exclusive of most industrial laboratories) recently published (1968 [4]), a discussion of the problems pertaining to aircraft interiors [5], the toxicity and fire resistance of foams for aerospace application [6], and the use of fire retardants in plastics and fibers. Rhys's paper [7] states that the use of fire-retardant chemicals in the United States in 1966 just for the plastics industry was

Phosphorus compounds	$46\bar{M}$ lb
Halogenated compounds	27
Antimony oxide	22
Others	22
	$\overline{117\bar{M}}$ lb

and projects 130$\bar{M}$ lb for 1967. Such a tabulation omits new synthetic materials made flame-resistant by their chemical structure rather than by added elements especially tailored to add fire safety. It also omits fire-safe coatings—paints and mastics—some of which are finding wide use on the inside of appliance housings. And these numbers are just for plastics; they do not include the retardants used in wood, in paints, for paper, on forest fires, and so on.

SOME DEFINITIONS

A number of terms in use today are confusing to some. For example, what do we mean by fire retardant, fire resistant, and fireproof? What is the difference, if any, between a flame retardant and a fire retardant? More often than not the meaning is clear from the context. Occasionally the author himself has not defined in his own mind exactly what he wishes the term to convey. In this book no attempt has been made to split hairs over definitions. Several of the foregoing terms have been used interchangeably just to avoid monotony in the writing and, it is hoped, in the reading. Thus fire retardant, fire resistant, flame retardant, and flame resistant are used interchangeably in many sections. Fireproof is not used loosely; indeed it is rarely used because there are so few substances that are fire or flameproof. Thereby arises a definition: a material that is fireproof is totally unaffected in its properties by fire. It is not consumed or altered in any important respect by fire. Thus the material cannot contribute to the fire or propagate it; it is inert. On the other hand, fire-retardant or fire-resistant materials are much reduced in flammability, burn slowly, and do not propagate fire rapidly, but they are substantially altered and eventually are consumed by a fire. Quantitative definitions are provided in the many tests that have been devised (see below) and vary over a wide range.

An example or two may help. Steel is not fireproof; it softens near 540°C and yields under stress at higher temperatures. It is not necessarily consumed in a fire but important properties are altered. Similarly, concrete may crack and break up under the thermal stresses of a fire. One has to define the intensity and duration of the fire before the terms have real meaning. A substance that withstands completely an oil fire may be totally consumed in an oxyacetylene flame.

"Fire retardant" is used as a modifying phrase to describe the property of resisting fire; it is also used in the nominative sense to describe a chemical species. Thus we talk of "fire-retardant lumber," meaning lumber which resists fire; but we say that ammonium phosphates are "fire retardants," that is, they are chemicals useful for retarding fire in a given system.

The words "resist" and "retard" imply a partial but not complete barrier to fire, whereas "proof" is an absolute term.

FIRE TESTING

In the literature search conducted for this book we did not deliberately search out papers on test methods per se. However, many of the papers were principally directed to the development of suitable methods and only incidentally provided data on chemicals and materials. Some of these findings are discussed here.

The tests may be classified in a number of ways. One such is by substrate: tests for wood, paper coatings, textiles, and polymers. Another is by end use: coatings for wood, steel, paper, and plastics. A third is by type of test: ease of ignition, surface flammability (flame spread and rate of flame spread), smoke generation, self-extinguishing time, and so forth. A fourth is by test geometry in relation to the combustion zone: horizontal flame under, horizontal flame above, vertical burning up from bottom, vertical burning down from top, at an angle toward or away, and so forth. Another variant is the predominant mode of heat transfer to the sample: convection versus radiant heating. There are many ways to test for flammability; in fact, there are so many tests that no attempt is made here to discuss them all. A partial list of tests encountered most frequently in the literature is shown in Table 1-1.

Some of these tests are helpful; others are not. Certainly the basic tests are all very much to the point and serve to provide a better understanding of the processes in pyrolysis or combustion. The tunnel tests give some indication not only of flammability, smoke generation, and fuel contribution but also of the danger of spreading fire to adjacent materials—a measure of the role in fire propagation that a material may play. Tunnel testing is becoming more prevalent as the small tunnels increase in reliability and correlate more and more closely with the basic 25-ft tunnel used in listing tests.

The small horizontal tests (D-635, D-1692, etc.) are frequently used for plastics. D-635 and D-1692 are most often used in screening work. In the polymer literature on fire retardency one finds more often than not that the reference is a patent and the test data are from D-635 or D-1692. These are not especially severe tests and materials receiving good ratings in them may be only so-so in more drastic testing, such as the Hooker HLT-15 test or the Butler chimney.

Hilado has reviewed these tests [8] and he concludes that the following laboratory tests are valid and useful for plastics and fabrics:

Oxygen index test
Setchkin test for ignition temperature
NBS, Rohm and Haas XP2 smoke tests
Two-foot tunnel test
ASTM D-635 and D-1692 (horizontal)
Hooker HLT-15, Butler chimney, IBM test, and ASTM D-626 or AATCC-34 (vertical)
ASTM D-1433 and D-1230 or AATCC-32 (inclined)

Table 1-1 Some Tests Used with Fire Retardants

Substrate	Test Type	Test No.	Sample Size	Geometry	Test Results
Wood	Fire tube	ASTM E-69	$\frac{3}{8}'' \times \frac{3}{4}'' \times 40''$	Vertical, flame under	Weight loss final and at intervals
	Schlyter panel Crib test	ASTM E-160	$\frac{1}{2}'' \times \frac{1}{2}'' \times 3''$ 24 pieces	Stacked in crib	Weight loss, observe afterflame and afterglow
	Tunnel test	ASTM E-286	$13\frac{3}{4}'' \times 8'$	Horizontal, flame under	Flame spread, smoke density, fuel contribution
	Tunnel test	ASTM E-84	$20'' \times 25'$	Horizontal, flame under	Flame spread, smoke density, fuel contribution
Insulation board	Inclined plane	ASTM C-209	$6'' \times 6''$	45°, flame under (1-ml EtOH)	Char area, duration of flaming
Paper	Vertical strip	ASTM D-777 (TAPPI T-461)	$2\frac{3}{4}'' \times 8\frac{1}{4}''$	Vertical, flame under	Char length (in.), afterglow (sec)
Textiles	Vertical strip	ASTM D-626 (AATCC 34; Fed. 5902, 5903)	$2'' \times 12\frac{1}{2}''$	Vertical, flame under	Char length (in.), duration of flame, afterglow time
	45° strip	ASTM D-1230 (AATCC 33; Fed. 5908)	$2'' \times 6''$	45°, flame at bottom edge; 1-sec ignition	Time of flame spread
	Horizontal strip Pill test (carpets)	Fed. 5900, 5906 ...	$>8'' \times 8''$	Pill (hexamethylene tetramine-"methenamine") centered in 8" steel circle	Burns to 1" from steel ring or does not

Material	Test	ASTM	Specimen size	Test conditions	Measurements
Paints	Cabinet	ASTM D-1360	$\frac{1}{4}"\times 6"\times 12"$	45°, alcohol flame under	Area and volume of char, weight loss, leach resistance
	Stick and wick	ASTM D-1361	$1"\times 1"\times 16"$	Vertical, alcohol wick under	Weight loss, flame spread rate, char height
	Radiant panel	ASTM E-162	$6"\times 18"$	60° angle panel at at 670°C	Flame spread, smoke density
	Inclined panel (see insulation board)				
	Tunnel tests—2', 8', 16', 25' (see wood)				
Mastics	Furnace test	ASTM E-119	Actual column or or beam	In use position	Time to reach specified temp (e.g., 585°C)
Plastics	Roofing asphalt	ASTM D-1167	$6"\times 6"$	Vertical, 10 sec under	Visual observations
	Flexible plastics	ASTM D-568	$1"\times 18"$	Vertical, 15 sec under	Nonburning, self-extinguishing, burn rate (in./min)
	Self-supporting plastics	ASTM D-635	$\frac{1}{2}"\times 5"$	Horizontal on wire mesh, end ignited	As for D-568
	Self-extinguishing plastics	ASTM D-757	$\frac{1}{8}"\times\frac{1}{2}"\times 4\frac{3}{4}"$	Horizontal on red-hot igniting bar	Burning rate (in./min)
	Sheeting and foam	ASTM D-1692	$\frac{1}{2}"\times 2"\times 6"$	Horizontal on wire mesh, ignite end	Burning rate or self-extinguishing time
	Thin sheet	ASTM D-1433	$3"\times 9"$	45°, ignite lower edge	Burning rate or self-extinguishing
	Vertical bar	...	$2"\times 2"\times 18"$	Vertical, 60-sec ignition	Extinguishing time, weight loss flame height, burn rate, char fraction
	Butler chimney	...	$\frac{3}{4}"\times\frac{3}{4}"\times 10"$	Vertical, 10-sec ignition	Weight loss, extinguishing time, flame height
	Hooker HLT-15	...	$\frac{1}{8}"\times\frac{1}{2}"\times 8"$	Vertical, 5 ignitions	Ratings based on surviving all 5 ignitions

Continued overleaf)

Table 1-1 (Continued)

Substrate	Test Type	Test No.	Sample Size	Geometry	Test Results
Plastics (continued)	Bureau of Mines flame penetration test	...	$1'' \times 6'' \times 6''$	Flame impinging on center of sample	Time to burn through
	Schlyter, inclined plane, tunnel tests, radiant panel (see wood, paints)				
General	Differential thermal analysis	...	Micro	...	Basic thermal parameters
	Thermogravimetric analysis	...	Micro	...	Basic thermal parameters
	Oxygen index test (candle test)	...	$\frac{1}{8}'' \times \frac{1}{4}'' \times 3''$	Vertical, flame on top	Mole fraction oxygen to just support flame
	NBS, Rohm and Haas XP-2 smoke test,	...	$1'' \times 1'' \times 1''$ up to $3'' \times 3'' \times 3''$	...	Obscuration time, max smoke density, smoke accumulation rate
	Ease of ignition (Setchkin)	ASTM D-1929	$\frac{3}{4}'' \times \frac{3}{4}''$ $(3 \pm 0.5 \text{ g})$	Place in holder in furnace	Flash ignition and self-ignition temp
	Test for non-combustibility	ASTM E-136	$1\frac{1}{2}'' \times 1\frac{1}{2}'' \times 2''$	Set in furnace at $750°C$	Sample temp must remain within $30°C$ of furnace temp

Table 1-2   Oxygen Index Method (After Fenimore and Martin [11])

Material	n*
Polyoxymethylene	
Delrin, Du Pont	0.150 ± 0.003
Celcon, Celanese	
0.1 cm thick	0.148 ± 0.001
0.3 cm thick	0.149
At 14 cm of Hg pressure	0.190
Polyethylene oxide, Polyox WSR-35, Union Carbide	0.150
Kitchen candle (wick in paraffin)	0.16
Polymethyl methacrylate, Plexiglas, Rohm and Haas	0.173 ± 0.001
Polypropylene, Profax 6505, Hercules	0.174 ± 0.002
Asbestos-filled, "slow-burning" JMDC-4400, Union Carbide	0.205 ± 0.005
"Self-extinguishing" JMDA-9490, Union Carbide (probably	
contains chlorine and antimony oxide)	0.282 ± 0.003
Polyethylene	
1220, Allied Chemical	0.174 ± 0.001
Marlex-5002, Phillips	0.175 ± 0.001
Marlex at 14 cm of Hg pressure	0.210
Polystyrene	
0.1 cm thick from stock	0.181
0.6 cm thick, Westlake high-temp sheet	0.183
Polybutadiene, cross-linked with 2 parts of dicumyl peroxide	0.183
Polyvinyl alcohol, Elvanol 70-05, Du Pont	0.225 ± 0.004
Chlorinated polyether, Penton, Hercules	
[—CH_2—$C(CH_2Cl)_2$—CH_2—O—]	0.232
Polycarbonates, various clear Lexan resins, General Electric	
[—C_6H_4—$C(CH_3)_2$—C_6H_4—$OC(O)O$—]	0.26 to 0.28
Polyphenylene oxide, General Electric	
[—$C_6H_2(CH_3)_2$—O—]	0.28 to 0.29
Carbon	
Porous carbon, PC-25, National Carbon	0.559 ± 0.003
Carbon electrode, L8109, National Carbon	0.635
Carbon electrode at 50 cm of Hg pressure	0.80
Carbon electrode at 24 cm of Hg pressure	1.00
Silicone rubber, General Electric	0.30 to 0.33
Polyvinyl chloride, Geon-101 without plasticizer, Goodrich	0.45
Polyvinylidene chloride, Saran 281S905, Dow	0.60
Teflon, Du Pont	0.95

* Limiting oxygen indices (n) of various materials (at atmospheric pressure unless otherwise stated).

It would appear that small-scale furnaces can be constructed to correlate with ASTM E-119, too, for testing mastics and other materials designed for long-term fire endurance.

A few specific articles in this area are worthy of note. Reichberger has reviewed the various tests, including some of the German methods [9]. A Russian group has studied a muffle furnace technique on 24 materials [10].

Table 1-3   Limiting Oxygen Index Rating of Some Materials (Goldblum
[12]) (Table courtesy of Society of Plastics Engineers, Inc.)

Material	% Oxygen	Material	% Oxygen
Acetal	16.2	Nylon 6-6	28.7
		Lexan* 3414-131 resin	
PMMA	17.3	plus 40% glass	29.3
		Lexan* 3412-131 resin	
PE	17.4	plus 20% glass	29.8
PP	17.4	PPO* 534 resin	30.0
SAN	18.1	Polysulfone P1700	30.4
PS	18.3	Lexan* 2014 resin	31.5
ABS/PC alloy	20.2	Lexan* DL444 resin	39.7
		Lexan* DL444 resin	
Noryl* 731 resin	24.0	plus 20% glass	42.0
Lexan* 101-111,		Polyvinylidene	
-112 resin	25.0	fluoride	43.7
Lexan* 141-111,			
-112 resin	25.0	Polyvinyl chloride	47.0
		Polyvinylidene	
Lexan* glass resin	27.8	chloride	60.0
		Polytetrafluoro-	
SE-1 Noryl* resin	28.0	ethylene	95.0

* Tradenames General Electric Company.

Fenimore and Martin describe their candle test or oxygen index method [11] and provide the results shown in Table 1-2. The oxygen index is defined as

$$n = \frac{[O_2]}{[O_2] + [N_2]}$$

where [ ] is the volume concentration of each gas in the air stream. The material is placed in an upright position and ignited at the top. The Limiting Oxygen Index (L.O.I.) is that value of n at which the material just fails to support flaming combustion. Goldblum [12] has provided additional data, as shown in Table 1-3. The test tells one a lot about the intrinsic flammability of a material but very little about its role in propagating a fire from one place to another. The heat in this test comes, after ignition, solely from the combustion of the sample itself. A radiant panel test such as that used at the National Bureau of Standards provides strong continuous heating (also see ref. 14). This tends to level out differences arising from differing heats of combustion. Some results are shown in Table 1-4.

Small tunnels as a laboratory tool have been described by Vandersall [15] and Levy [16]. Some results and their correlation to the large 25-ft tunnel are

Table 1-4 Radiant Panel Test (Robertson, Gross, and Loftus [13])

Name	Nominal Thickness, in.	Bulk Density, lb/ft³	Moisture Content, %	No. of Tests	Flame Spread Factor, F_s	Heat Evolution Factor, Q	Flame Spread Index, I_s	Coefficient of Variation, v, %	Underwriters' Laboratory Flame Spread Factor	SS-A-118b Classification	Flashover Time, min	Deposit, mg	Based on Red Oak	Underwriters' Laboratory Smoke Factor
Fiberboard perforated tile, finish removed	1/2	15.7	5.4	5	23.7	14.1	336	14.7				0.1	30	
Fiberboard A (FPL)[c]	1/2	20.0	8.0	6	15.3	18.2	278	5.1	285		5.00	0.4	130	357
Fiberboard J unfinished	1/2	18.0	7.0	9	18.2	12.8	236	5.0				0.2	70	
Plywood exterior fir unfinished	1/4	42.3	6.8	4	8.5	17.0	143	32.2				0.3	100	
Hardboard Common H	7/32	59.8	3.8	5	4.5	30.2	136	12.4				4.1	1400	
Hardboard Common M (FPL)	1/8	63.5	3.8	6	4.1	29.7	121	6.9		D		5.2	1700	
Fiberboard perforated tile	1/2	16.7	5.8	5	10.0	11.6	116	13.6	113	D	6.15	0.3	100	1900
Fiberboard J + paint A, 250 ft²/gal	1/2	18.3	6.5	5	8.4	12.2	107	44.0				0.4	130	
Red oak (FPL)	1/2	40.0	6.5	5	7.0	14.1	99	10.9	100	D		0.3	100	100
Fiberboard C (FPL)	1/2	19.7	6.0	5	10.3	9.0	94	18.3	77	D		0.2	70	94
Fiberboard J factory finish	1/2	20.2	6.6	5	9.9	8.4	83	15.9				0.2	70	
Fiberboard J + paint B, 250 ft²/gal	1/2	17.0	6.2	5	6.0	9.8	59	17.2				0.6	200	
Plastic C	1/8	81.5		5	4.9	11.5	56	8.4	130	D		9.7	3200	752
Plastic B	1/8	98.0		5	3.7	8.4	30	19.5	60			5.7	1900	368
Fiberboard J + paint B, 125 ft²/gal	1/2	17.8	6.9	5	4.1	6.4	27	26.0				0.9	300	
Gypsum board G (FPL)	1/2	50.5		5	4.2	3.6	14.3	44.5	10	C	10.35	0.0	0	0
Gypsum board K	3/8	47.4		5	4.4	3.1	13.3	29.9	10			0.1	30	
Mineral base tile	3/8	19.3	3.4	5	7.8	1.5	10.5	60.0	12.8			0.0	0	
Gypsum board L	1/2	51.2		4	3.0	2.3	7.9	105.0		D	16.36	0.1	30	17
Fiberboard I fire retardant finish (FPL)	1/2	19.0	6.4	5	1.8	2.6	4.5	53.8	14	D		0.5	170	14
Plastic A	1/8	89.4		5	1.0	4.0	4.0	42.3	20			3.3	1100	556
Fiberboard J + mineral spray	1/2	21.9	7.7	5	1.0	3.4	3.4	28.9				0.2	70	
Glass fiber batt	1	4.3	1.2	5	1.0	1.8	1.8	60.5	9–10			0.1	70	
Cellulose mineral board	1/2	47.8	7.5	5	1.0	1.3	1.3	48.6	10–15			0.3	30	
Glass fiber tile	1/16	10.3	0.6	5	1.0	1.2	1.2	77.7				0.2	100	
Fiberboard J + 1/8-in. mineral surface	5/8	42.2	7.5	5	1.0	1.2	1.2	50.7		C	16.45	0.1	70	0
Plywood exterior fir + aluminum foil	1/4	41.6	6.3	3	1.0	0.9	0.9	65.7		D		0.2	~30	
Glass fiberboard	2	11.0	0.4	5	1.0	0.6	0.6	37.3		B		0.1	70	
Asbestos cement board	3/16	117.0	0.4	5	1.0	0.0	0.0		0	A	17.10	0.0	0	

Flame spread index $\equiv I_s = F_s Q$
where F_s = flame time factor
Q = thermal factor
I_s = 0 for asbestos, 100 for red oak (arbitrarily).

Table 1-5   Two-foot Tunnel Test (Levy [16])

Item	Flame Spread Rating		Comment
	30'-30° Inc'd. Tunnel	25-ft Tunnel	
Asbestos—cement $\frac{1}{4}$ in.	0	0	No smoke (0 rating)
Cellular glass—9 pcf	3–5	5–8	No smoke
Acoustical tile—mineral type	13–16	10–15	Very little smoke
Tectum—1 in	15–20	15	Very little smoke
Phenolic foam—3.5 pcf	22	<25	Little smoke, but smolders continuously
Acoustical tile—treated cellulose	24–33	25	Little smoke
Intumescent paint	28–30	25	Low smoke
Redwood	71–78	65–80	Low smoke
Rigid vinyl foam—2 pcf	83–86	—	Surface flash and melted in 1 min
Hardboard—$\frac{1}{8}$ in.	87	90–105	Very little smoke
Red oak—nom. 1 in.	100	100	Smoke (100 rating)
Birch—nom. 1 in	100–105	105–110	Smoky
Urethane—SE 2 pcf	100–109	· · ·	Very smoky. High initial surface flash
Urethane—NB 2 pcf	102–105	· · ·	Very smoky. High initial surface flash
Urethane—NB 2 pcf	111–116	· · ·	Very smoky. High initial surface flash
Douglas fir plywood—$\frac{1}{2}$ in.	113–127	100–169	Low smoke
Northern white pine—nom. 1 in.	145–150	165	Smoky
Epoxy foam—2.5 pcf	150+	Very large	Very smoky. Burned completely
Urethane—6.5 pcf	150+	Very large	Very smoky. High surface flash
Polystyrene—1.5 pcf	150+	Very large	Burned completely in less than $1\frac{1}{2}$ min
Polystyrene FR—2 pcf	Melted when flame applied in less than 2 min, test interval		

shown in Table 1-5 and the correlation for paints in Figure 1-1. This work represents a significant advance in that expensive and time-consuming tests in the 25-ft tunnel may be deferred until the final, optimum system is ready for test. The small tunnels provide rapid, inexpensive screening tools.

Hilado [8, 17] has presented a penetrating analysis of flame tests in terms

of geometry and relation of heat source to flame direction. The configurations
are:

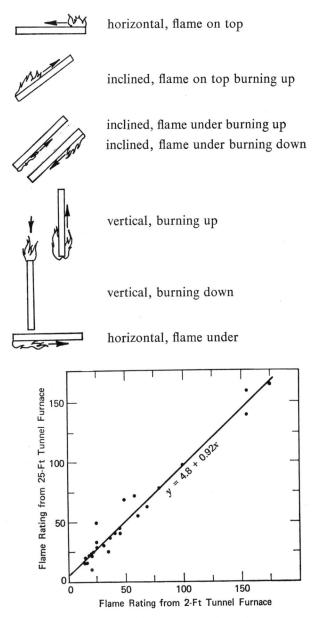

horizontal, flame on top

inclined, flame on top burning up

inclined, flame under burning up
inclined, flame under burning down

vertical, burning up

vertical, burning down

horizontal, flame under

Figure 1-1   Correlation of results from 25- and 2-ft flame tunnels.

Table 1-6　Smoke Density Values for Various Plastics (Gaskill and Veith [25])

	Time to Max Density, min	Max Density	Obscur. Time, min	Max Rate, min	Time for Max Rate, min
Smoldering–No Ventilation					
Acrylic FR UV-abs	29	380	3.8	44	9
Acrylic FR UV-abs[a]	26	480	3.7	56	23
Acrylic HR UV-abs	34	195	6.3	12	10
Acrylic HR UV-trans	33	190	6.5	14	14
Acrylic HR UV-trans[b]	18	200	6.5	50	16
Polyethylene	17	470	5.5	93	9
Polystyrene	26	345	4.0	32	9
Polystyrene[c]	6	460	4.0	222	5
Polytetrafluoroethylene	36	0	NR[d]	⋯	⋯
Polystyrene foam FR	23	10	NR	2	1
Phenolic canvas lam.	28	450	2.5	41	7
Phenolic canvas lam. FR	35	460	2.4	44	11
Rigid polyvinyl chloride-filled	30	490	1.6	33	5
Rigid polyvinyl chloride-unf.[e]	14	272	2.1	28	6
Rigid polyvinyl Chloride-unf.[f]	33	470	2.1	46	8
Range	6–35	0–490	11.6–00	2–220	1–23
Flaming–No Ventilation					
Acrylic FR UV-abs	7	480	1.8	151	4
Acrylic HR UV-abs	13	90	2.2	21	3
Acrylic HR UV-trans	6	140	2.3	39	4
Polyethylene	9	150±	4±	59±	4
Polystyrene	4	470	1.2	199	2
Polytetrafluoro-ethylene	30	55	11	4	9
Polystyrene foam FR	11	260	0.9	50	8
Phenolic canvas lam.	20	460	1.7	89	4
Phenolic canvas lam. FR	24	210	3.7	34	14
Rigid polyvinyl chloride-filled	11	530	0.5	250	1
Rigid polyvinyl chloride-unf.[e]	5	525	0.5	195	2
Rigid polyvinyl chloride-unf.[f]	10	535	0.6	170	2
Range	4–30	90–530	0.5–11	4–250	1–24

[a] Auto ignition at 20–24 min　　[d] Not reached
[b] Auto ignition at 14 min　　[e] $\frac{1}{8}$ in. thick
[c] Auto ignition at 4 min　　[f] $\frac{1}{4}$ in. thick

The most severe cases are those where the direction of gas flow is in the direction of flame spread so that convective heating is into the burning area rather than away. In this sense a vertical test burning up from the bottom is about as severe an exposure as one can design. The textile vertical test is therefore a very rigorous exposure.

Textile tests have been the subject of several papers [18] and are at present the center of some controversy. The 45° strip test AATCC 33 (ASTM D-1230) [19] employs a 1-sec ignition time in a manner that many workers feel is unrealistic. Indeed the test was designed to screen out only explosively flammable fabrics. The vertical test is perhaps too severe. A test that combines features of both is a semicircular device over which the fabric is stretched. The cloth ignites at the bottom on one side and, if it burns all the way, will burn up vertically, then on an incline, then horizontally, and finally downward. The angle goes from +90° to −90° and the conditions become progressively less severe. All sorts of observations can be made with such a test but probably no definitive ratings will come from it. It appears to be a versatile laboratory screening test of some promise. A variation is the vertical quarter-ellipse test with ignition from the top [20].

Tests for carpets and upholstery may have to be different from those for lighter fabrics. A study of the response of various materials to specified thermal fluxers has been reported the results of which will be useful in classifying materials but not in flammability comparisons. This work was directed to the behavior of various materials in the thermal exposure expected in a nuclear attack [21]. Carpet combustibility was regulated in a preliminary manner in the United States in late 1969. The test is the "pill test" in which a pellet of methenimine (hexamethylenetetramine) is ignited by a match and allowed to burn out on the carpet. If the area of burned carpet is less than 3″ from the point of ignition the carpet will pass [22].

A recent paper by Behnke and Seaman [23] discusses methods of measuring heat transfer through fabrics in terms of comfort indices or pain measurements and the tendency to produce skin blisters. This is, of course, for protective clothing and is a little outside the usual concern in fabric flammability.

Smoke generation continues to receive less attention than it deserves. The tunnel test (ASTM E-84) measures smoke secondarily but the Rohm and Haas XP2 test is designed solely for smoke determinations. The latter test has been modified, first by the Bureau of Standards [24] and then by the Lawrence Radiation Laboratory [25]. The results of the Lawrence work are shown in Tables 1-6 and 7. The index used in Table 1-7 is a measure of severity in that those materials that generate the greatest smoke in the shortest time are penalized most; the materials that generate the same smoke but at a slower rate receive a more favorable rating. Rigid, filled polyvinyl

Table 1-7   Smoke Observation Indices for Various Materials
(Gaskill and Veith [25])

	No Ventilation		6 Air Changes/Hr		20 Air Changes/Hr	
Material	Smold.	Flaming	Smold.	Flaming	Smold.	Flaming
Hardboard	25	5	7	1	2	0
Marine plywood	25	1	10	0	2	0
Red oak	35	1	15	1	3	0
Redwood	30	8	10	3	3	0
Black walnut	80	1	60	2	8	0
White oak	80	4	30	1	7	0
Acrylic FR UV-abs	25	360	10	350	1	240
Acrylic HR UV-abs	3	10	1	20	0	15
Acrylic HR UV-trans	3	35	1	40	0	20
Polyethylene	65	20	· · ·	· · ·	· · ·	· · ·
Phenolic lam. NFR	60	190	20	90	2	35
Phenolic lam. FR	45	10	25	9	1	3
Rigid polyvinyl chloride-filled	70	2400	15	525	5	165
Polystyrene	21	900	5	1175	1	1165

Index = maximum smoke density times average rate of accumulation time to reach critical level

chloride is a serious offender according to this test; some woods offend relatively little. Since smoke often obscures the route to safety and displaces oxygen in the lungs, both leading to death, it is often more hazardous than any other factor in a fire. Research to minimize smoke is sorely needed. Unfortunately, many flame retardants increase the tendency to smoke; see the tables for examples.

MECHANISMS OF FIRE RETARDATION

The following paragraphs briefly touch on current theories as to the actual function of flame retardants in altering combustion. The discussion is brief and is intended only to guide the reader to the literature. It is not the purpose of this book to delve deeply into mechanisms of thermal and oxidative degradation and the effect of fire-retardant elements thereon.

There are several distinct stages in the burning process: heating, degradation and decomposition, volatilization, and oxidation. Relatively few flames feed directly on solid or liquid phases; rather the flaming feeds on the gas phase. Flames are either diffusion- or premixed, depending on whether the air mixes with the gaseous fuel by diffusion, as in spontaneous fires, or via a mixing apparatus, as in a burner. There are a variety of ways to extinguish a flame: cooling the solid, altering the degradation and decomposition process to produce nonflammable volatiles, quenching the volatilization process, adding too much air or eliminating air from the combustion zone, or interfering with the oxidation reactions in the gas phase. Most techniques involve more than one of these ways and all of them have been and continue to be used on various occasions. The simple act of blowing out a match involves cooling the substrate and the vapor space and increasing the volume fraction of air in the combustion zone. Chemical fire retardants operate both in the solid phase (altering thermal degradation processes or forming barriers at the surface) and in the vapor space (interfering with oxidation). Let us deal with the latter first.

Oxidation in the vapor phase is a free radical process involving H, OH, O, and HO_2 radicals, to name but a few. There is substantial literature on the addition of substances to flames to achieve suppression. The action is often attributed to the ability of the added substance to trap these radicals. Bromine compounds are a good example of this:

$$RBr + H\cdot \rightarrow HBr + R$$

The generated R· must be a less active radical than the H· which is removed in order for the overall result to be flame inhibition. In general H·, OH·, and the like will be the most active species and their removal snuffs the flame.

Friedman and Levy [26] have prepared an extensive review of this literature through 1957, which includes discussion of both organic and inorganic compounds of the halogens and certain other materials known to affect flames. Table 1-8 lists some compounds and the minimum volume concentration required to snuff out a flame. From this, some interesting comparisons

Table 1-8   Summary of Compounds Evaluated as Fire-Extinguishing Agents for n-Heptane-Air
Room Temperature, 300–500 mm Hg Absolute Pressure [27]

Compound Formula	Compound Name	Peak in Flammability Curve, vol %
CBr_2F_2	Dibromodifluoromethane	4.2
CBr_3F	Tribromofluoromethane	4.3
$CF_3CHBrCH_3$	2-Bromo-1,1,1-trifluoro-propane	4.9
$CBrF_2CBrF_2$	1,2-Dibromotetrafluoroethane	4.9
CF_2ICF_2I	Tetrafluoro-1,2-diiodoethane	5.0
CH_2Br_2	Dibromomethane	5.2
CF_3CF_2I	Pentafluoroiodoethane	5.3
$CF_3CH_2CH_2Br$	3-Bromo-1,1,1-trifluoropropane	5.4
CH_3CH_2I	Ethyl iodide	5.6
CF_3CF_2Br	Bromopentafluoroethane	6.1
CH_3I	Methyl iodide	6.1
$CBrF_3$	Bromotrifluoromethane	6.1

Table 1-8   (continued)

Compound Formula	Compound Name	Peak in Flammability Curve, vol %
CH_3CH_2Br	Ethyl bromide	6.2
$CH_2BrCF_2CH_3$	1-Bromo-2,2-difluoropropane	6.3
$CClF_2CHBrCH_3$	2-Bromo-1-chloro-1,1-difluoro-propane	6.4
$CHBr_2F$	Dibromofluoromethane	6.4
$CBrF_2CH_2Br$	1,2-Dibromo-1,1-difluoroethane	6.8
CF_3CH_2Br	2-Bromo-1,1,1-trifluoroethane	6.8
$C_6F_{11}C_2F_5$	Perfluoro(ethylcyclohexane)	6.8
$1,3\text{-}C_6F_{10}(CF_3)_2$	Perfluoro(1,3-dimethylcyclohexane)	6.8
$1,4\text{-}C_6F_{10}(CF_3)_2$	Perfluoro(1,4-dimethylcyclohexane)	6.8
CF_3I	Trifluoroiodomethane	6.8
CH_2BrCH_2Cl	1-Bromo-2-chloroethane	7.2
$CClF_2CH_2Br$	2-Bromo-1-chloro-1,1-difluoro-ethane	7.2
$C_6F_{11}CF_3$	Perfluoro(methylcyclohexane)	7.5
C_7F_{16}	Perfluoroheptane	7.5
CH_2BrCl	Bromochloromethane	7.6
$CHBrF_2$	Bromodifluoromethane	8.4
$CClF_2CCl_2F$	1,1,2-trichlorotrifluoroethane	9.0
$CBrClF_2$	Bromochlorodifluoromethane	9.3
HBr	Hydrogen bromide	9.3
CH_3Br	Methyl bromide	9.7
$CF_2\text{-}CHBr$	2,2-Difluorovinyl bromide	9.7
C_4F_{10}	Perfluorobutane	9.8
$SiCl_4$	Silicon tetrachloride	9.9
$CBrF_2CBrClF$	1,2-Dibromo-2-chloro-1,1,2-trifluoroethane	10.8
$CClF_2CClF_2$	1,2-dichlorotetrafluoroethane	10.8
CCl_4	Carbon tetrachloride	11.5
$CF_3CHClCH_3$	2-chloro-1,1,1-trifluoropropane	12.0
$CF_3CH_2CH_2Cl$	3-chloro-1,1,1-trifluoropropane	12.2
$CClF_3$	Chlorotrifluoromethane	12.3
CF_3CF_3	Hexafluoroethane	13.4
CCl_2F_2	Dichlorodifluoromethane	14.9
$CHCl_3$	Chloroform	17.5
CHF_3	Trifluoromethane	17.8
$CHClF_2$	Chlorodifluoromethane	17.9
C_4F_8	Octafluorocyclobutane	18.1
SF_6	Sulfur hexafluoride	20.5
BF_3	Boron trifluoride	20.5
PCl_3	Phosphorus trichloride	22.5
HCl	Hydrogen chloride	25.5
CF_4	Carbon tetrafluoride	26
CO_2	Carbon dioxide	29.5
H_2O	Water	>8
$(C_2F_5)_2NC_3F_7$	Heptadecafluoro(N,N-diethyl-propylamine)	>8.5
CH_2Cl_2	Dichloromethane	>11

can be made. These are given in gas volumes; the results are, therefore, essentially on a molar rather than weight basis. Clearly the following is true:

$$I > Br > Cl > F$$

Note these pairs:

CH_3Br	9.7	vs.	CH_3I	6.1
CH_3CH_2Br	6.2	vs.	CH_3CH_2I	5.6
CH_2Br_2	5.2	vs.	CH_2Cl_2	>11
HBr	9.3	vs.	HCl	25.5

The effectiveness ratio for a pair of halogens varies with the compound type. As we shall see later, the two important halogens are chlorine and bromine and the latter is about twice as efficient on a weight basis as the former or four times on a mole basis. Thus the data on HCl and HBr take on added meaning as they relate to performance of halogenated retardants added to flammable substrates.

It should be added that fluorine binds so tightly to carbon that it does not serve as a radical trap. Fluorinated organics are themselves flame-resistant—Teflon burns only in 95% oxygen—but by virtue of their extreme stability they often cannot confer this property on substances with which they may be mixed.

Iodine compounds are too unstable to be of use as additives. This instability creates difficulties with other important properties of systems such as resistance to uv light and so on. Only bromine and chlorine compounds offer the right combination of properties and even these present some problems.

There are a good many references to flame inhibition. Some of these may be found in reference 28.

Antimony oxide is not effective by itself. But combined with a halogen compound it is a good retardant. The mechanism here is believed to be formation, on heating, of a volatile halide:

$$Sb_4O_6 + 4RCl \rightarrow 2SbOCl + 2SbCl_3$$
$$dec\ 170°\quad bp\ 223°$$

The antimony halide will therefore be found in the gas phase during burning. Studies have shown that most of the antimony is volatilized during burning. The volatile antimony compounds are effective free radical traps and will snuff the flame (see Chapter 5, p. 209). It appears that enough halogen to convert the antimony almost completely to SbX_3 is required. Two moles of chlorine per mole of antimony are not enough [29].

Both antimony and chlorine compounds also have an effect on the degradation reaction in the solid phase. Antimony promotes the charring reactions. Chlorine added to the solid is more effective than when it is simply fed to the flame zone separately [29] and one concludes it is doing some of its work in the solid, though exactly what is not clear. The role of antimony in char formation is undoubtedly the same as that of phosphorus and the latter has been studied extensively.

First, it is well known that most organophosphorus compounds are effective and only phosphorus salts of nonvolatile metals are not. Fenimore and Martin [29] show that, in polyethylene, triphenylphosphate, -phosphite, and -phosphine are all equivalent on a phosphorus basis. Even red phosphorus is claimed in one or two instances. The phosphorus compounds tend to create a great deal more char and less flammable volatiles. Weight loss on combustion is much reduced (see Chapter 4, p. 130; Chapter 5, p. 227). Smoke densities may be increased. Similar results are found with antimony compounds, with nitrogen, arsenic, and bismuth compositions, and even with sulfuric acid or organic sulfates. The key seems to be that an acid is required that is not too volatile. Substances that give rise to acid fragments on mild heating appear to have the desired effect. Phosphoric and sulfuric acids have high boiling points and therefore remain in the solid or liquid phases long enough to do their job. Indeed, after burning a substance treated with a phosphorus compound a sticky viscous liquid is usually found which is a polymeric acid containing phosphorus.

This subject has been carefully explored with regard to the thermal degradation of cellulose. Many papers have appeared on the basic phenomena, including the effect of a variety of chemical retardants [30]. The latest studies [31–35] have confirmed that the first step in cellulose degradation is

the production of levoglucosan [34, 35] but there is no agreement as to how this occurs: some suggest a free radical mechanism; others, various molecular rearrangements catalyzed by acids. The levoglucosan must then either decompose to volatile fragments or react further to produce char. The char presumably is highly cross-linked and reduced almost to pure carbon. The phosphorus-type retardants, as well as halogens to a lesser extent, greatly increase the carbon content. Byrne and his colleagues [32] have offered speculation as to how carbonium ions may form and aldol-type condensation may follow to produce cross-linking and char formation. The phosphorus acid participates probably by promoting carbonium ion formation either simply by donating a proton or by actually phosphorylating the levoglucosan first.

(a) $R_2CH-CH_2OH \xrightarrow{H^+} R_2CH-CH_2OH_2^+ \longrightarrow R_2CH-CH_2^+ \longrightarrow$

$$R_2C{=}CH_2 + H^+$$

(b) $R_2CHCH_2OH + \text{-}\ \overset{\displaystyle O}{\underset{\displaystyle OH}{\overset{\displaystyle \|}{P}}}-O-\overset{\displaystyle O}{\underset{\displaystyle OH}{\overset{\displaystyle \|}{P}}}-OH \longrightarrow$

$$R_2CHCH_2OPO_3H_2 + \text{-}\ \overset{\displaystyle O}{\underset{\displaystyle OH}{\overset{\displaystyle \|}{P}}}-OH \longrightarrow R_2C{\cdots}CH_2 + H_3PO_4$$

In (a) the acid merely protonates the alcohol to give a carbonium ion which can lead to production of an olefin plus another carbonium ion. In (b) phosphorylation is postulated via alcoholysis of an acidic POP linkage. This reaction is known to proceed rapidly at low temperatures [36]; polyphosphoric acid is also known to be a decomposition (thermal) product of various nonmetallic phosphates. There is as yet no clear evidence pointing to one or the other mechanism. Many acids affect the decomposition in the same way but these acids also form esters readily—sulfuric, boric, sulfamic, and so forth. And there the matter rests.

The results are clear enough, however. Table 1-9 presents some thermal

Table 1-9   Thermal Data for α-Cellulose

(After Tang and Neill [33])

	Activation Energy of Pyrolysis,[a] kcal/mole	Heat of Pyrolysis, cal/g	Max Rate of Heat Generation, cal/g min
2% $Na_2B_4O_7 \cdot 10H_2O$	30–32	58	730
2% $AlCl_3 \cdot 6H_2O$	33–33.5	57	665
2% $KHCO_3$	19–21	72	588
2% $NH_4H_2PO_4$	17–19	78	635
8% $NH_4H_2PO_4$	⋯	64	498
Blank	33–35	88	870

[a] First stage, zero to first order

data for various additives with cellulose and shows that the activation energy barrier is reduced (the temperature at the onset of pyrolysis is reduced, too) and the total heat generation process is slowed markedly. Details of work of this type are given in Chapters 4–6.

Another important but little understood aspect of fire retardancy is glow-proofing. It is often required that not only should the flame be extinguished but also afterglow should be prevented. Some flame retardants are also glow inhibitors; others are not. The phosphates are probably best at preventing afterglow; the halogens and borates are intermediate; and some retardants, for example, sulfamate, are ineffective [37]. (See reference 192, Chapter 5.) The glowing reaction may proceed in one of at least two ways:

(a)

$$C + \tfrac{1}{2}O_2 \rightarrow CO \qquad \text{(surface)}$$
$$CO + \tfrac{1}{2}O_2 \rightarrow CO_2 \qquad \text{(gas reaction)}$$
$$CO_2 + C \;\; \rightarrow 2CO \qquad \text{(surface)}$$

(b)

$$C + O_2 \;\; \rightarrow CO_2 \qquad \text{(surface reaction)}$$
$$CO_2 + C \;\; \rightarrow 2CO \qquad \text{(surface)}$$

The evidence points to mechanism (a) above [26]—glowing occurs by reactions in the surface and in the gas phase just above the surface. The reaction to CO_2 is much more exothermic than that to CO (a factor of 3.56) so that prevention of the CO_2 reaction substantially reduces the heat generation and may itself cause glow elimination. Many authors consider that a necessary part of the explanation is the fact that condensed or polymeric acids will form a film over the surface, thereby blocking the diffusion of oxygen to the burning site. The acid layer also will absorb some of the heat necessary for combustion. The situation is far from clear.

THERMAL DEGRADATION—GENERAL

There have been a great many papers on this subject in recent years, and there will be many more. Aspects relating to flame retardants are discussed in the pertinent sections of succeeding chapters. Here, two references should be underscored. The first is a published symposium at the University of London in 1961 which covers the subject in some breadth and should be consulted at least by the new investigator [38]. The second is the book by Madorsky [35], which the present author has found invaluable in writing this book. When Madorsky, of the National Bureau of Standards, retired he was prevailed upon to put into book form the results of many years of specialization in this field. This book is an absolute must for the research man interested in the thermal behavior of materials.

SYNERGISMS

A synergism may be defined as a case in which the effect of two components taken together is greater than the sum of their effects taken separately. There is much talk today of a synergism between nitrogen and phosphorus but this is only one way of viewing the evidence available thus far. The data, discussed in more detail in Chapters 5 and 8, show that nitrogen does play a role in fire retardance and that adding nitrogen often reduces the need for phosphorus. However, it takes a relatively large amount of nitrogen to reduce the phosphorus level noticeably. [One can only assume a synergism if one asserts that nitrogen in the absence of phosphorus is not at all a fire retardant. That this is not the case is attested to by the fact that nitrogen compounds are effective in nylon and in aminoplasts (see Chapter 8).] The effect in cotton is somewhat less than additive. Thus the following combinations produce about the same degree of fire retardancy [39]:

% P	% N
3.5	0
2	2.5
1.4	4
0.9	5

The effect of nitrogen on phosphorus requirements is also seen in polyurethanes. In these resins only about 1.5% P is required, whereas polyester resins require about 5% P. The principal difference between the two polymers is the presence of some 10% N in the urethane. The aromaticity and heterogeneous nature of the backbone are similar for both systems.

There are true synergisms in the phosphorus-halogen and antimony trioxide-halogen systems. Some examples are:

Polyolefins	5% P $\equiv 0.5\%$ P $+ 7\%$ Br $\equiv 20\%$ Br
Acrylates	5% P $\equiv 1\%$ P $+ 3\%$ Br $\equiv 16\%$ Br
Polyacrylonitrile	$10-12\%$ Br $\equiv 2\%$ Sb$_4$O$_6$ $+ 6\%$ Br
Epoxies	$13-15\%$ Br $\equiv 3\%$ Sb$_4$O$_6$ $+ 5\%$ Br

Antimony trioxide alone is ineffective. These are substantial effects over and above the additive components and reflect a chemical interaction which alters behavior in a way not possible with either component. The Sb$_4$O$_6$ is inert and may remain totally unaffected by burning of the polymer. When a halogen compound is present, the antimony halide is produced:

$$Sb_4O_6 + 12RCl \rightarrow 4SbCl_3 + 6(R_2O)$$

SbCl$_3$ (or SbBr$_3$) is volatile and reactive. It will participate in halogen transfer and char formation reactions within the solid phase and should act as an effective free radical trap in the vapor phase. Both kinds of reactions assist in flame suppression.

The explanation for the phosphorus-halogen synergism is less clear but probably follows similar lines. Thus combustion of a polymer containing these two elements could leads to production of some volatile phosphorus halides which alter combustion by trapping free radicals in the vapor and which facilitate dehydration to char in the solid. The results are of prime importance in lowering the total amount of fire retardant required. Often the amounts needed are so large as to affect adversely the physical properties of the system and only very painstaking adjustments overcome the detrimental effects. The much lower requirements afforded by these mixtures circumvent many such difficulties.

A startling reduction in requirements has been achieved experimentally by adding free radical precursors. These are molecules which readily produce free radicals on the input of energy, for example, dicumyl peroxide, ethanes with bulky substituents, xylyl disulfides, chelated paramagnetic ions of transition metals, N-nitrosoamines, and N-dichloroamides. Requirements for styrene are reduced as follows:

$$
\begin{array}{lll}
\text{Without free radical precursor} & 10\text{–}15\% & \text{Cl} \\
 & \text{or} \quad 4\text{–}5\% & \text{Br} \\
\text{With radical source} & 4\text{–}8\% & \text{Cl} \\
 & \text{or } 0.5\text{–}3\% & \text{Br}
\end{array}
$$

With 0.2% dicumyl peroxide, 0.15% P and 1.6% Br sufficed to produce a flame-resistant styrene. The rationale for this behavior is offered in Chapter 7, pp. 324–325; it is not altogether clear. Although addition of free radical sources undoubtedly creates as many problems as it solves, the technique indicates that there are breakthroughs that may be made to improve fire-retardant performance markedly. In this sense the development is most encouraging.

FIRE-RETARDANT REQUIREMENTS

Table 1-10 presents retardant requirements for the common polymers where reasonably good data are available. Each entry represents an average over the results of several papers or patents so that the uncertainty represented by, for example, one or another patent is removed in part. The values given should be taken in the spirit in which they were gathered, namely, for purposes of guidance rather than as absolute statements of fact. Certain obvious trends are apparent. Phosphorus is more effective at lower levels than any of the other elements taken alone. Bromine is superior to chlorine on a weight basis by about a factor of 2. Both phosphorus and antimony oxide reduce the halogen requirement substantially. Phosphorus does this more effectively than the antimony compound. Phosphorus-bromine combinations are perhaps the most efficient in the fire suppression sense of all the retardants evaluated. (Boric acid-borax and nitrogen or nitrogen/phosphorus

Table 1-10 Average Requirements for Fire-Retardant Elements to Render Common Polymers Self-Extinguishing

Polymer	Chapter	% P	% Cl	% Br	% P + % Cl	% P + % Br	% Sb_4O_6 + % Cl	% Sb_4O_6 + % Br
Cellulose	4, 5	2.5–3.5	>24	...	...	1 + 9	12–15 + 9–12	...
Polyolefins	7	5	40	20	2.5 + 9	0.5 + 7	5 + 8	3 + 6
Polyvinyl chloride	7	2–4	40	...	NA	...	5–15% Sb_4O_6	...
Acrylates	7	5	20	16	2 + 4	1 + 3	...	7 + 5
Polyacrylonitrile	7	5	10–15	10–12	1–2 + 10–12	1–2 + 5–10	2 + 8	2 + 6
Styrene	7	...	10–15	4–5	0.5 + 5	0.2 + 3	7 + 7–8	7 + 7–8
Acrylonitrile-butadiene-styrene	7	...	23	3	...	...	5 + 7	...
Urethane	8	1.5	18–20	12–14	1 + 10–15	0.5 + 4–7	4 + 4	2.5 + 2.5
Polyester	8	5	25	12–15	1 + 15–20	2 + 6	2 + 16–18	2 + 8–9
Nylon	8	3.5	3.5–7	...	...	...	10 + 6	...
Epoxies	8	5–6	26–30	13–15	2 + 6	2 + 5	...	3 + 5
Phenolics	8	6	16	...	...	...	...	...

combinations were omitted because there were not enough data points. These are, however, commented on for specific polymers in the appropriate chapters.) The compounds by which the elements are introduced into the polymers are, by and large, organic in nature. Nonvolatile metals are to be avoided because they tend to interfere with the ability of the system to participate in reactions leading to charring and to trapping of free radicals in the flame.

FUTURE TRENDS

Today flame retardance is achieved by incorporating one or more of the following elements: P, N, Sb, Cl, Br, or B. However, there is an alternative, and that is the use of structural modifications which inherently lead to reduced flammability. This approach is being used in altering present polymers and in the creation of entirely new types.

It is known from basic studies that the mode of polymer degradation is markedly affected by substituents on the backbone. Thus polyethylene degrades in a random manner to yield less than 1% monomeric ethylene in the volatiles but polyisobutylene depolymerizers to produce 18% monomer in the volatile fraction. Other examples are even more startling [35]:

$$% \text{ Monomer in Volatiles}$$

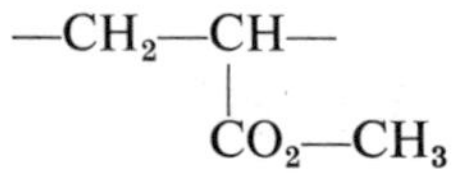

$$\begin{array}{cc}
-CH_2-CH- \\
\quad\quad | \\
\quad CO_2-CH_3
\end{array} \qquad 0.7$$

$$\begin{array}{cc}
\quad CH_3 \\
\quad | \\
-CH_2-C- \\
\quad\quad | \\
\quad CO_2-CH_3
\end{array} \qquad 91$$

$$\begin{array}{cc}
-CH_2-CH- \\
\quad | \\
\quad CN
\end{array} \qquad 5$$

$$\begin{array}{cc}
\quad CH_3 \\
\quad | \\
-CH_2-C- \\
\quad | \\
\quad CN
\end{array} \qquad \to 100$$

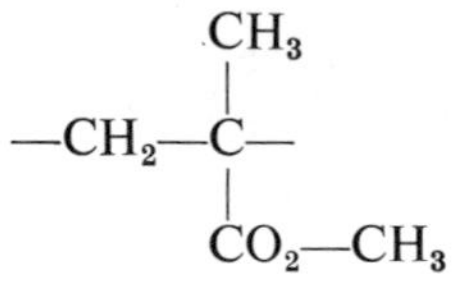

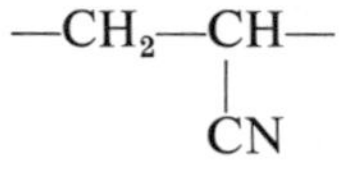

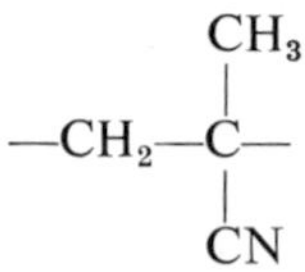

The monomers are highly flammable; reduction of the amount of monomer normally gives a reduction in flammability. Armed with such knowledge the chemist can readily avoid the more obvious pitfalls. A partial list of behavior in pyrolysis is offered in Table 1-11.

Table 1-11 Behavior of Some Polymers in Pyrolysis
(After Madorsky [35])

Polyethylene	Random scission	Small
Polymethacrylate	Random	Small
Polymethylmethacrylate	Unzips	High
Polyacrylonitrile	Chars-cyclizes	Small
Polystyrene	Unzips	High
Poly α-methylstyrene	Unzips $\approx 100\%$	Very high
Polyester	Random	Small
Polyurethane	Both	Moderate to high
Nylon	Random	Very low
Phenolics	Random	Moderate

Polyacrylonitrile degrades in an interesting manner, ultimately to a graphite-like resin. The ability to form internal cyclic structures of this type

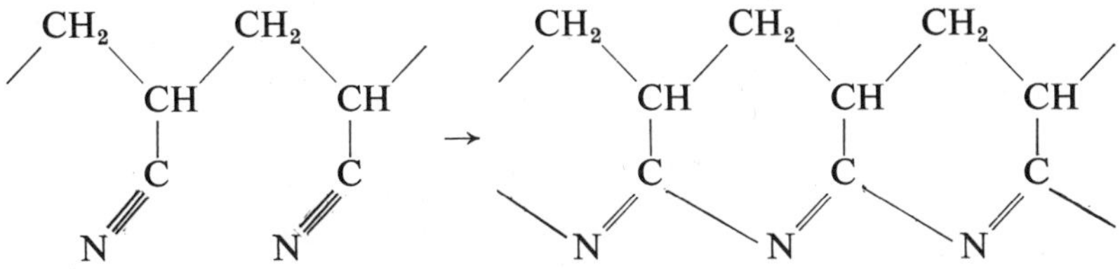

effectively prevents unzipping to monomer and reduces flammability.

Ring structures lend stability to every system studied. Aromatic structures are preferred, but even heterocycles like the glucosides confer greatly improved flame resistance, for example, in polyurethanes. By trimerizing isocyanates to form the triazine ring, resistance to burning of urethanes has been increased tenfold. A new development is the use of polyester polyols based on trimellitic anhydride, which thereby adds aromaticity to the final urethane.

A new generation of highly flame-resistant polymers is unfolding based on aromatic systems. A few of these are shown in Table 1-12. Some can be made almost fireproof; indeed, samples have been prepared which resist an oxyacetylene flame. The fewer pendant hydrogen on atoms outside the rings, the higher the flame resistance. Some of these materials are discussed in Chapters 7 and 8. New developments in polymerization techniques have only recently made these materials commercial possibilities. Their development, now spurred by the space program and other special applications, may be expected to proceed rapidly. It is conceivable that within a decade many uses for fibers and polymers with a moderate to high degree of fire risk will be served routinely by these new high-performance materials.

Table 1-12 Some New High-Performance Polymers

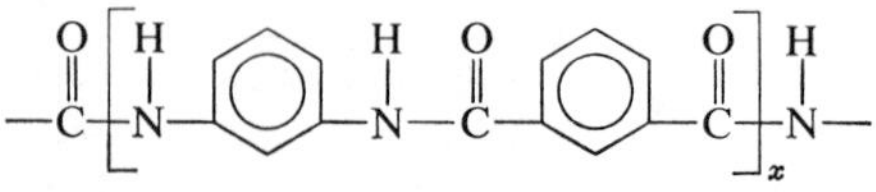

Nomex

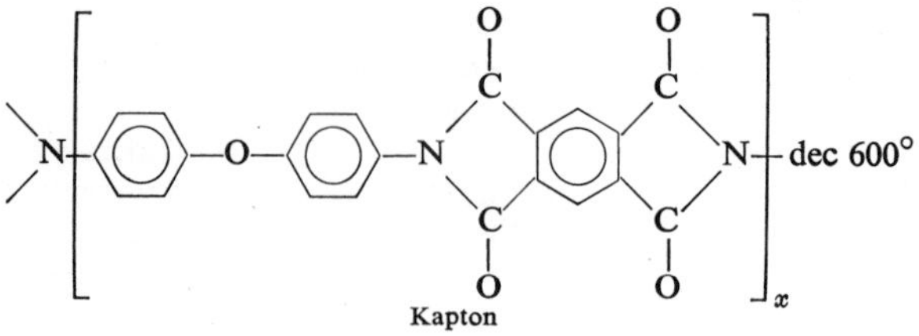

Kapton — dec 600°

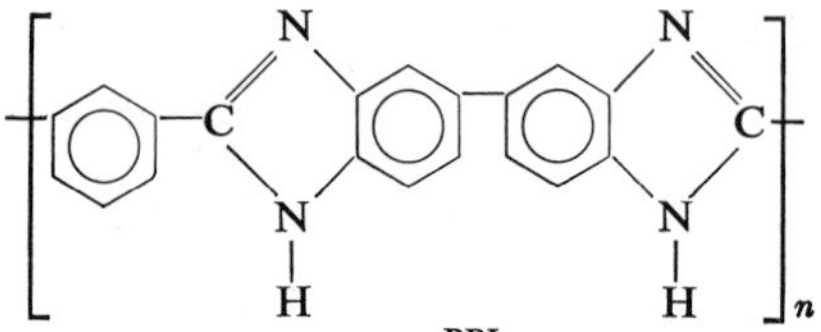

PBI

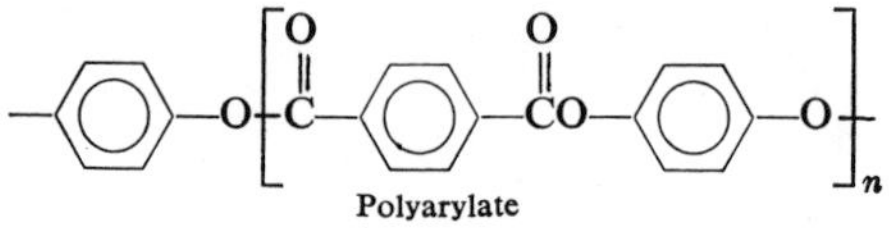

Polyarylate

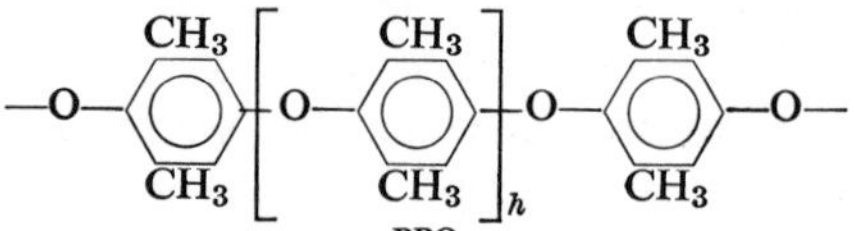

PPO

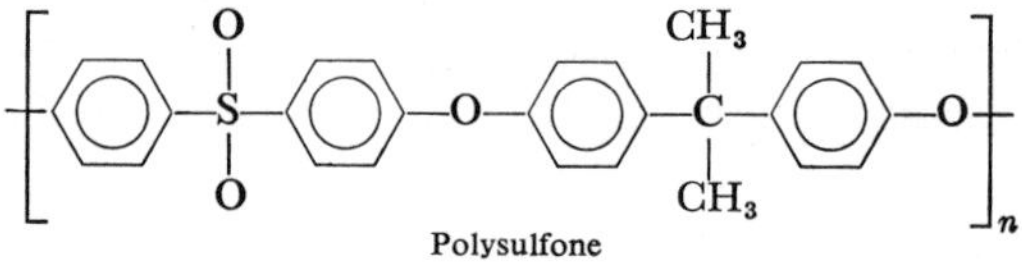

Polysulfone

REFERENCES

[1] C. I. Babcock, *Kirk-Othmer Encyclopedia of Chemical Technology*, 2nd ed., Vol. 9, Wiley-Interscience, New York, 1966, pp. 286–99.

[2] C. H. Vervalin, *Hydrocarbon Proc.*, **45**, 5, 227 (1966).

[3] D. I. Lawson, *J. Roy. Soc. Arts*, **112** (5090), 86 (1964).

[4] H. W. Emmons, "Fire Research—A Trip Report," *Rept. No. 1*, National Science Found. Grant GK 771, Div. Eng. Appl. Phys., Eng. Sci. Lab, Harvard Univ., Cambridge, Mass., 1968.

[5] H. W. Emmons, *Fire J.*, **1969**, 24 (May).

[6] B. Arden, *SAMPE J.*, **1968**, 33 (April/May).

[7] J. A. Rhys, *Chem. & Ind.*, **1969**, 187 (Feb. 15); N. E. Boyer, *Plast. Technol.*, **8**, 33 (1962); J. DiPietro, Paper presented at the 155th Annual Meeting of the Amer. Chem. Soc., San Francisco, April, 1968; L. W. Sayers, *Test. Inst. Ind.*, **3**, 168 (1965).

[8] C. J. Hilado, *Flammability Handbook for Plastics*, Technomic, Stamford, Conn., 1969, 164 pp.

[9] R. Reichberger, *Kunstst. Rundsch.*, **13** (9), 482 (1966).

[10] E. S. Khoroshaya, K. D. Korol'kova, L. F. Gdutviner, K. I. Friedgeim, I. V. Plotnikov, A. E. Rishin, and Yu. I. Kuznetsov, *Nauch.-Issled. Tr.*, *Vses. Nauch. Issled. Inst. Plenochnykh Mater. Iskusstv. Kozki*, No. 17, 187 (1966).

[11] C. P. Fenimore and F. J. Martin, *Mod. Plast.* **44**, 141 (1966); C. P. Fenimore, G. W. Jones, *Combustion and Flame, 1966* p. 295 (Sep.); also see ASTM D-2863.

[12] K. B. Goldblum, *SPE J.*, **25**, 50 (1969).

[13] A. F. Robertson, D. Gross, and J. Loftus, *ASTM Preprint 87*, 1956, 17 pp.

[14] J. D. Downing and R. M. Anderson, *J. Cell. Plast.*, **3**, 236 (1967).

[15] H. L. Vandersall, *J. Paint Technol.*, **39**, 494 (1967).

[16] M. M. Levy, *J. Cell. Plast.*, **3**, 168 (1967).

[17] C. J. Hilado, Paper presented at the 155th Annual Meeting of the Amer. Chem. Soc., San Francisco, April, 1968.

[18] J. M. Church, R. W. Little, and S. Coppick, *Ind. Eng. Chem.*, **42**, 418 (1950); H. J. Reese, *Melliand Textilber.*, **37**, 324 (1956); *ibid.*, **41**, 1403 (1968); F. Ward, *Text. Recorder*, **75**, 74 (1958).

[19] G. S. Buck, Jr., *Text. Inds.*, **114**, 101 (1950); G. Schoen, *Melliand Textilber.*, **48**, 215 (1967).

[20] P. M. Hay, *Amer. Dyestuff Reptr.*, **53**, 23 (1964).

[21] R. M. Stephenson and L. F. Stephenson, *Amer. Dyestuff Reptr.*, **57**, 623 (1968).

[22] Proposed tentative method of test for ease of ignition of finished textile floor covering materials, ASTM Committee D-13, Subcommittee A-11. Draft form; P. Juillard, *Teintex*, **32**, 102 (1967).

[23] W. P. Behnke and R. E. Seaman, *Mod. Text. Mag.*, **1969**, 19 (April).

[24] D. Gross, J. J. Loftus, and A. F. Robertson, *ASTM Pub. STP-422*, 1967.

[25] J. R. Gaskill and C. R. Veith, Paper presented at the 155th Annual Meeting of the Amer. Chem. Soc., San Francisco, April, 1968. Also see ASTM Special Technical Publication No. 422.

[26] R. Friedman and J. B. Levy, *WADC Tech. Rept. 56–568, ASTIA Doc. No. AD 110 685*, Wright Patterson Air Force Base, Ohio.

[27] "Final Report on Fire Extinguishing Agents for the Period 1 Sep. 1947 to 30 Jan. 1950," Purdue Univ. Found. and Dept. Chem., Purdue Univ., Lafayette, Ind., July, 1950.

[28] Products Chimiques Pechiney-Saint-Gobain, *Neth. Appl. 6,602,825*, Sept. 6, 1966; T. G. Lee, *J. Phys. Chem.* **67**, 360 (1963); N. Ando, *Zairyo*, **14**, 600 (1965); A. N. Baratov, *Zh. Vses. Khim. Obshchest.*, **12**, 276 (1967); M. C. Abrams, *Western States Combus. Inst.*, *Paper (USS) CI 61–26*, 1961, 15 pp., J. H. Burgoyne and J. F. Richardson, *Fuel*, **29**, 93 (1950); M. Friedrich, *Chem.-Ztg.*, **81**, 526 (1957); *ibid.*, **84**, 560 (1960); R. K. Sharma and J. Bardwell, *Combust. Flame*, **9**, 106 (1965); V. Zvonar, *Chem. Prum.*, **12**, 321 (1962); P. Volans, *Plast. Inst.*, *Trans. J.*, *Conf. Suppl. No. 2*, 47 (1967).

[29] C. P. Fenimore and F. J. Martin, *Combust. Flame*, **10**, 135 (1966).

[30] W. G. Parks, J. G. Erhardt, Jr., and D. R. Roberts, *Amer. Dyestuff Reptr.*, **39**, 294 (1950); G. J. Taupe and R. L. Stoker, *Trans. Amer. Soc. Mech. Eng.*, **73**, 1005 (1951); J. E. Dolan, *Chem. & Ind.*, **1952**, 368; H. A. Schuyten, J. W. Weaver, and J. D. Reid,

Advan. Chem., Ser. No. 9, 7–20 (1954); R. C. Laible, *Amer. Dyestuff Reptr.*, **47**, 173 (1958); S. K. Coburn and K. J. Morris, *Proc. Amer. Wood Preservers Assoc.*, **55**, 70 (1959); M. L. Amy and M. Phicot, *Chim. & Ind. (Paris)*, **83**, 411 (1960); A. Broido and S. B. Martin, *U.S. Dept. Com., Office Tech. Serv. AD 268, 729*, 1961, 20 pp.; A. F. Childs, *Dyer*, **122**, 585 (1959).

[31] C. M. Conrad and P. Harbrink, *Text. Res. J.*, **38**, 366 (1969); D. F. Arseneau, *Can. Wood Chem. Symp., Proc. 1st Symp.*, Toronto, 1963 (publ. 1965); D. J. Rasbash and B. Langford, *Chem. Eng. (London)*, **12**, 34 (1968).

[32] G. A. Byrne, D. Gardiner, and F. H. Holmes, *J. Appl. Chem.*, **16**, 81 (1966).

[33] W. K. Tang and W. K. Neill, *J. Polym. Sci., Part C*, 65 (1964).

[34] J. B. Berkowitz-Mattuck and T. Nogurchi, *J. Appl. Polym. Sci.*, **7**, 709 (1963).

[35] S. L. Madorsky, *Thermal Degradation of Organic Polymers*, Wiley-Interscience, New York, 1964, pp. 238 ff.

[36] F. B. Clarke and J. W. Lyons, *J. Amer. Chem. Soc.*, **88**, 4401 (1966).

[37] R. W. Little, *Text. Res. J.*, **21**, 901 (1951).

[38] "High Temperature Resistance and Thermal Degradation of Polymers," *S.C.I. Monograph No. 13*, Soc. Chem. Ind., London, 1961.

[39] G. C. Tesoro, S. B. Sello, and J. J. Willard, *Text. Res. J.*, **39**, 180 (1969); G. C. Tesoro, *Textilveredlung*, **2**, 435 (1967).

Chapter **2**

CHEMISTRY OF FIRE RETARDANTS BASED ON PHOSPHORUS

The chemistry of fire retardants centers around six elements: phosphorus, antimony, chlorine, bromine, boron, and nitrogen. Hundreds, perhaps thousands, of compounds of these elements have been synthesized for the purpose of conferring fire retardance on one or another substrate. Although it is true that there are occasional references to other elements, for example, the other group Va elements—As, Bi—and to other less likely materials—Se, Pb salts, Zn salts—it is becoming very clear that these six elements are the key to effective fire control. In this chapter and the next one the essentials of the chemistry of these elements are presented to enable the reader to become reasonably familiar with the nomenclature and some of the reactions to be expected. The references will enable him to gain further, more detailed insights.

Of the six, phosphorus has the most complex and perhaps the most fully developed chemistry. In all cases phosphorus is the central element in a compound; there may be an almost infinite variety of substituents in several oxidation states. Though antimony could in theory enter into the same multiplicity of reactions, the chemistry of this element is in fact but little developed. Nitrogen, in the same group, has a much simpler chemistry by virtue of its electronic makeup (there is no possibility for *spd* hybridization) and, for fire retardance, the number of compounds of nitrogen of interest is limited. In contrast to P, Sb, and N, the halogens are not found as central atoms in fire-retardant compounds. Rather, they are substituents primarily on organic compounds. So the chemistry of the halogens for our purposes is largely the chemistry of halogenation reactions, that is, how a halogen is best introduced to a given system. Boron has a well-developed chemistry but almost none of it has been applied in the field of fire retardants. Therefore, not a great deal will be said of its more complex and more interesting aspects,

Table 2-1 Representative Structures for Odd Oxidation States

Item No.	Oxidation No.	Structure	Name	Structure	Name
1	+5	$(HO)_3PO$	Phosphoric acid	$(RO)_3PO$	Trialkyl *phosphate* (R = alkyl)
2		...	...	$(NH_4)H_2PO_4$	Monoammonium dihydrogen phosphate
3		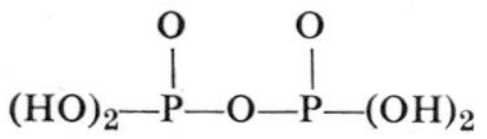 $(HO)_2{-}P{-}O{-}P{-}(OH)_2$	*Pyro*phosphoric acid	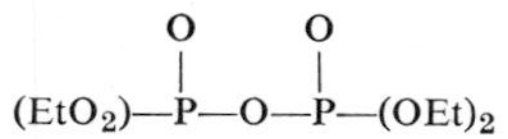 $(EtO_2){-}P{-}O{-}P{-}(OEt)_2$	Tetraethyl *pyro*phosphate
4		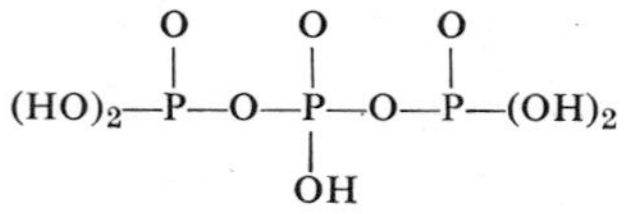 $(HO)_2{-}P{-}O{-}P{-}O{-}P{-}(OH)_2$	*Tri*phosphoric acid	$(NH_4)_5P_3O_{10}$	Penta ammonium triphosphate
5		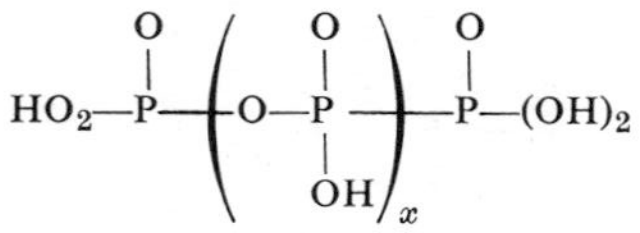 $HO_2{-}P{-}O{-}P{-}(OH/)_x$	*Poly*phosphoric acid	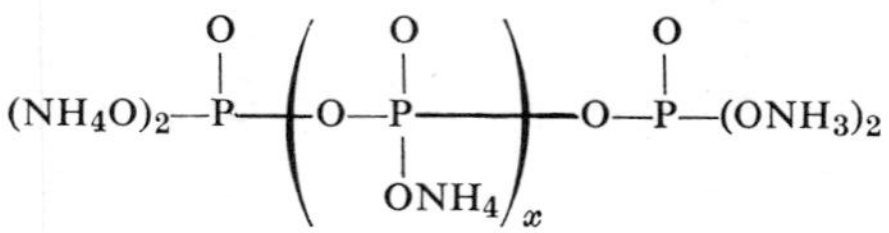 $(NH_4O)_2{-}P{-}O{-}P{-}O{-}P{-}(ONH_3)_2$ $(ONH_4/)_x$	Ammonium *poly*phosphate
6		$POCl_3$	Phosphorus oxychloride, phosphoryl trichloride	...	...
7	+3	$(HO)_2P(H)O$	Phosph*orous* acid	$(EtO)_3P$	Triethyl phosph*ite*
8		PCl_3	Phosphorus trichloride	$(RO)_2POH$, or, $(RO)_2P(H)O$	Diethyl phosph*ite*
9		PBr_3	Phosphorus tribromide	...	...

No.	Ox.				
10		$EtP(O)(OH)_2$	Ethylphospho*nic* acid	$CH_3P(O)(OEt)_2$	Diethylmethyl-phospho*nate* or Diethyl methane phospho*nate*
11	+1	$H_2P(O)(OH)$	Hypophosphorous acid	...	...
12		$Et_2P(O)OH$	Diethylphosphinic acid	$Et_2P(O)OCH_3$	Methyldiethyl phosphin*ate*
13		$EtP(O)(H)(OH)$	Ethyl phosphon*ous* acid or Ethane phos-phon*ous* acid	$EtP(O)(H)(OCH_3)$	Methylethyl phos-phon*ite* or Methyl ethane phosphon*ite*
14	−1	$[H_3PO]$ not known	[Trihydrogen phos-phine oxide]	Et_3PO	Triethyl phos-phine oxide
15		...	...	$CH_3(Et)_2PO$	Methyl diethyl phosphine oxide
16		$[H_2POH]$ not known	[Phosphin*ous* acid]	Et_2POH	Diethyl phos-phin*ous* acid
17		...	...	Et_2POCH_2	Methyl diethyl phosphin*ite*
18	−3	H_3P	Phosphine	EtH_2P	Monoethyl phos-phine
19		...	...	Et_2HP	Diethyl phosphine
20		...	...	Et_3P	Triethyl phosphine
21		H_4P^+	Phosphon*ium* ion	$[Et_3H_2P]Cl$	Diethyldihydrogen phosphonium chloride
22		...	...	$[Et_4P]Cl$	Tetraethyl phos-phonium chloride

ATOMIC STRUCTURE AND BONDING FEATURES

Phosphorus has an atomic number of 15 and an atomic weight, for the single naturally occurring isotope, of 30.98 [1]. There are 15 protons and 16 electrons in the nucleus. It has a nuclear spin of $\frac{1}{2}$ and no quadropole moment; this makes identification of its electronic environment by nuclear magnetic resonance (nmr) straightforward and, indeed, P^{31} nmr is now perhaps the single most powerful tool for studying transformations of phosphorus compounds [2]. The electronic structure is $1s^2 2s^2 2p^6 3s^2 3p^3$ for a total of 15. Bonding in phosphorus compounds can therefore include simple sharing of $3p$ orbitals or various $3s$-$3p$ hybrids. More important is the ready availability of $3d$ orbitals which allow for spd hybrids. Most phosphorus-ligand single bonds are now thought to be hybrids; for example, the PO single bonds in PO_4^{3-} are sp^3 hybrids, the Cl—P bonds in $POCl_3$ are largely sp^3, and the Cl—P bonds in PCl_3 are a mixture of p^3 and sp^3 character. The pentavalent and hexavalent compounds are, of course, spd hybrids, for example, PCl_5-sp^3d, PCl_6^- ion-sp^3d^2. Many P-ligand bonds have considerable π bond character. π bonds are formed in addition to single or σ bonds when additional orbitals of the P and the ligand overlap significantly. This is another way of describing multiple bond formation. For example, for the molecules $POCl_3$ and H_3PO_4, the formulas are often written

<pre>
        Cl   OH
        | |
  Cl—P=O   HO—P=O
        | |
        Cl   OH
</pre>

showing the PO bond as a double bond—one σ and one π. Calculations from X-ray data show that the double bond is indeed localized as shown for $POCl_3$. On the other hand, for $PO(OH)_3$ the π character is spread over all the PO bonds in a true molecular orbital so that the structural formula is misleading [1].

TYPICAL COMPOUNDS

Almost all fire retardants that contain phosphorus are in odd oxidation states. Table 2-1 lists these states in various forms and gives the presently accepted nomenclature (that used in the current literature—generally that used by Kosolapoff [3] and summarized by Van Wazer in Chapter 7 of reference 1). The phosphates and polyphosphates all contain quadruply connected phosphorus surrounded by oxygen. The phosphites can be rearranged to phosphonates containing one P—C bond. The phosphonates and phosphites have identical empirical formulas but very different chemical properties. A similar situation obtains for the phosphine oxides and the dialkyl phosphinites (items 14–17). Some confusion in the nomenclature of these compounds remains; it is best to use a structural representation where possible. This can usually be done on one line; for example, $(RO)_3P$ and $(RO)_2P(O)R$ are clearly differentiated.

ILLUSTRATIVE EXAMPLES AND PREPARATIONS

Phosphoric Acids and Their Salts

Solutions of monomeric or orthophosphoric acid are made either by reacting H_2SO_4 with phosphate rock

$$10H_2SO_4 + Ca_{10}(OH)_2(PO_4)_6 \rightarrow 6H_3PO_4 + 10CaSO_4 \cdot 2H_2O(c) \qquad (1)$$

to give what is known as "wet process" or "green" acid or by first producing elemental phosphorus followed by oxidation and hydration

$$Ca_{10}(OH)_2(PO_4)_6 + 15C + 10SiO_2 \rightarrow$$
$$\tfrac{3}{2}P_4 + 15CO + 10CaSiO_3 + H_2O \qquad (2)$$

$$P_4 + 5O_2 \longrightarrow P_4O_{10} \xrightarrow{6H_2O} 4H_3PO_4 \xrightarrow{xsH_2O} H_3PO_{4(soln)} \qquad (3)$$

to give what is called "furnace" acid. Wet process acid contains many impurities and is not used in significant amounts in industry. Furnace acid is made in high purity and finds wide use in such diverse end products as cola drinks, leavening agents, meat products, detergents, and metal finishing baths. For most fire-safety uses, derivatives of furnace acid are used.

If the amount of water used to absorb P_4O_{10} vapor is carefully controlled, incomplete hydrolysis occurs and polyphosphoric acids result; for example,

$$P_4O_{10} + 4H_2O \rightarrow 2H_2P_2O_7 \qquad (4)$$

In fact, polyphosphoric acids do not contain single species in the liquid state but rather they are mixtures for which the species distribution can be shown to be described by a Gaussian law [4]. For example, polyphosphoric acid containing 79.7% P_2O_5 averages the formula $H_4P_2O_7$, but actually contains ca. 18% H_3PO_4, ca. 42% $H_4P_2O_7$, ca. 23% $H_5P_3O_4$, and ca. 17% higher polymers. The polyphosphoric acids are significant as reactants to make phosphate esters and to prepare ammonium polyphosphates of low molecular weight. (See subsequent sections.)

Direct esterification of any of the phosphoric acids is difficult. It is not feasible, for example, to prepare a trialkyl phosphate by heating H_3PO_4 with an alcohol. The reaction is sluggish and requires high temperatures for lengthy periods. Commercially, the triesters are prepared, rather, from

$POCl_3$. Alcohols attack polyphosphoric acid randomly along the chain to produce, ultimately, monoalkyl orthophosphates and free phosphoric acid [5]:

$$(HO)_2-\overset{\overset{\displaystyle O}{\|}}{P}-O\text{-}(\overset{\overset{\displaystyle O}{\|}}{\underset{\underset{\displaystyle OH}{|}}{P}}-O)_{\!x}\text{-}\overset{\overset{\displaystyle O}{\|}}{P}-(OH)_2 + (x+1)ROH \xrightarrow[1\text{-}2\,\text{hr}]{80°C}$$

$$(x+1)ROP(O)(OH)_2 + H_3PO_4 \quad (5)$$

This reaction is used currently to phosphorylate polyols for use in polyurethane foam. Note that some free phosphoric acid is always produced; the lower x the higher the proportion of free acid. This can be minimized by using the highest possible P_2O_5 content in the acid.

Two crystalline salts are readily isolated from ammoniated solutions of orthophosphoric acid:

$NH_4H_2PO_4$	$(NH_4)_2HPO_4$
Monoammonium phosphate	Diammonium phosphate
MAP mol wt 115	DAP mol wt 132

These are very soluble in water with maximum solubility coming for a 50 mole % MAP:50 mole % DAP mixture. Even higher solubility can be achieved by ammoniating solutions of condensed phosphoric acid. Some properties of these species are shown in Table 2-2. The ammonium phosphates

Table 2-2 Properties of Ammonium Phosphates

	$NN_4H_2PO_4$, MAP	$(NH_4)_2HPO_4$, DAP	50:50 Mole Ratio, MAP:DAP	Ammoniated Acid at 79.7% P_2O_5 (to pH 6)^a
% N	12.2	21.1	17.0	18.3
% P	27.0	23.4	25.1	27.0
pH 1% soln	4.7	ca. 8	· · ·	6
g/100 g H_2O at 25°C	30	42	50–51	150
g(N)/100 g H_2O	3.6	8.8	8.5	27.5
g(P)/100 g H_2O	8.1	9.8	12.5	40.4
g(N)/100 g soln	2.8	6.2	5.7	11.0
g(P)/100 g soln	6.2	6.9	8.3	16.2

	Insol $(NH_4PO_3)_x$
% N	14.4
% P	32
pH, slurry	6.5
Insolubility in water at 25°C, %	85–90

^{a} Nominally $(NH_4)_3\,HP_2O_7$; solubility data at 0°C

lose ammonia at elevated temperatures [there is a noticeable odor of NH_3 over $(NH_4)_2HPO_4$ solutions even at room temperature]. This is described by the following equations [6]:

$$NH_4H_2PO_4(c) \rightleftharpoons NH_3(g) + H_3PO_4(s)$$
$$P_{mm} = 0.05 \text{ at } 125°C$$
$$(NH_4)_2HPO_4(c) \rightleftharpoons NH_4H_2PO_4(c) + NH_3(g)$$
$$\log P_{mm} = -\frac{3063}{T} + 1.75 \log T + 3.3$$

where P is in millimeters of Hg and T is the absolute temperature in degrees Kelvin.

Long-chain polymeric ammonium polyphosphates analogous to potassium Kurrol's salt have been reported and at least one is a commercial product. Two patents describe the preparation of $(NH_4PO_3)_x$, one by ammoniating polyphosphoric acid under pressure at high temperature followed by slow cooling and crystallizing in water [7], the other by polymerizing short-chain phosphates using urea to remove the chemical water as NH_3 and CO_2 [8]. The former process gives the polymer in low yield ($<50\%$) and will require further research to perfect. (The urea-polyphosphoric acid system has been studied in some detail—urea dissolves in the acid—and has itself been suggested as a fire retardant [9].) $(NH_4PO_3)_x$ is a crystalline compound that is relatively insoluble in water. The commercial product contains about 85% material that is insoluble in water at $25°C$ and 15% that is soluble. It is a mixture of soluble and insoluble species such that a true solubility cannot be given. It is essentially neutral in pH. When heated, it gradually gives off NH_3, the evolution becoming rapid at about $250°C$. Careful study of analytical data, ir and P^{31} nuclear magnetic resonance spectra, and viscometric and light scattering information have shown conclusively that the structure is that of a long-chain, unbranched polymer:

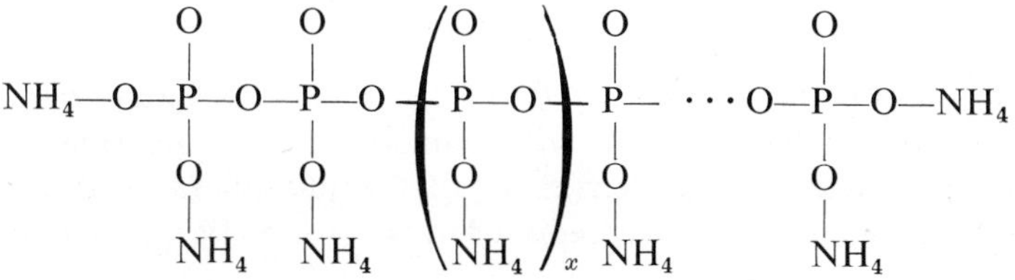

where x has been observed to be as high as 20,000. There are several crystalline forms; all have similar low solubilities in water [10]. This type of material finds use in intumescent paints and mastics and as an additive for plastic products, paper, and so forth. Table 2-2 summarizes some of its more salient features.

Whereas the ammonium orthophosphates are among the oldest of fire retardants [11], other P—N compounds having P in the $+5$ oxidation state are of much more recent origin. Urea-phosphoric acid solutions have been known as fire retardants for some years [12]. Guanidine phosphate is well known [13], as are guanylurea phosphate [14] and melamine phosphate:

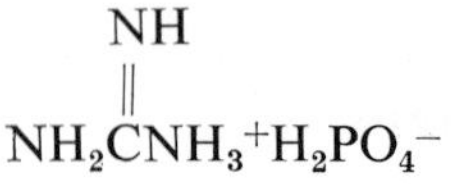

$$\overset{\displaystyle NH}{\underset{\displaystyle NH_2\overset{\|}{C}NH_3^+H_2PO_4^-}{\|}} \qquad \overset{\displaystyle O \quad NH}{\underset{\displaystyle NH_2\overset{\|}{C}NH\overset{\|}{C}{-}NH_3^+ \; H_2PO_4^-}{}}$$

Guanidine phosphate 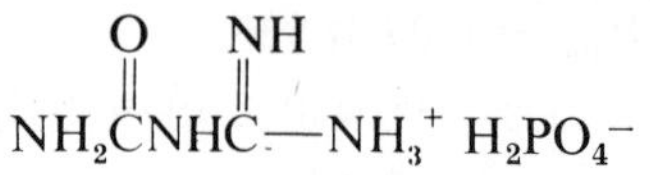   Guanylurea phosphate

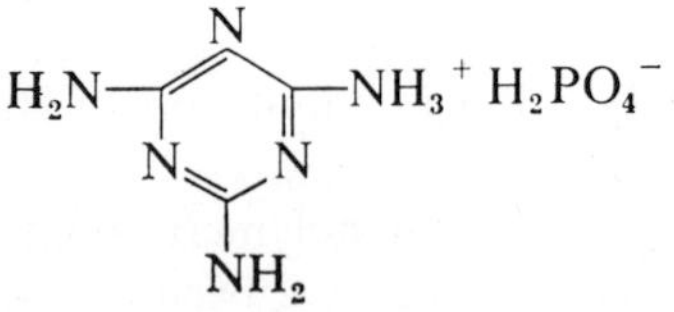

Melamine monophosphate

These materials are of interest in systems where a degree of oil solubility or, at least, oil compatibility is desired. Reaction products of dicyandiamide and H_3PO_4 have been disclosed in the patent literature [15] and are of current interest in fireproofing wood (see Chapter 4). Ammonia-phosphorus pentoxide reaction products and derivatives have been described [16], the solubility of which can be reduced with tempering. These products presumably contain P—N bonds and are polyphosphoramides. A similar type of P—N bonding is found for reaction products of melamine and P_4O_{10} in the absence of moisture [17]:

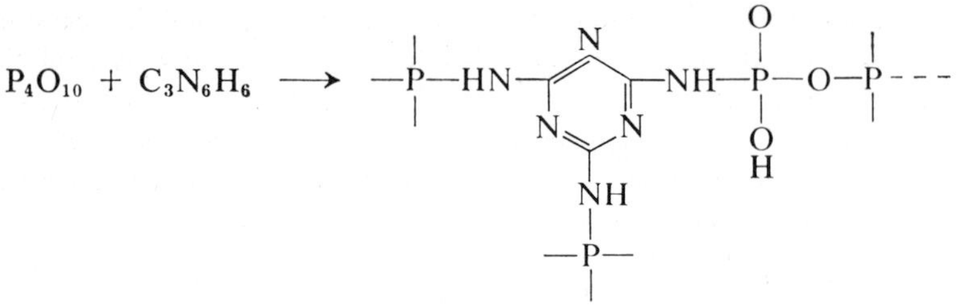

There are two more families of PN compounds of interest: the phosphorylamides and the phosphonitrilic compounds. The phosphorylamides can be regarded as simple derivatives of the phosphoric acids in which hydroxyl groups are successively replaced by NH_2 (or NHR, or NRR′) groups:

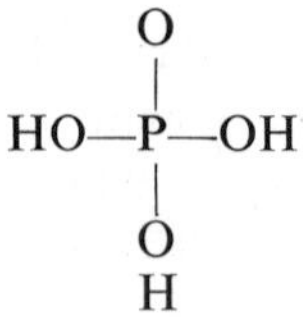

Phosphoric acid 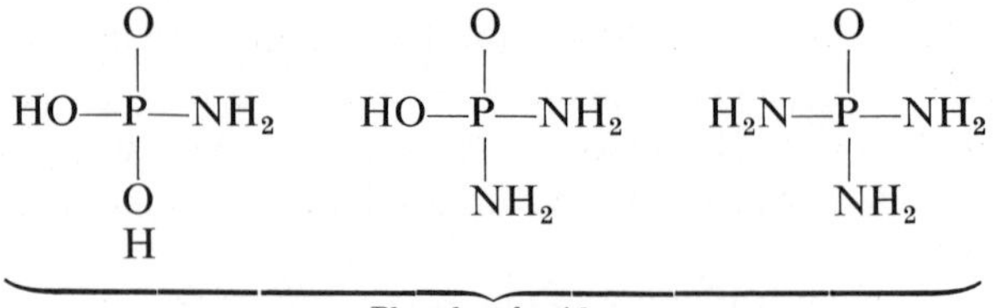   Phosphorylamides

Some idea of the variation that can be achieved can be gained from Table 2-3. Considerable attention has been devoted to the polyphosphorylamides [19] as fire retardants as ingredients in phenolic resins or aminoplasts [27] and in combination with polyols in intumescent systems [28]. Monomeric phosphoramides substituted variously on the N atoms have been proposed as fire-resistant functional fluids [29] and an amide derived from $POCl_3$ and aniline is mentioned for fire-resistant urethane coatings [30]. Note in Table 2-3 the reaction product of ethyleneimine and $POCl_3$ to yield tris-aziridinyl phosphorus oxide, usually referred to in the literature as APO (and incorrectly termed a phosphine oxide). In this compound the phosphorus remains in the +5 state and the product is properly termed a phosphorylamide. The aziridine ring is very reactive and will react with hydroxyl groups and the like to form interesting polymers. In combination with tetrakishydroxymethyl-phosphonium chloride (THPC), it has been studied extensively for conferring H_2O-resistant fire-retardant coatings on cotton textiles. (For more on THPC, see the section entitled "Phosphines, Phosphonium Salts, and Phosphine Oxides"; see also Chapter 5.) Most of the effort on these compounds has focused on the symmetrical trisubstituted amides or on unsymmetrical products made by simple, straightforward processes and it appears unlikely that materials from more complex syntheses will be commercialized. For details of P—N chemistry, the reader is referred to the recent thorough review by Fluck [31].

Fluck [31] also covers the phosphonitrilic compounds. Representative of this group, often cited in literature on fire retardents, is the polymeric product resulting from the reaction of a phosphorus pentahalide and ammonia:

$$PCl_5 + NH_4Cl \rightarrow \frac{1}{x}(NPCl_2)_x + 4HCl$$

The polymer obtained in this way is usually either a cyclic trimer or tetramer:

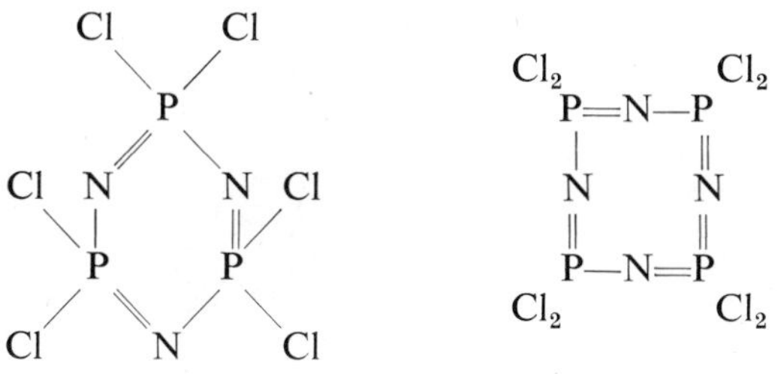

Trimer   Tetramer

A variety of substituted products can be produced by reacting $(NPCl_2)_x$ with amines to replace the chlorine atoms [32]. Chlorine can also be replaced by phenyl groups, by fluorine, by alcohols, and by phenols to give products

Table 2-3 Some Phosphorylamides

Reactants	Conditions	Product	References
$POCl_3$, NH_3	Chloroform, $-10°C$	$OP(NH_2)_3$ dec on heating	18
$POCl_3$, NH_3	Kerosene, $200°C$	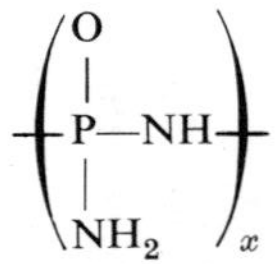 $\left(\begin{array}{c}\text{O}\\\|\\\text{P—NH}\\\|\\\text{NH}_2\end{array}\right)_x$ dec on heating	19
$POCl_3$, 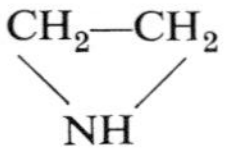 $CH_2{-}CH_2$ with NH	...	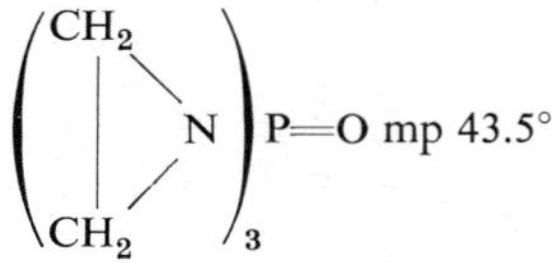 $\left(\begin{array}{c}\text{CH}_2\\ \\\text{N}\\ \\\text{CH}_2\end{array}\right)_3\!\!\text{P=O}$ mp 43.5°	20
$Et_2NP(O)Cl_2$, NH_3	Benzene, $20°C$	$Et_2NPO(NH_2)_2$ polymerizes on heating to 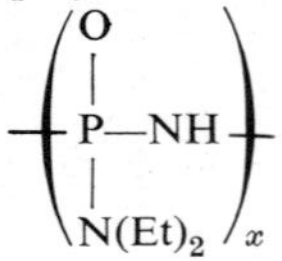 $\left(\begin{array}{c}\text{O}\\\|\\\text{P—NH}\\\|\\\text{N(Et)}_2\end{array}\right)_x$	21
$P_4O_4Cl_{10}$, $PhNH_2$	Chloroform, $<20°C$, then heat at $200°C$	$[P_4O_4(NHPh)_6]_x$ mp 120°	22

Ph_2NPOCl_2, HO–⟨⟩–OH 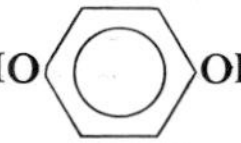	Under nitrogen, 205°C	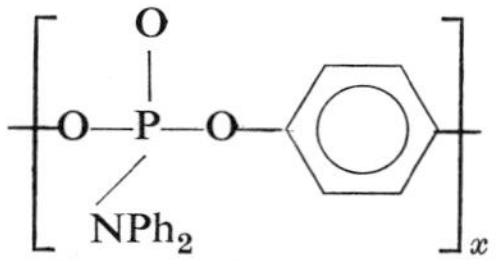 rubbery polymer	23
$CH_2=CHCH_2OH$, $POCl_3$, NH_3	Toluene, 0°C	$(CH_2=CHCH_2O)_2P(O)NH_2$	24
$POCl_3$, HOROH, $CH_2\!\!-\!\!CH_2$ (N)	2 steps: (1) glycol + $POCl_3$, 2 hr at 25°C; (2) in Et_3N, 1 hr at 5 NH	$(CH_2CH_2N)_2\!-\!P\!-\!O\!-\!R\!-\!O\!-\!P\!-\!(NCH_2CH_2)_2$ liquids and low melting solids	25
$(ClCH_2CH_2O)_2\!-\!P\!-\!NH_2$, Cl_3CCHO	Et_3N, 2.5 hr, 70°	$(ClCH_2CH_2O)_2$-$PNHCH(OH)CCl_3$ mp 91.2°	26
$[(CH_3)_2N]_2\!-\!P\!-\!Cl$, glycol	Et_3N, 6.5 hr, 55–80°	$[(CH_3)_2N]_2\!-\!P\!-\!O\!-\!R\!-\!O\!-\!H$	26a
$(RO)_2\!-\!P\!-\!NHR'$, $ClCH_2CH\!\!-\!\!CH_2$ (O)	Na, ϕ, $\tfrac{1}{2}$ hr, 50°	$(RO)_2\!-\!P\!-\!N(R')CH_2\!-\!CH\!-\!CH_2$ (O)	26b

Table 2-4 Phosphonitrilic Compounds Useful as Fire Retardants

Reactants	Conditions	Product	Reference
NH_4Cl, PCl_5	ϕCl, $1\frac{1}{2}$ hr, $125°$	$(PNCl_2)_{3-4}$ trimer: bp $124°$ (10 mm) tetramer: bp $185°$ (10 mm)	33
$(PNCl_2)_3$, NH_3, MeOH	MeOH, $25-40°$, $\frac{1}{2}$ hr	$(PN(NH_2)_2)_3$; $[PN(NH_2)(OCH_3)]_3$	33a
$(PNCl_2)_3$, NH_3, $COCl_2$	2 steps: (1) NH_3 in either (2) $COCl_2$ in $\phi\,Cl_2$, $1\frac{1}{2}$ hr at $170°$	$P_3N_3Cl_4(NCO)_2$ solid	34
$(PNCl_2)_3$, ϕ, $AlCl_3$	Reflux in N_2, 72 hr	$(PN(Cl)\phi)_3$	34a
$(PNCl_2)_3$, NH_2RNH_2 ($R \geq$ 3C atoms)	THF, 7 hr, $100°$ in vacuo	$-(P_3N_3)-NHRNH-_x$	35
$(PNCl_2)_3$, ϕONa, NH_3	2 steps in $CHCl_3$	$-[N{=}P(O\phi)(NH_2)]-_3$	36
$(PNCl_2)_3$, $\phi(OH)_3$	1 hr, $160°$	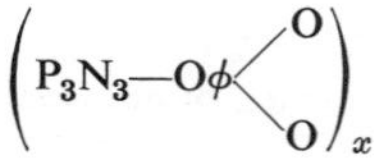 $\left(P_3N_3-O\phi\overset{\displaystyle O}{\underset{\displaystyle O}{\bigtriangleup}}\right)_x$	37
$(PNCl_2)_{3-4}$, $NaOCH_2CF_3$ $NaOCH_2C_3F_7$	2 steps: (1) polymerize $(PNCl_2)_x$, 2–4 hr, $300°$, vac., (2) NaOR in C_6H_6, 18–48 hr, $65°$	$P_xN_x(OCH_2CF_3)_{2x}$; $P_xN_x(CCH_2C_3F_7)_{7/2x}$	38
$(PNCl_2)_3$, $\phi(F)(OH)$, CF_3CH_2OH	NaH, KOH, xylene, 16 hr, $60°$	$P_3N_3(O\phi F)(OCH_2CF_3)$ bp $170-240°$ (0.03 mm)	39
$(PNCl_2)_{2x}$, $(CH_3)_2Zn$	Xylene, 48 hr, $120°$	$[PN(CH_3)_2]_x$ stable to $300°$	40

with high or low resistance to hydrolytic scission of the P—N bonds and to produce the desired reactivity and solubility. Table 2-4 lists some compounds of interest. The principal reactants have been amines and alcohols or phenols. Of note are the fluoroalkoxy derivatives which are often nonflammable as well as highly resistant to hydrolysis (unlike many other members of this family). Kosolapoff summarizes products that can be obtained directly from PCl_5 and amines [41]. The basic chemistry of these systems has received considerable attention from Moeller and his coworkers [42].

Phosphorus isocyanates have received a little attention [42a] for specialty

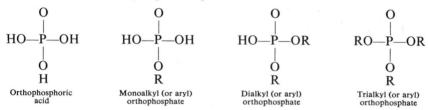

urethanes and the like.

Fluorophosphates, for example, NH_4PF_6, have also been claimed as fire retardants [42b].

Organic Esters of Phosphoric Acid

By substituting for one or two or three hydroxyl groups on orthophosphoric acid with RO^- groups, one gets a series of esters:

O ‖ HO—P—OH | O | H	O ‖ HO—P—OH | O | R	O ‖ HO—P—OR | O | R	O ‖ RO—P—OR | O | R
Orthophosphoric acid	Monoalkyl (or aryl) orthophosphate	Dialkyl (or aryl) orthophosphate	Trialkyl (or aryl) orthophosphate

The monoesters can be prepared from the corresponding alcohol or phenol and polyphosphoric acid as shown in Equation 5 [5]. Purification entails removal of the free H_3PO_4 which is always generated. In practice, this scheme is likely to be used only when the excess acid is unobjectionable. The principal use is in phosphorylating polyols for use in urethanes (see Chapter 8). Monoesters with alkyl groups containing less than about six carbon atoms are readily soluble in water. Higher homologs are dispersible in water and are excellent surfactants.

Diesters are most frequently produced in mixture with monoesters from the reaction of alcohols with phosphorus pentoxide:

$$6ROH + P_4O_{10} \rightarrow 2ROP(O)(OH)_2 + 2(RO)_2P(O)OH$$
$$\text{Monoester} \qquad\qquad \text{Diester}$$

The reaction also produces very small amounts of triester and free H_3PO_4. The diester can be isolated from the monoester by virtue of the fact that the

Table 2-5 Some Phosphate Esters of Interest

Reactants	Conditions	Product	References
EtOH, polyphosphoric acid (80%, P_2O_5) mole ratio 1EtOH/2P	60°, 8 hr	EtOP(O)(OH)$_2$, H_3PO_4 liquid	5
6EtOH, P_4O_{10}	60°, 8 hr	EtPO(O)(OH)$_2$ + (EtO)$_2$P(O)OH liquids	43, 44
3EtOH, POCl$_3$	Room temp, trace PCl$_3$	(EtO)$_3$PO bp 210–220°	44
ROH, H_3PO_4	120°, 10 mm	Alkyl phosphoric acids	44
ROH = sorbitol, mannitol, pentaerythritol			
Phenol, POCl$_3$	150–300°C, catalyst (metal halide), 6–9 hr	(PhO)$_3$PO mp 50°	44
CH$_2$—CH$_2$, POCl$_3$ (epoxide)	Catalyst (PCl$_3$, AlCl$_3$, etc.)	(Cl—CH$_2$CH$_2$O)$_3$PO bp 180° (5 mm)	44
CH$_2$—CH$_2$, (ϕO)$_2$P(O)Cl (epoxide)	25–40°	(ϕO)$_2$P(O)(OCH$_2$CH$_2$Cl)	44a
RCH—CH$_2$, H_3PO_4 (epoxide)	...	$\left(\text{RCHCH}_2\text{O}\right)_3$ PO with OH	45
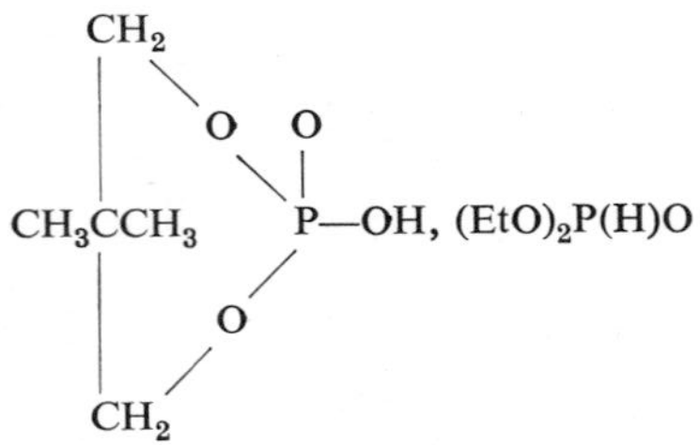 cyclic structure with CH$_2$, CH$_3$CCH$_3$, CH$_2$, P—OH, (EtO)$_2$P(H)O	Transesterification 165°C, 7 hr	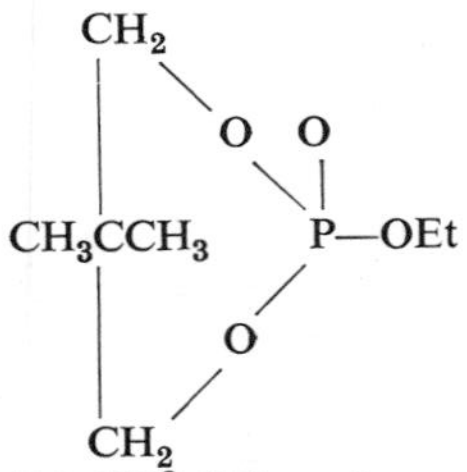 cyclic structure with P—OEt, CH$_3$CCH$_3$, CH$_2$ 109–112° (0.8 mm)	46

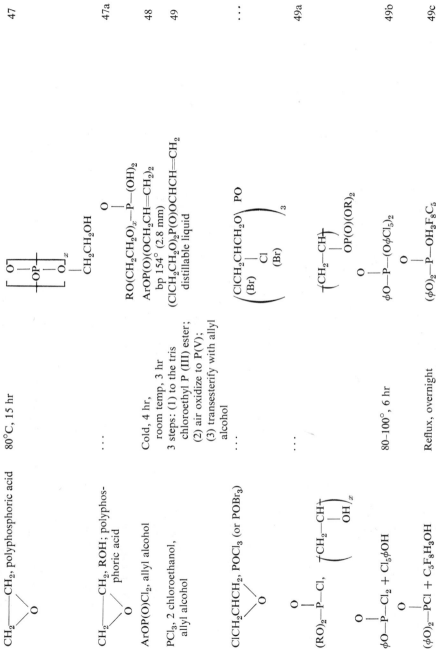

Reactants	Conditions	Product	Ref.
CH_2—CH_2 epoxide, polyphosphoric acid	80°C, 15 hr	[polymer structure]	47
CH_2—CH_2, ROH; polyphosphoric acid	...	$RO(CH_2CH_2O)_x$—P—$(OH)_2$ ... CH_2CH_2OH	47a
$ArOP(O)Cl_2$, allyl alcohol	Cold, 4 hr, room temp, 3 hr	$ArOP(O)(OCH_2CH=CH_2)_2$ bp 154° (2.8 mm)	48
PCl_3, 2 chloroethanol, allyl alcohol	3 steps: (1) to the tris chloroethyl P (III) ester; (2) air oxidize to P(V); (3) transesterify with allyl alcohol	$(ClCH_2CH_2O)_2P(O)OCHCH=CH_2$ distillable liquid	49
$ClCH_2CHCH_2$, $POCl_3$ (or $POBr_3$)	...	$(ClCH_2CHCH_2O)_3$ PO (Br)	...
$(RO)_2$—P—Cl, $[CH_2$—CH—$OH]_x$		$[CH_2$—CH—$OP(O)(OR)_2]$	49a
ϕO—P—Cl_2 + $Cl_5\phi OH$	80–100°, 6 hr	ϕO—P—$(O\phi Cl_5)_2$	49b
$(\phi O)_2$—PCl + $C_5F_8H_3OH$	Reflux, overnight	$(\phi O)_2$—P—$OH_3F_8C_5$	49c

Table 2-5 (Continued)

Reactants	Conditions	Product	References
$(RO)_2P(O)OH$, $RCHCH_2$ (epoxide)	2 steps: (1) to the triester; (2) heat to polymerize losing 2ROH	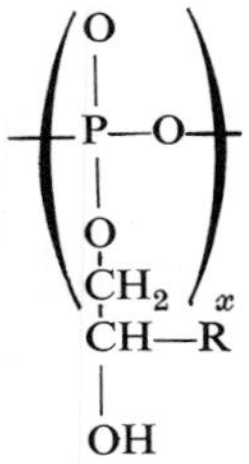 polymer: $\left(-P(O)(-O-)-O-CH_2-CH(R)-OH\right)_x$	50
$(EtO)_3PO$, P_4O_{10}	Chloroform, to 130°C, 2 hr	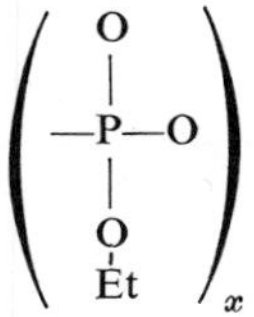 polymer: $\left(-P(=O)(-O-)-O-Et\right)_x$, OH	51
$ClCH_2CHCH_2$ (epoxide), 100% H_3PO_4	2 hr, 95°, inert atmos	$(ClCH_2CHCH_2O)_3PO$ liquid, OH	52
$CH_2C(CH_2Cl)_2CH_2$, $POCl_3$ (epoxide)	1 hr, 100°, strip at 1-mm Hg	$[(ClCH_2)_3CCH_2O]_3PO$ viscous oil	53
$(CH_2{=}CHCH_2O)_3PO + CCl_4$	Benzoyl peroxide, bromine reflux, stand 12 hr	$[CH_2(Br)CH(Br)CH_2O]_2P(O)OCH_2CH(Cl)CH_2CCl_3$	54
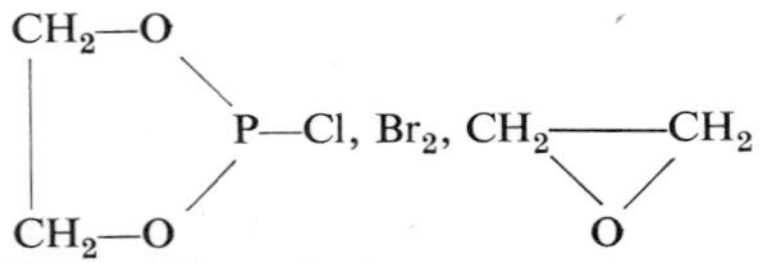 cyclic structure: $CH_2{-}O$, $CH_2{-}O$, $P{-}Cl$, Br_2, $CH_2{-}CH_2$ (epoxide)	2 steps: (1) Br_2 at 10–20°; (2) EO, 50–120°	$(BrCH_2CH_2O)_2{-}P(=O){-}OCH_2CH_2Cl$	54a

barium salts have very different solubilities in water, the diester salt being more soluble [43]. Commercially, this separation is not often practiced since both esters are equally useful in the principal applications: gasoline additives (as transition metal salts), plasticizers, specialty fire retardants, and anti-corrosion additives (as NH_4 salts).

Both mono- and diesters can be prepared via oxyhalophosphates and ROH:

$$ROH + xsPOCl_3 \rightarrow ROP(O)Cl_2 \xrightarrow[H_2O]{\text{room temp}} ROP(O)(OH)_2$$

$$2ROH + POCl_3 \rightarrow (RO)_2P(O)Cl \xrightarrow[\substack{\text{warm} \\ \text{MOH}}]{H_2O} (RO)_2P(O)OM$$

When R = alkyl, the alcoholysis proceeds smoothly at or near room temperature. When phenols are used, higher temperatures are required and the substitution cannot be cleanly held to mono- or di- but rather a mixture may be obtained [43]. Use of magnesium, iron, or aluminum chloride catalysts often gets around these problems [44]. To prevent hydrolysis of ester linkages, base is used in the hydrolysis of the diester monochloride to produce, cleanly, the dialkyl phosphate.

Triesters are afforded through the action of alcohols on phosphoryl trichloride in the presence of a tertiary base:

$$3ROH + POCl_3 + 3\text{ Base} \xrightarrow[\text{solvent}]{\text{inert}} (RO)_3PO + 3\text{ Base·HCl}$$

The base can be regenerated and recycled. In the case of phenols, the base and solvent can be eliminated and higher temperatures used. A variety of catalysts have been employed along with excess phenol to afford high yields of triaryl phosphates [44]. These compounds (e.g., tricresyl, triphenyl, or mixed esters) have been used for many years as fire-retardant plasticizers; the literature is replete with references to these in a wide variety of substrates.

Table 2-5 has entries of interest in fire retardance culled from the literature. Specific applications of many of these esters are described in the appropriate chapters. It is important to note the ease with which halogens can be introduced into these esters; for example, by using ethylene oxide the halogen in $POCl_3$ can be retained in the product, or by using a halogenated alcohol one obtains a halogenated ester. A problem with phosphate esters is that, in certain systems such as polyurethanes, they do not resist hydrolysis sufficiently. The triesters tend to split into acidic fragments plus alcohols and the acid deactivates the basic polymerization catalyst.

Phosphites and Phosphonates

Most practical routes to phosphites and phosphonates are via phosphorus trihalides. The reaction of PCl_3 with various reagents has been very carefully studied [55]. PCl_3 is somewhat more reactive with alcohols than is $POCl_3$ and there is less difference in the ease of replacement of successive chlorine atoms. By the same token, the phosphite esters are relatively reactive, once formed. Thus, primary alcohols and phenols readily afford the primary esters:

$$ROH + PCl_3 \rightarrow ROPCl_2 + HCl$$

at or near room temperature and without added base. Dialkyl and diaryl phosphites are not obtained in good yield by the use of two moles of alcohol (or phenol) per mole of phosphorus halide. Rather, the diesters can be prepared by use of a tertiary base

$$2ROH + PCl_3 + 2B \rightarrow (RO)_2PCl + 2B \cdot Cl$$

or by using an excess of ROH. If 3 moles of ROH are used, the result depends on R. If R is alkyl, the diester is produced directly [56]:

$$3ROH + PCl_3 \rightarrow (RO)_2P(H)O + RCl + 2HCl$$

If R is aryl, the triester results:

$$3ArOH + PCl_3 \rightarrow (ArO)_3P + 3HCl$$

The diaryl ester can be prepared in fair yield by the direct reaction of 2 moles of aryl-OH with 1 of phosphorus halide and in high yield in this ratio in the presence of a tertiary base (or possibly NH_3 at low temperature to avoid formation of P—N moieties). The trialkyl phosphites must be prepared at low temperature in the presence of a base; otherwise the diester results.

Hydrolysis of the halophosphites produced in the above reactions, that is, $ROPCl_2$, $(RO)_2PCl$, is somewhat tricky. If one is careful to use only just enough water to produce HCl and no excess, the desired product can be achieved. It should be remembered that the monoesters, in particular, are sensitive to hydrolysis to orthophosphorous acid. The triesters are the most stable and are most often encountered.

Cyclic phosphites are often met in the literature. They are prepared from diols:

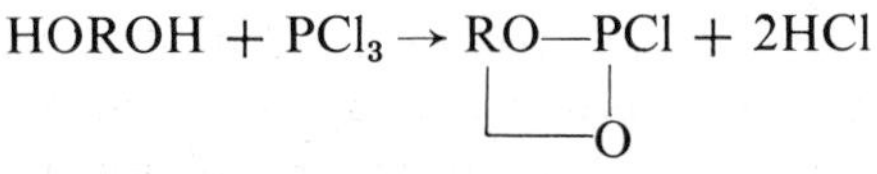

$$HOROH + PCl_3 \rightarrow \begin{array}{c} RO\!\!-\!\!PCl \\ |\qquad| \\ \rule{0.6cm}{0.4pt}O \end{array} + 2HCl$$

A reaction of interest for fire-retardant work is

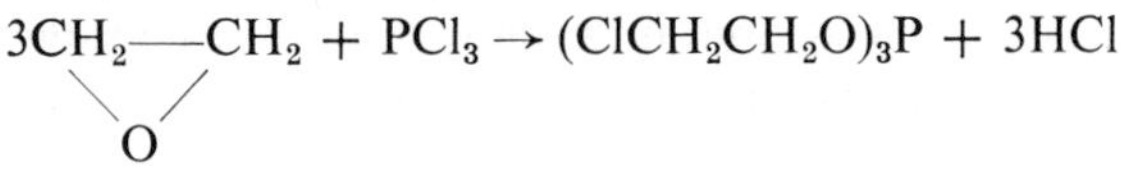

$$3CH_2\!\!-\!\!CH_2 + PCl_3 \rightarrow (ClCH_2CH_2O)_3P + 3HCl$$
$$\diagdown\!\diagup$$
$$O$$

This goes very smoothly at or below room temperature.

A selection of phosphites mentioned in the literature on fire retardants is given in Table 2-6.

Phosphonates contain one carbon-phosphorus bond. The most common are the diesters, which can be regarded as isomerized phosphite esters in which the coordination number of the phosphorus is increased from 3 to 4

$$
\begin{array}{cc}
\text{CH}_3 & \text{CH}_3 \\
| & | \\
\text{O} & \text{O} \\
| & | \\
\text{CH}_3\text{O}\text{---P---OCH}_3 & \text{CH}_3\text{---P---OCH}_3 \\
& | \\
& \text{O} \\
\text{Trimethylphosphite bp 111°C} & \text{Dimethyl methylphosphonate} \\
& \text{bp 181°C}
\end{array}
$$

(or the valency from 3 to 5). That these are truly two different compounds may be ascertained from the difference in boiling points. There are equally dramatic differences in refractive indices, densities, solubilities, and so forth. The phosphonate calls upon orbitals in the $3d$ shell of the phosphorus atom; the phosphite does not [57].

The most common preparation of phosphonate esters is by the action of a halide on the corresponding phosphite (Michaelis-Arbuzov reaction) [57]:

$$
\begin{array}{c}
\text{O} \\
|| \\
(\text{RO})_3\text{P} + \text{R}'\text{X} \rightarrow (\text{RO})_2\text{---P---R}' + \text{RX}
\end{array}
$$

The reaction is best when R′ is primary aliphatic and poorest when it is aryl. The reaction is run at elevated temperatures with provision for rapid removal of the RX by-product so as to prevent its participation, in turn, in a similar reaction with unreacted triester. When R′ = R, we have a simple isomerization and only a trace of R′X is needed. An interesting case is that of the haloalkyl phosphite:

$$
\begin{array}{c}
\text{O} \\
|| \\
(\text{ClCH}_2\text{CH}_2\text{O})_3\text{P} \rightarrow (\text{ClCH}_2\text{CH}_2\text{O})_2\text{---P---CH}_2\text{CH}_2\text{Cl}
\end{array}
$$

This requires no added halide: it is built into the starting material.

An alternative method makes use of an alkali metal salt of a diester of phosphorous acid:

$$
\begin{array}{c}
\text{O} \\
|| \\
(\text{RO})_2\text{POM} + \text{R}'\text{X} \rightarrow (\text{RO})_2\text{---P---R}' + \text{MX}
\end{array}
$$

This synthetic route is often used when R′ ≠ R; the regular Michaelis-Arbuzov reaction, when R = R′. Free phosphonic acids may often be obtained by the action of PCl_5 on olefins followed by hydrolysis:

$$
\begin{array}{c}
\text{O} \\
|| \\
\text{---RCH}=\text{CH}_2 + \text{PCl}_5 \rightarrow \text{RHC(Cl)CH}_2\text{PCl}_4 \rightarrow \text{RCH}=\text{CH---P---(OH)}_2
\end{array}
$$

Table 2-6 Some Phosphites and Phosphonates of Interest

Reactants	Conditions	Product	Reference
		Phosphites	
PCl_3, phenol	70° gradually to 150°, N_2 stripping	$(\phi O)_3 P$ bp 360°	44
PCl_3, EtOH	Low boiling solvent, 80° or so. Strip and distil under red. press.	$(EtO)_2 P(H)O$ bp 72° (8–10 mm)	44
PCl_3, $C(CH_2OH)_4$	Reflux 3 hr; H_2O	$C(CH_2OPO_2H_2)_4$	44
PCl_3, ethylene oxide	Keep below 20°	$(ClCH_2CH_2O)_3 P$ bp 112° (2.5 mm)	44
PCl_3, ethylene oxide	Isopropyl ether, trace $ClCH_2CH_2OH$, 70°	$(ClCH_2CH_2O)_3 P$	60
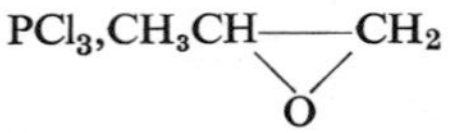 PCl_3, $CH_3CH{-\!-}CH_2$ with O bridge	Et_3N, N_2, 5 hr, 35°	$(CH_3CHCH_2O)_3 P$ with Cl substituent	61
PCl_3, methanol, R_3N	Cool to room temp. filter	$(CH_3O)_3 P$ bp 111°	55
PCl_3, polyolefin glycols	...	Polyolefin glycol polyphosphites	55a
$(HORO)_2POR'$ + Br(Cl), OH, Br(Cl), Br(Cl), Br(Cl) ring	$NaO\phi$, 4 torr, 130°	$(HORO)_2PO$—aryl with Br, Br, Br, Br	63
$R' = H$, allkyl, aryl		bp (R = dipropyl ether) 130°/4 torr	
$(RO)_3P$, HOR'OH	$NaO\phi$ (cat.), 120–150°, 15 mm Hg	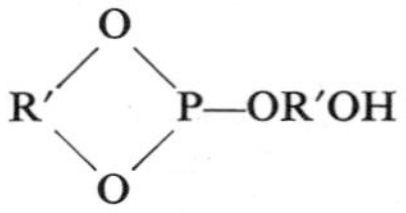 R'—P—OR'OH ring with O, O	63a

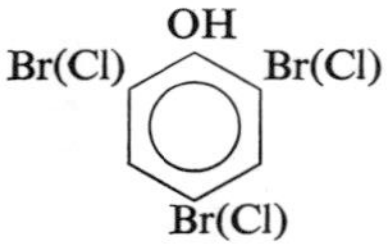

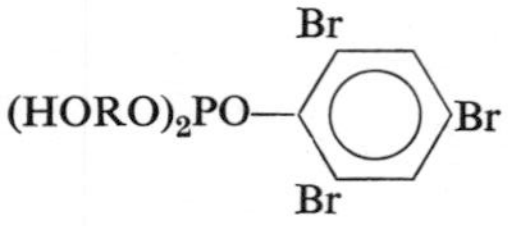

ϕPCl_2, NaCNO	ϕ, $CH_3CO(CN)$, reflux 22 hr	$\phi P(NCO)_2$	63b
		Phosphonates	
PCl_3, RCHO	$ZnCl_2$, H_2O	$ClCH{-}P(OH)_2$ R, O	63c
PBr_3, $CH_2\!\!-\!\!CHCH_2Br$ (epoxide O)	$125°$, isomerize $155°$	$BrCH_2CHCH_2{-}P{-}(OCH_2CH{-}CH_2Br)_2$, Br, Br, O	63d
PCl_3, CH_2O, ROH, Et_3N	$0{-}5°$, 3 hr	$ClCH_2P(O)(OR)_2$ bp (R = propyl) $78{-}90°$ (1 mm) R = allyl, methallyl	44, 62
$HP(O)(OR)_2$, CH_2O, $HN(CH_2CH_2OH)_2$	Add phosphite to other reactants slowly at room temp; finish at $50°$	$(RO)_2P(O)CH_2N(CH_2CH_2OH)_2$ liquid	64, 65
$HP(O)(OEt)_2$, $HOCH_2CH_2OH$	$140{-}160°$, $20{-}25$ mm Hg	$HOCH_2CH_2O{-}P(H)(OEt)_2$ viscous liquid	66
$HP(O)(OR)_2$, $R'CH{=}CHR''$	Ketone solvent, 7 hr, $25°$, uv light	$R'{-}CH_2{-}C(H)(R'')P(O)(OR)_2$	67
$HP(O)(OR)_2$, $R'CH{=}CHR''$	$1\frac{1}{4}$ hr, $50°$, $NaNH_2$	$R'{-}CH_2{-}C(H)(R'')P(O)(OR)_2$	68
$HP(O)(OR)_2$, $R'CH{=}CHR''$	$15{-}30°$, Na	$R'{-}CH_2{-}C(H)(R'')P(O)(OR)_2$ high boiling liquids	69
$HP(O)(OEt)_2$, $CH_2{=}CHCONH_2$	NaOEt, $80{-}90°$, filter, wash	$(EtO)_2P(O)CH_2CH_2CONH_2$ mp $74°$	70
$HP(O)(OR)_2$, $CH_2{=}CHCl$	Per cpd or uv light	$ClCH_2CHP(O)(OR)_2$	70a
$HP(O)(OEt)_2$, (CH_2CH_2) $HC{=}O$ (epoxide)	25 hr, $27°$	$(CH_2{-}CH_2)$, $H{-}C{-}OH$, $EtOPO$, EtO moldable at $180°$	71

(Continued overleaf)

Table 2-6 (Continued)

Reactants	Conditions	Product	Reference
HP(O)(OR)$_2$, RCH$_2$CHO —Br	2 steps: (1) form α-hydroxy phosphonate; (2) NaOH, H$_2$O, 2–3 hr (filter out NaBr)	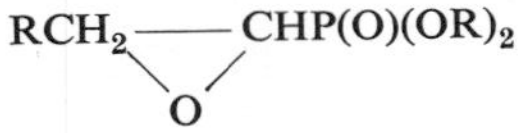 RCH$_2$—CHP(O)(OR)$_2$ bp 78–79° (0.9 mm)	72
HP(O)(OR)$_2$, 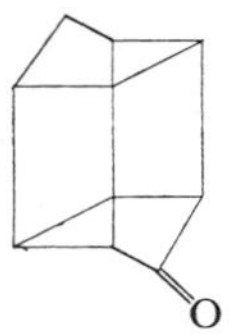 fully chlorinated	Xylene, 16 hr, 90°	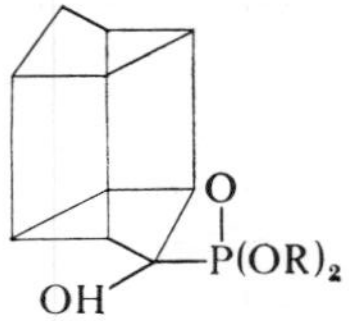 fully chlorinated solid	73
ClCH$_2$POCl$_2$, (HO)$_2$POCH$_2$OH	Vac, 2 hr, 110°	$\left[(HO)_2-P-CH_2O \right]_2-P-CH_2Cl$ liquid	74
ClCH$_2$POCl$_2$, CH$_2$=CHCH$_2$OH	Benzene, 1 hr, 0°	$(CH_2=CHCH_2O)_2-P-CH_2Cl$ bp 94–98° (1 mm)	75
(EtO)$_3$P, BrCH$_2$CH—CH$_2$ O 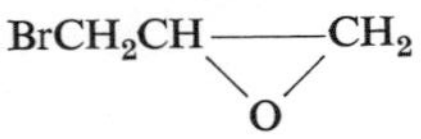	Reflux, 140°	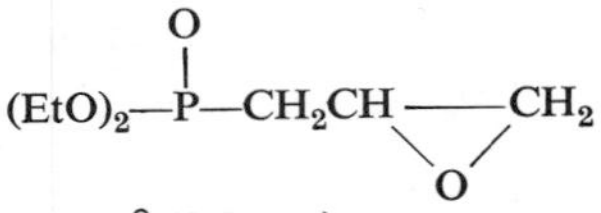 $(EtO)_2-P-CH_2CH-CH_2$ O bp 101° (1.5 mm)	76

Reactants	Conditions	Product	No.
(EtO)₃P, OHC–⬡–CHO 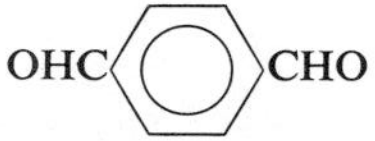	Dioxane, HCl, 1 hr, 25°	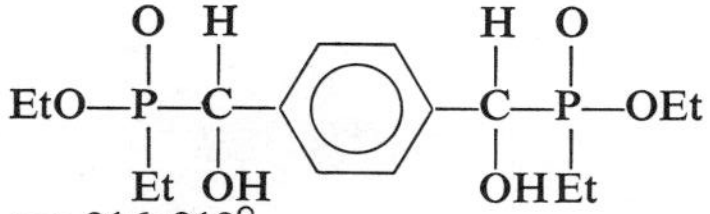 $EtO-\overset{O}{\underset{}{P}}-\overset{H}{\underset{Et\ OH}{C}}-⬡-\overset{H}{\underset{}{C}}-\overset{O}{\underset{OHEt}{P}}-OEt$ mp 216–218°	77
(EtO)₃P, 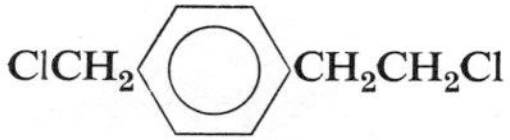 ClCH₂–⬡–CH₂CH₂Cl	2 steps: (1) 20 hr, 90°; (2) KOH, EtoH, 6 hr, reflux	 $CH_2=CH-⬡-CH_2-\overset{O}{\underset{}{P}}-(OEt)_2$ bp 120° (0.2 mm)	78
(EtO)₃P, CH₂=CHOCH₂CH₂X	5 hr, 170–190°, press.	$CH_2=CHOCH_2CH_2-\overset{O}{\underset{}{P}}-(OEt)_2$	78a
(EtO)₃P, $(CH_2-CH)_x$ OOCCH₂Cl	…	$(CH_2-CH)_x$ COOCCH₂$-\overset{O}{\underset{}{P}}-(OEt)_2$	79
(RO)₃P, 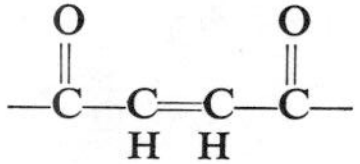 $-\overset{O}{\underset{}{C}}-\overset{}{\underset{H}{C}}=\overset{}{\underset{H}{C}}-\overset{O}{\underset{}{C}}-$	…	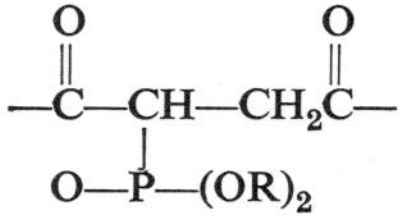 $-\overset{O}{\underset{}{C}}-\overset{}{\underset{O-\overset{O}{P}-(OR)_2}{CH}}-CH_2-\overset{O}{\underset{}{C}}-$	79a
(iso-C₄H₉O)₃P, CH₂BrCH₂Br	2 steps: (1) 2–3 hr, 170–190 (Arbuzov); (2) KOH, ethanol	$(C_4H_9O)_2-\overset{O}{\underset{}{P}}-CH=CH_2$ bp 115–116° (5 mm)	80
(φO)₃P, octakis (2-hydroxypropyl) sucrose	Trace (φO)₂PHO, C₁₀H₂₁OH, vac, 150°	Sucrose polyoxypropylene phosphite, diphenyl ester (solid)	81
φOP(OR)₂, XφOH (X = Cl, Br)	NaOφ, transesterify	XφOP(OR)₂	81a
(HORO)₃P, CCl₄	Reflux 7 hr	$(HORO)_2-\overset{O}{\underset{}{P}}-CCl_3$	82

(Continued overleaf)

Table 2-6 (Continued)

Reactants	Conditions	Product	Reference
$(CH_2=CHCH_2O)_2P$, $Cl\phi CH_2Cl$	10 hr, 90°	$Cl\phi CH_2-\overset{\displaystyle O}{P}-(OCH_2CH=CH_2)$	82a 82b
$(CH_2=CHCH_2O)_3P$, 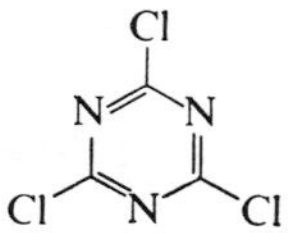 (2,4,6-trichlorotriazine)	Per cpd, 85–90°	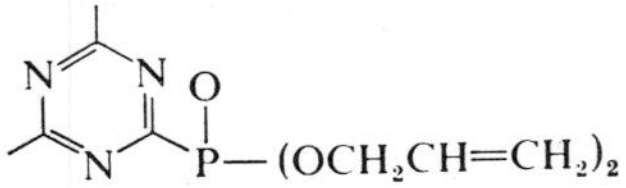 triazine$-\overset{\displaystyle O}{P}-(OCH_2CH=CH_2)_2$	83
$RCH=CH_2-\overset{\displaystyle O}{P}-Cl_2$, ROH	...	$RCH=CH_2-\overset{\displaystyle O}{P}-(OR)_2$	83
$(XCH_2CH_2O)_3P$ ($X = Cl, Br$)	2 steps: (1) Arbuzov rearr. at 160°; (2) NaOH, 20 hr, 50°	$(XCH_2CH_2O)_2-\overset{\displaystyle O}{P}-CH=CH_2$ bp 137–139° (4 mm)	83
$[(XCH_2)_2CHO]_3P$	105°	$[(XCH_2)_2CHO]_2-\overset{\displaystyle O}{P}-CH(CH_2X)_2$	84
$EtP(OEt)_2$, $BrCH_2CH_2CH_2Br$	2½ hr, 160°	$\overset{Et}{\underset{\displaystyle O}{\overbrace{CH_2CH_2CH_2PO}}}$ bp 83° (2 mm)	85
$CH_2=\overset{\displaystyle R}{C}-\overset{\displaystyle O}{P}-(OH)_2$, $R'-CH-CH_2$ (epoxide)		$CH_2=\overset{\displaystyle R}{C}-\overset{\displaystyle O}{\underset{OH}{P}}-OCH_2\overset{}{\underset{OH}{CH}}-R'$	85a

Amidophosphonates

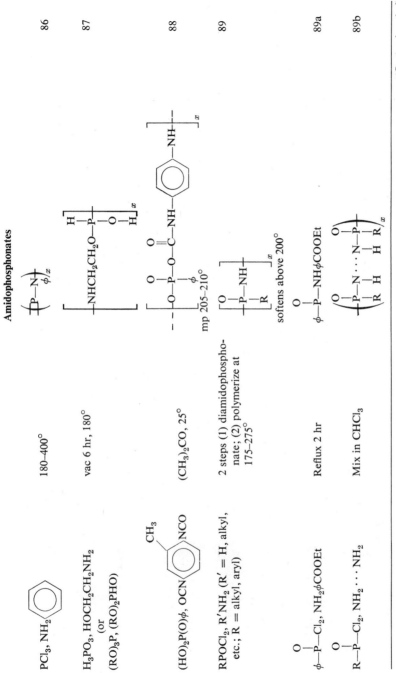

Reactants	Conditions		Ref.
PCl₃, NH₂ (phenyl)	180–400°	$\left[\!\!\begin{array}{c}\text{P—N}\\\phi\end{array}\!\!\right]_x$	86
H₃PO₃, HOCH₂CH₂NH₂ (or (RO)₃P, (RO)₂PHO)	vac 6 hr, 180°	structure	87
(HO)₂P(O)φ, OCN—(CH₃ aryl)—NCO	(CH₃)₂CO, 25°	structure, mp 205–210°	88
RPOCl₂, R'NH₂ (R' = H, alkyl, etc.; R = alkyl, aryl)	2 steps (1) diamidophosphonate; (2) polymerize at 175–275°	structure, softens above 200°	89
φ—P(O)—Cl₂, NH₂φCOOEt	Reflux 2 hr	φ—P—NHφCOOEt	89a
R—P(O)—Cl₂, NH₂···NH₂	Mix in CHCl₃	structure	89b

(Continued overleaf)

Table 2-6 (Continued)

Reactants	Conditions	Product	Reference
$(R_2N)_2POM$, RX	0°	$R-\overset{O}{\underset{\Vert}{P}}-(NR_2)_2$	90
$CH_2{=}CH-\overset{O}{\underset{\Vert}{P}}-Cl_2$, H_2O	4 hr, 10–20°	$CH_2{=}CH-\overset{O}{\underset{\Vert}{P}}-(OH)_2$ liquid, dist *in vacuo*	91
		Polymeric Phosphonates	
$CH_2{=}CHR-\overset{O}{\underset{\Vert}{P}}-(OR')_2$ (R = O or 1CH₂)	Per cpd as initiators	$-CH_2-CH-$ $\;\;\;\;\;\;\;\; R$ $\;\;\;\;\;\;\;\; OP(OR')_2$ glassy polymers	92
RPOCl₂, HOR'OH [R = alkyl, alkenyl; R' = φ, φC(CH₃)₂φ]	4–8 hr, 175–200°, (MgCl₂)	$\left[\overset{O}{\underset{R}{\underset{\Vert}{P}}}-O-R'-O\right]_x$ soften 100–130°	93
$(\phi O)_2PHO$, $HO\phi C(CH_2)_2\phi OH$	NaAlO₂, 1–2 hr, 190–240°, vac	high polymer waxy solids to viscous oils	94
CH₃POCl₂, CH₃PO(OCH₃)₂	25–60°	$\left[\overset{O}{\underset{CH_3}{\underset{\Vert}{P}}}-O\right]_x$ liquids	95

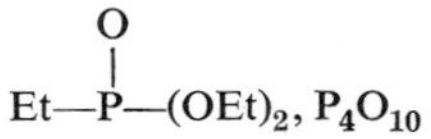

96

$$O \atop Et—P—(OEt)_2$$, P_4O_{10}

CHCl₃, heat slowly 6 hr to 90°, vac

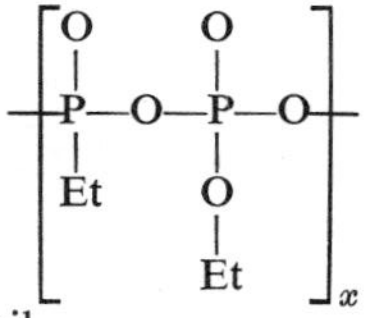

oil

97

$$PCl, (CH_3)_2CO$$

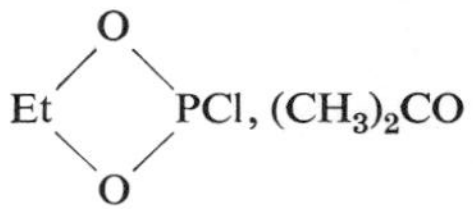

CH₃Cl₂, trace H₂O, 20–40 hr at reflux

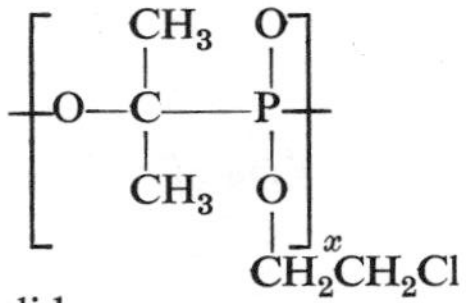

oil

98

PH, NCO (with OCH₃, CH₃ substituents)

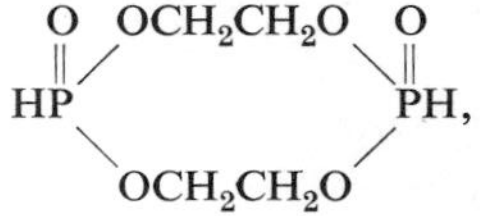

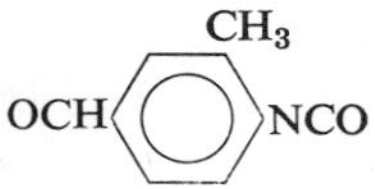

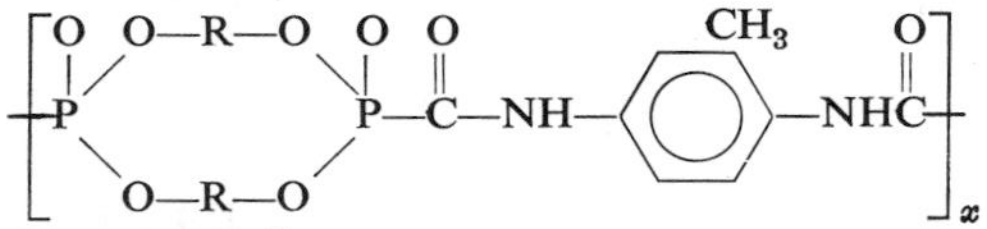

3½ hr, 130°

softens: 75–125°

solid

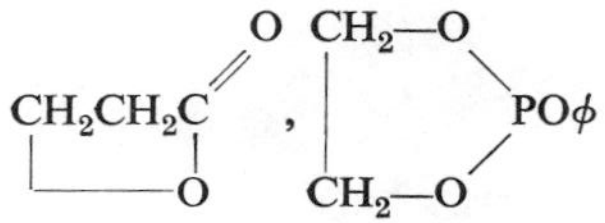

99

POϕ (cyclic structure)

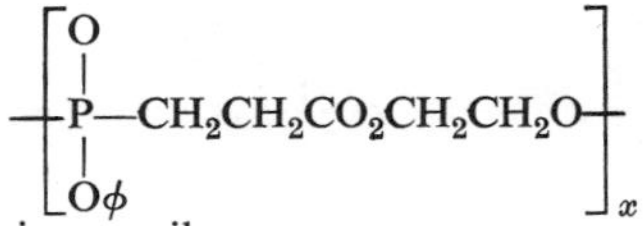

10 min, 150°

viscous oil

(Continued overleaf)

Table 2-6 (Continued)

Reactants	Conditions	Product	Reference
PCl_3, 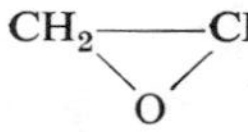 CH_2—CH_2—O (epoxide), CH_3CHO	3 steps: (1) react PCl_3, 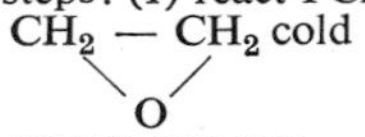 CH_2—CH_2 (ethylene oxide) cold ($ClCH_2CH_2OH$ present in traces); (2) add CH_3CHO cold, heat and remove $ClCH_2CH_2Cl$; (3) isomerize to phosphonate $\frac{1}{2}$ hr, 190–200°	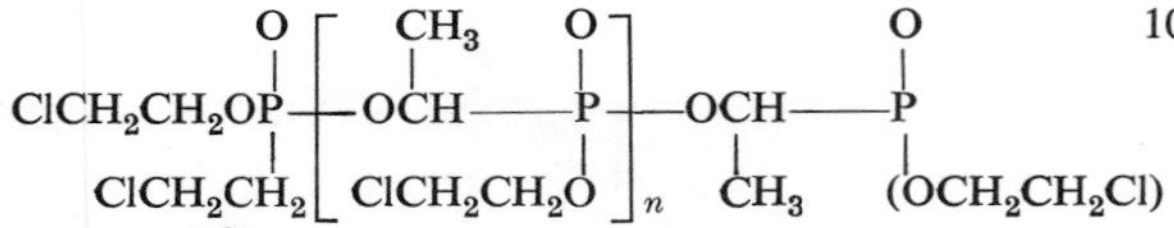 $ClCH_2CH_2OP\left[\begin{array}{c}O\\\parallel\end{array}\begin{array}{c}CH_3\\ \mid\\OCH\\ \mid\\ CH_2CH_2\end{array}\begin{array}{c}O\\\parallel\end{array}\begin{array}{c}\\P\\ \mid\\ OCH \\ \mid\\ ClCH_2CH_2O\end{array}\right]_n\begin{array}{c}O\\\parallel\\P\\ \mid\\ CH_3\end{array}(OCH_2CH_2Cl)$ n averages 2	100
$+CH_2$—$CH\!\!+_x$, $(RO)_2$—$\overset{\displaystyle O}{\overset{\|}{P}}$H	Per cpd	$+CH_2$—$CH\!\!+_x$ ϕ $PO\binom{O}{R}_2$	100a
PCl_3, $RCOOH$ ($R = ClCH_2$, CH_3)	50°–160°C, 1–48 hr, then hydrolyze	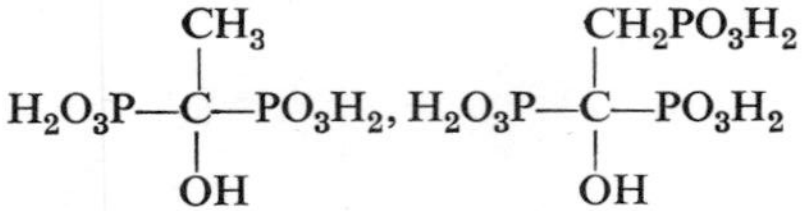 $\begin{array}{c}CH_3 \quad\quad\quad CH_2PO_3H_2\\ \mid \quad\quad\quad\quad\quad \mid\\ H_2O_3P-C-PO_3H, H_2O_3P-C-PO_3H_2\\ \mid \quad\quad\quad\quad\quad\quad \mid\\ OH \quad\quad\quad\quad OH\end{array}$	100b
Phosphite-polyphosphonate, pentaerythritol ($EtO)_3P$, $OHC\phi CHO$	$\frac{3}{4}$ hr, 100°–200°, 35 mm	Phosphite-polyphosphonate/pentaerythritol rxn. prod.	100c
	Dioxane, HCl	$\phi(CH-\overset{\displaystyle O}{\overset{\|}{P}}-(OEt)_2)_2$ $\mid$ OH	100d

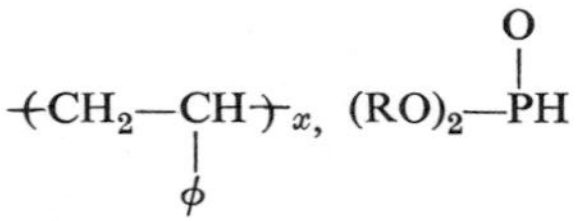

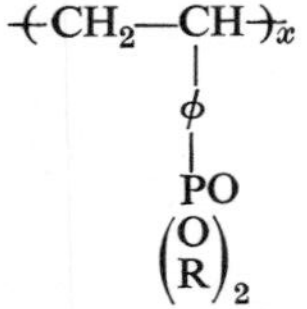

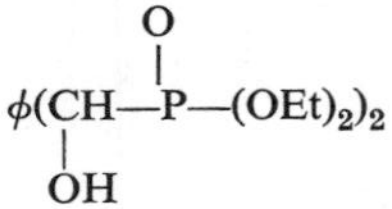

Another possibility [58] is:

$$PCl_3 + RCl \xrightarrow{AlCl_3} RP(O)Cl_2 \longrightarrow RP(O)(OH)_2$$

This method has some advantages in raw material costs if the free phosphonic acid is required. Recently, the addition of dialkyl phosphite to terminal unsaturated double bonds has gained commercial significance:

$$(RO)_2P(H)O + CH_2\!\!=\!\!CH\!-\!- \rightarrow (RO)_2\overset{\displaystyle O}{\overset{\|}{-}}\!\!P\!\!-\!\!CH_2\!\!-\!\!CH_2\!\!-$$

This requires strong base but goes in good yield (see items 10–13 in Table 2-6). Carbonyl groups reacted with the hydrogen of a P—H bond afford α-hydroxy phosphonates. This is most commonly carried out with phosphorous acid or one of its diesters, although hypophosphorous acid or one of its esters reacts similarly [58]:

$$(RO)_2P(H)O + R'\overset{\displaystyle R}{\overset{\|}{C}}\!\!=\!\!O \rightarrow (RO)_2\!\!-\!\!\underset{\underset{O}{\,}}{P}\!\!-\!\!\underset{\underset{OH}{\,}}{C}R(R')$$

where R is H, alkyl.

A commercially significant enlargement of this reaction is a Mannich-type condensation of the P—H moiety with an aldehyde and ammonia or a primary or secondary amine:

$$(RO)_2P(H)O + R'CHO + R''(R''')NH \rightarrow (RO)_2\overset{\displaystyle O}{\overset{\|}{-}}\!\!P\!\!-\!\!CH(R')N(R'')R'''$$

where R, R', R'', R''' can be H, alkyl. With ammonia and formaldehyde, nitrilotris(methylenephosphonic acid) or its esters result. Many variants on this reaction have been reported [59]. The products are high in phosphorus content and are useful fire retardants.

A novel anhydride of one of these has been claimed as a versatile reactive or additive fire retardant [59a]:

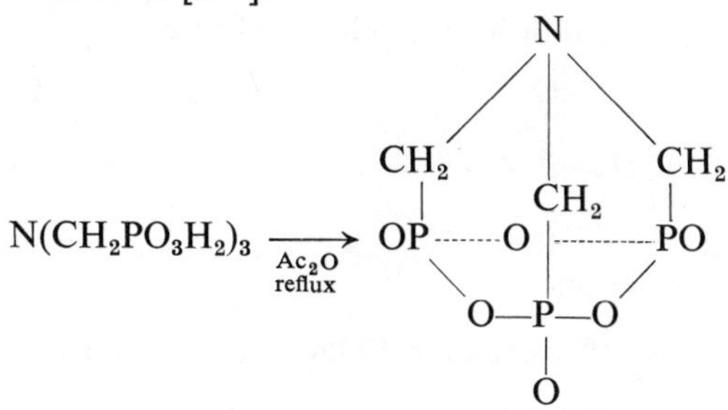

Trianhydride

The phosphonates are an important class of fire retardants because of the stability of these compounds. The P—C bond is very resistant to chemical attack and it confers stability against oxidation of the phosphorus atom. The synthesis possibilities are varied and flexible, leading easily to products with desired functionality.

Table 2-6 is a fairly lengthy presentation of phosphites and phosphonates of interest as fire retardants. The number of entries, though not exhaustive, gives some idea of the importance of this group of phosphorus chemicals. The table gives some preparation schemes for several phosphites of interest primarily as reactants to prepare more stable phosphonates or polyphosphonates.

The schemes based on Mannich condensations are used commercially for an important group of fire retardants. The poly(vinyl phosphonates) are illustrative of various ways vinyl derivatives might be used as polymers or copolymers. Many of the entries demonstrate how one can combine halogens with phosphorus in a single molecule. Others are examples of preparation of unsaturated compounds which can subsequently be polymerized or copolymerized. Interesting polymers in which the phosphorus compounds act as the polyol in urethane polymerization are given. The Arbuzov rearrangement or reaction appears repeatedly in Table 2-6; it is a commonly used technique. Polymeric phosphonates may be grouped into three classes: those with the phosphorus in a subsidiary pendant position, those with O—P—C groupings in the backbone, those with O—P—O groupings in the backbone, the P—C bond being in a substituent position; namely

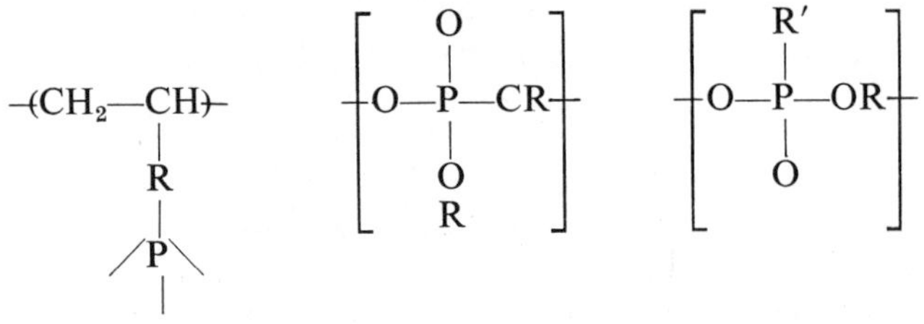

Very few phosphinic acids have been discussed as flame retardants. Compounds such as the following might be of interest:

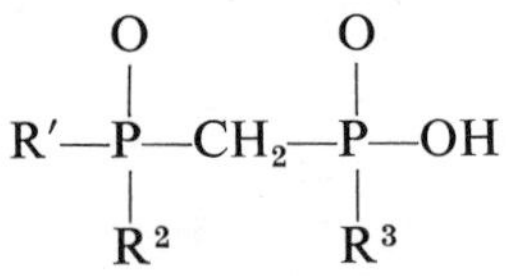

Reference 100e 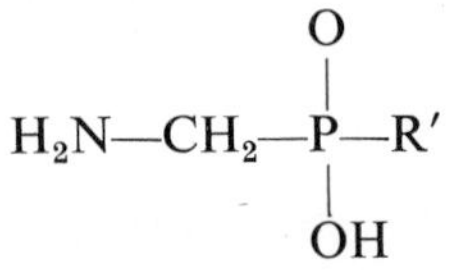   Reference 100f

Phosphines, Phosphonium Salts, and Phosphine Oxides

The chemistry of these oxidation states of phosphorus is not as well worked out as it is for the positive oxidation states. Phosphines and phosphonium salts are nominally in the -3 state, phosphine oxides in the -1 state. The literature [101] suggests a number of ways of making these materials, almost all of which must start with phosphine itself, that is, PH_3. It is appropriate then to consider how one obtains phosphine. Phosphorus and hydrogen do not combine under ordinary conditions to produce monomeric phosphine; indeed, phosphine yields hydrogen at elevated temperatures:

$$PH_3 \xrightarrow{\Delta} H_2 + \text{polymeric phosphines}$$

For laboratory work phosphine of high purity is obtained from the alkaline hydrolysis of phosphonium iodide which, in turn is produced from phosphorus and iodine:

$$P_4 + 4I_2 \longrightarrow 2P_2I_4$$

$$10P_2I_4 + 13P_4 + 128H_2O \longrightarrow 40PH_4I + 32H_3PO_4$$

$$PH_4I + MOH \xrightarrow{H_2O} PH_3(g) + MI + H_2O$$

Hydrolysis of phosphides has long been a laboratory favorite:

$$Ca_3P_2 + 6H_2O \rightarrow 2PH_3 + 3Ca(OH)_2$$

The product usually contains some hydrogen and some diphosphine: $H_2P—PH_2$. Of course, the difficulty lies in obtaining the phosphide.

Commercially, phosphine is produced in small volume by only one or two manufacturers. Processes which have been or are now being used include pyrolysis of phosphites and/or hypophosphites to give PH_3 and phosphate [102], reaction of phosphorus with amalgams of Zn or Cd in an aqueous acid medium [103], electrolysis of phosphorus in an acid aqueous medium using a special lead cathode and graphite anode and operating at $85°$ and 24–32 A/ft^2 [104], and a steaming process wherein phosphorus is hydrolyzed at $250°$ and higher in a medium consisting essentially of very strong phosphoric acid, obtaining red phosphorus and strong phosphoric acids as by-products [105]:

$$8P + 12H_2O \xrightarrow{\Delta} 3H_3PO_4 + 5PH_3$$

$$P_{white} \xrightarrow{\Delta} P_{red}$$

These latter two processes are believed to be the ones operated commercially.

Organophosphines can be synthesized in a variety of ways [101]. It is difficult to obtain mono- or diphosphines in high yield—one gets mixtures, the composition of which is often controlled by equilibrium laws. Tertiary phosphines are somewhat easier to isolate in high purity. Primary phosphines may be obtained by addition to unsaturated compounds:

$$PH_3 + RCH{=}CH_2 \rightarrow RCH_2—CH_2PH_2$$

This reaction may be either radical or acid catalyzed, the acid being of the type represented by $AlCl_3$. Reaction with Grignard reagents produces in turn other Grignards, which may then be reacted with halides to give a variety of products usually of low purity. Thus [101]:

$$RPH_2 + 2R'MgX \rightarrow RP(MgX)_2 + 2R'H$$
$$RP(MgX)_2 + 2R''X \rightarrow RPR''_2 + 2MgX_2$$

Phosphorus halides react with Grignard reagents to produce tertiary phosphines in good yield:

$$PX_3 + 3RMgX \rightarrow R_3P + 3MgX_2$$

Primary and secondary alkyl phosphine halides react similarly to give unsymmetrical tertiary phosphines. Halides of phosphorus react with aryl and alkyl halides in the presence of sodium (Wurtz reaction)

$$PX_3 + 3RX + 6Na \rightarrow R_3P + 6NaX$$

with yields of tertiary phosphine usually in excess of 50%. When dialkyl anilines are reacted with PX_3, para substitution results, giving

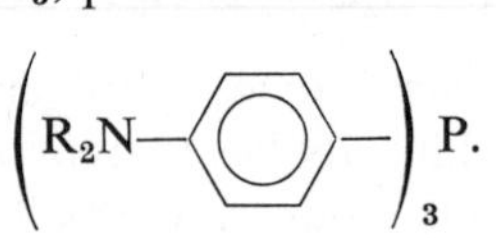

$$\left(R_2N-\!\!\left\langle\bigcirc\right\rangle\!\!- \right)_3 P.$$

The reaction is taken to completion either by using excess aniline derivative or by using pyridine to capture the HCl [101].

Halophosphines may at times be required as intermediates. Various pyrolysis routes have been described [101]: heating benzene and PCl_3 affords phenyl dichlorophosphine. Further heating yields diphenyl chlorophosphine. Primary and secondary phosphines react with halogens at low temperatures:

$$RPH_2 + 2X_2 \rightarrow RPX_2 + 2HX$$

Alkali metals bonded to phosphorus provide a convenient reaction site for organic halides:

$$NaPH_2 + RX \rightarrow RPH_2 + NaX$$
$$Na_3P + 3RX \rightarrow R_3P + 3NaX$$
$$\text{etc.}$$

The degree of substitution is controlled by the amount of alkali metal used. Triphenyl methyl sodium is suggested as a sodium source using ether as the medium. Alkyl halides react smoothly; aryl halides, not so well.

Though olefin oxides react only sluggishly with PH_3 (requiring autoclaving at 100° to give hydroxyalkyl phosphines in low yield), carbonyl compounds will add phosphine under certain conditions. The reaction is difficult to control in steps and most readily yields phosphonium salts. Thus,

$$4CH_2O + PH_3 + HCl \xrightarrow{\text{ether}} (HOCH_2)_4P^+Cl^-$$

Polymeric aldehydes may be used and PH_4I may be substituted for PH_3 and HCl. The product is cleaved in base so workup should avoid alkaline reagents. Phosphonium salts are readily obtained, too, simply by mixing tertiary phosphines and a halide and heating. Small radicals react better than large ones, with very small radicals reacting spontaneously at room temperature. Iodides react better than bromide and better than chloride. In the presence of $AlCl_3$, replacement occurs in aryl compounds and can be complete:

$$\phi_3P + xs\ CH_3\phi X \xrightarrow[250°]{AlCl_3} (CH_3\phi)_4PX$$

Strong heating of phosphonium compounds ($>200°$) may cause reversion to the tertiary phosphine and the halide; this is not a particularly good synthesis route.

Primary and secondary phosphines are very sensitive to oxidation and, in general, must be protected from oxygen. All phosphines exhibit a characteristic, offensive, garlic-like odor, and many are extremely toxic. Care must be exercised in handling these materials. Tertiary phosphines and phosphonium compounds are more stable and somewhat less toxic than phosphine and its mono- and disubstituted derivatives.

In contrast, the phosphine oxides are perhaps the most stable of organophosphorus compounds. These materials are generally crystalline solids of sufficient stability that a succession of reactions can be carried out on the substituents without disrupting the geometry around the phosphorus. Preparation is straightforward: simply contact the tertiary phosphine with air and heat or use controlled amounts of nitric acid, SO_2, permanganate, and so forth:

$$R_3P + \tfrac{1}{2}O_2 \rightarrow R_3PO$$

Particularly smooth oxidation is afforded by mercuric oxide and ferric chloride. The stability of this configuration is reflected by the ease with which certain phosphinous esters rearrange:

$$R_2POR' \rightarrow R_2R'PO$$

Other phosphinous esters yield the phosphine oxide by heating with a halide (Arbuzov reaction):

$$R_2POR' + R''X \rightarrow R_2R''PO + R'X$$

Another route to the oxide is provided by the decomposition of phosphonium

Table 2-7 Some Phosphines, Phosphonium Salts, and Phosphine Oxides Suggested as Fire Retardants

Reactants	Conditions	Product	Reference
PH_3, ethylene oxide (CH_2—CH_2, O)	Dry ether, 20 hr, 100°, autoclave	$HOCH_2CH_2PH_2$ bp 139–140°	106
$(ClCH_2)_4PCl$	Aq alkali	$(ClCH_2)_3P$ bp 100° (7 mm)	107
PH_3, CH_2O	P + Cl_2 cat., 5 hr, cold	$(HOCH_2)_3P$	107a
PH_3, ethylene oxide (CH_2—CH_2, O)	See above, use excess ethylene oxide (CH_2—CH_2, O)	$(HOCH_2CH_2)_3P$ bp 183–185°	106
PH_3, $CH_2{=}CHCN$	$NiCl_2$	$P(CH_2CH_2CN)_3$	106a
$CH_2{=}CHCH_2MgBr$, PCl_3	Add PCl_3 in ether, boil ½ hr, cool to 0°, add NH_4Cl	$(CH_2{=}CHCH_2)_3P$ bp 69° (13 mm)	108
$CH_2{=}CHCH_2MgBr$, $PCl_2(\phi Br)$	Add $PCl_2(\phi Br)$, boil ½ hr, cool to 0°, add NH_4Cl	$(CH_2{=}CHCH_2)_2P(4\text{-}BrC_6H_4)$ bp 186° (37 mm)	108
$CH_2{=}CHMgBr$, PCl_3	THF, reflux 6 hr, cool	$(CH_2{=}CH)_3P$	108
PH_3, CH_2O, HCl	Hg catalyst, 7 hr, 35°	$(HOCH_2)_4PCl$ mp 151°	109
$P(CH_2OH)_3$, $CH_2{=}CHCN$	Mix cold	$P(CH_2CH_2CN)_3$	109a
$(HOCH_2)_4PCl$, PCl_5	CCl_4, heat	$(ClCH_2)_4PCl$ mp 192–193°	110
$4\text{-}Cl\text{-}C_6H_4\text{-}CH_2\text{-}CH{=}CH_2$, $\phi\text{-}P(H)Na$	Et_2O, liq NH_3, −60°	$CH_2{=}CH\text{-}C_6H_4\text{-}P(H)\phi$ 	110
$(\phi O)_3P$, Cl–R, Na	2 hr, 30–50°	R_3P	110a
$(HOCH_2)_4PCl$	Air blow, 120–180°, 5½ hr	Polymeric product	111
PH_3, CH_2O, H_2O	H_2O, $HgCl_2$ (or other transition metal halides), 10 hr, 35°	$(HOCH_2)_4P^+OH^-$ oil	112
$(HOCH_2)_3P$, MeI	Mix, cool, add EtOH	$[(HOCH_2)_3PCH_3]^+I^-$	112a

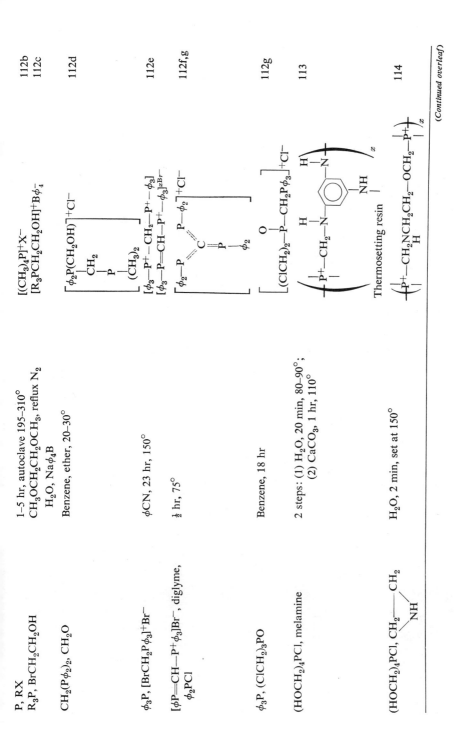

P, RX	Conditions	Product	Ref
R_3P, $BrCH_2CH_2OH$	1–5 hr, autoclave 195–310°; $CH_3OCH_2CH_2OCH_3$, reflux N_2; H_2O, $Na\phi_4B$	$[(CH_3)_4P]^+X^-$; $[R_3PCH_2CH_2OH]^+B\phi_4^-$	112b, 112c
$CH_2(P\phi_2)_2$, CH_2O	Benzene, ether, 20–30°		112d
ϕ_3P, $[BrCH_2P\phi_3]^+Br^-$	ϕCN, 23 hr, 150°		112e
$[\phi P=CH-P^+\phi_3]Br^-$, diglyme, ϕ_2PCl	$\tfrac{1}{2}$ hr, 75°		112f,g
ϕ_3P, $(ClCH_2)_3PO$	Benzene, 18 hr		112g
$(HOCH_2)_4PCl$, melamine	2 steps: (1) H_2O, 20 min, 80–90°; (2) $CaCO_3$, 1 hr, 110°	Thermosetting resin	113
$(HOCH_2)_4PCl$, $CH_2{-}CH_2$ NH	H_2O, 2 min, set at 150°		114

(Continued overleaf)

Table 2-7 (Continued)

Reactants	Conditions	Product	Reference
$(HOCH_2)_4PCl$, ϕOH	Heat on hot plate	Polymer	115
Bu_3P, $BrCH_2CH_2OH$	Heat	$Bu_3(HOCH_2CH_2)PBr$	116
$Bu_3(HOCH_2CH_2)PBr$	ROAc, Na_2CO_3, reflux 8 hr in dimethoxy ethane	$Bu_3(CH_2\!\!=\!\!CH)PBr$ mp 148–150° polymerizes with benzoyl peroxide, 72 hr, 80°	116
$(HOCH_2)_3P$	EtOH, 5 hr, air stream, 50°	$(HOCH_2)_3PO$	116a
$(HOCH_2)_4P^+Cl^-$	H_2O, Na_2SO_3, NaOH, 2 hr, 75°	$(HOCH_2)_3PO$	116b
$(HOCH_2)_3PO$, $SOCl_2$	2 hr, 50°; H_2O	$(ClCH_2)_3PO$	116c
$(ClCH_2)_3PO$, ϕ [imide ring, $O\!\!=\!\!C$ ... $C\!\!=\!\!O$, N], NK, H_2NNH_2	2 steps	$(NH_2CH_2)_3PO$	116c
$(CNCH_2CH_2)_2\!-\!\overset{\|}{\underset{O}{P}}\!-\!H$, $Cl\phi NCO$	Et_3N, $CHCl_3$, $\tfrac{1}{2}$ hr, reflux	$(CNCH_2CH_2)_2\!-\!\overset{O}{\underset{O}{\|}}\!P\!-\!CNH\phi Cl$	116d
ϕ_2PCl, $(CH_3)_2CO$	Mg, N_2, 18 hr, reflux	$\phi_2\!-\!\overset{O}{\underset{CH_3}{P}}\!-\!\overset{CH_3}{\underset{}{C}}\!-\!OH$	116e
$(AcOCH_2)_3PO$, $HBr(g)$	3 hr, 180°C	$(BrCH_2)_3PO$ mp 128–129°	117
ϕ_2PCl, $ClCH_2CH_2OH$	2 steps: (1) Et_2O, pyridine, N_2 10 min, −1–4°; (2) EtOH, KOH, 15 min, reflux	$CH_2\!\!=\!\!CH\!-\!\overset{O}{\underset{}{P}}\!-\!\phi_2$ mp 117–118°	118

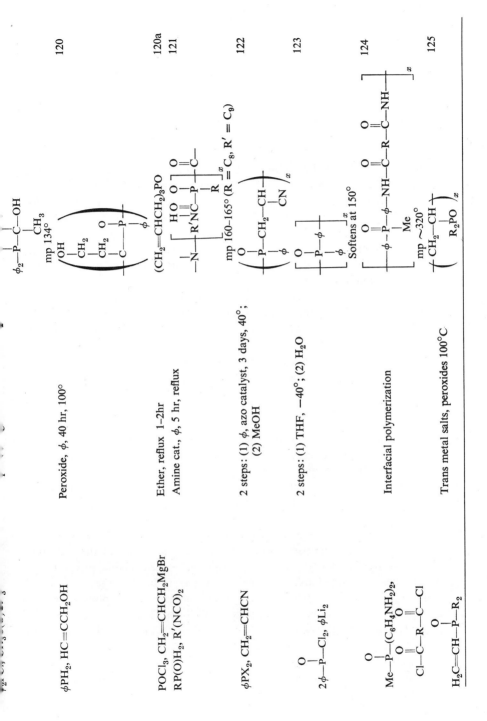

φPH₂, HC≡CCH₂OH — Peroxide, φ, 40 hr, 100° — $\phi_2\text{—P—C—OH}$ (CH₃), mp 134° — 120

POCl₃, CH₂=CHCH₂MgBr; RP(O)H₂, R'(NCO)₂ — Ether, reflux 1–2hr; Amine cat., φ, 5 hr, reflux — (CH₂=CHCH₂)₃PO — 120a; 121

ϕPX_2, CH₂=CHCN — 2 steps: (1) φ, azo catalyst, 3 days, 40°; (2) MeOH — mp 160–165° (R = C₈, R' = C₉) — 122

$2\phi\text{—P(=O)—Cl}_2$, φLi₂ — 2 steps: (1) THF, −40°; (2) H₂O — Softens at 150° — 123

Me—P(=O)—(C₆H₄NH₂)₂, Cl—C(=O)—R—C(=O)—Cl — Interfacial polymerization — mp ~320° — 124

H₂C=CH—P(=O)—R₂ — Trans metal salts, peroxides 100°C — $\left(\text{—CH}_2\text{—CH}\big(\text{R}_2\text{PO}\big)\text{—}\right)_x$ — 125 alkoxides or hydroxides. An example is the treatment of tetraalkyl phosphonium halide with strong alkali to give the hydroxide followed by heating:

$$R_4PCl \xrightarrow[base]{} R_4POH \xrightarrow{\Delta} R_3PO + ROH$$

Grignard reagents afford the oxides with phosphoryl halides:

$$3RMgX + POX_3 \rightarrow R_3PO + 3MgX_2$$

Best yields are obtained by adding the phosphoryl halide to excess Grignard.

Table 2-7 summarizes some of the chemistry that has been explored in the search for fire retardants. The scant number of items involving primary and secondary phosphines reflects their instability. Phosphonium salts and phosphine oxides show the most promise and the hydroxymethylphosphonium compounds most of all. "Tetrakis"—$(HOCH_2)_4PCl$—is currently receiving commercial attention for treating cotton textiles and its use is reviewed in detail in Chapter 5.

The hydroxylalkyl and the haloalkyl phosphines, phosphonium salts, and phosphine oxides have received considerable attention because the hydroxyl or halo group provides a convenient reactive site for copolymerization or further reaction, for example, substitution or dehydrohalogenation to olefin derivatives. Some of this work has recently been reviewed in three articles [126]. Consult Table 2-7 for specific syntheses.

GENERAL COMMENTS

As the foregoing text should indicate, there is a very well developed chemistry of phosphorus already available and more is steadily being introduced into the literature. Because of its versatility, it is clear that phosphorus will continue to be of central interest as a fire retardant, especially in view of the increasing interest in fire-safe systems in which the phosphorus is chemically integrated.

The commercial realities are that phosphorus compounds become more expensive as one progresses from the $+5$ oxidation state to the -3 state. Thus, phosphates are most readily available at reasonable prices; many phosphines are very hard to come by at any price. On the other hand, stability of organophosphorus compounds increases from phosphates and phosphites to phosphonates to phosphine oxides. Thus we find the phosphonates are often a compromise in cost and performance and this, in turn, may explain the focus of so much research on these compounds. Perhaps future efforts will lead to improved techniques for producing phosphine and thus lower the cost of phosphonium salts and phosphine oxides. If costs were not a factor, reactive compounds in the phosphine oxide family would probably be desired in many applications where stability is required, for example, adding to fiber spinning melts.

REFERENCES

[1] J. R. Van Wazer, *Phosphorus and Its Compounds*, Vol. I, Wiley-Interscience, New York, 1958, p. 1. This book is required reading for the researcher in phosphorus chemistry. In addition, *Phosphorus and Its Compounds*, Vol. II, J. R. Van Wazer, ed., Wiley-Interscience, New York, 1961, is useful in terms of commercial routes to various end products; it is cited several times in the references of this chapter. For an updating and enlargement of several of the topics covered by Van Wazer, the interested reader should consult the excellent series titled *Topics in Phosphorus Chemistry*, M. Grayson and E. J. Griffith, 5 vols., Wiley-Interscience, New York, 1964–67; it is cited at several places in the references of this chapter. For a broad view of what can be done in organophosphorus chemistry, the landmark work of Kosolapoff is a must (see reference 3 below). Three recent review series in the Russian literature will be helpful on organophosphorus compounds. They are: N. A. Pudovik et al., *J. Gen. Chem. USSR*, **37**, 2277 (p. 2393 in Russian) (November, 1967), and articles by Evdahov et al., Gladshtein et al., and Valetdinov et al., *ibid.*, 2385, 2391, 2400; N. A. Pudovik and V. K. Khairullin, *Russ. Chem. Rev.*, **37** (5), 317 (May, 1968); V. K. Promovienkov and S. Z. Ivin, *ibid.* (9), 670 (September, 1968). All these page numbers are given as of the English translation. There are other source works on phosphorus chemistry but most are more specialized. One may consider the following on the subject of P-containing polymers (primarily inorganic): J. R. Van Wazer and C. F. Callis, in *Inorganic Polymers*, F. G. A. Stone and W. A. G. Graham, eds., Academic, New York, 1962, Chapter 2.
[2] M. M. Crutchfield, C. H. Dungan, J. H. Letcher, V. Mark, and J. R. Van Wazer, "P^{31} Nuclear Magnetic Resonance," in Grayson and Griffith, eds., *op. cit.*, Vol. 5, 1967.
[3] G. M. Kosolapoff, *Organophosphorus Compounds*, J. Wiley, New York, 1950, pp. 4–6.
[4] Reference 1, Chapter 12.
[5] F. B. Clarke and J. W. Lyons, *J. Amer. Chem. Soc.*, **88**, 4401 (1966).
[6] A. DePassille, *C. R. Acad. Sci., Paris*, **199**, 356 (1934); see also T. E. Warren, *J. Amer. Chem. Soc.*, **49**, 1904 (1927); Van Wazer, ed., *op. cit.*, p. 503.
[7] A. W. Frazier, *U.S. 3,342,579* (to Tennessee Valley Authority), Sept. 19, 1967
[8] Monsanto Co., *Belg. 651,782, 674,161, 674,205; 677,866*, Dec. 12, 1965; Dec. 22, 1965, Sept. 9, 1966; C. Y. Shen and N. E. Stahlheber, *U.S. 3,397,035* (to Monsanto Co.), Aug. 13, 1968 (also *Can. 794,769*, Sept. 17, 1968); H. A. Rohlfs and H. Schmidt, *U.S. 3,419,349* (to Chemische Werke Albert), Dec. 31, 1968 (also *Can. 797,138*, Oct. 22, 1968); Soc. Industrielle et Financière de Lens Finales, *Belg. 711,361*, Feb. 27, 1968.
[9] I. S. I. Vol'fkovich and A. I. Chekhovskikh, *Prikl. Khim.*, **39** (2), 249 (1966).
[10] C. Y. Shen, N. E. Stahlheber, and D. R. Dyroff, *J. Amer. Chem. Soc.*, **91**, 62 (1969).
[11] For example, in Vol. 1 of *Chem. Abstr.*, we find Hugo Becker, *Brit. 20,460*, Sept. 14, 1906, claiming ammonium phosphates for fireproofing fabrics. Much earlier references have been cited by, for example, R. L. Little, ed., *Flameproofing Textile Fabrics*, Reinhold, New York, 1947.
[12] R. Groebe, *U.S. 2,089,697* (to General Electric Co.), Aug. 10, 1937.
[13] N. Taguchi et al., *Jap. 9300* (to Oshika Shinko Co.), 1961.
[14] "Guanylurea Salts," *New Product Bull. No. 35*, American Cyanamid Co., New York, 1952.
[15] Farb. Bayer A.-G., *Brit. 786,736*, Nov. 27, 2957; Farb. Bayer A.-G. (by K. W. Muller), *Ger. 1,061,069*, July 9, 1959.

[16] Monsanto Chemical Co., *Brit. 729,243*, May 4, 1955; P. G. Arvan, *U.S. 2,958,577* (to Monsanto Chemical Co.), Nov. 1, 1960; E. W. Snyder, *U.S. 3,413,380* (to Union Carbide Corp.), Nov. 26, 1968; *U.S. 2,122,122* (to Victor Chemical Works), June 28, 1938.

[17] J. E. Malowan, *U.S. 2,544,706* (to Monsanto Chemical Co.), Mar. 13, 1951; M. L. Nielsen and H. K. Nason, *U.S. 2,603,614* (to Monsanto Chemical Co.), July 15, 1952.

[18] R. Klement and O. Koch, *Chem. Ber.*, **87**, 333 (1954); R. Klement and L. Benek, *Z. Anorg. Allg. Chem.*, **287**, 12 (1956).

[19] J. E. Malowan and F. R. Hurley, *U.S. 2,596,935* (to Monsanto Chemical Co.), May 13, 1952; J. E. Malowan, *U.S. 2,749,233* (to Monsanto Chemical Co.), June 5, 1956; E. H. Rossin and M. J. Scott, *U.S. 2,816,004* (to Monsanto Chemical Co.), sec. 10, 1957.

[20] H. Bestian, J. Heyna, A. Bauer, G. Ehlers, B. Hirskour, T. Jacobs, W. Noll, Weibezahn, and F. Römer, *Ann.*, **566**, 210 (1950); M. Lidaks, S. Hillers, and A. Medne, *Latv. PSR Zinat. Akad. Vestis*, **1959** 87–90 (*Chem. Abstr.*, **53**, 17992a); R. A. Pizzarello and A. F. Schneid, *U.S. 3,335,131* (to Interchemical Corp.), Aug. 8, 1967.

[21] Farb. Bayer A.-G., *Brit. 830,800*, Mar. 23, 1961.

[22] Chemische Fabrik Joh. A. Benckiser G.m.b.H. (by M. Becke-Goehring), *Ger. 1,042,228*, Oct. 30, 1958.

[23] H. W. Coover, *U.S. 2,682,521* (to Eastman Kodak Co.), June 29, 1954.

[24] G. E. Walter, I. Hornstein, and G. M. Steinberg, *U.S. 2,574,516* (to Glenn L. Martin Co.), Nov. 13, 1951.

[25] Dow Chemical Co. (by A. P. Ingram, Jr.), *Belg. 662,571*, Oct. 15, 1965.

[26] A. Maeder, *U.S. 2,894,019* (to CIBA, Ltd.), July 7, 1959.

[26a] H. C. Vogt and J. T. Patton, Jr., *U.S. 3,335,129* (to Wyandotte Chemicals Corp.), Aug. 8, 1967.

[26b] J. C. Frick, Jr., and R. L. Arceneaux, *U.S. 2,939,849* (to U.S. Dept. Agr.), June 7, 1960.

[27] M. L. Nielsen and H. K. Nason, *U.S. 2,596,936* (to Monsanto Chemical Co.), May 13, 1952; M. L. Nielsen, *U.S. 2,596,937* (to Monsanto Chemical Co.), May 13, 1952.

[28] M. L. Nielsen and R. W. Arnold, *U.S. 2,680,077* (to Monsanto Chemical Co.), June 1, 1954.

[29] Monsanto Co., *Belg. 685,385*, Feb. 13, 1967.

[30] Hooker Chemical Corp., *Belg. 685,268*, Feb. 9, 1967.

[31] E. Fluck, in Grayson and Griffith, eds., *op. cit.*, Vol. 4, 1967.

[32] R. A. Shaw, *Endeavour*, **27**, 74 (1968).

[33] Compagnie Française des Matières Colorantes (by X. Bilger), *Fr. 1,327,119*, May 17, 1963; Chemische Fabrik Joh. A. Benckiser G.m.b.H. (by M. Becke-Goehring), *Ger. 1,059,186*, June 11, 1959; C. A. Redfarn and H. Coates, *Brit. 788,735* (to Albright & Wilson, Ltd.), Jan. 8, 1958; Imperial Chemical Industries, Ltd. (by E. Hofmann), *Brit. 876,035*, Appl. Nov. 19, 1958; T. Bieniek, G. Burnie, J. M. Maselli, *U.S. 3,443,913* (to W. R. Grace & Co.) May 13, 1969; B. Gruskin, *U.S. 3,454,409* (to W. R. Grace & Co.) Jul. 8, 1969.

[33a] Compagnie Française des Matières Colorantes (by X. Bilger), *Fr. 1,327,120*, *1,225,073*, May, 7, 1963, June 29, 1960; Establissements Kuhlmann (by X. Bilger), *Belg. 630,055*, Oct. 21, 1963, *Fr. 1,375,029*, Oct. 16, 1964.

[34] Imperial Chemical Industries, Ltd. (by E. Hofmann), *Brit. 888,662*, Jan. 31, 1962.

[34a] G. J. Klender, *U.S. 3,419,504* (to Uniroyal, Inc.), Dec. 31, 1968.

[35] Olin Mathieson Chemical Corp. (by M. Becke-Goehring and Dieter Neubauer), *Ger. 1,143,027*, Jan. 31, 1963.

[36] Albright & Wilson, Ltd., *Fr. 1,343,907*, Nov. 22, 1963; Compagnie Française des Matières Colorantes (by X. Bilger), *Fr. 1,157,097*, May 27, 1958; D. J. Jaszka, *U.S. 3,442,629* (to Hooker Chemical Corp.), May 6, 1969.

[37] C. A. Redfarn, *U.S. 2,866,773* (to Walker Extract and Chemical Co.), Dec. 30, 1958; H.-C. Wu. and M.-C. Kao, *Ko Fen Tau Hsun*, **7**, 229 (1965) (in Chinese) (see *Chem. Abstr.*, **64**, 9897*b*); R. G. Rice and M. V. Ernest, *Can. 781,991* (to W. R. Grace Co.), Apr. 2, 1968.

[38] M. E. Mirhej and J. F. Henderson, *J. Macromol. Chem.*, **1**, 187 (1966); *Chem. Eng. News*, **34** (Jan. 13, 1969).

[39] H. F. Lederle, E. H. Kober, and G. F. Ottmann, *J. Chem. Eng. Data*, **11**, 221 (1966); R. F. W. Ratz and C. J. Grundman, *U.S. 2,876,248* (to Olin Mathieson Chemical Corp.), Mar. 3, 1959.

[40] C. F. Lie and R. L. Evans, *U.S. 3,169,933* (to Minnesota Mining and Manufacturing Co.), Feb. 16, 1965.

[41] Reference 3, Chapter 10.

[42] T. Moeller and P. Nannelli, *Inorg. Chem.*, **1**, 721 (1962); T. Moeller and S. Lanoux, *ibid.*, **2**, 1061 (1963); T. Moeller and S. G. Kokalis, *J. Inorg. Nucl. Chem.*, **25**, 875, 1397 (1963), and references contained therein.

[42a] H. P. Latscha and P. B. Hormuth, *Z. Anorg. Allg. Chem.*, **359**, 78 (1968); L. I. Samaraj, O. I. Kolodjaznij, and G. I. Derkatsch, *Angew. Chem. (Int. Ed.)*, **7**, 618 (1968); G. Tomaschewski, B. Breittfeld., *J. prakt. Chem*; *311* 256 (1969).

[42b] S. B. Monroe, *U.S. 3,216,968*, Nov. 9, 1965.

[43] Reference 3, Chapter 9.

[44] D. H. Chadwick and R. S. Watt, in Van Wazer, ed., *op. cit.*, Chapter 19. A listing of commercial esters may be found in this chapter; C. R. Gleason and G. H. Slack, *U.S. 3,328,492* (to Hexionic Acid Corp.), June 27, 1967; H. Grabhofer, H. Muller, R.-F. Posse, and H. Ulrich, *U.S. 3,342,903* (to Agfa A.-G.), Sept. 19, 1967; H. R. Guest and B. W. Kiff, *U.S. 2,957,856* (to Union Carbide Corp.), Oct. 25, 1960.

[44a] I. K. Rultsova and S. M. Shuer, *Plast. Massy*, **1962** (12), 23.

[45] Richardson Co. (by T. E. Ronay and R. D. Dexheimer), *Ger. 1,251,450*, Oct. 5, 1967; B. R. Franko-Filipasie, *U.S. 3,324,202* (to FMC Corp.), June 6, 1967; The Lubrizol Corp. (by T. C. Jennings), *Can. 805,177*, Jan. 28, 1969; P. J. Apice, *U.S. 3,433,854* (to Allied Chemical Corp.), Mar. 18, 1969; A. L. Austin, R. J. Hartman, and J. T. Patton, Jr., *U.S. 3,439,067* (to Wyandotte Chemicals Corp.), Apr. 15, 1969.

[46] R. L. McConnell and H. W. Coover, Jr., *U.S. 2,960,528* (to Eastman Kodak Co.), Nov. 15, 1960.

[47] V. Trescher, G. Braun, and H. Nordt, *Can. 759,469* (to Farb. Bayer A.-G.), May 23, 1967.

[47a] F. S. Eiseman, Jr., and L. M. Schenck, *U.S. 3,331,896* (to General Aniline & Film Corp.), July 18, 1967.

[48] T. Yanagawa, S. Hashimoto, and I. Furukawa, *Doshisha Kogaku Kaishi*, **8**, 69 (1957).

[49] British Celanese, Ltd. (by F. E. King, V. F. G. Cooke, and J. Lincoln), *Brit. 872,206*, July 5, 1961.

[49a] D. E. Kvalnes and N. O. Brace, *U.S. 2,691,567* (to E. I. duPont de Nemours & Co.), Oct. 12, 1954.

[49b] Dynamit Nobel A.-G. (by R. M. Ismail), *Can. 805,183*, Jan. 28, 1969.

[49c] C. A. Seil, R. H. Boschan, and J. P. Holder, *U.S. 3,308,207, 3,308,208* (to Douglas Aircraft Co.), Mar. 7, 1967.

[50] W. M. Lanham, *Brit. 812,390* (to Union Carbide Corp.), Apr. 22, 1959; *Ger. 1,061,763*, July 23, 1959; Hercules Inc., *Belg. 725,477*, Dec. 13, 1968.

[51] H. Fernholz and F. Wunder, *U.S. 3,378,502* (to Farb. Hoechst A.-G.), Apr. 16, 1968.

[52] P. E. Pelletier and F. Pelletier, *U.S. 3,281,502* (to Wyandotte Chemicals Corp.), Oct. 25, 1966.

[53] S. Carpenter and E. R. Witt, *U.S. 3,324,205* (to Celanese Corp.), June 6, 1967; *U.S. 3,403,049*, Sept. 24, 1968.

[54] G. Palethorpe, *U.S. 3,318,978* (to Monsanto Co.), May 9, 1967.

[54a] G. H. Birum, J. L. Schwendeman, and R. M. Anderson, *U.S. 3,334,112* (to Monsanto Co.), Sept. 26, 1967.

[55] See reference 3, Chapter 8; reference 1, Chapter 7; reference 44.

[55a] L. Friedman, *U.S. 3,330,888* (to Union Carbide Corp.), July 11, 1967.

[56] Various commercial processes for lower alkyl diesters are discussed in H. E. Sorstokke, *Can. 787,946* (to Stauffer Chemical Co.), June 18, 1968, and in the patents referred to therein.

[57] A detailed review of the Michaelis-Arbuzov reaction has appeared: R. G. Harvey and E. R. de Sombre, in Grayson and Griffith, eds., *op. cit.*, Vol. 1, 1964.

[58] Reference 3, Chapter 7, pp. 129–30; K. L. Freeman and M. J. Gallagher, *Aust. J. Chem.*, **21**, 2297 (1968); V. S. Abramov, A. L. Shalman, and A. P. Bulgakova, *J. Gen. Chem.*, *USSR*, **38** (6), Eng. trans, p. 1266 (June 1968).

[59] K. Moedritzer and R. R. Irani, *J. Org. Chem.*, **31**, 1603 (1966); *U.S. 3,257,479, 3,269,812, 3,288,846* (to Monsanto Co.), June 21, 1966, Aug. 30, 1966, Nov. 29, 1966.

[59a] R. R. Irani and R. S. Mitchell, *U.S. 3,395,113* (to Monsanto Co.), July 30, 1968; *U.S. 3,470, 112* (to Monsanto Co.) Sep. 30, 1969; Paper presented at the 155th Annual Meeting of the Amer. Chem. Soc., San Francisco, April, 1968.

[60] Deutsche Advance Produkton G.m.b.H., *Brit. 1,048,070*, Nov. 9, 1966.

[61] A. J. Kolka, *U.S. 2,866,808* (to Ethyl Corp.), Dec. 30, 1958.

[62] A. D. F. Toy and K. H. Rattenbury, *U.S. 2,841,604* (to Victor Chemical Works), July 1, 1958; Victor Chemical Works (by A. D. F. Toy and K. H. Rattenbury), *Ger. 1,041,251*, Oct. 16, 1958.

[63] M. S. Larrison, *U.S. 3,333,026, 3,333,027* (to Weston Chemical Corp.), July 25, 1967.

[63a] L. Friedman, *U.S. 3,433,856* (to Weston Chemical Corp.), Mar. 18, 1969; *Text. World*, **1968**, 161 (July).

[63b] J. J. Pitts, M. A. Robison, and S. I. Trotz, *Inorg. Nucl. Chem. Lett.*, **4**, 483 (1968).

[63c] Electric Reduction Co. of Canada, Ltd. (by H. Coates and J. D. Collins) *Can. 794,378*, Sept. 10, 1968.

[63d] B. S. Taylor and M. R. Lutz, *U.S. 3,412,052* (to FMC Corp.), Nov. 19, 1968.

[64] W. M. Ramsey and C. Kezerian, *U.S. 2,917,528, 2,964,549* (to Victor Chemical Works), Dec. 15, 1959, Dec. 13, 1960; T. M. Beck and E. N. Walsh, *U.S. 3,235,517* (to Stauffer Chemical Co.), Feb. 15, 1966; see also Stauffer Chemical Co., *Belg. 687,956*, Apr. 7, 1967; T. M. Beck and E. N. Walsh, *U.S. 3,076,010* (to Stauffer Chemical Co.), Jan. 29, 1963; W. M. Ramsey and C. Kezerian, *U.S. 2,917,528* (to Victor Chemical Works), Dec. 15, 1959.

[65] R. R. Hindersinn and M. I. Iliopulos, *U.S. 3,385,914* (to Hooker Chemical Corp.), May 28, 1968.

[66] Imperial Chemical Industries, Ltd. (by Harry McGrath), *Brit. 1,002,929*, Sept. 2, 1965; see also E. E. Nifantev, A. I. Zavalishina, I. S. Nasonovskii, I. V. Komlev, *Zh. Obshch. Khim*, *38*, 2453 (Nov. 1968)

[67] N. V. de Bataafsche Petroleum Maatschappij, *Dutch 69,357*, Jan. 15, 1952.

[68] F. Johnston, *U.S. 2,754,319* (to Union Carbide and Carbon Corp.), July 10, 1956.

[69] Victor Chemical Works, *Brit. 766,722*, Jan. 23, 1957.

[70] S. A. Zahir, *U.S. 3,374,292* (to CIBA, Ltd.), Mar. 19, 1968; *U.S. 3,381,063* (to CIBA, Ltd.), Apr. 30, 1968. For derivatives of this material, see CIBA, Ltd., *Belg. 653,179*, Mar. 17, 1965, and CIBA, Ltd., *Neth. Appl. 6,410,849*, Mar. 19, 1965.

[70a] Farb. Hoechst A.-G. (by K. Schimmelschmidt and H. J. Kleiner), *Can. 804,618*, Jan. 21, 1969.

[71] F. J. Welch and H. J. Paxton, Jr., *U.S. 3,183,214* (to Union Carbide Corp.), May 11, 1965.

[72] Farb. Hoechst A.-G. (by M. Reuter and E. Wolf), *Ger. 1,046,047*, Dec. 11, 1958; E. Klanke, E. Kuhle, I. Hammann, and W. Lorenz, *U.S. 3,407,248* (to Farb. Bayer A.-G.), Oct. 22, 1968.

[73] E. D. Weil and K. J. Smith, *U.S. 3,202,692* (to Hooker Chemical Corp.), Aug. 24, 1965.

[74] S. C. Ternin, *U.S. 3,065,183* (to Koppers Co., Inc.), Nov. 20, 1960; *U.S. 3,179,522* (to Koppers Co., Inc.), Apr. 20, 1965.

[75] A. D. F. Toy and K. H. Rattenbury, *U.S. 2,714,100* (to Victor Chemical Works), July 26, 1955.

[76] E. E. Hardy and T. Reetz, *U.S. 2,770,620* (to Monsanto Chemical Co.), Nov. 13, 1956.

[77] G. H. Birum and R. B. Clampitt, *U.S. 3,372,209* (to Monsanto Co.), Mar. 5, 1968.

[78] J. G. Abramo, A. Y. Garner, and E. C. Chapin, *U.S. 3,161,667* (to Monsanto Co.), Dec. 15, 1964; J. G. Abramo, A. Y. Garner, and E. C. Chapin, *U.S. 3,051,740* (to Monsanto Chemical Co.), Aug. 28, 1962.

[78a] A. I. Polyakov and L. A. Il'ina, *Siberian Chem. J.*, *Eng. Trans.*, p. 467 (July–August, 1968).

[79] E. N. Rostovskii, O. V. Shehelkunova, and N. S. Bondareva, *Vysokomol. Soedin. Khim. Svoistva i Modifikatsuja Polimerov, Sb. Statei*, **1964,** 151; Carlisle Chemical Works (by I. Heckenbleikner and K. R. Molt) *Can. 804,619*, Jan. 21, 1969; R. H. Wiley, *U.S. 2,478,441* (to E. I. duPont de Nemours & Co.), Aug. 9, 1949.

[79a] N. E. Boyer and R. R. Hindersinn *Can. 802,990* (to Hooker Chemical Corp.), Dec. 31, 1968.

[80] V. S. Abramov and V. S. Tsivunin, *Tr. Kazansk. Khim. Tekhnol. Inst.*, **26,** 96 (1959).

[81] A. Guttag, *U.S. 3,382,236* (to Weston Chemical Corp.), May 7, 1968.

[81a] M. S. Larrison, *U.S. 3,333,027, 3,333,028* (to Weston Chemical Corp.), July 25, 1967.

[82] Pure Chemicals, Ltd. (by L. Friedman), *Brit. 1,012,630*, Dec. 8, 1965.

[82a] G. F. D'Alelio, *U.S. 3,325,569*, June 13, 1967.

[82b] A. D. F. Toy and R. S. Cooper, *J. Amer. Chem. Soc.*, *76*, 2191, (1954); G. F. D'Alelio, *U.S. 3,062,774* (to Dal Mon Research Co.), Nov. 6, 1962.

[83] A. D. B. Graham, *U.S. 3,255,145* (to Ethyl Corp.), June 7, 1966; Stauffer Chemical Co., *Belg. 713,066*, Apr. 1, 1968.

[84] FMC Corp. (by B. S. Taylor and M. R. Lutz), *Fr. 1,372,907*, Sept. 18, 1964; G. E. Schroll, *U.S. 3,250,827* (to Ethyl Corp.), May 10, 1966.

[85] A. Y. Garner, *U.S. 2,916,510* (to Monsanto Chemical Co.), Dec. 8, 1959.

[85a] S. Z. Ivin, I. D. Shelakova, V. K. Promonenkov, B. B. Levin, and I. N. Fetin, *Zh. Obshch. Khim.*, **37** (7), Eng. trans., p. 1556 (July, 1967).

[86] Badische Anilin- & Soda-Fabrik A.-G. (by K. H. Koenig and H. Pommer), *Belg.* *609,477,* Apr. 24, 1962.

[87] Albright & Wilson, Ltd., *Brit. 931,146,* July 10, 1963; R. T. McConnell and H. W. Coover, Jr., *U.S. 3,030,340* (to Eastman Kodak Co.), Apr. 17, 1962; Bakelite and O.P.C.I.A. (by G. L. Quesnel, J. Girard, and A. Thiot), *Fr. 1,213,894,* Apr. 5, 1960. See also *U.S. 3,121,697,* Feb. 18, 1964.

[88] C. H. Campbell, *U.S. 3,153,013* (to Monsanto Co.), Oct. 13, 1964.

[89] H. W. Coover, *U.S. 2,642,413* (to Eastman Kodak Co.), June 15, 1953; H. W. Coover, Jr., R. L. McConnell, and N. H. Shearer, Jr., *Ind. Eng. Chem.,* **52,** 412 (1960).

[89a] J. R. Caldwell and J. C. Martin, *U.S. 2,882,294* (to Eastman Kodak Co.), Apr. 14, 1959.

[89b] J. Farago, *U.S. 3,116,268* (to E. I. duPont de Nemours & Co.), Dec. 31, 1963.

[90] Rhone Poulenc *Belg. 691,377,* June 16, 1967.

[91] Farb. Hoechst A.-G. vorm. M. Lucius and Brüning (by K. Schimmelschmidt and W. Denk), *Ger. 1,023,033,* Jan. 23, 1958.

[92] I. A. Krivosheeva, A. I. Razumov, and G. S. Kolesnikov, *Vysokomol. Soedin.,* **3,** 1247 (1961); I. A. Krivosheeva, A. I. Razumov, B. Ya. Teitel'baum, and T. A. Yagfarova, *Vysokomol. Soedin., Karbot. Vysokomol. Soedin., Sb. Statei,* **1963,** 160; Etablissements Kuhlmann (by M. C. Demarq and J. Sleziona), *Fr. 1,403,732,* June 25, 1965; S. Hashimoto, I. Furukawa, and T. Yanagawa, *Kogyo Kagaku Zasshi,* **64,** 1682 (1961); A. D. F. Toy and R. S. Cooper, *J. Amer. Chem. Soc.,* **76,** 2191 (1954); Farb. Hoechst A.-G. (by H. Krämer, G. Messwarb, and W. Denk) *Ger. 1,032,537,* June 19, 1958.

[93] D. Israelov and L. A. Rodivilova, *Polym. Sci., USSR.,* **8,** 1557 (1966) (Eng. trans.); H. W. Coover, Jr., *U.S. 2,743,258* (to Eastman Kodak Co.), Apr. 24, 1956; W. B. McCormack and H. E. Schroeder, *U.S. 2,891,915* (to E. I. duPont de Nemours & Co.), June 23, 1959.

[94] H. W. Coover, Jr., and R. L. McConnell, *U.S. 3,271,329* (to Eastman Kodak Co.), Sept. 6, 1966.

[95] D. Grant, J. R. Van Wazer, and C. H. Dungan, *J. Polym. Sci., Part A-1,* **5,** 57 (1967); A. N. Pudovik, A. A. Muratova, F. F. Sushentsova, and M. M. Zoreva, *Vysokomol. Soedin.,* **6,** 258 (1964).

[96] Farb. Hoechst. A.-G., *Belg. 671,561,* Oct. 28, 1965.

[97] A. Carson, W. E. Feely, and M. J. Hurwitz, *U.S. 3,371,131* (to Rohm and Haas Co.), Feb. 27, 1968.

[98] A. N. Pudovik and M. A. Pudovik, *Dokl. Akad. Nauk SSSR,* **168,** 354 (1966) (p. 492 in Eng. trans.).

[99] R. L. McConnell and H. W. Coover, Jr., *U.S. 3,062,788* (to Eastman Kodak Co.), Nov. 6, 1962.

[100] G. H. Birum, *U.S. 3,014,956, 3,042,701, 3,058,941, 3,192,242, 3,317,510* (to Monsanto Co.), Dec. 26, 1961, July 3, 1962, Oct. 16, 1962, June 29, 1965, May 2, 1967; R. M. Anderson and G. H. Birum, *U.S. 3,157,613* (to Monsanto Co.), Nov. 17, 1964; G. H. Birum, *U.S. 3,363,031* Jan. 9, 1968; *U.S. 3,391,226,* July 2, 1968 (both to Monsanto Co.).

[100a] R. J. Rolih and I. K. Fields, *U.S. 3,220,989* [to Standard Oil Co. (Ind.)], Nov. 30, 1965.

[100b] Proctor & Gamble Co., *Belg. 712,772,* Mar. 26, 1968.

[100c] G. H. Birum, *U.S. 3,317,510* (to Monsanto Co.), May 2, 1967.

[100d] G. H. Birum and R. B. Clampitt, *U.S. 3,372,209* (to Monsanto Co.), Mar. 5, 1968.

[100e] B. P. Block, I. C. Popoff, J. P. King, and L. K. Huber, *U.S. 3,332,986* (to Pennsalt Chemicals Corp.), Aug. 1, 1967.

[100f] I. C. Popoff, B. P. Block, and L. K. Huber *U.S. 3,332,987* (to Pennsalt Chemicals Corp.), Aug. 1, 1967.

[101] See reference 3, Chapters 2–6; reference 1, Chapter 5.

[102] J. H. Brown and I. Gordon, *Can. 777,708* (to Hooker Chemical Corp.), Feb. 6, 1968.

[103] M. G. Palmer, *U.S. 3,375,074* (to Hooker Chemical Corp.), Mar. 26, 1968.

[104] G. T. Miller, *U.S. 3,251,756* (to Hooker Chemical Corp.), May 17, 1966.

[105] E. J. Love and F. A. Ridgway, *U.S. 3,371,994* (to Hooker Chemical Corp.), Mar. 5, 1968.

[106] I. L. Knunyants and R. N. Sterlin, *Dokl. Akad. Nauk SSSR*, **56**, 49 (1947); W. J. Vullo, *U.S. 3,452,098* (to Hooker Chemical Corp.) Jun. 24, 1969.

[106a] Farb. Bayer A.-G., *Belg. 677,189*, Mar. 1, 1966; Farb. Hoechst A.-G. (by M. Reuter and E. Wolf), *Ger. 1,078,574*, Mar. 31, 1960.

[107] A. Hoffman, *J. Amer. Chem. Soc.*, **43**, 1684 (1921); *ibid.*, **52**, 2995 (1930).

[107a] Farb. Hoechst A.-G. (by M. Reuter and F. Jacob), *Ger. 1,040,549*, Oct. 5, 1958; (by M. Reuter and L. Orthner), *Ger. 1,035,135*, Jan. 8, 1959; *Ger. 1,035,135*, July 31, 1958.

[108] W. J. Jones, W. C. Davies, S. T. Bowden, C. Edwards, V. E. Davis, and L. H. Thomas, *J. Amer. Chem. Soc.*, **1947**, 1446: L. Maier, D. Seyferth, F. G. A. Stone, and E. G. Rochow, *ibid.*, **79**, 5884 (1957). *Swiss Pat. Appl. Ser. No. 15. 143/69.*

[109] M. Reuter, *U.S. 2,912,466* (to Farb. Hoechst A.-G.), Nov. 10, 1959; W. A. Reeves, F. F. Flynn, and J. D. Guthrie, *J. Amer. Chem. Soc.*, **77**, 3923 (1955).

[109a] Farb. Hoechst A.-G. (by E. Wolf and M. Reuter), *Ger. 1,082,910*, June 9, 1960.

[110] A. Y. Garner, *U.S. 3,010,998* (to Monsanto Co.), Nov. 28, 1961.

[110a] Carlisle Chemical Works, Inc. (by I. Hechenbleikner and K. R. Mott), *Can. 811,190*, Apr. 22, 1969; M. Grayson, P. T. Keough, and M. M. Ranhut, *U.S. 3,409,707* (to American Cyanamid Co.), Nov. 5, 1968.

[111] Hooker Chemical Corp. (by I. Gordon and J. H. Brown), *Ger. 1,201,066*, Sept. 16, 1965.

[112] Farb. Hoechst A.-G. (by Martin Reuter and by Reuter and L. Orthuer), *Ger. 1,064,061, 1,041,957*, Aug. 27, 1959, Oct. 30, 1958.

[112a] M. Reuter, L. Orthner, F. Jacob, and E. Wolf, *U.S. 2,937,207* (to Farb. Hoechst A.-G.), May 17, 1960.

[112b] L. Maier, *U.S. 3,432,559* (to Monsanto Co.), Mar. 11, 1969.

[112c] M. M. Ranhut, N. B. Borowitz, and M. Grayson, *U.S. 3,422,149* (to American Cyanamid Co.), Jan. 14, 1969; M. Grayson, P. T. Keough, M. M. Rauhut, *U.S. 3,452,099* (to American Cyanamid Co.) Jun. 24, 1969.

[112d] D. W. Grisley, Jr., *U.S. 3,334,145* (to Monsanto Research Corp.), Aug. 1, 1967.

[112e] D. W. Grisley, Jr., *U.S. 3,341,605* (to Monsanto Research Corp.), Sept. 12, 1967.

[112f] G. H. Birum and C. N. Mathews, *U.S. 3,426,073, 3,426,074*, Feb. 4, 1969; G. H. Birum, *U.S. 3,446,852*, May 27, 1969 (all to Monsanto Co.).

[112g] D. J. Daigle, L. H. Chance, and G. L. Drake, Jr., *J. Chem. Eng. Data*, **13**, 585 (1968).

[113] Albright & Wilson, Ltd., *Brit. 740,269, 790,641*, Nov. 9, 1955, Feb. 12, 1958.

[114] Albright & Wilson, Ltd., *Brit. 764,313*, Dec. 28, 1956.

[115] W. A. Reeves and J. D. Guthrie, *U.S. 2,846,413* (to USA), Aug. 5, 1958; L. M. Kindley, H. E. Podall, and N. Filipescu, *SPE Trans.*, **2**, 122 (1962).

[116] M. Grayson, P. T. Keough, and M. M. Rauhut, *U.S. 3,320,321* (to American Cyanamid Co.), May 16, 1967; *U.S. 3,332,962*, July 25, 1967; P. J. Pare and M. Hauser, *U.S. 3,125,555* (to American Cyanamid Co.), Mar. 17, 1964.

[116a] Farb. Hoechst A.-G. (by M. Reuter and F. Jacob), *Ger. 1,040,549*, Oct. 9, 1958.

[116b] Literature from Hooker Chemical Corp., Niagara Falls, N.Y., May 11, 1961.

[116c] E. B. Trostyanskaya, E. S. Venkova, and Yu. A. Mikhailin, *Zh. Obshch. Khim.*, **37** (77), Eng. trans., p. 1572 (July, 1967).

[116d] American Cyanamid Co. (by M. M. Ranhut), *Fr. 1,324,166*, Apr. 12, 1963.

[116e] A. Y. Garner, *U.S. 3,346,647* (to Monsanto Research Corp.), Oct. 10, 1967.

[117] R. B. Clampitt, G. H. Birum, and R. M. Anderson, *U.S. 3,306,937* (to Monsanto Co.), Feb. 28, 1967; *U.S. 3,468,678*, Sep. 23, 1969.

[118] R. C. Miller, *U.S. 3,299,015* (to E. I. duPont de Nemours & Co.), Jan. 17, 1967.

[119] A. Y. Garner, *U.S. 3,346,647* (to Monsanto Research Corp.), Oct. 10, 1967.

[120] E. C. Chapin and A. Y. Garner, *U.S. 3,158,642* (to Monsanto Co.), Nov. 24, 1964. See also A. Y. Garner, *U.S. 3,235,536* (to Monsanto Co.), Feb. 15, 1966.

[120a] L. H. Chance and J. D. Guthrie, *J. Appl. Chem. (London)*, **10**, 395 (1960).

[121] S. A. Buckler and M. Epstein, *U.S. 3,213,042* (to American Cyanamid Co.), Oct. 19, 1965.

[122] W. B. McCormack, *U.S. 2,671,077, 2,671,078* (to E. I. duPont de Nemours & Co.), Mar. 2, 1954.

[123] P. R. Bloomfield, *U.S. 3,044,984* (to Microcell, Ltd.), July 17, 1962.

[124] T. Y. Medved, T. M. Frunze, C. M. Hu, V. V. Kurashev, V. V. Korshak, and M. I. Kabochnik, *Vyokomol. Soedin.*, **5**, 1309 (1963).

[125] F. J. Welch and H. J. Paxton, *J. Polym. Sci., Part A*, **3**, 3427, 3439 (1965).

[126] K. A. Petrov and V. A. Parshina, *Russ. Chem. Rev.*, **37** (7), Eng. trans., p. 532 (1968); H. Hellmann, J. Bader, H. Birkner, and O. Schumacher, *Amer. Chem.*, **659**, 49 (1962); A. R. Hands and A. J. H. Mercer, *J. Chem. Soc.*, *C* (11), 1331 (1968).

Chapter 3

SOME CHEMISTRY OF ANTIMONY, BORON, CHLORINE, AND BROMINE

Research on chemical fire retardants has focused on P, Sb, B, Cl, and Br. Scattered reports and claims have been made for systems containing other elements, for example, silicate glasses and compounds of zirconium, selenium, and arsenic. In this chapter we develop the chemistry of Sb, B, Cl, and Br around the type of compounds reported most frequently in the literature on fire retardants. This leads, in the case of boron at least, to emphasis on the inorganic compounds and to a very sketchy treatment of the new and very interesting chemistry of the boranes and their derivatives. The reader should bear this bias in mind as he proceeds and should, in any case, consult the references for details.

The chemistry of antimony is much like that of phosphorus but has not been developed in nearly as much detail. Boron chemistry is much more sophisticated but little of it has been utilized for fire retardants, attention in this area being devoted almost exclusively to the inorganic salts. The halogens have, of course, been the target of a great many researches. We are interested primarily in those halogen compounds that have the halogen as a substituent in an organic molecule. The chemistry of these substituent halogen atoms reduces primarily to the halogenation reactions and an understanding of the chemical and physical properties of the products.

ANTIMONY

Antimony, symbol Sb, is a member of the Group V family which includes, in order of increasing atomic number, nitrogen, phosphorus, arsenic, antimony, and bismuth. Antimony has an atomic number of 51 and an atomic weight of 121.76. There are two abundant isotopes: one with mass number 121 and 70 neutrons, the other with mass number 123 and 72 neutrons in the nucleus. Both have 51 protons in the nucleus and 51 electrons in 5 electron

75

shells. The valence shell configuration is $5s^2 5p^3$ (cf. phosphorus $3s^2 3p^3$) [1]. The nuclear spin is $\frac{5}{2}$ for mass number 121 and $\frac{7}{2}$ for mass number 123. Both isotopes, therefore, have quadropole moments and may exhibit complex spectra in high-resolution nmr spectrometry, both from effects owing to the quadropole and from excessive multiplet formation from spin-spin splitting in, for example, proton resonance studies. Little has been reported on nmr studies on antimony compounds.

There are many similarities to phosphorus in bond formation: bonds in triply connected antimony compounds will be p^3, sp^2 hybrids, or sp^3 hybrids (with one filled orbital not engaged in bonding); quadruply connected compounds will be largely sp^3 hybrids with substituents at the apices of a tetrahedron; and penta- and hexavalent compounds must use $5d$ orbitals—$sp^3 d$, $sp^3 d^2$ hybrids. Measured bond angles [2] on $SbCl_3$, $SbBr_3$, and SbI_3 indicate a mixture of hybrids: probably p^3 and sp^2 in analogy with the corresponding phosphorus compounds. Whereas phosphorus accommodates six substituents with difficulty, antimony does so with ease and a number of SbX_6 species have been isolated. As one goes from nitrogen to bismuth, the elements become more metallic and the higher oxidation states become less favored. Thus, the +5 state is the most common for phosphorus and reduction to lower states requires considerable energy. Antimony compounds in the +5 state are fairly unstable; the +3 state is favored. The antimony compounds melt and boil at very much higher temperatures than the corresponding phosphorus species [3–5]:

	Melting Point, °C	Boiling Point, °C		Melting Point, °C	Boiling Point, °C
PH_3	−134	−88	SbH_3	−88	−17
PCl_3	−94	76	$SbCl_3$	73	220
P_4O_6	24	175	Sb_4O_6	656	1425

Antimony is relatively rare in nature, comprising only 0.0001 % of the igneous rocks of the earth [3], and ore bodies are relatively small and scattered, and are principally located outside the United States. In recent years the United States has imported some 95 % of its total antimony requirements but produced most of its needs for Sb_2O_3. The ore is largely Sb_2S_3 and Sb_2O_3 and is roasted to Sb_2O_3, purified by distillation (volatilization):

$$Sb_4S_6 + 9O_2 \rightarrow Sb_4O_6 + 6SO_2$$

Note that modern usage refers to the trioxide more properly as Sb_4O_6.

Antimony trihalides are prepared readily by direct halogenation of the oxide:

$$Sb_4O_6 + 6X_2 \rightarrow 4SbX_3 + 3O_2$$

The pentahalides are more difficult to prepare and only the fluoride and chloride have been isolated [2]. The sulfides are commonly employed but not as fire retardants, and they will not be discussed here.

Apart from the difference in atomic size and number of inner shell electrons leading to more metal-like properties, antimony behaves much like phosphorus. Indeed, the nomenclature of derivatives is very similar and should, in fact, be identical. It is unfortunate that the ending -ic is now being employed for Sb(III) compounds when it is so well entrenched for P(V) species. A perusal of reviews of antimony chemistry quickly shows that, when the techniques of the phosphorus chemist are applied, antimony more often than not responds as would phosphorus. It would appear that more deliberate application of these methods would yield dividends quickly. For example, triesters of Sb(III) have been made from Sb_4O_6 but $SbCl_3$ is a much more likely starting material, and so on. The current nomenclature [2] is:

Compound	Current	Old Form	P Cpd Name
Sb(V) salts	Antimonate(V)	Antimonate	Phosphate
Sb(III) salts	Antimonate(III)	Antimonite	Phosphite
Sb(V) acid	Antimonic(V)	Antimonic	Phosphoric
Sb(III) acid	Antimonic(III)	Antimonous	Phosphorous

The more general form of nomenclature established by the IUPAC Committee [6] has taken hold for antimony, not for phosphorus. Naming habits for the latter are unlikely to change; it may be wiser to alter the naming of the other Group V elements to conform to that for phosphorus.

The most common antimony compounds used as fire retardants are Sb_4O_6 and $SbCl_3$ [7]. Some of their properties are listed in Table 3-1. Antimony(III)

Table 3-1 Properties of Two Trivalent Antimony Compounds [5]

	Sb_4O_6 (Rhombic)	$SbCl_3$
Melting point, °C	656	73.4
Boiling point, °C	1425	220.2
Specific gravity (25°), g/ml	5.67	3.140
Solubility	0.002 g/100 g, H_2O (15°)	Sol ϕ, CS_2, hydrolysis in H_2O^a
Refractive index	2.18, 2.35, 2.35	$\cdots^b$

^{a} $SbCl_3 + H_2O \rightarrow SbOCl + 2HCl$
^{b} For $SbBr_3$ and SbI_3, $n > 1.7$

compounds are amphoteric. Thus, the antimonyl ion—SbO^+—forms in acid solutions of Sb_4O_6 but the antimonite anion forms in base [5]:

$$Sb_4O_6 + 2H_2SO_4 \rightarrow 2(SbO)_2SO_4 + 2H_2O$$

and

$$Sb_4O_6 + 12NaOH \rightarrow 4Na_3SbO_3 + 6H_2O$$

Organic compounds of antimony have received attention only very recently in fields other than medical research. Salts of organic acids, esters, and the like are of considerable interest in clear, transparent polymer systems where Sb_4O_6 or $SbCl_3$ may yield opaque products. To this end certain antimony esters have been suggested and a fledgling field is developing.

Analogs of phosphite esters are easy to prepare, even from Sb_4O_6, for example [5]:

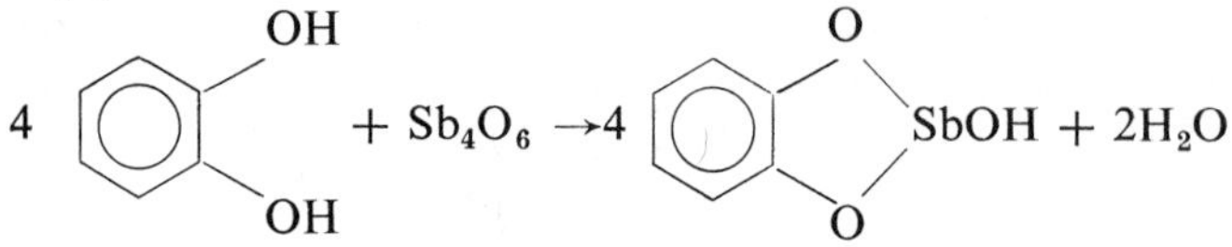

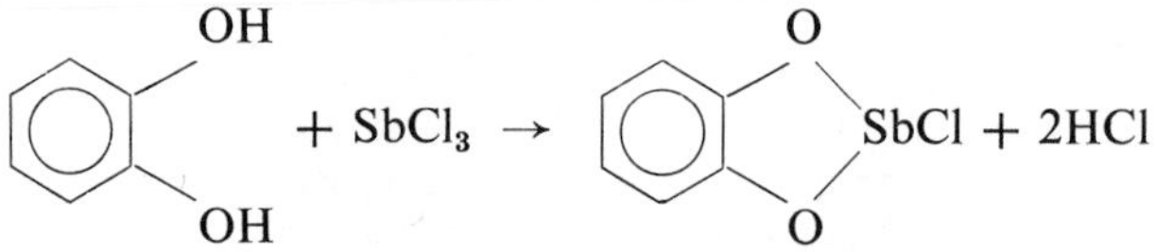

Butyl antimonite has been obtained in an autoclave reaction at 240° and 20 atmospheres [8]:

$$Sb_4O_6 + 12CH_2(CH_2)_3OH \rightarrow 4[CH_3(CH_2)_3O]_3Sb + 6H_2O$$
$$\text{bp } 102° \text{ (0.3 mm)}$$

Cyclic alkyl antimonites have been prepared [9]:

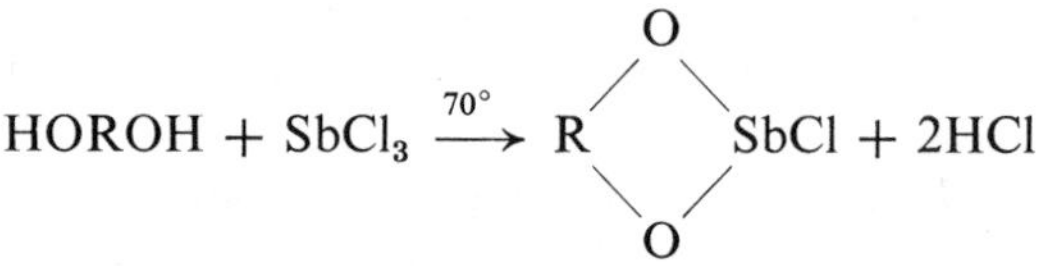

The reaction has been run with allyl alcohol in benzene with excess ammonia to remove the liberated HCl [10]. At 79° a yield of 75% was obtained. By running at $-10°$, conversion to a brominated product was achieved:

$$(CH_2{=}CHCH_2O)_3Sb + 2Br_2 \xrightarrow{CH_2Cl_2}$$
$$\text{bp160°(30 mm)}$$
$$(CH_2{=}CHCH_2O)Sb[OCH_2CH(Br)CH_2Br]_2$$

Isomerization to the antimonate (Arbuzov rearrangement) has not been reported but should be possible.

Antimonic acids (or, less preferably, stibonic acids) analogous to phosphonic acids have been prepared but largely by complex methods. For example, aryl antimonic acids are prepared from aryl diazonium salts and Sb_4O_6 in HCl [6]:

$$ArN_2Cl + SbCl_3 + 3H_2O \xrightarrow[\text{2. reacidify}]{\text{1. base}} Ar—Sb(O)(OH)_2 + N_2 + 4HCl$$

It should be possible to make these compounds from antimonites in a much more straightforward manner

$$Sb(OR)_3 \xrightarrow[R'X]{\Delta} Sb(R')(O)(OR)_2$$

but this apparently remains to be attempted. A similar situation prevails for antiminic or stibinic esters: $R_2Sb(O)(OR')$.

Stibine, SbH_3, and mono-, di-, and trialkyl and arylstibine are known, though the mono- and di-compounds are very rare. Tertiary stibines have been made [2, 10a] via Grignard reagents and $SbCl_3$ (cf. tertiary phosphine synthesis), via a Wurtz reaction or via trialkyl-aluminum compounds.

Table 3-2   Some Additional Antimony Compounds Suggested for Fire-Retardant Use

Compound	Use	Reference
KSb tartrate: 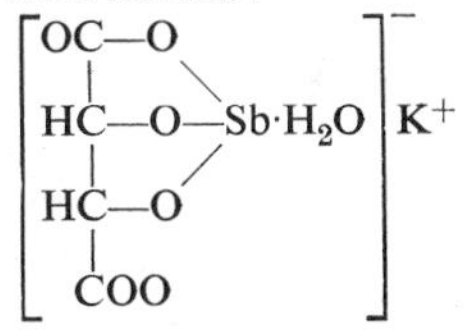 	Polyester	2, 11
Sb caproate	Polyester	12
$Sb(OCH_2CH_3)_3$	Polyester	13
Sb polymethylene glycolate: 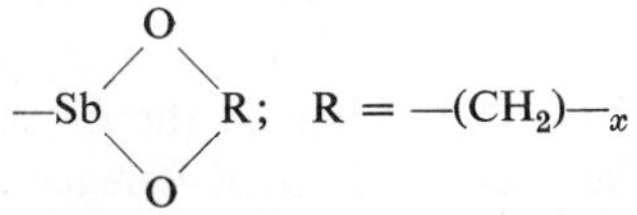 $R; \ R = —(CH_2)—_x$	Polyester	14
$Sb(OCH_2CH=CH_2)_3$	Polyester	15
	Polyester	16
Sb phosphate	Textiles	17
$KSb(OH)_6$	Textiles	1, 18
NH_4SbF_6	Polyolefins	19, 20
$Sb[OCH(CH_3)CH_2CH_3]_3$	Epoxies	21
Sb_2S_3	—	21a

Stibine oxides and stibonium compounds are known but are not common.

Table 3-2 lists some additional antimony compounds suggested in the literature as fire retardants for the indicated substrates. The largest category is that of organic derivatives for polyester resins. Note the salt formation demonstrating antimony's amphoteric nature. A 1962 review lists 225 references (mostly patents) to uses of antimony compounds as fire retardants [11].

At this writing, it appears that the chemistry of antimony is just beginning to unfold. If the element were more abundant, we might expect this subject to become as sophisticated as it is for phosphorus. However, it seems unlikely that the element will ever be as inexpensive as phosphorus and thus there will not be the incentive to use it. Further, there are but few advantages one can cite for antimony over phosphorus (the lower water solubility of Sb_4O_6 is one). Thus, it is likely that antimony compounds will remain rather high-priced specialties finding use in a rather limited way.

BORON

Boron, symbol B, is the lightest member of the Group III elements, which also include Al, Ga, In, Tl, Sc, and Y. Boron has two natural isotopes with mass number 10 and 11 (18.3 and 81.7%, respectively) [22–24]. There are 5 protons and 5 and 6 neutrons, respectively, in the nucleus. The nuclear spin numbers are 3 and $\frac{3}{2}$. Both isotopes have quadrupole moments and these, along with the large number of spin states, serve to complicate nmr spectra of protons and other magnetic nuclei when bonded to boron. Since B has spin $\frac{3}{2}$, useful nmr spectra are obtained by probing the boron nucleus directly. Additional definitive data on structure have been obtained by measuring the interaction of the quadrupole moment of B^{11} (and B^{10}) with electric field gradients and related effects [22].

The electronic configuration is $1s^2 2s^2 2p$. A $2s$ electron in the outer shell may readily be promoted to $2p$ to yield $2s2p_x2p_y$; these may then form sp^2 hybrids and molecules with trigonal planar symmetry. That this is, in fact, the case has been shown, for example, for $(CH_3)_3B$ in which the C—B—C bond angles are just 120° [25]. Because there are but three valence shell electrons, compounds with three substituents will readily form; compounds with four may only form if the substituent brings with it an unshared electron pair to donate. Such compounds are common: the BF_4^- ion, for example.

Boron is present in the earth's crust at a level of only 3 ppm but there are areas where it is concentrated in very substantial volume for mining. In the United States the major deposits are in western Nevada and southern California. The principal forms are borax, $Na_2B_4O_7 \cdot 10H_2O$, and boronite, $Na_2B_4O_7 \cdot 4H_2O$, in the Kramer deposits near Boron, Calif., and a variety of hydrous forms in the Searles Lake brine [23] from which borax is crystallized. The sediments overlying the Kramer deposit are principally

$$Na_2O \cdot 2CaO \cdot 5B_2O_3 \cdot 16H_2O \qquad and \qquad 2CaO:3B_2O_35 \cdot H_2O$$

The mixed sodium-calcium borate has found use in the past as a fire retardant for combating forest fires; sodium borates are used in a variety of resins [25a] and in celluloses.

Comercially, boron is isolated as borax. This, in turn, may be recrystallized as less hydrated species:

$$Na_2O \cdot 2B_2O_3 \cdot 10H_2O \rightarrow Na_2O \cdot 2B_2O_3 \cdot 5H_2O \rightarrow Na_2O \cdot 2B_2O_3 \cdot 4H_2O$$

$$Na_2O \cdot 2B_2O_3 \xleftarrow{\text{calcine}} \rceil$$

By altering the Na_2O/B_2O_3 ratio in solution, the metaborates crystallize:

$$Na_2O \cdot B_2O_3 \cdot 8H_2O \rightarrow Na_2O \cdot B_2O_3 \cdot 4H_2O$$

The phase diagrams for several other $M_2O \cdot B_2O_3 \cdot H_2O$ systems have been worked out and it is possible to prepare by crystallization a variety of alkali and alkaline earth salts. Of these the potassium pentaborate, $K_2O \cdot 5B_2O_3 \cdot 8H_2O$, is notable for its reduced water solubility and the ammonium salts $(NH_4)_2O \cdot 2B_2O_3 \cdot 4H_2O$ and $(NH_4)_2O \cdot 5B_2O_3 \cdot 8H_2O$ are of interest both for somewhat enhanced solubility *vis-à-vis* borax and for their ability to decompose to acids on heating.

Orthoboric acid, H_3BO_3 (mp 170.9°, sp gr$_4^{14}$ 1.5172), is prepared by the action of strong acids on borax followed by crystallization of the solid acid. Colemanite, $2CaO \cdot 3B_2O_3 \cdot 5H_2O$, may be used instead of borax [23]. H_3BO_3 dissolves in water to form saturated solutions of the following compositions: 0°—2.48% H_3BO_3, 20°—4.8%, 80°—19.1%, and 100°—28.2%. It dissolves in various polyols by formation of esters (see below); for example, at 20° one can prepare a 19.9% solution in glycerine (98.5%). The apparent pK_a in dilute aqueous solutions is 5.8×10^{-10}. Determination by titration with base is complicated by formation of polyanions as the alkalinity increases, especially in concentrated solutions. This can be avoided by adding a polyol to release the titratable proton with formation of a polyol—$B(OH)_4^-$ complex.

On heating, H_3BO_3 loses water in two stages:

$$H_3BO_3 \xrightarrow[\text{Dec 185°}]{130–200°} HBO_2 \xrightarrow[\substack{102 \text{ mm Hg} \\ 6 \text{ hr}}]{260–270°} B_2O_3$$

Metaboric acid
I. mp 236°
II. mp 200.9°
III. mp 176°

B_2O_3 Softens at 325°

Since most fire-retardant applications of boron compounds utilize borax or boric acid, it is worthwhile to consider in some detail the physical and

Table 3-3 Solubilities of Alkali Metal and Ammonium Borates at Various Temperatures [23]

Compound	Solubility, % Anhydrous Salt by Weight, at Temperature, °C											
	0	10	20	25	30	40	50	60	70	80	90	100
Li$_2$O·5B$_2$O$_3$·10H$_2$O[a]	2.2–2.5	2.55	2.81	2.90	3.01	3.26	20.88	24.34	27.98	31.79	36.2	41.2[b]
Li$_2$O·2B$_2$O$_3$·4H$_2$O	0.88	1.42	2.51	3.34	4.63	9.40	3.50	3.76	4.08	4.35	4.75	5.17
Li$_2$B$_2$O$_4$·16H$_2$O[c]												
Li$_2$B$_2$O$_4$·4H$_2$O						7.40	7.84	8.43	9.43	{10.58 / 9.75[b]	11.8 / 9.7[b]	13.4[d] / 9.70[b]
Na$_2$O·5B$_2$O$_3$·10H$_2$O	6.28	8.10	10.55	12.20	13.75	17.40	21.80	26.90	32.25	37.84	43.80	50.30
Na$_2$O·2B$_2$O$_3$·10H$_2$O	1.18	1.76	2.58	3.13	3.85	6.00	9.55	15.90				
Na$_2$O·2B$_2$O$_3$·5H$_2$O[e]								16.40	19.49	23.38	28.37	34.63
Na$_2$O·2B$_2$O$_3$·4H$_2$O[f]								14.82	17.12	19.88	23.31	28.22
Na$_2$O·B$_2$O$_3$·8H$_2$O[g]	14.5	17.0	20.0	21.7	23.6	27.9	34.1					
Na$_2$O·B$_2$O$_3$·4H$_2$O								38.3	40.7	43.7	47.4	52.4
K$_2$O·5B$_2$O$_3$·8H$_2$O	1.56	2.11	2.82	3.28	3.80	5.12	6.88	9.05	11.7	14.7	18.3	22.3
K$_2$O·2B$_2$O$_3$·4H$_2$O		9.02	12.1	13.6	15.6	19.4	24.0	28.4	33.3	38.2	43.2	48.4
K$_2$O·B$_2$O$_3$·2.5H$_2$O		42.3	43.0	43.3	44.0	45.0	46.1	47.2	48.2	49.3	50.3	
Rb$_2$O·5B$_2$O$_3$·8H$_2$O	1.58	2.0	2.67	3.10	3.58	4.82	6.52	8.69	11.4	14.3	18.1	23.75[h]
Cs$_2$O·5B$_2$O$_3$·8H$_2$O	1.6	1.85	2.5	2.97	3.52	4.8	6.4	8.31	10.5	13.8	18.0	23.45[i]
Cs$_2$O·2B$_2$O$_3$·5H$_2$O[j]												
Cs$_2$O·B$_2$O$_3$·7H$_2$O			36.8[k]									
(NH$_4$)$_2$O·2B$_2$O$_3$·4H$_2$O	3.75	5.26	7.63	9.00	10.8	15.8	21.2	27.2	34.4	43.1	52.7	
(NH$_4$)$_2$O·5B$_2$O$_3$·8H$_2$O	4.00	5.38	7.07	8.03	9.10	11.4	14.4	18.2	22.4	26.4	30.3	

a Incongruent solubility below 37.5 or 40.5°C
b According to Bouaziz
c Transition point to 4-hydrate, 36.9 or 40°C
d At 101.2°C

e Transition point to 10-hydrate, 60.7°C, 16.6% Na$_2$B$_4$O$_7$
f Transition point to 10-hydrate, 58.2°C, 14.55% Na$_2$B$_4$O$_7$
g Transition point to 4-hydrate, 53.6°C, 36.9% Na$_2$B$_2$O$_4$
h At 102°C

i At l01.65°C
j Incongruent solubility
k At 18°C

chemical properties of H_3BO_3—$Na_2B_2O_7 \cdot 10H_2O$ systems and their analogs. Table 3-3 lists solubilities in water for some borate salts from 0 to 100°. Observe the relatively high solubility of sodium metaborate, $Na_2O \cdot B_2O_3 \cdot 8H_2O$, at low temperatures and its moderate temperature dependence in contrast to borax. The highest solubility is achieved at elevated temperatures with the ammonium borate, $(NH_4)_2O \cdot 2B_2O_3 \cdot 4H_2O$. In the borax-boric acid system shown in Figure 3-1, there is a sharp maximum in solubility, the location of which shifts to higher Na_2O/B_2O_3 ratios with

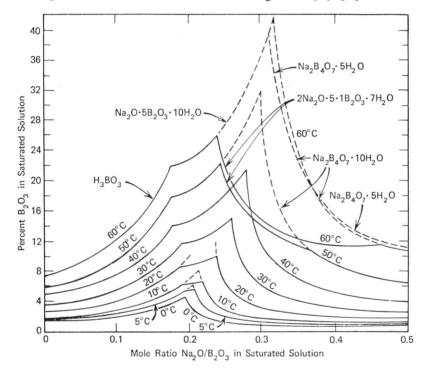

Figure 3-1 Solubility isotherms for the system $Na_2B_4O_7$-B_2O_3–H_2O at 0–60°C. The compound $2Na_2O \cdot 5.1B_2O_3 \cdot 7H_2O$, Suhr's borate, usually does not appear, since it crystallizes very slowly in the absence of seed. [23]

increasing temperature. This behavior and the earlier discussed difficulty with titration is attributed to formation in the intermediate region of polyanions and reversion to a monomeric structure at full neutralization to metaborate. The structures shown in Table 3-4 have been deduced [23, 26] from colligative properties, Raman spectroscopy, and potentiometric titrations of solutions combined with X-ray diffraction data on a variety of crystalline borates.

The effect of this polyion formation is to remove from solution the starting

Table 3-4 Some Polyions Postulated for Aqueous Solutions, M₂O/B₂O₃ Systems

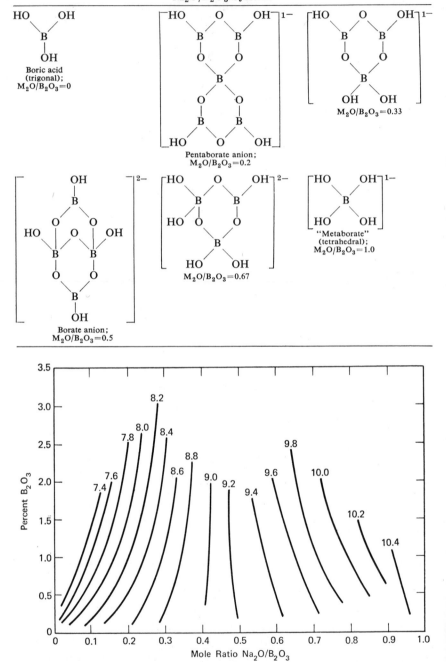

Figure 3-2 Values of pH in the system $Na_2O-B_2O_3-H_2O$ at 25°C.

Table 3-5   Melting Points for Various Compounds
in the Borate Family (23)

H_3BO_3	$170.9°$
HBO_2(III, II, I)	$176°, 200.9°, 236°$
B_2O_3	Softens at $325°$ (pours at $500°$)
$2Na_2O \cdot B_2O_3$	$625°$
$Na_2O \cdot B_2O_3 \cdot 8H_2O$	Begins at $54°$
$Na_2O \cdot B_2O_3 \cdot 4H_2O$	Begins at $90–95°$
$Na_2O \cdot B_2O_3$	$966°$
$Na_2O \cdot 2B_2O_3 \cdot 10H_2O$	$75°$
$Na_2O \cdot 2B_2O_3 \cdot 5H_2O$	$120°$
$Na_2O \cdot 2B_2O_3$	$644°$ (β-form), $742.5°$ (α-form)
$Na_2O \cdot 3B_2O_3$	$728°$ (A), $766°$ (B)
$Na_2O \cdot 4B_2O_3$	$816°$
NaCl	$801°$
$K_2O \cdot B_2O_3$	$950°$
$K_2O \cdot 2B_2O_3$	$815°$
$K_2O \cdot 3B_2O_3$	$825°$
$K_2O \cdot 4B_2O_3$	$857°$
$K_2O \cdot 5B_2O_3$	$760°$ (A), $780°$ (B)
$(NH_4)_2O \cdot 2B_2O_3 \cdot 4H_2O$	NH_3 press. at room temp
$(NH_4)_2O \cdot 5B_2O_3 \cdot 8H_2O$	Stable to $150°$

material, for example, H_3BO_3, borax, or metaborate, thereby increasing the amount of these materials that can be added before their solubility limit is reached. At low temperature this maximizes at or near the region where the double-ringed pentaborate ion should predominate, but it shifts at higher temperature to the area where a trimer is favored. Knowledge of these factors enables one to select the proper combination of temperature and alkalinity for preparing the most concentrated borate solutions. Figure 3-2 presents the results of pH determinations on these solutions for use if a given pH is desired [23].

Since one of the important mechanisms of fire retardance with borates is believed to be the formation of glassy films, the melting point data in Table 3-5 are pertinent. As it turns out, only the boric acids and the hydrates of the borates are low melting. Anhydrous borates melt at temperatures considerably above those at which most natural and synthetic polymers normally decompose to give flammable volatiles. Indeed, NaCl melts at a lower temperature than many of the salts listed.

Organic boron compounds are most frequently derived from H_3BO_3, BCl_3, or BH_3 and its polymeric homologs [23, 27]. Boron trichloride is prepared commercially by chlorinating B_2O_3 or a borate at high temperature in the presence of carbon:

$$B_2O_3 + 3C + 3Cl_2 \xrightarrow{900-1000°} 2BCl_3 + 3CO$$

The tribromide may be prepared in a similar manner. BCl_3 (mp $-107°$, bp 12.5°) is a gas at room temperature and hydrolyzes readily to boric acid and HCl. The tribromide (bp 90.8°) also hydrolyzes readily.

Boron-hydrogen compounds are extremely reactive and have become very useful laboratory reagents in recent years for the production of organoboron compounds. Diborane is readily prepared:

$$NaH + B(OCH_3)_3 \xrightarrow{225-275°} NaBH_4 + 3NaOCH_3$$

$$3NaBH_4 + BF_3 \xrightarrow[35°]{diglyme} \underset{\text{Diborane}}{2B_2H_6} + 3NaF$$

Higher boranes are obtained by heating diborane.

Esters of boric acid may be prepared either by direct reaction or by the action of alcohol on a trihalide:

$$H_3BO_3 + 3ROH \rightarrow B(OR)_3 + 3H_2O$$

$$BX_3 + 3ROH \rightarrow B(OR)_3 + 3HX$$

The water formed in the first reaction is usually removed by azeotropic distillation; the second reaction goes smoothly at room temperature. The intermediate mono- and dialkyl and aryl esters can be made by varying the proportions of reactants. The borate esters are very susceptible to hydrolysis to boric acid, many having half-lives at a room temperature of only a few minutes. This can be minimized somewhat by esterifying with bulky, sterically awkward groups. This susceptibility to hydrolysis severely limits the borate esters as fire retardants. Polyols react with boric acid and boron trihalides to yield a variety of polymers as well as monomeric products:

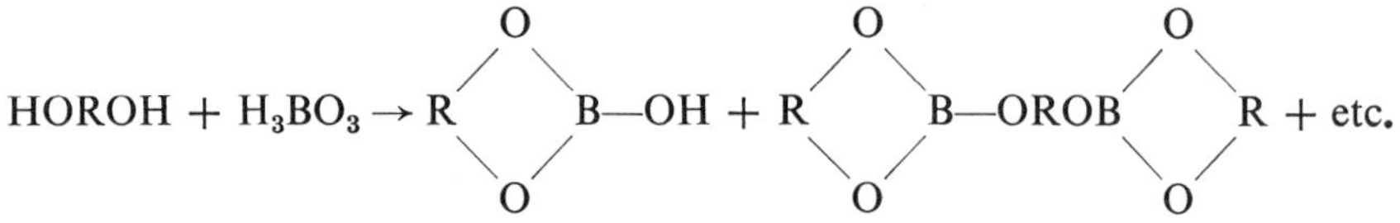

$$HOROH + H_3BO_3 \rightarrow R\begin{matrix} O \\ / \ \backslash \\ \backslash \ / \\ O \end{matrix}B-OH + R\begin{matrix} O \\ / \ \backslash \\ \backslash \ / \\ O \end{matrix}B-OROB\begin{matrix} O \\ / \ \backslash \\ \backslash \ / \\ O \end{matrix}R + \text{etc.}$$

Glycerine reacts to give both a monomer and polymers. These, too, are moisture sensitive. However, resistance to attack by water is much increased by inclusion of a nitrogen atom in the alcohol. The nitrogen bonds with the empty boron bonding orbital to form rather stable tetrahedral compounds, for example:

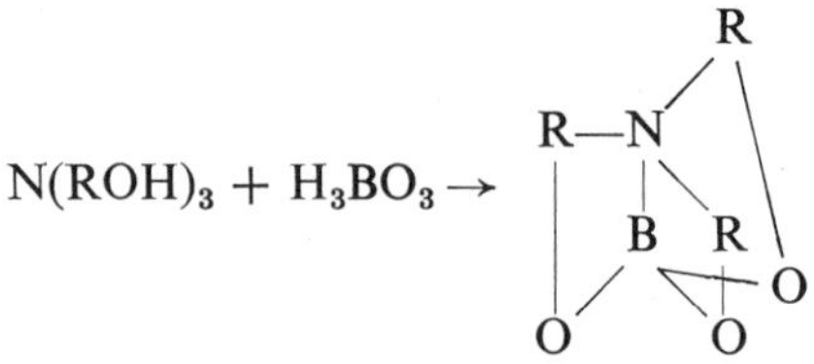

$$N(ROH)_3 + H_3BO_3 \rightarrow$$

Hydrolysis half-lives of the order of months have been reported for such derivatives. Esters of metaboric acid have been studied in epoxy resin systems. These are fairly unstable compounds with poor moisture resistance. They may be made from boric or metaboric acids, or even from boron trioxide, by controlling the ratio of alcohol to boron:

$$3ROH + 3HBO_2 \rightarrow$$

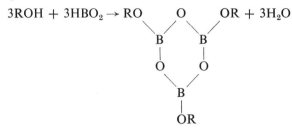

The water formed is removed by azeotropic distillation. Alternatively, an orthoborate can be heated with boron trioxide to give the trialkoxy or triaryloxyboroxines.

Compounds with boron-carbon bonds are termed either as substituted boranes or as boronates (one B—C bond), borinates (2 B—C bonds), and borines. Boronates may be synthesized from borate esters and Grignard reagents; borinates from boronate esters and Grignard reagents:

$$(RO)_3B + R'MgX \rightarrow (RO)_2R'BMgX \xrightarrow[H_2O]{}$$
$$R'B(OH)_2 + ROMgX + ROH \longleftarrow$$
$$R'B(OR)_2 + R'MgX \rightarrow R'_2BOH$$

Other methods make use of alkali metal alkyls with borate esters. Boron trihalides will add to olefins under mild conditions to give boronic dihalides which afford the acid on hydrolysis. Heating of the boronic and borinic acids yields the trimeric and dimeric anhydrides, respectively. The boron-carbon bond is stable but the BOR ester bonds are not particularly resistant to hydrolysis. Again, reaction of the acid with a β-aminoalcohol affords a very stable tetrahedral ester.

Symmetrical trihydrocarbyl boranes may be prepared by the reaction of a variety of boron compounds with Grignard reagents, with aluminum alkyls, and other metal alkyls:

$$BF_3 + 3R'MgX \rightarrow BR'_3 + 3MgXF$$
$$B(OR)_3 + 3R'X + 2Al \rightarrow BR'_3 + AlX_3 + Al(OR)_3$$

Borane derivatives are derived from the hydroboration reaction [28]:

$$\text{B—H} + \text{C}=\text{C} \rightarrow \text{B—C—CH}$$

If the reactant is unsubstituted borane, the symmetrical trialkyl borane results. The alkyl and aryl boranes are very reactive materials useful as intermediates. They are readily oxidized to borinic and, slower, to boronic esters via peroxy intermediates. The lower alkylboranes are very flammable and are somewhat difficult to handle. Their main uses are as polymerization catalysts and as intermediates.

A few miscellaneous structures specifically mentioned in the literature or claimed as flame retardants are presented in Tables 3-6 and 3-7. These are

Table 3-6   Some Additional Inorganic Boron Compounds Suggested as Flame Retardants

Compound	Use	Reference
$ZnO \cdot 2B_2O_3$ or $ZnO \cdot B_2O_3$	Paint	29
$PbO \cdot 2B_2O_3$		30
$2CaO \cdot 3B_2O_3 \cdot 5H_2O$ colemanite	Forest fire fighting	31
$Na_2O \cdot 2CaO \cdot 5B_2O_3 \cdot 12H_2O$ ulexite		
Various metal borates	Wood impregnant	32
$4NaF \cdot 5B_2O_3 \cdot 5H_2O$	Wood	33
NH_4BF_4	Polyolefins	34
Zn·tetramine·BF_4 complex	Polyolefins	35

over and above references to boric acid, borax, and other materials in the sodium, potassium, and ammonium borate families.

It is evident from the preponderance of references in the literature to inorganic borates and boric acid as fire retardants that these are the preferred materials. Indeed, the references in Table 3-7 represent the sum total of the citations to organoboron compounds for use against fire, whereas there are hundreds of references to boric acid and borax. This is undoubtedly because the organic derivatives with good stability are very new and many are still too costly. It may also be because the glass-forming mechanism can only operate with the inorganic hydrates and acids. The acid properties of many of the trisubstituted organoboron compounds (Lewis acids) may simply not be good enough in comparison with H_3PO_4 and the other acidic Group V materials. This remains to be tested by careful experimentation; we can look to such studies for guidance as to further research on organic fire retardants based on boron.

CHLORINE AND BROMINE

Most of the halogens have, at one time or another, been studied as flame retardants. Astatine has not. Iodine has received but little attention. Fluorine

Table 3-7 Some Organoboron Compounds as Fire Retardants

Reactants	Conditions	Product	Use	Reference
BCl_3, ROH	Room temperature	$(CH_3O)_3B$ bp $68.5°$	Wood	36
H_3BO_3, glycerine	(1) Dioxane, reflux, (2) azeotrope dist 4 hr hold $120°$, low vacuum	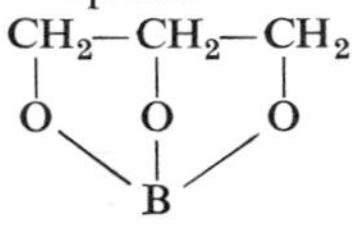 CH_2—CH_2—CH_2 (O—B—O ring) viscous liquid	Urethanes	37
H_3BO_3, $N(CH_2CH_2OH)_3$	...	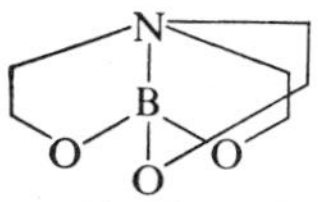 (N—B—O bicyclic structure)	Textiles	38
$(CH_2{=}CHCH_2O)_3B$, Br_2	CCl_4, $<18°$, 2 hr, 2–4 mm	$(BrCH_2CHBrCH_2$—$O)_3B$ liquid $n_D^{25} = 1.5666$	Urethanes	39
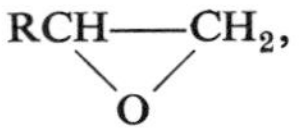 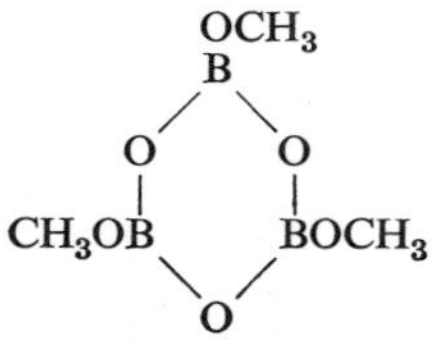 RCH—CH_2 (epoxide O), $R'NH_2$; (boroxine ring: OCH_3, B, $BOCH_3$, CH_3OB, O—O—O—O)	Mix at $100°$, cure at $200°$	Boron-containing epoxy foam (soften $320°$)	Epoxies	40
H_3BO_3, CH_3—C—CH_2—$N(CH_3)_2$ (OH), CH_3CH—$CHCH_3$ (OH OH OH)	Benzene, reflux	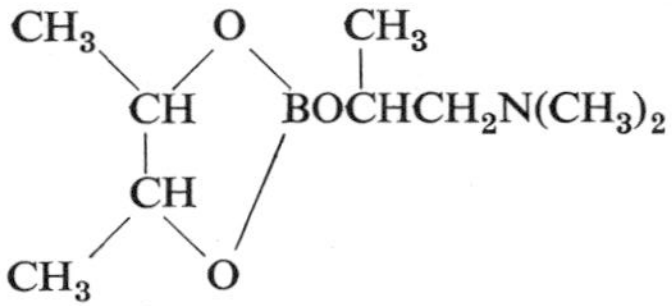 (dioxaborolane ring) CH_3, O, CH_3, CH, CH, O, CH_3, $BOCHCH_2N(CH_3)_2$ bp $132°$ (8.5–33 mm)	Epoxies	41
$HOCH_2CH_2OH$, $C(CH_2OH)_4$, $HOOC$—$CH{=}CHCOOH$, H_3BO_3	$150°$	Borated alkyd resin	Glass, fiber, bonding	42
H_3BO_3, CH_3—$\phi(NCO)_2$	Dry N_2, 110, $\tfrac{1}{2}$ hr	Borated isocyanate polymer	Miscellaneous	43

has only recently been evaluated and has shown limited utility in certain areas [44]. The vast majority of papers on the halogens in fire retardation have dealt with chlorine and bromine and it is on these two elements that the following discussion is focused.

Chlorine, symbol Cl, atomic weight 35.457, has two principal isotopes of mass number 35 and 37 present in the earth's crust in a ratio of roughly 3 to 1 [45]. There are 17 electrons in the electron shells and therefore 17 protons in the nucleus. Isotope Cl^{35} has 18 neutrons in the nucleus; Cl^{37} has 20 neutrons. Both isotopes have a nuclear spin of $\frac{3}{2}$; both have quadropole moments. The electron configuration in the valence shell is $3s^23p^5$. Chlorine lacks one electron of having the inert gas configuration and has a very strong tendency to add an electron to complete the shell. Elemental chlorine, Cl_2, is a greenish-yellow gas at room temperature, is highly toxic, and has a characteristic pungent, suffocating odor.

Bromine, symbol Br, atomic weight 79.916, also has two stable isotopes of mass number 79 and 81 apportioned in the earth's surface layer, 50.5% isotope 79 and 49.5% isotope 81 [46]. There are 35 electrons in the electron shells, 35 protons in the nucleus, and 44 and 46 neutrons in the nuclei of the two stable isotopes. Both isotopes have nuclear spins of $\frac{3}{2}$ and exhibit quadropole moments. The electronic configuration in the valence shell is $4s^24p^5$, lacking one of having the inert gas configuration. Bromine, too, has a strong affinity for an electron to complete the outer shell, but this is somewhat weaker than for chlorine. This is because the nuclear attraction for the valence electrons is more screened in bromine than in chlorine. The effect of this may be seen in the values for ionization potentials, for the dissociation constant of Br_2 and Cl_2 at a given temperature, for the heats of dissociation, and so on [47]. Bromine is the only nonmetal which is a liquid in its elemental form at room temperature.

	Chlorine	Bromine
Ionization potential, eV	13.01	11.84
Dissociation constant at 1000°C	10^{-8}	8×10^{-3}
Heat of dissociation, kcal/mole	56.9	45.2

Some physical and chemical properties of the two elements are given in Table 3-8.

Both elements are powerful oxidizing agents, chlorine being the stronger of the two. Indeed, the commercial method of isolating bromine is to oxidize bromide with chlorine and then to recover the gaseous bromine [46, 48].

Table 3-8   Properties of Chlorine and Bromine

	Chlorine	Bromine
Atomic weight	35.457	79.909
Melting point, °C	−102	−7.3
Boiling point, °C	−34.6	58.78
Density, g/cm³	3.209×10^{-3}	3.1226 (20°)
	(0°, 1 atm)	
Heat of fusion, kcal/mole	1.615	2.580
Heat of vaporization, kcal/mole	4.420	7.418
Solubility in water at 20°C:		
moles/l	0.090	0.210
g/l	6.38	33.56
Organic solvents	Sol in alkali	Alcohol, chloroform, ether, CS_2

The half-cell potentials are strongly negative at 25°C:

	E_0, (v)
$Cl^- \rightleftharpoons \frac{1}{2}Cl_2 + e^-$	−1.358
$Br^- \rightleftharpoons \frac{1}{2}Br_2 + e^-$	−1.065

For example, both chlorine and bromine will, in theory at least, oxidize copper, mercury, silver, and palladium to their positive ions and chlorine will oxidize platinum, again in theory. In fact, many of these metals become passivated and oxidize but slowly.

A few inorganic halides have been suggested as fire retardants. Some of these are listed in Table 3-9 and will be discussed further in the appropriate

Table 3-9   Some Inorganic Halides and Suggested
Fire-Retardant Applications

$ZnCl_2$	
$MgCl_2$	
$MgBr_2$	Wood treating
Br^-, BrO_3^- together	
$ZnBr_2$	Cellulosic derivatives
$TiOCl_2, TiCl_4$	Textiles
NH_4Br	Polystyrene

chapters on the four substrates listed. The preparation of the salts shown is straightforward and needs no discussion.

Nearly all the flame-retardant applications for chlorine and bromine utilize organohalogen compounds. They may be used as inert additives, as reactive substituents, or as integral parts of the finished system, for example, as halogenated monomers. Indeed, in certain instances completed synthetic polymer systems have been halogenated as a final finishing step. That this can be done is largely because of the highly reactive nature of the halogens.

Cl_2 and Br_2 react with organic molecules either by addition reactions at points of unsaturation (or at electron-rich centers as in phosphine adducts)

$$X_2 + \quad \overset{\diagdown}{\underset{\diagup}{C}}=\overset{\diagup}{\underset{\diagdown}{C}} \quad \rightarrow \quad \overset{\diagdown}{\underset{\diagup}{C}}X-\overset{\diagup}{\underset{\diagdown}{C}}X$$

where X is Cl or Br, or by substitution reactions for atoms or groups in the molecule

$$X_2 + RH \rightarrow RX + HX$$

Another common mode of addition of a halogen to unsaturated centers is by use of the acids:

$$\overset{\diagdown}{\underset{\diagup}{C}}=\overset{\diagup}{\underset{\diagdown}{C}} + HX \rightarrow \overset{\diagdown}{\underset{\diagup}{C}}H-\overset{\diagup}{\underset{\diagdown}{C}}X$$

Similarly, HX can be removed from halogenated molecules by treatment with base under appropriate conditions. If the molecule is polyhalogenated, this provides a way to remove one halogen at a time.

$$CHCl_2-CHCl_2 \overset{base}{\longrightarrow} CCl_2{=}CHCl + HX(\cdot base)$$

Substitution may be achieved with an unsaturated compound (rather than addition) by proper choice of conditions. Reaction with aromatic compounds commonly yields substitution but may instead be addition if conditions are right. Reaction with an alkyl aryl compound can proceed with either part of the molecule depending, again, on conditions. These concepts are best clarified by some examples.

Chlorine Compounds

Substitution of hydrogen by chlorine in alkanes is usually achieved by thermal means. (The specific data presented in the following discussion are taken primarily from references 49–52.) The high temperatures employed facilitate reaction by dissociating the Cl_2 molecule into two Cl radicals, which then attack the hydrocarbon. Thus

$$\underset{bp\ -161.5°}{CH_4} + Cl_2 \overset{400°}{\longrightarrow} \underset{bp\ -23.7°}{CH_3Cl} + HCl$$

and

$$\underset{bp\ -88.3°}{C_2H_6} + Cl_2 \overset{400°}{\longrightarrow} \underset{bp\ 12.4°}{C_2H_5Cl} + HCl$$

To achieve further substitution higher temperatures may be used or enhanced reactivity may be gained by irradiation with light in the 3000–5000 Å wavelength region from, for example, light from mercury lamps. Alternatively oxide catalysts, such as alumina or Fuller's earth, may be used:

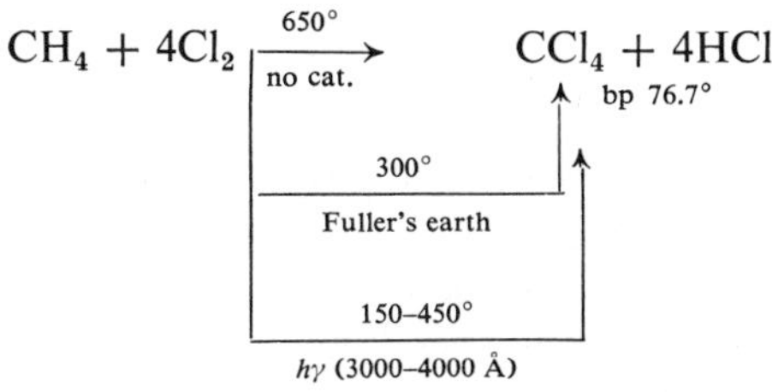

$$CH_4 + 4Cl_2 \xrightarrow[\text{no cat.}]{650°} CCl_4 + 4HCl$$
bp 76.7°

300°
Fuller's earth

150–450°
$h\gamma$ (3000–4000 Å)

$$CHCl_2CHCl_2 \xrightarrow[\text{uv light}]{>200°, \ Cl_2} CHCl_2CCl_3 + CCl_3CCl_3$$
bp 146.3°   bp 161°   Sublimes 187°

These reactions generally are not highly selective and a mixture of products results. In the chlorination of methane and ethane, multiproduct facilities are usually built and the product stream fractionated into its components for sale, for instance, as methyl chloride, methylene chloride, chloroform, and carbon tetrachloride from methane and chlorine feedstocks.

One fairly common fire-retardant series is the chlorinated paraffins known as the Chlorowaxes Paraffins having a carbon chain of from C_{10} to C_{30} or so are polychlorinated in the presence of light at 100° and no solid catalyst.

Addition of chlorine to unsaturated centers proceeds readily in the presence of halogen carriers such as $AlCl_3$ or $FeCl_3$, forming a complex anion and positive chlorine. The attack at the double bond is believed to proceed as follows:

$$AlCl_3 + Cl_2 \rightarrow [AlCl_4]^-Cl^+$$

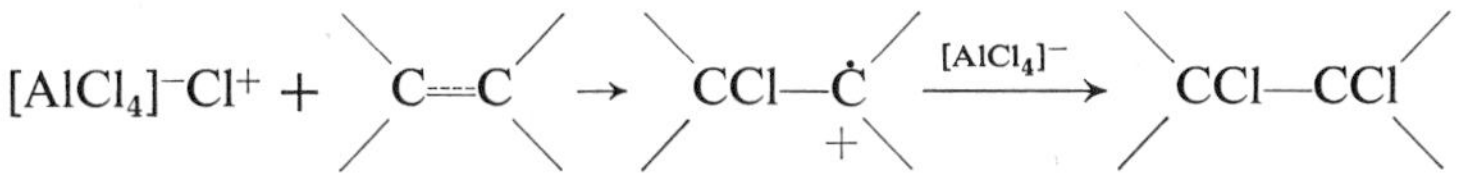

$$[AlCl_4]^-Cl^+ + \ \overset{}{C}{=}\overset{}{C} \rightarrow \overset{}{C}Cl{-}\overset{+}{C} \xrightarrow{[AlCl_4]^-} \overset{}{C}Cl{-}CCl$$

Addition will also occur in the presence of light but only if oxygen is rigorously excluded. Oxygen acts as a chain breaker. For this reason most additions to olefins are run with chloride catalysts. A few examples:

$$CH_2{=}CH_2 + Cl_2 \xrightarrow[\text{AlCl}_3]{100°} CH_2ClCH_2Cl$$
bp −103.9°   bp 83.5°

$$CHCl{=}CCl_2 + Cl_2 \xrightarrow[\text{FeCl}_3]{70°} CHCl_2CCl_3$$
bp 86.7°   bp 161°

$$CH{=}CH + 2Cl_2 \xrightarrow[\text{FeCl}_3]{70-80°} CHCl_2CHCl_2$$
Sublimes at −83.6°   bp 146.3°

$$CH{=}CH + Cl_2 \xrightarrow[\text{FeCl}_3{-}\text{HgCl}_2]{60-80°} CHCl{=}CHCl$$
bp 31.8°

Some interesting products can be obtained by careful selection of conditions. Propylene can be converted to allyl chloride in high yield with very little loss of the double bond content. Alternatively, a saturated product results. Temperature is the key variable:

$$CH_2\!\!=\!\!CHCH_3 + Cl_2 \xrightarrow{>300°} CH_2\!\!=\!\!CHCH_2Cl + HCl$$
$$\text{bp } 45.0°$$

$$\xrightarrow{<200°} CH_2Cl\!\!-\!\!CHCl\!\!-\!\!CH_3$$
$$\text{bp } 96.8°$$

Similarly, the fully chlorinated diene known as "hex," starting material for a host of insecticides and for a number of fire retardants, is obtained by using conditions favoring substitution:

$$C_5H_{12} + 9Cl_2 \xrightarrow{80-90°} C_5H_5Cl_7 \xrightarrow[Al_2O_3]{Cl_2, 300-400°}$$

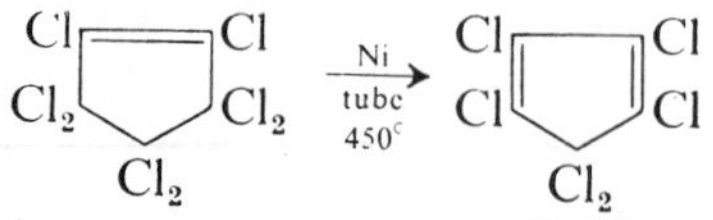

"hex"
mp 9.9°

"Hex" and maleic anhydride combine in a Diels-Alder reaction to yield chlorendic anhydride which is useful in urethanes, polyesters, and the like.

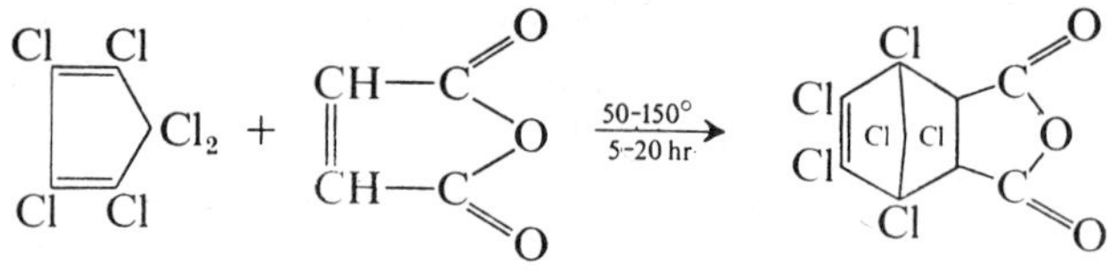

HCl adds to double and triple bonds in the presence of chloride catalysts:

$$CH_2\!\!=\!\!CH_2 + HCl \xrightarrow[AlCl_3]{200-250°} C_2H_5Cl$$
$$\text{bp } 12.2°$$

$$CH_2\!\!=\!\!CCl_2 + HCl \xrightarrow[AlCl_3]{30°} CH_3CCl_3$$
$$\text{bp } 74.1°$$

$$CH\!\!\equiv\!\!CH + HCl \xrightarrow[HgCl_2, C]{150°} CH_2\!\!=\!\!CHCl$$
$$\text{bp } -13.9°$$

The elements of HCl are readily removed from chlorocarbons by treatment with base. Thus, treatment of trichloroethane with lime at 50° affords vinylidene chloride:

$$CH_2Cl\!\!-\!\!CHCl_2 \xrightarrow[50°]{Ca(OH)_2} CH_2\!\!=\!\!CCl_2$$
$$\text{bp } 113.5° \qquad\qquad\qquad \text{bp } 31.8°$$

Chloroprene is obtained via the catalytic action of HCl, $CuCl_2$, and NH_4Cl at 60° on acetylene, causing a dimerization followed by addition of HCl to the remaining triple bond:

$$2CH{\equiv}CH \rightarrow CH_2{=}CH{-}C{\equiv}CH \rightarrow CH_2{=}CH{-}\underset{\underset{\textstyle Cl}{|}}{C}{=}CH_2$$

Chloroprene bp 59.4°

Chlorohydrins are formed by addition of aqueous chlorine to double bonds:

$$Cl_2 + H_2O \rightarrow HOCl + Cl^- + H^+$$

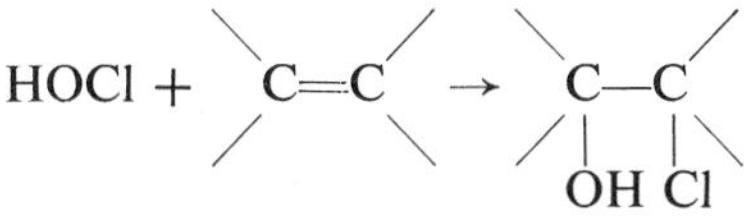

$$HOCl + \ \underset{/}{\overset{\backslash}{C}}{=}\underset{\backslash}{\overset{/}{C}} \ \rightarrow \ \underset{\underset{\textstyle OH}{/|}}{\overset{\backslash}{C}}{-}\underset{\underset{\textstyle Cl}{|\backslash}}{\overset{/}{C}}$$

The reaction is normally run in base to promote the formation of hypochlorite. Probably the most important of these materials is epichlorohydrin, an important intermediate for epoxy resins. This may be made from allyl chloride in two steps:

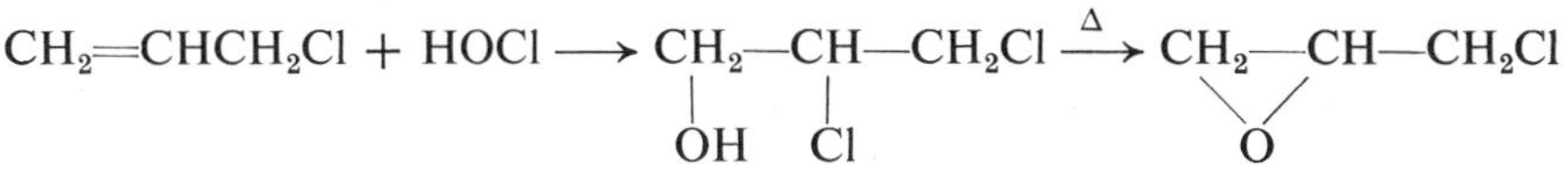

$$CH_2{=}CHCH_2Cl + HOCl \longrightarrow \underset{\underset{\textstyle OH}{|}}{CH_2}{-}\underset{\underset{\textstyle Cl}{|}}{CH}{-}CH_2Cl \overset{\Delta}{\longrightarrow} \underset{\underset{\textstyle O}{\backslash/}}{CH_2}{-}CH{-}CH_2Cl$$

Epichlorohydrin bp 116.1°

Aromatic substitution is afforded by the usual Friedel-Crafts catalysts. Polysubstitution requires longer times and higher temperatures.

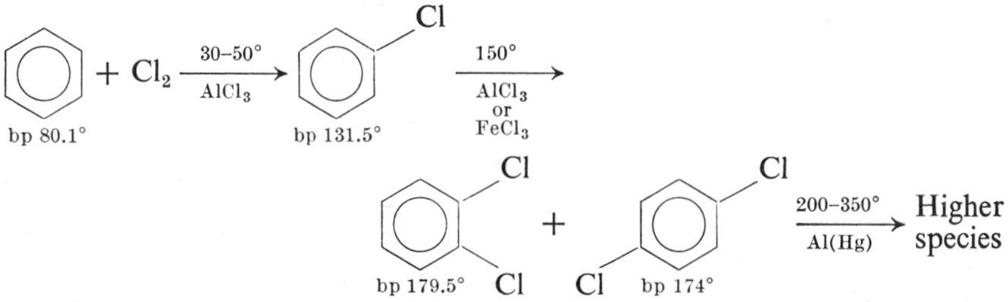

Poly(butene-1) has been chlorinated by using Cl_2 in a solvent at less than 60° with uv light. Up to 70% Cl has been incorporated in this manner [52a]. (For references to chlorinating polyolefins, see Chapter 7.)
Addition to the ring is facilitated by irradiation:

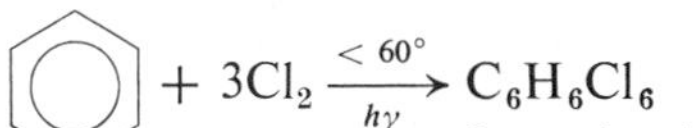

Benzene hexachloride
mp ca. 65° (mixed isomers)

Alkyl aryl compounds may be substituted on the alkyl group by treating at low temperatures in the presence of light, the alkyl group being more reactive than the ring in this case. Polynuclear compounds such as naphthalene give addition without catalysts but substitution with chloride catalysts at temperatures between 80° and 200°. Thus, biphenyl may be chlorinated with $FeCl_3$ below 150°. Phenol gives lower chlorophenols without catalysts, higher homologs with $FeCl_3$ or $AlCl_3$.

Phthalic anhydride affords tetrachlorophthalic anhydride on treatment with chlorine in the presence of carrier catalysts:

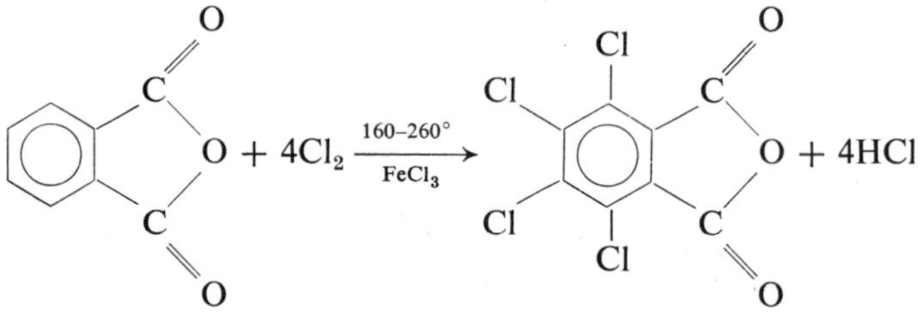

The product is of considerable importance in flame-retarded polyesters.

Some of the chlorohydrocarbons used as flame retardants are shown in Table 3-10.

Table 3-10   Chlorohydrocarbons Used as Fire Retardants
References Are for Data Sources Over and Above Those in
References 45 and 49–52

	Use	Reference
Alkanes		
Chloroethanes	Paint, styrene	
Chloropropanes	Polyesters	
Chlorinated C_{10}-C_{30} paraffins	Paints, wood, textiles, polyolefins	53
Chlorinated fish oil	Paint	
Chlorinated rubber	Rubber	
Chlorinated PVC	Vinyls, textiles	
Chlorinated polyisobutylene	Polyurethane	
Chlorinated polyolefins	Polyolefins	
Polyvinyl chloride	Textiles	
Benzene hexachloride-$C_6H_6Cl_6$	Cellulose derivatives	
Olefins		
Chloroethylenes, vinyl chloride	Textiles, styrene, acrylics	54
Chloropropene	Styrene, vinyls, acrylics	55
Chlorobutene and butadiene	Rubber, vinyls	56
Chloroprene	Polyolefins	
Vinyl chloroacetate	Polyesters, acrylics	
Allyl chloride	Epoxies	

Table 3.10 (continued)

	Use	Reference
Olefins (continued)		
Hexachlorocyclopentadiene and derivatives ("hex")	Paints, polyesters, polyurethanes, epoxies, styrene, acrylics	57
Chlorendic acid and derivatives	Polyesters, polyurethanes, epoxies, vinyls	58
Alcohols, acids, aldehydes, etc.		
Chloroalcohols C_2-C_{12}, polyols	Vinyls, polyesters	59
Pentaerythritol chlorohydrin	Polyesters, polyurethane	
Tetrachlorobutane-1,4-diol	Epoxies	
Epichlorohydrin	Epoxies	
1,1,1-Trichloro-2,3 epoxypropane	Polyesters, polyurethane	
Chloroadipic acid	Nylon, vinyls	
Vinyl chloroacetate	Polyesters	
Dichlorosuccinic acid	Polyurethane	60
Chlorinated fatty acids	Styrene	
Chloral	Polyurethane, epoxies, polyaldehydes	61
Chloroalkyl acrylonitrile	Acrylics	
Chlorinated aryldiamines	Epoxies	
CCl_4, alkyl acetate condensate	Resins, textiles	61a
Aromatics		
Alkoxychlorobenzenes	Vinyls	
Chlorinated hexamethylbenzene	Vinyls	
Chlorinated alkylaryl ethers	Polyesters	
Chlorophenols	Textiles, styrene, acrylics, wood phenolics, polyphenylenes	62
Pentachlorophenol glycidyl ether	Polyurethane, epoxies	
Chlorostyrenes	Polyesters, styrene, polyolefins	
Chlorothiophenol esters	Acrylics, vinyls	
Chlorinated 1,4-bis-hydroxymethyl benzene	Cellulosics, textiles	
Chlorophenyl isocyanate	Textiles	
Chlorobiphenyls and polyphenyls	Textiles, polyesters, polyurethane, styrene	63
Chlorinated 4,4'-bis-hydroxybiphenyl	Polyesters	
Chlorinated 3,3'-bis-isocyanatobiphenyl	Polyurethanes	
Chlorinated naphthalenes	Textiles, polyesters	
Chlorinated bisphenol A and glycidyl ethers	Polyesters, epoxies	44
Chlorinated diphenyl carbonate	Polycarbonates	
Tetrachlorophthalic acid and derivatives	Textiles, polyesters	64
Chlorinated alkyd resins	Paints	
Chloranil $\left(O=\!\!\left\langle\!\!\begin{array}{c} Cl\;\; Cl \\ \\ Cl\;\; Cl \end{array}\!\!\right\rangle\!\!=O \right)$	Vinyls	

Bromine Compounds

Bromine reacts with organic compounds in much the same way as chlorine does. It adds to double and triple bonds and it participates in similar substitution reactions. The principal differences are attributable to the larger size of the bromine atom with its greater number of electrons screening the attraction of the nucleus for valence shell electrons. Thus, the oxidizing power of bromine is less than that of chlorine and the stability of bromides is somewhat less than that of the corresponding chlorides. The larger size of the bromine atom makes some fully brominated structures more difficult to form. However, these differences are minor as compared to the great similarity between the reactions of bromine and of chlorine.

The lower monobromoalkanes are produced from the corresponding alcohols in good yield, using either bromine and a reducing agent to produce Br^- *in situ* (most of the specific data in the following discussion are taken from references 46, 48, 51, and 52)

$$H_2S + 4Br_2 + 8CH_3OH \rightarrow 8CH_3Br + H_2SO_4 + 4H_2O$$

$$S + 3Br_2 + 6CH_3OH \rightarrow 6CH_3Br + H_2SO_4 + 2H_2O$$

or by adding the bromide to an acid-alcohol mixture, thus

$$2NaBr + 2CH_3OH + H_2SO_4 \rightarrow 2CH_3Br + Na_2SO_4 + 2H_2O$$

These processes yield the monobromide without higher degrees of substitution and are carried out at low temperature, for example, the boiling point of the alcohol. Direct bromination of methane is apparently not practiced significantly on the industrial scale, perhaps because there is little demand for all the species produced: CH_3Br, CH_2Br_2, $CHBr_3$, and CBr_4. Ethyl bromide (bp 38.4°) and propyl bromide (bp 59.3°) through at least lauryl bromide (bp 175–180°) are usually prepared from the alcohol in sulfuric acid.

CH_2Br_2 is prepared from the chloride by use of an aluminum catalyst:

$$CH_2Cl_2 + AlBr_3 \xrightarrow{\;xs\,AlBr_3\;} CH_2Br_2 + AlCl_3$$

Bromine adds readily to unsaturated centers, often quantitatively.

Ethylene bromide is one of the most common organobromine compounds and is prepared by the addition reaction:

$$CH_2{=}CH_2 + Br_2 \xrightarrow{\;>30°\;} \underset{\text{bp } 131.4°}{CH_2BrCH_2Br}$$

The reaction is exothermic and requires cooling to maintain control over the rate.

Butadiene adds bromine at the 1,4 positions to give 1,4-dibrome-2-butene. Addition to the ethylene group in styrene is achieved in CCl_4 to yield 1,2-dibromoethyl benzene.

Vinyl bromide is of considerable interest for flame-proofing acrylics and is prepared by adding hydrogen bromide to acetylene

$$CH{\equiv}CH + HBr \rightarrow CH_2{=}CHBr$$

or by removing the same moiety from the appropriate dibromide

$$CH_2BrCH_2Br \xrightarrow{\text{base}} CH_2{=}CHBr + HBr$$

Both methods have been reported. Allyl bromide is derived either from allyl alcohol via NaBr and H_2SO_4 or, again, by removal of the elements of HBr from a dibromide, in this case at high temperature with no base:

$$CH_3CHBrCH_2Br \xrightarrow{\Delta} CH_2{=}CHCH_2Br + HBr$$

Carboxylic acids are readily brominated in the α-position with bromine at temperatures of $100°$ or so. Thus,

$$CH_3CH_2COOH + Br_2 \xrightarrow{100-130°} CH_3CHBrCOOH + HBr$$

Alkyl groups in alkyl-aryl compounds may be selectively brominated by use of low temperature and irradiation with light from tungsten or mercury lamps [65]. Thus, toluene and bromine afford benzyl bromide:

$$C_6H_5CH_3 + Br_2 \xrightarrow[h\gamma]{80-85°} \underset{\text{bp 198–202°}}{C_6H_5CH_2Br} + HBr$$

Aromatic structures are brominated with the aid of a catalyst such as $FeBr_3$. Higher temperatures favor higher degrees of substitution:

$$C_6H_6 + Br_2 \xrightarrow[20-40°]{FeBr_3} \underset{\text{bp 156°}}{C_6H_5Br} \xrightarrow[CCl_4, 90-100°]{FeBr_3, Br_2} C_6H_2Br_4, \text{ etc.}$$

The ring is brominated in preference to side chains by use of the catalyst rather than irradiation. Thus, toluene can be fully brominated on the ring at $80-90°$ in the presence of $FeBr_3$ in CCl_4 in high yield. In this regard chlorination and bromination are very similar.

Tetrabromophthalic anhydride may be obtained by brominating phthalic anhydride in oleum:

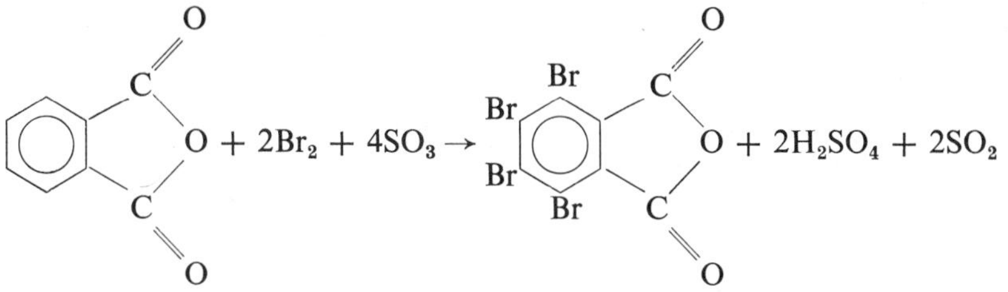

Table 3-11 Some Organobromine Compounds Which Have Been Studied as Fire Retardants

	Use	Reference
Alkanes		
Bromoethanes	Vinyls	
Bromocycloalkanes	Polyolefins	66
Bromocycloalkanol, acrylic acid ester	Acrylics	
Brominated polybutadiene	Vinyls, styrene	67
Olefins		
Vinyl bromide	Styrene, acrylics	
Tetrabromododecene	Polyesters	
Hexabromobicycloheptene derivatives	Acrylics	68
Brominated cyclododecatriene	Resins	68a
Hexabromo cyclopentadiene		68b
Alcohols, acids, aldehydes, etc.		
2,3,3-Tribromoallyl alcohol and esters (e.g., acrylate esters)	Styrene, vinyls	
2,2,3,3-Tetrabromobutane-1,4-diol	Styrene	
Brominated pentaerythritol	Polyesters	
Brominated polyols	Polyurethanes	
2,3-Dibromopropyl phthalate	Paper	
Brominated tall oil	Polyurethanes	
2,2-Bis(bromomethyl)-1,3-propanediol	Polyesters	68c
2-Bromoethyl itaconate	Styrene, acrylics	69
Dibromosuccinic acid	Polyesters	60
Bromoacetaldehyde, bromobenzaldehyde	Polyvinyl alcohol	70
$BrCH_2$-R-$CONR^1R^2$, where R = C_5-C_{21}; R^1, R^2 are lower alkyls	Polyurethanes	
Bis (2,3-dibromopropyl) malate	Polystyrene	70a
Aromatics		
Brominated polyphenyls	Polyolefins	71
Pentabromotoluene	Polyurethanes	
Bromophenyl vinyl ether	Polyesters	
Styrene dibromide	Styrene	
Bromophenols, acrylate esters	Paints, styrene	
Bromophenol, glycidyl ether	Polyesters	

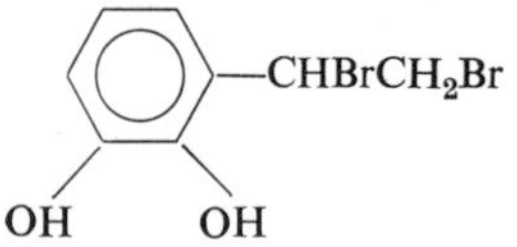 —CHBrCH$_2$Br / OH OH	Polyesters	
Bromotoluene di-isocyanate	Polyurethanes	
Tetrabromophthalic acid or anhydride	Polyesters	72, 73
Bromophthalimide	Nylon	
Tetrabromobisphenol A	Epoxies	44
Brominated salicylanilide		73a

Tetrabromobisphenol A(I) used extensively in flame-retardant epoxy resins is produced by treating bisphenol A with Br_2 in an alcohol solvent:

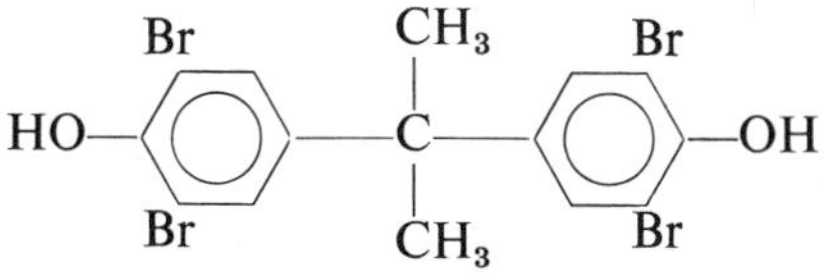

I. mp 181.2°

Table 3-11 contains a partial list of organobromine compounds discussed in the literature as flame retardants. Application of some of these will be considered in the appropriate chapters on the substrates indicated.

Comparisons of Chlorine and Bromine—Effects on Flammability

Addition of a halogen atom to a molecule in general reduces flammability. The mechanism of this was touched on briefly in Chapter 1. Table 3-12 lists

Table 3-12   Some Flash Points[a] Showing the Effect of Chlorine and Bromine

	X = Cl	X = Br
CH_3X	+333°C	None
CH_3CH_2X	−50°	None
CH_2XCH_2X	+17°	None
CH_2XCHX_2	None	None
$CH_2{=}CHX$	−78°	None
$CH_2{=}CHCH_2X$	−29°	−1°
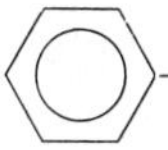 ⬡—X	+24°	+50°
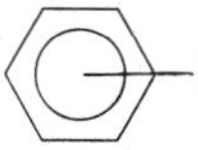 ⬡—X_2	+60°	· · ·

[a] Generally closed cup; handbook values or from reference 46

some flash points of chloro and bromo compounds; the greater effectiveness, on an atom by atom basis, of bromine is clearly demonstrated. It has been said [73] that one atom of bromine is as effective in reducing flammability as two of chlorine. The table lends, at least qualitatively, support to this statement. Of course, since bromine is more than twice as heavy as chlorine and more expensive by far on a weight basis, chlorine is at least as useful as bromine in flame-retardant applications. Experience confirms this for many systems, but not always.

The halogens are rendered more effective when used with phosphorus, antimony, and the like. Specific compounds, such as PCl_3 and $SbCl_3$, have been discussed earlier. Synergists for bromine and chlorine are phosphorus and antimony compounds. (See Chapter 1 and ref. 74.)

OTHER ELEMENTS

Many other elements have been studied as fire retardants in addition to those discussed in Chapter 2 and here in Chapter 3. However, none have become commercially significant in spite of fairly diligent efforts in some instances. The other members of Group V—N, As, and Bi—have received attention. Research is currently in progress to determine the extent to which nitrogen is effective, especially in combination with phosphorus. Indeed, considerable evidence is already in on this point. Nitrogen compounds are flame retardants themselves but this has not been quantified in many systems. When combined with other elements that yield acids on heating—as in $NH_4SO_3NH_2$, ammonium sulfamate [75], for example—flame retardance is certainly observed. However, it has not been shown how much of this retardance may be attributed to the nitrogen. Arsenic and bismuth are effective fire retardants, as might be expected by virtue of their chemical similarity to phosphorus and antimony. Relatively little has been done with these two elements for a variety of reasons. Their chemistry will not be discussed.

Some sulfur compounds have been found useful. One, ammonium sulfate, $(NH_4)_2SO_4$, finds considerable utility in paper and other cellulose products. Presumably, the material decomposes on heating to leave concentrated H_2SO_4 behind. This acid then participates much as does phosphoric acid to alter the route of thermal decomposition (see Chapters 1 and 4). The chemistry of $(NH_4)_2SO_4$ is so straightforward that it need not be discussed in detail here. Its principal properties are summarized in Table 3-13.

Table 3-13   Some Properties of Ammonium Sulfate

Molecular formula	$(NH_4)_2SO_4$	
Molecular weight	132.15	
Specific gravity	1.769 g/ml	
Melting point	Dec at 280°C	
Solubility		
g/100 g soln at 0°C		70.6
at 100°		103.8
moles/l at 0°		3.90
at 100°		4.85
g/l at 0°		515
at 100°		640
Insoluble in alcohol, NH_3, acetone		

Selenium has been studied as a flame retardant [76]. This use should be discouraged because of the toxicity of selenium. The research on selenium-based fire retardants was done largely in the early days of this century before the toxic character of this element was fully understood.

Molybdenum pentachloride is being claimed by a manufacturer as a flame retardant, but little is yet known in detail [77].

Tin salts are occasionally mentioned [78]; tin salts were part of Perkin's flame-retardant finish for textiles (see pp. 266 and 221).

Sodium silicate solutions have been used as impregnants for porous substrates to confer fire resistance. In a sense this is like the use of low-melting borates in that an impervious film is provided to resist the fire rather than chemical alteration of the burning substrate by the retardant. The use of silicates (water glass and the like) has been either in protective coatings or in immersion baths to fill the pores of the material to be protected. The process and its results are principally physical, not chemical, in nature and thus no detailed chemical discussion is warranted here.

Occasionally a reference to organosilicon compounds is found. Some of the silicones are only slightly flammable and are used in some cases for this property. Once in a while a derivative such as the reaction product of halophenols and $SiCl_4$ is disclosed in the literature specifically for its fire-resistant properties [79].

REFERENCES

[1] T. Moeller, *Inorganic Chemistry*, Wiley, New York, 1952.

[2] G. O. Doak, L. D. Freedman, and G. G. Long, "Antimony Compounds," in *Encyclopedia of Chemical Technology*, 2nd ed., Vol. 2, Wiley-Interscience, New York, 1963, pp. 570 ff.; C. C. Downie, *Paint Manufacture*, **16**, 13 (1946).

[3] S. C. Carapella, Jr., "Antimony and Antimony Alloys," in *Encyclopedia of Chemical Technology, op. cit.*, pp. 562 ff.

[4] J. R. Van Wazer, *Phosphorus and Its Compounds*, Vol. 1, Wiley-Interscience, New York, 1958.

[5] M. C. Sneed and R. C. Brasted, *Comprehensive Inorganic Chemistry*, Vol. 5, Van Nostrand, Princeton, N.J., 1956, Chapter 2.

[6] International Union of Pure and Applied Chemistry, *Nomenclature of Inorganic Chemistry*, Butterworths, London, 1959.

[7] For early references to the use of antimony as a fire retardant in a reasonably scientific manner see A. Arent, *Brit. 132,813, 138,641, 146,099*, Sept. 20, 1919, Feb. 5, 1920, June 12, 1920; *Can. 229,246*, Feb. 27, 1923; a more recent typical citation is W. E. Green, *U.S. 3,333,970* (to Associated Lead Manufacturers), Aug. 1, 1967.

[8] Associated Lead Manufacturers, Ltd., and F. B. Lewis, *Brit. 982,717*, Feb. 10, 1965.

[9] Allied Chemical Corp. (by E. A. Dickert), *Belg. 657,422*, Apr. 15, 1965.

[10] Diamond Alkali Co. (by F. F. Roselli), *Fr. 388,520*, Feb. 5, 1965.

[10a] R. S. Cooper, *U.S. 2,664,411* (to Diamond Alkali Co.), Dec. 29, 1953.

[11] M. K. Moran, ed., "Uses of Antimony Compounds as Fire and Flame Retardants," *Bibliography FR-3*, Metals and Thermit Corp., Rahway, N.J., 1962, 28 pp.

[12] Diamond Alkali Co. (J. E. Dereich), *Ger. 1,026,951*, Mar. 27, 1958; M. M. Gherson, *Pensez Plast.*, **1960** (19697), 55.

[13] B. O. Schoepfle and P. Robitschek, *U.S. 3,031,425* (to Hooker Chemical Corp.), Apr. 24, 1962.

[14] Farb. Bayer A.-G. (by E. Eimers and L. Goerden), *Ger. 1,089,967*, Sept. 29, 1960.

[15] Riken Synthetic Resin Co., Ltd. (by T. Akita and J. Okazawa), *Jap. 4,696*, Apr. 25, 1963; Peter Spence and Sons, Ltd., *Brit. 844,555*, Aug. 17, 1960 (addition to *Brit. 837,696*, reference 16); M. Worsley, B. N. Wilson, and B. O. Schoepfle, *U.S. 3,054,760* (to Hooker Chemical Corp.), Sept. 18, 1962.

[16] Peter Spence and Sons, Ltd. (by L. Williams and R. Sidlow), *Brit. 837,696*, June 15, 1960.

[17] Chemische Werke Witten G.m.b.H. (by E. Behnke and H. Wulff), *Ger. 1,109,886*, Appl. Aug. 13, 1958.

[18] J. D. Broatch, *U.S. 2,852,414* (to British Jute Trade Research Corp.), Sept. 16, 1958.

[19] R. S. Robe, *Brit. 795,133*, May 14, 1958; H. C. Rapp, *U.S. 3,239,482* (to Raychem Corp.), Mar. 8, 1966.

[20] H. C. Rapp, *U.S. 3,239,482* (to Raychem Corp.), Mar. 8, 1966.

[21] Peter Spence and Sons, Ltd. (by R. Sidlow and L. Williams), *Brit. 953,206*, Mar. 25, 1964.

[21a] A. G. Walker *British Plastics*, *42*, 131 (1969 Jul.).

[22] E. L. Muetterties, ed., *The Chemistry of Boron and Its Compounds*, Wiley, New York, 1967. This is a comprehensive monograph on the subject and is a prime source book.

[23] F. E. Bacon, N. P. Nies, H. C. Newsom, M. L. Iverson, S. M. Draganov, W. G. Woods, and G. W. Campbell, Jr., articles on "Boron," in *Encyclopedia of Chemical Technology*, 2nd ed., Vol. 3, Wiley-Interscience, New York, 1964. These articles give a reasonably complete view with a good balance of scientific and practical information.

[24] J. Kleinberg, W. J. Argersinger, Jr., and E. Griswald, *Inorganic Chemistry*, Heath, Boston, 1960, Chapter 12.

[25] See reference 24, pp. 182–83.

[25a] U.S. Borax Chemical Corp., *Belg. 716,757*, Dec. 18, 1968.

[26] V. F. Ross and J. O. Edwards, Chapter 3 of reference 22.

[27] M. F. Lappert, Chapter 8 of reference 22.

[28] H. C. Brown, *Hydroboration*, Benjamin, New York, 1962.

[29] N. G. Bonch-Osmolovskii and E. V. Bonch-Osmolovskaya, *USSR 138,298*, Appl. Aug. 4, 1960; J. J. Wachter, *U.S. 2,785,144* (to Westinghouse Electric Corp.), Mar. 12, 1957; C. H. Teesdale and R. E. Prince, *U.S. 1,265,540*, May 7, 1918; H. A. Gardner, *Paint Mfrs. Assoc. of U.S.*, *Circ. No. 280*, 107 (1926).

[30] N. M. Hopkins, *U.S. 1,706,733* (to Burnot Fireproofing Products Co.), Apr. 2, 1929.

[31] G. A. Connell and G. D. Holmes, *Forestry*, **36**, 91 (1963); G. A. Hesterberg, *Proc. World Forestry Congr.*, *5th*, Seattle, Wash., **2**, 999 (1960); G. A. Connell, *U.S. 2,858,895* (to U.S. Borax Chemical Corp.), Nov. 4, 1958.

[32] U.S. Borax Chemical Corp., *Brit. 1,023,640*, Mar. 23, 1966.

[33] N. D. Clare and A. J. Deyrup, *U.S. 2,823,145* (to E. I. duPont de Nemours & Co.), Feb. 11, 1958.

[34] T. H. Ling, *U.S. 3,287,312* (to Anaconda Wire and Cable Co.), Nov. 22, 1966.

[35] Rexall Drug and Chemical Co., *Belg. 669,948*, Mar. 22, 1966.

[36] G. A. Martin, *U.S. 3,342,629* (to Callery Chemical Co.), Sept. 19, 1967.

[37] I. S. Bengelsdorf and W. G. Woods, *U.S. 3,250,732* (to U.S. Borax Chemical Corp.), May 10, 1966.

[38] W. Posner, *Brit. 550,168*, Dec. 28, 1942.

[39] W. G. Woods, D. Laruccia, and I. S. Bengelsdorf, *U.S. 3,189,565, 3,250,797* (to U.S. Borax Chemical Corp.), June 15, 1965, May 10, 1966.

[40] H. H. Chen and A. C. Nixon, *SPE Trans.*, **5**, 90 (1965); L. E. Brown and J. B. Harshman, Jr., *AEC Accession No. 35935, Rept. No. MLM-CF-64-8-1* (Avail. OTS), 1964, 14 pp.

[41] W. G. Woods, W. D. English, and I. S. Bengelsdorf, *U.S. 3,257,347* (to U.S. Borax Chemical Corp.), June 21, 1966.

[42] J. P. Stalego, *U.S. 3,218,279* (to Owens-Corning Fiberglass Corp.), Nov. 16, 1965.

[43] R. S. Aries, *Fr. 1,295,905*, June 15, 1962.

[44] Organofluorine compounds usually require almost complete substitution of hydrogen by fluorine; perfluoro compounds are preferred. For examples, see the following: Fluorinated oils—P. J. Gaylor, *U.S. 2,370,787* (to Standard Oil Development Co.), Mar. 6, 1945; Fluorinated carboxylic acids—E. L. Kropa and J. J. Padbury, *U.S. 2,523,470* (to American Cyanamid Co.), Sept. 26, 1950; Perfluoroalkyd vinyl ethers— D. W. Codding (to Minnesota Mining and Manufacturing Co.), Jan. 24, 1956; Fluoroketone copolymers—E. I. duPont de Nemours & Co. (by E. G. Howard, Jr.), *Brit. 1,020,678*, Feb. 23, 1966; Bisphenol A fluorinated in the isopropylidene portion— Allied Chemical Corp., *Neth. Appl. 6,407,623*, Jan. 4, 1965.

[45] J. S. Sconce, ed., "Chlorine, Its Manufacture, Properties and Uses," *ACS Monograph 154*, Reinhold, New York, 1962.

[46] Z. E. Jolles, ed., *Bromine and Its Compounds*, Academic, New York, 1966. This is an excellent reference work on bromine chemistry.

[47] Reference 1, p. 419.

[48] V. A. Stenger and G. J. Atchison, articles on "Bromine" and "Bromine Compounds," in *Encyclopedia of Chemical Technology, op. cit.*, Vol. 3, pp. 750–83.

[49] D. G. Nicholson, "Chlorine," D. W. F. Hardie, J. C. Gage, B. H. Pilorz, P. S. Bauchwitz, R. R. Whetstone, H. Sidi, H. C. Hubbard, G. D. Lichtenwalter, G. H. Riesser, and J. D. Doedens, "Chlorocarbons and Chlorohydrocarbons," in *Encyclopedia of Chemical Technology*, 2nd ed., Vol. 5, Wiley-Interscience, New York, 1964, pp. 1–6, 85–296. This lengthy series of articles gives details on practical routes to various compounds and is the best source of information on probable commercial practice. The coverage is broad and informative.

[50] E. H. Huntress, *Organic Chlorine Compounds*, Wiley, New York, 1948. This is a very thorough index similar to Beilstein—indeed cross-indexed to it—which is perhaps the quickest entry into the literature for compounds known as of 1948.

[51] P. B. D. de la More and J. H. Ridd, *Aromatic Substitution*, Academic, New York, 1959. A reference of limited applicability to the present task but helpful on specific points and on reaction theory.

[52] L. F. Fieser and M. Fieser, *Reagents for Organic Synthesis*, Wiley, New York, 1967. This book, already a classic, is especially helpful for laboratory synthesis guidance.

[52a] Ethylene Plastique, S.A., *Belg. 706,168*, Nov. 7, 1967.

[53] K. M. Bell, B. W. McAdam, and H. T. Wallington, *Plastics*, **31**, 1439 (1966).

[54] Badische Anilin- & Soda-Fabrik A.-G. (by K. Herrle and M. Herner), *Belg. 644,644*, Sept. 3, 1964.

[55] A. P. Suprun, T. A. Soboleva and G. P. Lopatina, *Vysokomol. Soedin.*, *Karbot. Vysokomol. Soedin., Sb. Statei*, **1963**, 128.

[56] C. W. Seelbach and D. L. Cottle, *U.S. 3,058,966* (to Esso Research and Engineering Co.), Oct. 16, 1962; N. G. Karapetyan, I. S. Boshniakov, and A. S. Margarian, *Vysokomol. Soedin.*, **7**, 1993 (1965).

[57] S. G. Gelfand, *U.S. 3,313,857* (to Hooker Chemical Corp.), Apr. 11, 1967; W. H. Chang, *U.S. 3,385,867* (to Pittsburgh Plate Glass Co.), May 28, 1968; Badische Anilin- & Soda-Fabrik A.-G. (by M. Minsinger, H. H. Friederich, and K. J. Fust), *Ger. 1,074,573*, Feb. 4, 1960; J. R. Caldwell, *U.S. 3,085,885* (to Eastman Kodak Co.),

Apr. 16, 1963; P. E. Hoch, *U.S. 3,040,106* (to Hooker Chemical Corp.), June 19, 1962; W. J. Jackson, Jr., and J. R. Caldwell, *Brit. 952,152*, Mar. 11, 1964; Hooker Chemical Corp., *Neth. Appl. 6,515,354*, May 26, 1966; H. Bluestone, *U.S. 2,925,445* (to Shell Development Co.), Feb. 16, 1960 (also see *U.S. 2,676,132*); Chemische Werke Albert, *Neth. Appl. 6,608,460*, Dec. 19, 1966; Hooker Chemical Corp., *Neth. Appl. 6,612,481*, Mar. 21, 1967; R. E. Lidov, *U.S. 2,733,248* (to Shell Development Co.), Jan. 31, 1956; P. Robitschek and C. T. Bean, *U.S. 2,863,848* (to Hooker Chemical Corp.), Dec. 9, 1958; J. Hyman, *Brit. 618,432*, Feb. 22, 1949; M. Kleiman, *U.S. 3,288,813* (to Velsicol Chemical Corp.), Nov. 29, 1966; Dynamit Nobel A.-G., *Belg. 707,726*, Dec. 12, 1967; H. S. Bloch and L. Schmerling, *U.S. 3,409,386* (to Universal Oil Products Co.), Nov. 5, 1968; Borg-Warner Corp., *Belg. 697,459*, Oct. 24, 1967; Dynamit Nobel A.-G., *Belg. 717,305, 718,569*, Dec. 2, 1968, Dec. 31, 1968; R. R. Hindersinn and J. F. Porter, *U.S. 3,403,036* (to Hooker Chemical Corp.), Sept. 24, 1968; R. R. Hindersinn and H. W. Marciniak, *U.S. 3,418,263* (to Hooker Chemical Corp.), Dec. 24, 1968; R. R. Hindersinn and C. S. Hilado, *U.S. 3,449,279* (to Hooker Chemical Corp.), June 10, 1969.

[58] J. Kovacs and C. S. Marvel, *J. Polym. Sci., Part A-1*, **5**, 1279 (1967); J. L. Thomas, *SPE J.*, **23**, 30 (1967).

[59] M. H. Earing, *U.S. 3,318,960* (to Wyandotte Chemical Corp.), May 9, 1967; W. W. Levis, Jr., *U.S. 3,325,545* (to Wyandotte Chemical Corp.), June 13, 1967; J. T. Baker Chemical Co., *Belg. 690,806*, Dec. 7, 1966; Dynamit Nobel A.-G., *Belg. 690,819*, Dec. 12, 1966; R. A. Grimm, J. E. Menting, A. J. Stirton, and J. K. Weil, *J. Amer. Oil Chem. Soc.*, **45**, 895, 897, 901 (1968).

[60] J. C. Wygant, R. M. Anderson, and E. J. Prill, *U.S. 3,317,568* (to Monsanto Co.), May 2, 1967.

[61] H. J. Dietrich, J. V. Karabinos, and M. C. Raes, *J. Polym. Sci., Part A-1*, **5**, 1395 (1967); H. J. Dietrich, *Amer. Chem. Soc., Div. Polym. Chem., Preprints*, **8**, 481 (1967).

[61a] T. M. Patrick, Jr., *U.S. 2,775,615* (to Monsanto Chemical Co.), Dec. 25, 1956.

[62] Deutsche Solvay Werke G.m.b.H. (by G. Faerber), *Ger. 1,051,503*, Feb. 26, 1959.

[63] C. H. Penning, *Ind. Eng. Chem.*, **22**, 1180 (1930); Soc. D'Ugine, *Belg. 671,608*, Oct. 29, 1965.

[64] E. C. Stivers, *U.S. 3,340,226* (to Raychem Corp.), Sept. 5, 1967; B. W. Nardlander and W. E. Cass, *U.S. 2,455,873* (to General Electric Co.), Dec. 7, 1948; P. E. Hoch, *U.S. 3,337,624* (to Hooker Chemical Corp.), Aug. 22, 1967; G. J. Bohrer, *U.S. 2,443,887* (to General Electric Co.), June 22, 1948.

[65] Reference 46, p. 356.

[66] Gelfand, reference 57; Dow Chemical Co., *Belg. 698,808*, Apr. 22, 1967; H. A. Wright, *U.S. 3,456,022* (to Koppers Co., Inc.), July 15, 1969.

[67] Badische Anilin- & Soda-Fabrik A.-G., *Belg. 691,886*, Dec. 28, 1966.

[68] Kovacs and Marvel, reference 58.

[68a] Monsanto Chemicals, Ltd. (by R. J. Stephenson and D. G. Hare), *Brit. 1,051,944*, Dec. 21, 1966.

[68b] R. R. Hindersinn and H. W. Marciniak, *U.S. 3,392,136* (to Hooker Chemical Corp.), July 9, 1968.

[68c] Dow Chemical Co., *Belg. 711,564*, Mar. 1, 1968.

[69] L. H. Lee, *U.S. 3,297,658* (to Dow Chemical Co.), Jan. 10, 1967.

[70] K. Matsubayashi, *Sen-i Gakkaishi*, **19**, 27 (1963) (in Japanese).

[70a] R. J. Stephenson, *U.S. 3,465,031* (to Monsanto Co.) Sep. 2, 1969.

[71] Chemische Fabrik Kalk G.m.b.H., *Fr. 1,371,629*, Sept. 4, 1964.

[72] H. Jenker and R. Straug, *U.S. 3,382,254* (to Chemische Fabrik Kalk G.m.b.H.), May 7, 1968; Raychem Corp., *Brit. 1,061,372*, Mar. 8, 1967.

[73] Reference 48, p. 771.

[73a] Z. E. Jolles, *U.S. 3,432,450* (to Berk, Ltd.), Mar. 11, 1969.

[74] Reference 46, p. 669.

[75] M. H. Cupery, *Ind. Eng. Chem.*, **30**, 627 (1938); P. F. Tryon, *U.S. 2,337,693* (to Commercial Solvents Corp.), Dec. 28, 1943.

[76] E. C. Crocker, *Ind. Eng. Chem.*, **17**, 163 (1925); N. M. London and the London Electric Wire Co. and Smiths, Ltd., *Brit. 365,936*, Nov. 14, 1930.

[77] Private communication from Sylvania Electric Products, Inc., Sept. 4, 1968.

[78] G. R. Hill and D. G. Needham, *U.S. 3,432,461* (to Phillips Petroleum Co.), Mar. 11, 1969.

[79] Dynamit Nobel A.-G., *Belg. 717,515*, Dec. 16, 1968.

Chapter 4

CELLULOSE: CHEMISTRY, WOOD, PAPER, AND FOREST FIRE CONTROL

Cellulose, in various forms, is one of the commonest materials used for construction and furnishings. Buildings, especially residences, are most frequently made of wood and contain within them cotton fabrics and paper products of many types. Most building fires feed on cellulosic fuels. This fact, in turn, has stimulated extensive research on fire retardants for cellulose. As much as one-third to two-fifths of all references on fire retardants deal with cellulosic products. Of these citations perhaps half are on cotton textiles, one-third on wood, and the balance on paper and chemically modified cellulose. In this chapter the pertinent chemistry of cellulose is presented, followed by a discussion of fire retardants for wood and paper and compositions for fighting forest fires. Chapter 5 is devoted primarily to cotton textiles.

STRUCTURAL CHEMISTRY OF CELLULOSE [1–4]

Natural cellulose is a polysaccharide composed of β-glucose units. Each glucose unit has the cyclic, pyranoside structure. These rings are connected by loss of a mole of water between C_1 and C_4 of adjacent rings to form ether linkages. A segment of a cellulose chain is shown in the Haworth formula: There are one primary and two secondary alcohol groups for each anhydroglucose residue. In natural, or native, celluloses there are usually more than

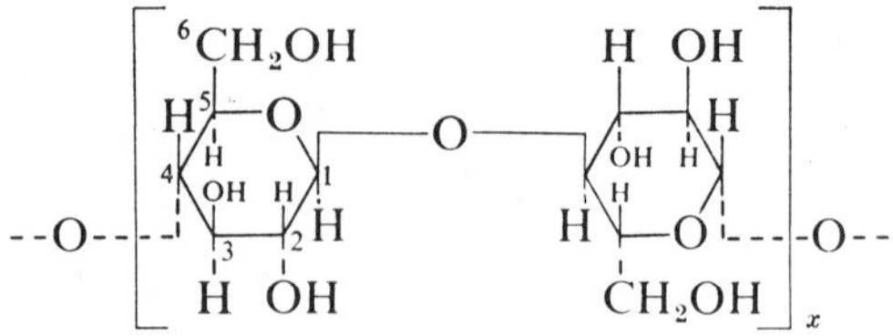

108

1000 anhydroglucose units in a single chain molecule; some may contain up to 10,000 segments per molecule. These are nonbranching or straight chains, and this arrangement of atoms gives the so-called primary structure of the cellulose molecule. Two additional structural modes are required to characterize cellulose in its usual form: the secondary structure, usually relating to an ordered spatial orientation of each polymer chain, and the tertiary structure, denoting either special configurations such as irregular but specific folds or turns of a chain (as in an enzyme molecule arrayed spatially around its "active site") or ordered packing of the individual chains to produce crystallites.

The three hydroxyl groups in each repeating unit provide sites for strong hydrogen bonding either to adjacent units on the same chain or to neighboring chains. Intramolecular bonding will determine the secondary structure; intermolecular bonds produce tertiary structure. Native cellulose is only partially crystalline; there are crystallite regions interspersed throughout an amorphous matrix. Because of the imperfect crystallinity, structure determinations have not been produced with the same precision as for many other polymer systems and details are still being argued.

A recent paper [5] has summarized the situation. First, the pyranoside rings are in the chair form. Second, the most reasonable intramolecular hydrogen bonding appears to be that of two pairs: one between the OH on C_2 and the acetal O in the ether link between rings, the other between the OH on C_3 and the O in the ring adjacent:

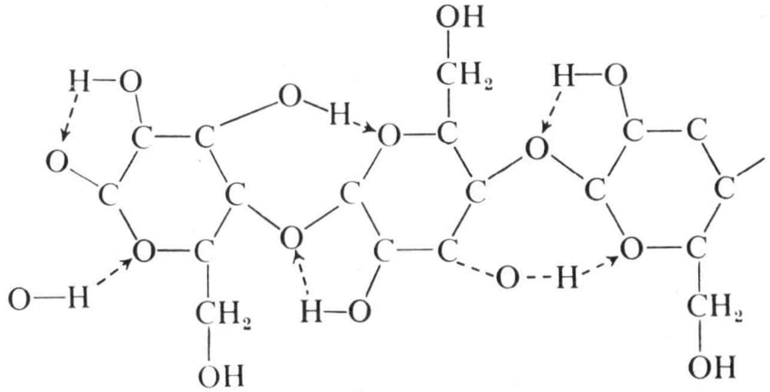

The primary hydroxyl groups appear not to participate in the intramolecular bonding but are available for bonding to adjacent chains. This intermolecular bonding provides the basis for the high strength of cellulose molecular aggregates (fibrils) and is at the heart of many questions concerning the strength of cellulose fibers. A recent study of reactivity showed that the sites at C_6 and C_2 are accessible to reagents; C_3 is much less so [5a]. The site at C_2 appears to be the most available for reaction.

The simplest tertiary structure indicates strong hydrogen bonding in one plane and merely van der Waals attractions in the other. This would mean that cellulose is a layered crystal and should therefore be noticeably weaker along certain directions. This is indeed the case [5b]. In the light of the present knowledge of polymer configuration in crystals, it is likely that the inter-molecularly, hydrogen-bonded chains assume a secondary structure of a flat, ribbonlike helix with perhaps a very long repeating period. Such a secondary structure would erase any plane of tertiary bonding in the crystallites.

One may visualize the cellulose helices packed together spirally around the fiber axis, held by hydrogen bonds. These fibrils lie not parallel to the axis but at an angle varying with the species, from 25 to 45°. This structure is consistent with the observation that many natural cellulose crystallites are known to have a spiral structure in which the tiny fibrils assume a helical configuration in the direction of the fiber axis [6].

Disordered or amorphous regions of cellulose may be assumed to have less perfectly paried intermolecular hydrogen bonds. These regions are known to swell more readily in certain solutions and to be more accessible to certain reagents. The cellulose reacts as though it were composed of sheets rather than of individual chains [6a]. But even in NaOH 50% or more of the cellulosic sites are unavailable in cotton [5a]. Different sources of cellulose have different percentages of amorphous regions. The order of decreasing crystallinity or increasing accessibility is: cotton, wood pulps, mercerized cotton, textile rayons, and high-tenacity rayons [7]. Thus we may expect somewhat easier, smoother reactions with wood cellulose or processed cellulose than with cotton fibers.

Reactions of Cellulose—Degradation by Chemical and Thermal Effects

The cellulose structure may be disrupted by (a) breaking the hydrogen bonds between chains and within chains or (b) cleaving the acetal linkage between rings on a given chain. If the acetal bonds are broken, the molecules depolymerize and all semblance of fiber behavior is lost. If the hydrogen bonds are broken, the polymer chains will be intact but fiber strength will be greatly decreased; the molecules become water-soluble or subject to swelling by solvents. In the terminology of the biochemists, rupture of the hydrogen bonds is denaturation in the sense that the natural secondary and tertiary configurations are eliminated.

The hydrogen bonds can be ruptured by reagents that bond more tightly to the hydrogen and oxygen atoms participating in the bond. In other words, these are materials that will compete favorably for a given site. When this occurs, the effects derived from the secondary and tertiary structures are lost.

An example is the ability of a solution of copper-ammonia complex (ammine complex) [8] to dissolve cellulose with apparently little or no cleavage of the primary structure. This may be considered as a denaturant, just as are a great many other metals, salts, and miscellaneous reagents that are known denaturants for a variety of native structures in biochemistry [9].

Presumably, treatment of cellulose with alkali (mercerization) [10] causes a loss of hydrogen bonding by the interposition of the base. This in turn loosens the structure and permits the invasion of water molecules, which results in swelling. Temporary treatment with alkali allows the rearrangement of secondary and tertiary structures by weakening the hydrogen bonds. Following this rearrangement, the original strength of the hydrogen bonds can be restored by washing out the alkali. In this way improvements in luster, dye affinity, and fiber strength are achieved [11]. The final crystal form may be attained by such a treatment [11a].

The cellulose primary chain is degraded by hydrolysis, oxidation, microbial attack, or mechanical attrition. Of interest to us is hydrolytic degradation, which is particularly sensitive to acid catalysis but not to base catalysis. The reaction is:

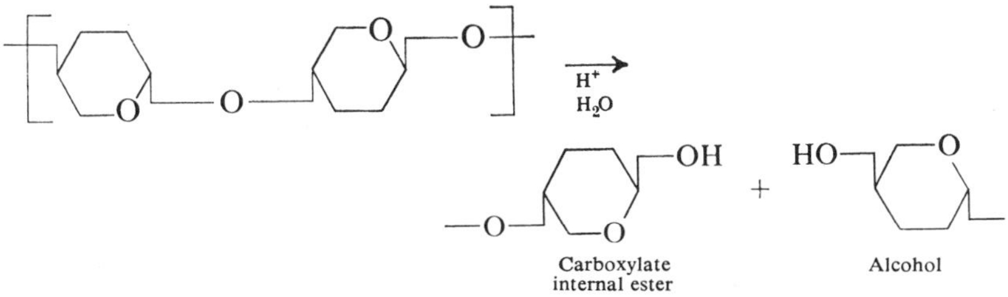

Carboxylate
internal ester   Alcohol

It is believed that the acetal oxygen linking the rings together is first protonated in a fast reaction and that the chain then is cleaved in a slower,

Table 4-1 Hydrolysis of
Cotton Cellulose in H_3PO_4 at
25°C [12]

Time, hr	Wt Av Chain Length
0	1600
15	1410
60	1055
135	760
210	612
330	468
750	258
1140	182
2100	101

rate-determining step [12]. Table 4-1 shows hydrolysis in H_3PO_4 at room temperature as a function of time. This is for homogeneous hydrolysis; that is, the cellulose was first dissolved in H_3PO_4 and the subsequent reduction in chain length followed by viscometry. Heterogeneous hydrolysis or hydrolysis of cellulose in a separate, insoluble phase is less straightforward. For this case, the apparent rate is influenced by both intrinsic chemical factors and surface, diffusion-controlled factors. The important facts are the marked influence of hydrogen ion on the rate and the apparent uniformity of cellulose in its behavior in hydrolysis. On the latter point, there has been much discussion with regard to possible weak links in the cellulose chain. The evidence for such weak links is unconvincing.

Above 140°C or so thermal degradation of cellulose occurs [13]. This appears to be a combination of oxidative changes in the glucose rings combined with cleavage of the chain and catalyzed by acidic products of the oxidation. The presence of moisture facilitates the process. Clues to the mechanism of oxidative degradation may be found in details of experiments on specific oxidants [14]. It has been shown that periodate selectively oxidizes the secondary alcohol groups to aldehydes:

$$
\begin{array}{ccc}
\overset{\displaystyle |}{\underset{\displaystyle |}{O}}\;\;\;CHOH & & \overset{\displaystyle |}{\underset{\displaystyle |}{O}}\;\;\;CHO \\
HOCH_2CH\;\;CHOH & \xrightarrow{\text{periodate}} & HOCH_2CH\;\;CHO \\
\end{array}
$$

Further oxidation will produce the corresponding dicarboxylic acid. On the other hand, nitrogen dioxide acts mostly on the primary alcohol:

$$
\begin{array}{ccc}
\overset{\displaystyle |}{\underset{\displaystyle |}{O}}\;\;\;CHOH & & \overset{\displaystyle |}{\underset{\displaystyle |}{O}}\;\;\;CHOH \\
HOCH_2CH\;\;CHOH & \xrightarrow{N_2O_4} & HOOCCH\;\;CHOH \\
\end{array}
$$

Chromic acid is nonspecific and produces general effects, such as loss of fibrous structure and overall development of oxidation products. Hypohalite solutions produce a small amount of carboxylic acids in a manner similar to nitrogen dioxide.

Pyrolysis or nonoxidative thermal degradation has been studied for cellulose by many workers and results for specific type of cellulosics are discussed in appropriate sections of this book. However, some general results may be useful at this juncture. Madorsky [15] summarizes work done at the National Bureau of Standards in the last few years. In this work fractions coming over from the pyrolysis chambers are condensed at various temperatures to separate products into groups as a function of boiling point. Further separations are then carried out within groups by other techniques. The condensing temperatures are (a) that at pyrolysis, (b) 25°C, (c) −80°C, and (d) −190°C. The fractions are then those that are volatile only at pyrolysis temperature, those volatile at 25° but not at −80° or −190°, those volatile at −80° but not at −190°, and those volatile at −190°. Results for cotton cellulose are shown in Table 4-2. Analysis by the mass spectrograph

Table 4-2   Pyrolysis of Cellulose in a Vacuum [15]

Temp, °C	Duration, min		Volatilization, %		Volatile Fractions, % of Total Volatiles/Step			
	For Step	Cum	For Step	Cum	V_{pyr}	V_{25}	V_{-80}	V_{-190}
250	300	· · ·	2.7	· · ·	26.3	53.9	12.0	7.8
280	38	38	3.3	3.3	33.9	50.7	10.5	4.9
280	58	96	7.0	10.3	55.9	36.1	6.0	2.0
280	53	149	9.1	19.4	63.8	29.6	5.4	1.2
280	100	249	15.9	35.3	· · ·	· · ·	· · ·	· · ·
280	77	326	9.8	45.1	70.2	23.6	4.7	1.5
280	180	506	16.7	61.8	· · ·	· · ·	· · ·	· · ·
280	155	661	7.8	69.6	70.8	21.9	5.5	1.8
321	9	· · ·	22.0	· · ·	65.2	28.2	4.3	2.3
397	159	· · ·	91.4	· · ·	69.6	22.5	6.2	1.7

indicated that V_{25} is largely water with a little acetaldehyde, V_{-80} is CO_2, and V_{-190} is CO. V_{pyr} is a tarry substance constituting some two-thirds of the products at the higher temperatures.

Data from both mass spectrograms and infrared spectra suggest that the tars are based on the levoglucosan structure:

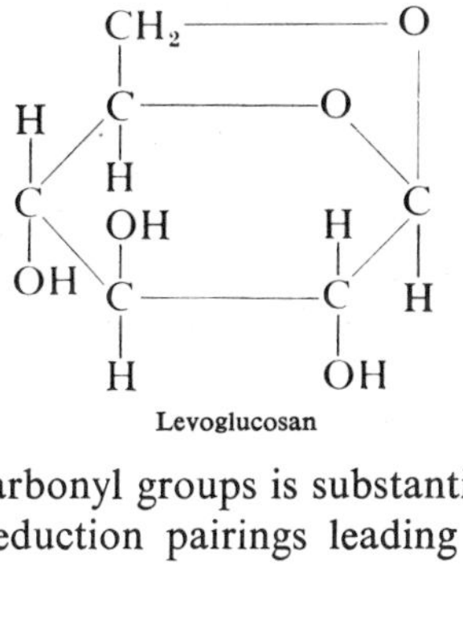

Levoglucosan

However, evidence for carbonyl groups is substantial, probably attributable to internal oxidation—reduction pairings leading to acids and aldehydes along with the recorded amounts of CO generated (V_{-190}). It is of interest that when pyrolysis is carried out in nitrogen instead of in a vacuum the amount of tar is greatly decreased with an attendant increase in V_{25}. This is apparently due to secondary degradation reactions in the gas phase which yield smaller, more volatile fragments. The activation energy for this pyrolysis in a vacuum is about 50 kcal/mole. The effects of fire retardants on this behavior are discussed later in some detail. Suffice it to say here that the fire retardants affect the ratio of volatile to nonvolatile combustibles and the relative ease with which they are formed; that is, the activation energy is often reduced so that events occur at lower temperatures or more readily at the same temperature. It is important to realize that most fire situations include at best only partial pyrolysis. Oxygen is present, to a greater or lesser extent, although it may be argued that there are many geometrical configurations where the substrate is exposed primarily to pyrolysis with oxidation occurring at a considerable distance. In any case, data from controlled pyrolysis offer many insights into the degradation process and into its alteration by chemical treatments.

Introduction of Specific Substituents

Cellulose derivatives are usually prepared by the introduction of substituents at one or more of the three hydroxyl groups on each of the repeat units. Some common examples are cellulose nitrate, ethyl cellulose, hydroxyethyl cellulose, carboxymethyl cellulose, cellulose acetate, and cellulose xanthate. It is useful to review briefly how these commercially important materials are produced and then to see how this and related technology has been applied to the problem of introducing fire-retardant properties.

Cellulose nitrate [16, 17] is prepared by means of HNO_3—H_2SO_4—H_2O mixtures at low temperature:

$$\text{Cell—(OH)}_3 + HNO_3 \xrightarrow[\substack{\frac{1}{2} \text{ hr} \\ 20\text{–}40°C}]{H_2SO_4} \text{Cell—(ONO}_2)_3$$

The mixed acid contains about 20% H_2O at the end and some 60% H_2SO_4 and 20% HNO_3. Large excesses of the acid are used to avoid large temperature changes and to assure the proper equilibrium between hydrates of the acid and free water in the system. The usual product contains between 10 and 13.5% N (mononitrate $\sim$7% N, di- 11.2% N, tri- 14.1% N). The reaction proceeds smoothly and with very little chain depolymerization. Although there are conflicting theories on the nature of this reaction, the data seem to support the idea that the reaction with cellulose fibers is homogeneous; that is, the rate is not diffusion-controlled and the reactants penetrate the system very rapidly and uniformly.

Cellulose sulfate forms over a period of about 2 hr at 0–5°C in either alcoholic H_2SO_4 or chlorosulfonic acid in pyridine. It is important to minimize the depolymerization of the chains by using carefully controlled levels of water (dilute with alcohol) or by using pyridine.

Cellulose acetate is prepared from dry cellulose and acetic anhydride:

$$\text{Cell—}(OH)_3 + 3Ac_2O \xrightarrow[\substack{H_2SO_4 \\ 5-45° \\ 1-2\ hr}]{AcOH} \text{Cell—}(OAc)_{2.5-3.0} + 3AcOH$$

The product is washed free of residual acids to prevent depolymerization.

The foregoing reactions are esterifications. Of equal significance are etherifications [17, 18]. In this case NaOH is added to swell the cellulose to make it more accessible and to enhance the reactivity of the hydroxyl groups. The strength of the caustic varies from about 30% for low degrees of substitution to as high as 70% for high degrees. At least 3 moles of NaOH are needed per glucose residue for high degrees of substitution, which suggests that formation of a sodium alkoxide species at the cellulose is a prerequisite. Ethyl cellulose is prepared from ethyl chloride

$$\text{Cell—}(OH)_3\cdot 3NaOH + C_2H_5Cl \xrightarrow[6-12\ hr]{90-150°} \text{Cell—}(OEt)_3 + NaCl$$

(or from ethanol in sulfuric acid [18a]). Carboxymethyl cellulose is prepared from chloroacetate:

$$\text{Cell—}(OH)_3\cdot 3NaOH + ClCH_2COONa \xrightarrow[to\ 13\ hr]{25-100°C}$$
$$\text{Cell—}(OCH_2COONa)_{0.4-1.4} + NaCl$$

Cyanoethyl cellulose is made from acrylonitrile:

$$\text{Cell—}(OH)_3\cdot 3NaOH + CH_2{=}CHCN \xrightarrow[reflux]{\frac{1}{2}\ hr}$$
$$\text{Cell—}(OCH_2CH_2CN)_{0.5-2.5} + NaOH$$

Hydroxymethyl cellulose is prepared from ethylene oxide:

$$\text{Cell—}(OH)_3\cdot 3NaOH + CH_2{-}CH_2 \xrightarrow[30°C]{15\ hr}$$
$$\underset{O}{\diagdown\diagup}$$
$$\text{Cell—}(OH)_2OCH_2CH_2OH + 3NaOH)$$

All the reactions suffer from side reactions with formation of simple ethers, hydrolysis products of the alkyl halides, and so on. The efficiency based on the halide is therefore often low and this raises processing costs.

"Viscose" is probably the most important commercial derivative of cellulose. It is formed by the action of alkali and carbon disulfide on cellulose:

$$\text{Cell—}(OH)_3\cdot 3NaOH + CS_2 \xrightarrow[20-35°]{1-3\ hr} \text{Cell—}(OCSSNa)_3$$

The product is unstable and is, of course, not used as such but is converted into end products as regenerated cellulose.

Many studies have been made of the modification of cellulose by chemical treatments, cross-linking reactions (e.g., by dimethylol urea[18b]), and grafting by chemical or radiation treatment. These have been reviewed in abstract form recently; the interested reader can enter the literature readily from reference 3.

Reactions with Phosphorus Compounds

One of the simplest phosphorus-cellulose reaction products is cellulose orthophosphate ester and its derivatives. Direct esterification with ortho-phosphoric acid is difficult and leads to extensive depolymerization. A full triester can be produced by means of a mixture of phosphoric and sulfuric acids [19, 20]. Phosphorylation is easier and milder with condensed phosphoric acid in a nonaqueous medium. Mixtures of nearly anhydrous phosphoric acid and phosphorus oxychloride have been used [21], but degradation will often result from these treatments with strongly acidic systems. Unfortunately, the more basic the reaction system, the less reactive are the phosphates. In one study [22] sodium polyphosphates effected phosphorylation about 50% greater at pH 5–6 than at pH 9–10:

$$Na_5P_3O_{10} + Cell\text{—}OH \xrightarrow[\substack{1\ hr \\ H_2O\ pH\ 9}]{160°} Cell\text{—}OPO_3Na_2 + Na_3HP_2O_7$$

Relatively little loss in strength was noted from this treatment and reasonable fastness to laundering was observed. However, the presence of the sodium in the product reduces the flame-retardant properties of the fabric and conversion to a heat-decomposable cation (ammonium ion or amines) must be done to achieve the desired protection. This has undoubtedly been the reason this approach has not been commercialized.

The need clearly is to achieve esterification while keeping the pH neutral or basic. An early approach was the use of $POCl_3$ in ammonia or in pyridine [23]:

$$Cell\text{—}OH + POCl_3 \xrightarrow[\substack{120° \\ 1-3\ hr}]{pyridine} Cell\text{—}O\overset{\displaystyle O}{\overset{\|}{\text{—}P\text{—}}}Cl_2 + pyridine \cdot HCl$$

$$Cell\text{—}O\overset{\displaystyle O}{\overset{\|}{\text{—}P\text{—}}}Cl_2 + 4NH_4OH \xrightarrow[H_2O]{} Cell\text{—}O\overset{\displaystyle O}{\overset{\|}{\text{—}P\text{—}}}(ONH_4)_2 + 2NH_4Cl$$

Up to 9% phosphorus can be introduced in this manner. Unfortunately, up to 50% loss in strength occurs despite the presence of pyridine, which suggests some direct attack by $POCl_3$ on the cellulose chain linkages. Analysis of the product indicates that there may be some disubstituted phosphorus present, that is, phosphorus that titrates as a monobasic acid.

Phosphorus pentoxide has been used in $CHCl_3$ or in a bath containing pyridine [23a]. Its action is more severe and some dehydration and charring may occur.

A mixture of urea and phosphoric acid (or urea phosphate which forms at ca. 110°) has been used to phosphorylate cellulose—on a commercial scale at one time. Of several references to this method [24] those to Little and Nuessle et al. are perhaps the best reviews. The cellulosic material is immersed in a solution of 40% urea and 20% $(NH_4)_2HPO_4$, squeezed, dried, and cured 10 min at 150°, and given a final wash. As much as 9% phosphorus can be so introduced but only about 3% is required for fire retardancy. Schiffner and Lange [24] report substantial strength loss but show how several of the process variables can be controlled so as to hold this loss to 20–30% of the untreated samples. Processing in the presence of pyridine, dimethyl formamide, aniline, and other nitrogeneous organic solvents is said to be beneficial in terms of minimizing strength loss and promoting the reaction. The process has not gained wide acceptance, perhaps because (a) the phosphate ester linkage may not be sufficiently resistant to hydrolysis in laundering; (b) the tendering of the fabric may still be too much for applications where strength is critical: in furniture fabrics, uniforms, tentage, and the like; and (c) the replacement of NH_4^+ by Na^+ in laundering reduces flame retardancy.

Variants on the process include the use of methyl vinyl pyridine as a grafted polymer followed by treatment with phosphoric acid [25]; a solution of cellulose in phosphoric acid followed by bubbling N_2O_4 gas through to promote esterification [26]; and the use of diethylenetriamine in place of urea [27].

Chlorophosphate esters have been used to introduce the phosphorus as an ester:

$$\text{Cell—OH} + (RO)_2PO(Cl) \xrightarrow{\text{base}} \text{Cell—OP(O)(OR)}_2 + HCl \cdot \text{base}$$

The base used [28] was pyridine, the temperature varied from 25° to 80°, and the time, from 5 min to 2 days. Some tendering of cotton fabric occurred with attendant loss of strength. Low levels of phosphorus were readily attached to give glow resistance, and up to 5% phosphorus has been incorporated. The technique has not been adopted commercially, perhaps because of the sensitivity of the chlorophosphate to moisture and hence the need to eliminate water carefully from the system by washing the cellulose first with pyridine. Some basic research on the structure of such materials has been done by ir spectroscopy and the results support the structures indicated above [29].

Phosphorus-nitrogen compounds have been studied for fire-retarding cellulose. Nielsen [30] found that polyphosphorylamide reacts with the hydroxyl groups to produce an ester linkage with a complex P—N chain:

$$\left[\begin{array}{c} O \quad H \\ | \quad\quad | \\ -P\!-\!\!-\!N- \\ | \\ NH_2 \end{array}\right] + HO\!-\!Cell \xrightarrow[150°]{5\ min} Cell\!-\!O\!\left[\begin{array}{c} O \quad H \\ | \quad\quad | \\ P\!-\!\!-\!N- \\ | \\ A \end{array}\right]_x$$

where, if $x = 4$, A represents $2ONH_4$, $1OH$, and $1NH_2$. The treatment confers excellent fire resistance, only a 5% loss of strength, but only fair resistance to hot laundering at the add-on needed to obtain the necessary phosphorus (14% to get 3.8% P).

On the other hand, the reaction product of ethyleneimine and phosphorus oxychloride reacts to form ethers with cellulose with two carbons and a nitrogen between phosphorus and the oxygen from the cellulose [31]:

$$\left(\begin{array}{c} CH_2 \\ | \quad\quad \diagdown \\ \quad\quad N \\ | \quad\quad \diagup \\ CH_2 \end{array}\right)_3\!\!PO + HO\!-\!Cell \xrightarrow[Zn(BF_4)_2\ as\ cat.]{3\ min\ 155°} \left(\begin{array}{c} CH_2 \\ | \quad\quad \diagdown \\ \quad\quad N \\ | \quad\quad \diagup \\ CH_2 \end{array}\right)_2\!\!\begin{array}{c}O\\ \|\\ P\!-\!N\!-\!CH_2CH_2O\!-\!Cell \end{array}$$

There is a serious loss in strength when APO is used alone in this fashion. This, plus the toxicity of APO, has apparently ruled out its use.

Forming ethers connected to phosphorus through one carbon atom (phosphonomethylation) has been studied extensively by two groups of workers [32]. The reaction is similar to that for preparing carboxymethyl cellulose:

$$Cell\!-\!OH + ClCH_2\overset{\overset{\displaystyle O}{\|}}{P}(OH)_2 + 3NaOH \xrightarrow[125°]{13\ min}$$

$$Cell\!-\!OCH_2\overset{\overset{\displaystyle O}{\|}}{P}(ONa)_2 + NaCl + 3H_2O$$

The product is not particularly fire-retardant as the sodium salt but becomes so on washing with a solution of an ammonia derivative. The phosphorus is firmly bound and will withstand boiling in 2% NaOH solution containing soap (of course, the NH_3 will be removed and the fabric converted back to the sodium form by such a treatment). The disadvantages are stiffening induced by the treatment, loss of the ammonium form through conversion to the sodium form on laundering, loss of strength at high phosphorus levels where best retardancy is found (no loss of strength occurs at 1% P), and the relative unavailability of the reagent at low cost (this could likely be overcome). The loss of strength at high phosphorus levels was subsequently partially overcome [33] by crosslinking prior to phosphonomethylation but at the cost of greatly increased stiffness. Schiffner and Lange [34] used urea with chloromethyl phosphonic acid, thereby maintaining a neutral to basic pH during treatment—urea decomposing to give NH_3 and other products—and yielding directly the ammonium salt of the treated cellulose. However,

Table 4-3   Reactions to Form Various Organic Phosphorus Esters of Cellulose

$$\text{Cell—OH} + (\text{ClCH}_2\text{CH}_2\text{O})_2\overset{\displaystyle O}{\underset{\displaystyle |}{P}}\text{—CH—CH}_2 \xrightarrow[100°]{3\text{ hr}}$$

$$\text{Cell—OCH}_2\text{CH}_2\text{O—P}\begin{smallmatrix} O & CH=CH_2 \\ | & \\ \\ \end{smallmatrix}\quad \text{OCH}_2\text{CH}_2\text{Cl}$$

35

$$2\text{ Cell—OH} + (\text{ClCH}_2\text{CH}_2\text{O})_3\text{P} \xrightarrow{\text{benzene}}$$

$$\text{Cell—OCH}_2\text{CH}_2\overset{\displaystyle O}{\underset{\displaystyle |}{\text{OP}}}\text{—OCH}_2\text{CH}_2\text{O Cell}$$
$$\qquad\qquad |\!\!-\!\!\text{OCH}_2\text{CH}_2\text{Cl}$$

36

$$\text{Cell—O}\overset{\displaystyle O}{\overset{\displaystyle ||}{\text{CCH}}}_2\text{Cl} + (\text{EtO})_3\text{P} \xrightarrow[\text{(Arbuzov)}]{\text{heat}} \text{Cell O}\overset{\displaystyle O}{\overset{\displaystyle ||}{\text{C}}}\text{CH}_2\overset{\displaystyle O}{\underset{\displaystyle |}{\text{—P}}}\text{—(OEt)}_2$$

37

$$\text{Cell}\begin{smallmatrix} \text{OH} \\ \diagup \\ \diagdown \\ \text{OH} \end{smallmatrix} + \text{CH}_3\overset{\displaystyle O}{\underset{\displaystyle |}{\text{—P}}}\text{—Cl}_2 \xrightarrow[\text{(2) base}]{\substack{(1)\ 10\ \text{min} \\ 70-80°}} \text{Cell}\begin{smallmatrix} \diagup \\ \diagdown \end{smallmatrix}\overset{\displaystyle O}{\underset{\displaystyle }{}}\text{P—CH}_3$$

38

$$\text{Cell—OH} + (\text{CH}_3\text{O})_2\overset{\displaystyle O}{\underset{\displaystyle |}{\text{—P}}}\text{—H} \xrightarrow[0.1\% \text{ NaOH}]{\substack{2-6\ \text{hr} \\ 100-180°}} \text{Cell—O}\overset{\displaystyle O}{\underset{\displaystyle |}{\text{—P}}}\text{—H (ester interchange)}$$
$$\qquad\qquad\qquad\text{OCH}_2$$

39

$$\text{Cell—OH} + (\text{HO})_2\overset{\displaystyle O}{\underset{\displaystyle |}{\text{—P}}}\text{—H} \xrightarrow[\substack{80-100° \\ \text{base}}]{10-15\ \text{hr}} \text{Cell—O}\overset{\displaystyle O}{\underset{\displaystyle |}{\text{—P}}}\text{—H (reacts with —CH}_2\text{OH)}$$
$$\qquad\qquad\qquad\underset{\text{Base}}{O}$$

40

$$\text{Oxidized Cell—CHO} + (\text{CH}_3\text{O})_2\overset{\displaystyle O}{\underset{\displaystyle |}{\text{—P}}}\text{—H} \rightarrow \text{Cell—CH}\overset{\displaystyle O}{\underset{\displaystyle |}{\text{—P}}}\text{—(OCH}_3)_2$$
$$\qquad\qquad\qquad\qquad\qquad |\!\text{OH}$$

41

$$\text{Cell—OH} + \text{HO}\overset{\displaystyle O}{\underset{\displaystyle |}{\text{—P}}}\text{—H} \xrightarrow[120°]{4-6\ \text{hr}} \text{Cell—O}\overset{\displaystyle O}{\underset{\displaystyle |}{\text{—P}}}\text{—H}$$
$$\qquad\quad |\!\text{H}\qquad\qquad\qquad |\!\text{H}$$

42

$$\text{Cell—OH} + \text{Cl}\overset{\displaystyle O}{\underset{\displaystyle |}{\text{—P}}}\text{—R}_2 \xrightarrow[110°]{\frac{1}{2}-2\ \text{hr}} \text{Cell O}\overset{\displaystyle O}{\underset{\displaystyle |}{\text{—P}}}\text{—R}_2$$

43

$$\text{Cell—OH} + \text{CH}_3\text{—P}\begin{smallmatrix} \diagup O \diagdown \\ \diagdown O \diagup \end{smallmatrix}\text{P—CH}_3 \xrightarrow[130°]{7-40\ \text{hr}} \text{Cell—O—P}\begin{smallmatrix} \diagup \text{CH}_3 \\ \diagdown \\ O \quad \text{OH} \end{smallmatrix}$$

44

$$\text{Cell—OH} + \text{R}_2\text{N—P}\begin{smallmatrix} \diagup \\ \diagdown \end{smallmatrix} \xrightarrow[120°]{20-40\ \text{hr}} \text{Cell—OP}\begin{smallmatrix} \diagup \\ \diagdown \end{smallmatrix} + \text{R}_2\text{NH}$$

45

the problem of conversion to the less effective sodium salt on laundering remains.

A great many papers have appeared on the production of various fully esterified phosphorus derivatives of cellulose. These products decompose on strong heating to produce the various phosphorus acids required to carry out the fire-retardant reactions. They are not convertible under ordinary laundering to sodium salts. On the other hand, since most of the treatments do not produce hydroxyl groups to replace those consumed in the reaction and some penetrate deeply the cellulose fiber structure, the strength from hydrogen bonding is sacrificed in part. The reagents are expensive and thus, though good fire retardance is often achieved, the processes have not received commercial attention. Table 4-3 summarizes some of these reactions. In certain cases the reaction products have been studied as to structure by ir techniques [46] and in terms of hydrolysis in alkaline solution [47]. Rogovin et al. have reviewed much of this in an attempt to draw conclusions from the work of that group [48].

As an alternative to direct reaction with cellulose, coupling agents have been employed with phosphorus compounds. The best known systems are based on tetrakishydroxymethyl phosphonium salts [49]. Coupling may be achieved by pretreating the cellulose with the agent:

$$\text{Cell—OCH}_2\text{CH}_2\text{NH}_2 + (\text{HOCH}_2)_4\text{P}^+\text{X}^- \rightarrow$$

$$\text{Cell—OCH}_2\text{CH}_2\text{N—}\overset{\overset{\displaystyle O}{\displaystyle |}}{\text{P}}(\text{CH}_2\text{OH})_2 + \text{H}_2\text{O} + \text{H}_2 + \text{CH}_2\text{O}$$

(Note the tendency to form the stable phosphine oxide during this reaction.) More often the phosphonium compound is not reacted with cellulose derivatives but is copolymerized with other substances [50] to form an insoluble polymer that is held to the cellulose by hydrogen bonding:

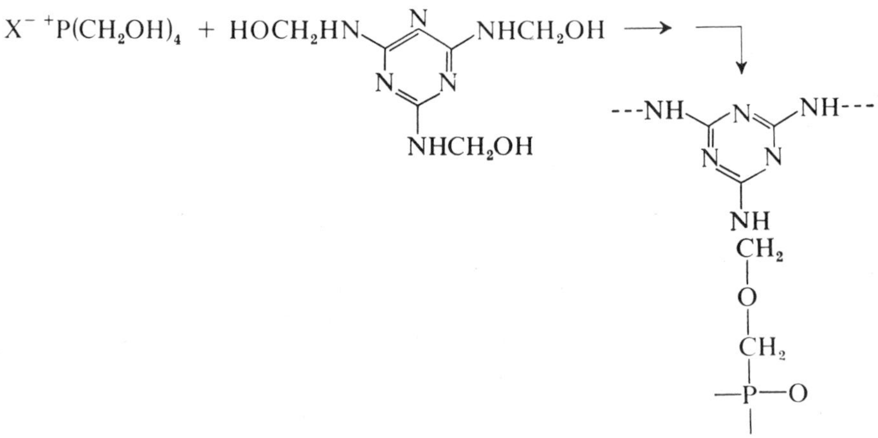

This kind of system is discussed further in Chapter 5 on cotton textiles.

Reactions with Nitrogen and Halogen Compounds

Very little in the way of basic studies has been done with these two materials and almost nothing with other fire-retardant elements. Animized cotton has been described and its production characterized [49, 50]. The reaction is simply:

$$\text{Cell—OH} + \text{HO—}\overset{\displaystyle O}{\underset{\displaystyle O}{\overset{|}{\underset{|}{S}}}}\text{—OCH}_2\text{CH}_2\text{NH}_2 \xrightarrow{\text{NaOH}}$$

$$\text{Cell—OCH}_2\text{CH}_2\text{NH}_2 + \text{H}_2\text{SO}_4 \text{ in NaOH}$$

This group in combination with phosphorus or sulfur obtained from acid-reacting azo dyes produces fire-retardant properties. The nitrogen will [51] assist in conferring flame retardance and is not merely a point of attachment for the dye. A fairly complicated nitrogen-containing cellulose derivative has been described in which the nitrogen is contained as the diethylaminoethyl ester of acrylic acid which in turn was grafted to cellulose by reaction with an hydroxyalkyl group placed on cellulose by prior reaction [52]. This kind of elegant synthesis is not likely to lead to commercial results because of cost. Crease resistance is obtained by use of methylol compounds which polymerize with themselves and which may also couple to the cellulose surface on heating or treatment with condensation catalysts [53]:

$$\text{Cell—OH} + \text{HOCH}_2\text{N}\Big\langle \xrightarrow[\text{cat}]{\text{heat}} \text{Cell—O—CH}_2\text{N}\Big\langle$$

Indeed, this probably occurs in the curing of systems of methylol phosphonium salts with cellulose and methylol nitrogen compounds (see above).

Halogenation has been studied primarily as a route to other derivatives. An early procedure [54] used as an intermediate step

$$\text{Cell—OH} + \text{CH}_3\text{SO}_2\text{Cl} \rightarrow \text{Cell—OSO}_2\text{CH}_3 \xrightarrow[\text{X=Cl, Br, F, I}]{\text{NaX}} \text{Cell—X}$$

The iodo- and bromo- derivatives exhibited good flame-retardant properties but poor glow resistance; a combination of phosphorylation and halogenation appeared promising. This system has not been explored fully. Over 10% chlorine can be introduced to cellulose by means of thionyl chloride in dimethylformamide [55]:

$$\text{Cell—OH} + \text{SOCl}_2 \xrightarrow[1\frac{1}{2}\text{ hr }80°]{\text{DMF}} \text{Cell—Cl} + \text{SO}_2 + \text{HCl}$$

The DMF plays a role in the reaction and is not simply an inert medium.

The chlorocellulose so produced is a good intermediate for aminocellulose or nitrilocellulose. Some work has also been done on reacting N-methylol haloacetamides to cellulose but the precise nature of the product has not been defined [56].

Other reactions of interest are considered in the following sections on paper, wood, and cotton substrates.

WOOD

Historical

References on fire-retardant studies date back at least 200 years and were initially directed to textiles. The earliest reference to fire retardants for wood reported in *Chemical Abstracts* occurs in Vol. 1 (1905) and refers to a mixture of ammonium phosphate and boric acid [57]—a system still in wide use today (see below). In the same year a patent was issued on the use of sodium silicate [58]. In 1907 a review [59] discussed ammonium phosphates, tungstates, borates, and chlorides of calcium, magnesium, zinc, tin (II), and sodium silicate, aluminum hydroxide, and boric acid. The review refers to the use of $CaSO_4$ in 1841. A variety of techniques were employed in the early years of this century, even including the application of ac electric current to fix the retardant. Mentioned in a British patent of 1912 [60] are zinc, mercury(II), and copper(II) salts and a technique of impregnation using evacuation before immersion. The proceedings of the American Wood Preservers Association have provided many substantial and useful status reports. One in 1914 summarized the important properties of fire retardant-wood and singled out NH_4Cl, $(NH_4)_2SO_4$, and NH_4 phosphates as being especially useful [61].

The most comprehensive of the early studies were reported over a period of six years (1930–35) in five progress reports from the Forest Products Laboratory, Forest Service, U.S. Department of Agriculture [62–66]. A very lengthy resumé of the work of a committee of the American Wood Preservers Association was issued in 1944 that updated the earlier work [67]. These references will provide the reader with a good picture of the work done up to the mid-1940s. There is also an early British review article [67a]. Much of this was on simple inorganic salts; almost nothing is recorded in these years on sophisticated *in situ* polymerizations or even on the use of urea or other weak nitrogenous bases to aid in reacting phosphoric acid with cellulose. These techniques were developed later. (The compounds evaluated in 1944 [67] were $Na_2Cr_2O_7$, $(NH_4)_2SO_4$, $(NH_4)_2HPO_4$, $NH_4H_2PO_4$, $Na_2B_4O_7$, H_3BO_3, $ZnCl_2$, and chromated $ZnCl_2$.) (For general references to pertinent wood chemistry and associated procedures, techniques, and tests, see Browning's work, reference 67.)

Table 4-4 summarizes the approaches taken to fire-retard wood. The

Table 4-4 Summary of Research on Fire Retardants for Wood

Chemicals	Treating Conditions	Reference
Phosphorus compounds		
Esters of H_3PO_4, $H_4P_2O_7$, $(HPO_3)_x$, H_3PO_3, H_3PO_2	Simple impregnation	68
H_3PO_4, melamine resin	Simple impregnation	69
$(\phi O)_3PO$	Simple impregnation in H_2O or alcohol	70
$(RO)_3P$ where R is alkyl, aryl, haloalkyl, haloaryl	20–80% in aliphatic solvent	71
$CH_2{=}CH{-}PO_3{-}R_2$ $CH_2{=}CH{-}CH_2{-}PO_3R_2$ $CH_2{=}CH{-}CH_2{-}PO_2R_2$ where R is alkyl, haloalkyl, allyl	Evacuate wood, add monomer, soak 8 hr, irradiate with cobalt-60 16 hr, hold 8 hr at 90°, evacuate 16 hr at 90°; add-on: 12–20% P	72
Alkyl arene phosphinic acids	Soak 24 hr at room temp, cure $\frac{1}{3}$–$\frac{2}{3}$ hr at 130–160°, wash with DMF	73
Phosphorus-nitrogen compositions		
$NH_4H_2PO_4$, $(NH_4)_2HPO_4$	(1) Soak wood in NH_3; (2) soak in H_2O solution of phosphate; (3) dry	74
$NH_4H_2PO_4$	(1) Steam wood; (2) evacuate; (3) add phosphate solution; (4) apply pressure; (5) kiln dry. (Basic description of process still in use)	75
$NH_4H_2PO_4$, $(NH_4)_2HPO_4$	(1) Impregnate in hot solution; (2) quench in cold phosphate solution	76
$NH_4H_2PO_4$, $(NH_4)_2SO_4$	Use in veneer; compatible with phenolic adhesives; use in wood with surfactant	77
$(NH_4)_2HPO_4$	Impregnate with small amount of Na_2SiF_6 and carboxymethyl cellulose	78
$NH_4H_2PO_4$, urea	Use 4–10 parts urea per part $NH_4H_2PO_4$; soak up to 100% wet add-on; dry; cure up to 170°, 2.5–3.0% $NH_4H_2PO_4$ Add-on; use on particle board, too	79
$NH_4H_2PO_4$, $(NH_4)_2HPO_4$, urea, CH_2O	Impregnate with H_2O solution, heat, cure; mix with wood particles and press into wood	80
$H_2NC({=}NH)NH_2$, H_3PO_4, CH_2O	Impregnate with H_2O solution, dry and cure	81
$H_2NC({=}NH)NHCN$, H_3PO_4 or $(NH_4)_2HPO_4$ or $(NH_4)_2SO_4$	Evacuate wood, treat with 20% H_2O solution at 160 psi and 50–60°, dry cure 24 hr at 70–100°; add-on, dry solids is ca. 20% of dry wood weight	82
$H_2NC({=}NH)NHCN$, H_3PO_4, CH_2O	As above	83
$H_2NC({=}NH)NHCN$, $NH_4H_2PO_4$, $NH_2(CH_2)_6NH_2$	Surface application	84

Table 4-4 (Continued)

Chemicals	Treating Conditions	Reference
$H_2NC(=NH)NHCN$, CH_2O, $(NaPO_3)_x$	Impregnated in H_2O solution	85
$NH_4H_2PO_4$, $(NH_4)_2HPO_4$, urea, $C(CH_2OH)_4$, CH_2O	Surface treat from H_2O solution, heat to polymerize; gives intumescent coating	86
Resin from $NH_2C(=NH)NHCN$ and CH_2O, P—N polymer	Impregnation in H_2O solution	87
Ammoniated polyphosphoric acid (acid at 79% P_2O_5)	Standard vacuum and pressure cycle, H_2O solution	88
NH_4 phosphates or $(NH_4)_2SO_4$, polyethylene glycols	Evacuate, then add H_2O solution, dry several days; claims improved leaching resistance	89
$Zn_3(PO_4)_2$, $Zn_3(AsO_4)_2$, NH_3	Treat with H_2O solution of the ammine complex; ppts of Zn polyphosphates and arsenates form on heat curing	90
$Zn(H_2PO_4)_2$, $NH_4H_2PO_4$ or $(NH_4)_2HPO_4$	Treat with Zn compound in H_2O, then H_2O solution of NH_4 phosphate, ppt insol Zn phosphate in wood; claims fire resistance	91
$\left(\begin{array}{c} CH_2 \\ \mid \quad \diagdown \\ \quad\quad N \\ \mid \quad \diagup \\ CH_2 \end{array}\right)_3 P=O$, also $(PNX_2)_x$ polymers	Impregnate in standard manner	92
$(\phi O)_2{-}\overset{\displaystyle O}{\overset{\displaystyle \|}{P}}{-}NHCH_2CH=CH_2$	Impregnate, polymerize by heat, free radical initiators or irradiation	93
$[P(CH_2OH)_4]^+X^-$, nitrogen base or amide, alkyd resin	Treat from H_2O solution, air dry, heat cure	94

Phosphorus-halogen compositions

$Cl_5\phi OH$, $(C_4H_9O)_3PO$ in aromatic petroleum oil with or without chlorinated paraffin	Impregnate in mineral spirits	95
$POCl_3$, with or without vinyl chloride	Evacuate, impregnate with dil solution in CH_2Cl_2, dry; add-on: 6–7%	96
$Cl_5\phi OH$, $Zn_3(PO_4)_2$, NH_3	2 steps: ammoniacal Zn phosphate solution, then phenol in chloro-paraffin solvent	97
$(ClCH_2CH_2O)_3PO$	Impregnation	98
$(ClCH_2CH_2O)_2PHO$ or $(CH_3CH_2O)_2PHO$, with or without NH_3	Evacuate, then impregnate with benzene solution, remove benzene (treat with dry NH_3 in one version of process)	99

Table 4-4   (Continued)

Chemicals	Treating Conditions	Reference
Bis(2-haloalkyl)alkenyl phosphonates; e.g., $(ClCH_2CH_2O)_2P(O)$ $\quad\quad CH{=}CH_2$ Bis(ω-monohaloalkyl) ω-monohaloalkane phosphonates; e.g., $(ClCH_2CH_2O)_2P(O)$ $\quad\quad CH_2CH_2Cl$	Impregnate in dichloro-benzene, aromatic mineral spirits, etc.; add-on: up to 2.5 lb/ft^3	100
Halophosphonate polymer	with methacrylate impregnant	100a
Halogen compositions		
NH_4Br	Simple immersion treatment	101
Br^-, BrO_3^- (yields Br_2) or $Br^- + Cl_2 \rightarrow Br_2$	Brief immersion in 23% H_2O solution at room temp, dry $\frac{1}{2}$ hr at 105°, then dip in $2NH_4SO_4$, then 5% aq borate; add-on: 6% Br	102
$CaCl_2$	Irradiate with Co-60 rays, then impregnate with solution of $CaCl_2$	103
$ZnCl_2$ with or without NH_3	Dissolve basic $ZnCl_2$ in aq NH_3, dil to washing strength (4.5–30%), use evacuation followed by pressure, then dry; add-on: 1–8 lb salts/ft^3	104
$ZnCl_2$, chromates	Evacuate, then apply pressure on 5% H_2O solution for 3–4 hr, 175 psi, 60°; add-on: 1 lb/ft^3	105
$MgCl_2$ or $MgBr_2$ with MgO; with or without other fillers	Essentially an oxychloride cement binder for wood particle board; used as slurry; board is hot pressed $\frac{1}{4}$ hr, 220 psi, 120°	106
$SbCl_3$	Impregnate in solution in CCl_4, amyl acetate, alcohol, etc.	107
$SbCl_3$, chlorinated rubber, or phosphates	Use with creosote	107a
Sb_2O_3, chloroparaffin	Coated wood chips with 5–7% Cl, 9–12% Sb_2O_3, then bond with resin into particle or chip board	108
$Cl_5\phi OH$, chlorocarbon	Use chlorocarbon residue from $CHCl{=}CCl_2$ mfr to replace half of usual mineral spirits as solvent	109
Chloroparaffin, Na_2SiO_3	Makes wood fiberboard using H_2O solution at high dilution, molded and cured 1 hr at 100–180°	110
Chlorolignin	Chlorinated lignin (13% Cl) in H_2O solution used to impregnate wood in usual manner	111

Table 4-4   (Continued)

Chemicals	Treating Conditions	Reference
Boric acid-borax compositions		
H_3BO_3, $Na_2B_4O_7$	Early references simple surface applications; later use vacuum-pressure cycle, 15–21% H_2O solution, $\frac{1}{2}$ hr, 145 psi, 65°; add-on: 5–8 lb salt/ft^3	112
H_3BO_3	Coat wood chips at 10% by weight, bond with resin, hot press to board	113
H_3BO_3, $(NH_4)_2SO_4$, dichromates, $CuSO_4$, $(NaPO_3)_x$	Impregnate from 16% H_2O solution	114
H_3BO_3, Sb_2O_3	Add wood chips to form particle board	115
$Na_2B_4O_7$, Na_2CO_3, Mg, Zn, Fe salt	Treat with H_2O solution of borax, then ppt borate by treating with solution of transition metal salt	116
$Na_2B_4O_7$, Na_2SiO_3	Impregnate with H_2O solution 1 M Na_2SiO_3, 0.5 M $Na_2B_4O_7$, then pressure treat with CO_2, 5 min, 500 psi	117
$Na_2B_4O_7$, HCl	Soak chips in H_2O solution at 75°, dry at 121°, then bond into board with resin	118
$Na_2B_4O_7$, H_3BO_3, $Cl_5\phi OH$	Treat with emulsion in petroleum oil	119
H_3BO_3, $ZnCl_2$, $CuSO_4$, $Cr_2O_7^=$	Treat with H_2O solution	120
H_3BO_3, $HOCH_2CH_2NH_2$, mole ratio 1:1	Treat chips with H_2O emulsion of bonding resins plus retardant composition, mold board under pressure	121
$(CH_3O)_3B$	Impregnate with methanol solution 24 hr at 25 psi, up to 65°; add-on; 1 lb/ft^3	122
Nitrogen, phosphorus, boron compositions		
NH_4 phosphates, H_3BO_3 chlorinated naphthalene	Form fiberboard admixing retardants, dry, treat with NH_3	123
$(NH_4)_2HPO_4$, $Na_2B_4O_7$, H_3BO_3	Treat with H_2O solution with weight ratio of 1:3:1 in order shown at left	124
$(NH_4)_2HPO_4$, $Na_2B_4O_7$, NH_4Cl	Treat in solution at neutral pH	125
$(NH_4)_2HPO_4$, $Na_2B_4O_7$, Na_3PO_4	Spray, immerse, or paint with H_2O solution	126
$NH_4H_2PO_4$, H_3BO_3, Na benzyl naphthalene sulfonate	Treat with 20–45% solution	127
Miscellaneous compositions		
Chromates with other metal salts, e.g., $CuSO_4$	Vacuum-pressure cycle or paint on with resinous binders	128
Sodium silicate	Used in alkaline solution in H_2O or as colloidal solution at low pH, bonding fibers with or without organic binders	129
$TiCl_2(OAc)_2$	Treat wood chips and bind with resins	130

Table 4-5   Some Commercial Fire-Retardant Mixtures

Chemicals	Proportion	Specification	Reference
$(NH_4)_2SO_4$	78	Type 1[a]	67
$NH_4H_2PO_4$ or $(NH_4)_2HPO_4$	19		
$(NH_4)_2SO_4$	60	$\cdots$	
H_3BO_3	20	$\cdots$	
$(NH_4)_2HPO_4$	10	Type 2	
$Na_2B_4O_7$	10	Minalith	
$Na_2B_4O_7$	60	Type 3	67
H_3BO_3	40	$\cdots$	
$ZnCl_2$	77.5	Type 4	132
$Na_2Cr_2O_7 \cdot 2H_2O$	17.5	CZC	
$ZnCl_2$	62	$\cdots$	132
$Na_2Cr_2O_7 \cdot 2H_2O$	15.5	Type 4	
$(NH_4)_2SO_4$	10	CZC(FR)	
H_3BO_3	10	$\cdots$	
$Na_2B_4O_7$	67–70	$\cdots$	64, 65
$NH_4H_2PO_4$	33–30	$\cdots$	
$ZnCl_2$	54	$\cdots$	65
$NH_4H_2PO_4$	46	$\cdots$	
$ZnCl_2$	35	$\cdots$	132
$(NH_4)_2SO_4$	35	Protexol Class D	
H_3BO_3	25	Pyresote	
$Na_2Cr_2O_7 \cdot 2H_2O$	5	$\cdots$	

[a] Bureau of Ships Ad Interim Specification 51C38 (INT), Apr. 1, 1943.

entries are classified by element and, in the case of phosphorus, in order of decreasing oxidation number. Certain literature citations are discussed individually in later paragraphs of this text. They are primarily papers containing good basic data or up-to-date comparisons among a variety of treatments.

Over the years various combinations of a handful of these chemicals have received close attention by researchers in the industry. In fact, the first commercial production of fire-retardant lumber in the United States was based on a mixture of $(NH_4)_2SO_4$ and $NH_4H_2PO_4$ (80:20) [131]. Some of the more popular mixtures are listed in Table 4-5 with proportions recommended for use. Although fire retardance can be achieved to a degree by simple surface application of the treating solution or by immersion of the wood, these methods are by and large unsatisfactory. In most cases the wood is placed in a treating vessel, the system evacuated to $\frac{1}{10}$ to $\frac{1}{4}$ of an atmosphere and held for $\frac{1}{4}$ to $1\frac{1}{2}$ hr. The treating solution is introduced and pressures up

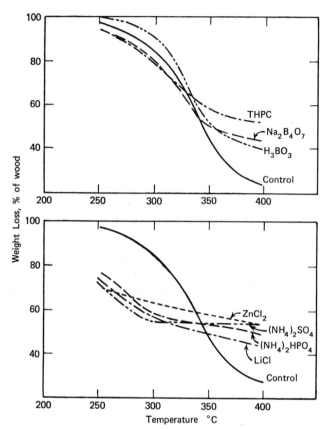

Figure 4-1 Thermogravimetric studies on ponderosa pine. Heated in N_2 at 6°C/min. Corrected for chemicals and moisture to show true effect of chemicals on the volatilization of wood [134].

to 175 psig are maintained for as long as 7 hr at temperatures up to 65°. The treated wood is then kiln-dried. Add-on or dry chemical retention varies depending on the wood species and the chemicals employed, but is generally 1–10 lb/ft³ of wood. Soft pine may weigh as little as 30 lb/ft³; certain oaks, over 70 lb [133]. Thus, the add-on on a weight percent basis ranges from 1.4 to 33%. Typical values for framing lumber would be in the middle of this range.

Fire-retardant lumber is evaluated not only for resistance to flame spread, afterglow, smoke generation, and the like, but also for permanence, effect on strength, resistance to decay, corrosive effects on nails, and so on. Let us now compare the performance of the various treatments from these several points of view. First are basic data obtained in recent years by quantitative thermal analysis, either TGA or DTA. This work was carried out in the 1960s at the Forest Products Laboratory. An examination of the weight loss versus temperature curves [134] illustrated in Figure 4-1 shows two types of

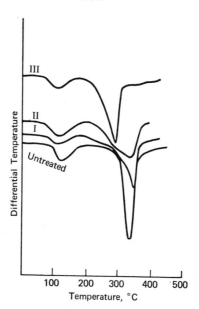

Figure 4-2 Differential thermal analysis of cellulose (Whatman). Helium atmosphere: I—2% Na$_2$B$_4$O$_7$·12H$_2$O· II—2% NaCl; III—2% NH$_4$H$_2$PO$_4$ [135].

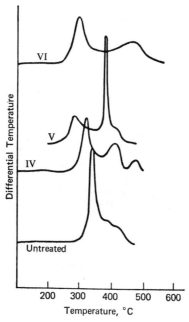

Figure 4-3 Differential thermal analysis of cellulose (Whatman). Oxygen atmosphere: IV—8% Na$_2$B$_4$O$_7$·10H$_2$O; V—8% NaCl; VI—8% NH$_4$H$_2$PO$_4$ [135].

behavior: one in which volatilization is significantly reduced at 400° but there is little effect at lower temperatures, and the other in which volatilization is reduced as much at 400° as for the first class but is noticeably increased at lower temperatures beginning at about 200°. DTA curves are shown in Figure 4-2 [135]. Note that $Na_2B_4O_7$ does not noticeably affect the location of the second endotherm but reduces the magnitude of it more than any other treatment. $NH_4H_2PO_4$, on the other hand, reduces the temperature at the second endotherm by almost 100° but does not reduce the magnitude of it nearly as much as does $Na_2B_4O_7$. Observe that the data in both Figures 4-1 and 4-2 are for pyrolysis; that is, the vapor space is filled with an inert gas. Very little quantitative work is reported in the literature on thermal decomposition in the presence of oxygen. The curves in Figure 4-3 are therefore rare entries. They show two exotherms, one at about 300°, another at 440–450°. The exotherm at 300° is combustion of volatiles; that at 440° is attributed to glowing combustion of the char remaining [135]. With boron compounds the flaming combustion is suppressed considerably but glowing of the char is, if anything, enhanced and a second glowing exotherm appears. Ammonium phosphates and sulfates reduce the flaming combustion at 300° and nearly eliminate the exotherms from glowing. All of this is in agreement with current belief that the sodium borates act primarily through formation of a physical barrier whereas the phosphates and sulfates act chemically to change the combustion pathways to less combustible products (see Chapter 1).

From quantitative analysis of data like those in Figures 4-1, 4-2, and 4-3, the effect of fire retardants on the activation energy and heat of pyrolysis or of combustion can be computed. Some results are shown in Table 4-6. Note

Table 4-6 Effect of Fire Retardants on Thermal Decomposition
Parameters for Wood (Ponderosa Pine) [134–136]

	Amount Salt, %	Temp Range, °C	Heat of Pyrolysis, cal/g	Amount Salt, %	Activation Energy, kcal/mole
Control	0	200–390	77	0	36
$Na_2B_4O_7$	2	200–395	31	17	25
$NH_4H_2PO_4$	2	180–315	45	15	33
$(NH_4)_2HPO_4$	⋯	⋯	⋯	11	32
$NH_4SO_3NH_2$	⋯	⋯	⋯	16	34

that sodium tetraborate sharply reduces both the activation energy and the heat of pyrolysis (no oxygen), whereas the phosphates and sulfamate have a somewhat lesser but still marked effect. The effective fire retardants therefore facilitate degradation reactions and lower the total heat required in the absence of air. The volatile products that are released in the presence of fire

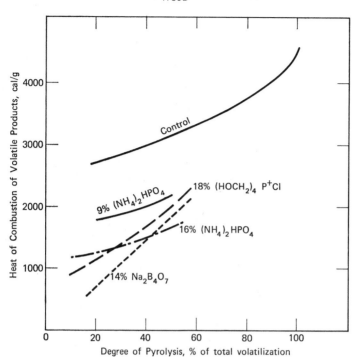

Figure 4-4 Heat of combustion of volatile pyrolysis products as function of degree of volatilization for ponderosa pine [137].

retardants provide less energy on combustion than the same weight of volatiles from untreated wood. This is shown in Figure 4-4 and is further substantiation for the theories that maintain that the action of the chemical fire retardants shown is to influence chemically the degradation reactions in the wood itself. The upshot of all this may be a great deal less heat released from a treated piece of wood. For one such case, the char from the treated piece was 37% of the original weight and only 22% for a comparable untreated piece.

Taking the total heat of combustion of wood as 4.610 cal/g (8300 Btu/lb) and of char as 7.230 cal/g (13,000 Btu/lb), the heat released is then [138]:

	Heat Remaining in Char, cal/g	Heat Released by Wood, cal/g
Untreated	1590	3020
Treated	2670	1940

Not only is the route of degradation altered but the amount of heat ultimately released may be reduced by 35% or so.

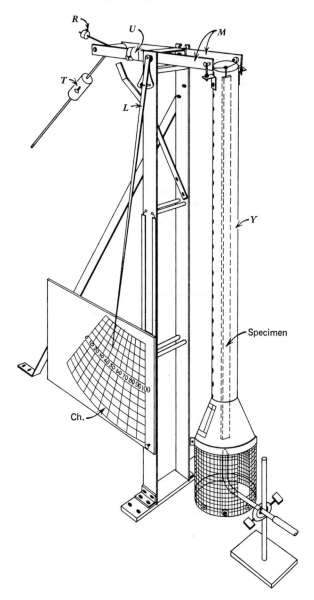

Figure 4-5 Drawing of the fire tube apparatus [62].

Flammability Ratings

In the 1930s and 1940s the fire tube test [139] was widely used to rate flame retardants for wood. Most of the screening studies employed this device. Since the 1950s the concept of flame spread has gained favor as measured in various flame tunnels leading ultimately to the 25-ft tunnel [140]. The Schlyter panel test [141] and the crib test [142] have also been used. Most of the data are from the fire tube test [143] which is, in essence, a type of thermobalance (see Figure 4-5). A key result is the weight loss at the end of the test. Observations are also made of the time at which flames reach the top of the tube, duration of flames, afterglow in seconds, and so on. In Figure 4-6 curves relating final weight loss in the fire tube to retention of fire retardant salts are shown. Note that most of this work was done on yellow pine but some was with oak and some with fir plywood. Figure 4-6 allows one to compare essentially all of the treatments that have commercial significance. If one assumes that the lower the weight loss, the better the flame retardant properties, then from Figure 4-6 it appears that the various ammonium

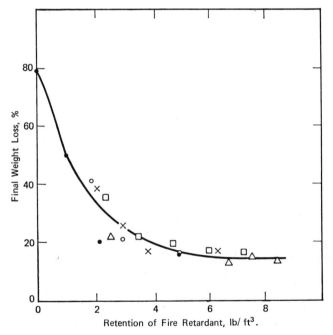

Figure 4-6a Final weight loss in the fire tube test as a function of the amount of fire retardant retained by treated wood. *Key:* ● $NH_4H_2PO_4$ fir plywood [145]; ○ $(NH_4)_2 HPO_4$ pine [67]; × $(NH_4)_2SO_4$ fir plywood [145]; △ 11-37-0 fir plywood [145]; □ Dicyandiamide-H_3PO_4 ponderosa pine [146].

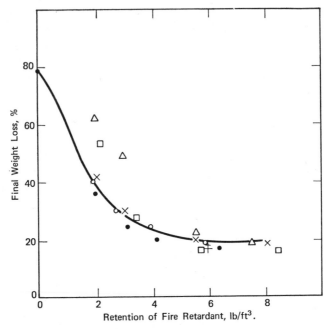

Figure 4-6b Retention of fire retardant, lb/ft³. *Key:* ● Na$_2$B$_4$O$_7$ fir plywood[145]; ○ H$_3$BO$_3$ fir plywood [145], × 60Na$_2$B$_4$O$_7$ 40H$_3$BO$_3$ yellow pine [67]; △ 40Na$_2$B$_4$O$_7$ 60H$_3$BO$_3$ yellow pine[67]; □ 67Na$_2$B$_4$O$_7$ 33NH$_4$H$_2$PO$_4$ yellow pine[65]; ‡ 40Na$_2$B$_4$O$_7$ 60NH$_4$H$_2$PO$_4$ yellow pine [65].

phosphates are all about the same with the one having the highest phosphorus content being the best, that is, NH$_4$H$_2$PO$_4$. (NH$_4$)$_2$SO$_4$ is about as good as (NH$_4$)$_2$HPO$_4$ but does not quite match NH$_4$H$_2$PO$_4$. Of the boron compounds, Na$_2$B$_4$O$_7$ and H$_3$BO$_3$ seem equally effective and about equal to the phosphates in terms of suppressing weight loss. Two of the mixtures are inferior at the lower retention levels. The ZnCl$_2$ combinations are only about two-thirds as effective as the phosphate, sulfate, or boron compounds. All the additives shown reduce weight loss by about a factor of 4 at the higher loadings.

Flame spread ratings in an 8-ft tunnel have recently been published for fir plywood and correlated with results in the fire tube and in the Schlyter test. The findings are shown in Figures 4-7, 4-8, and 4-9. This same work also dealt with smoke density and heat evolution. The results are in Figure 4-8. As might be expected, there is a strong correlation among the fire tube, flame tunnel, and Schlyter tests but it is not straightforward since these tests do not measure exactly the same properties. It is reasonable to assert that a fire-retardant treatment which gives excellent results in one of these tests will very likely be excellent in all three. Notice in Figure 4-7 that ZnCl$_2$, NH$_4$H$_2$PO$_4$, and H$_3$BO$_3$ tend to increase smoke density; in the case of the first two, very

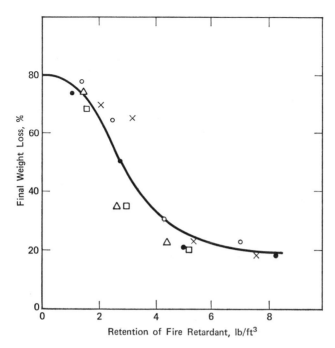

Figure 4-6c   Retention of fire retardant, lb/ft^3. *Key:* ● ZnCl$_2$ yellow pine [67]; ○ 77.5 min ZnCl$_2$, 17.5 min Na$_2$Cr$_2$O$_7$·2H$_2$O yellow pine [67]; × 54ZnCl$_2$:46NH$_4$H$_2$PO$_4$ yellow pine [65]; △ 80ZnCl$_2$-chromate:10H$_3$BO$_3$:10(NH$_4$)$_2$SO$_4$ oak [144]; □ 35ZnCl$_2$:35(NH$_4$)$_2$SO$_4$: 25H$_3$BO$_3$:5Na$_2$Cr$_2$O$_7$·2H$_2$O oak [144].

markedly. (NH$_4$)$_2$SO$_4$ and the borates decrease smoke compared to that from the untreated wood. All the treatments reduce the evolution of heat. The phosphates are perhaps the best of the agents at preventing afterglow. The borates and ZnCl$_2$ do not reduce afterglow except partially at higher levels of treatment. Thus, to achieve all the desired results—low weight loss, low flame spread, low smoke, minimum heat evolution, and no afterglow—a combination is needed. This explains the variety of commercial mixtures which have been successfully used over the years.

Effect on Strength, Moisture Resistance, Corrosion of Fasteners, and Decay Resistance

The particular end use of treated lumber dictates the relative importance of various physical and chemical properties affected by the addition of the fire retardant. Strength under various test conditions, corrosion of nails, screws, and the like, and resistance to decay are certainly of considerable

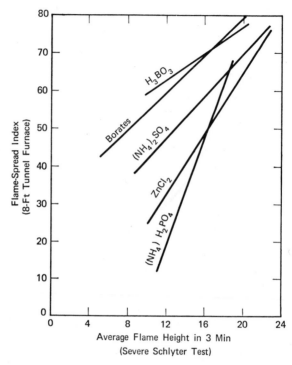

Figure 4-7a Correlation of flammability measurements made by the 8-ft tunnel furnace method and the severe Schlyter panel method on ⅜-in. Douglas fir plywood treated to different retention levels with elementary fire-retardant chemicals [145].

importance. The effect of typical loadings of fire-retardant salts on the modulus of rupture of fir is shown in Table 4-7 (p. 142) for the kiln-dried condition and after postheating to three temperatures. There is generally a reduction in MOR caused by all the treatments. It is least with borax and diammonium phosphate, most severe with AWPA Type D. In a detailed study of the effect of NON-COM™ fire retardant (trademark of the Koppers Co., Inc.), the U.S. Navy concluded [148] that the elastic modulus of treated wood may be reduced by 12% or so and the dynamic strength (impact) parameters may be reduced up to 20–40% for laminated beams and plywood. This reduction found with this proprietary, mixed-salt treatment was attributed primarily to the increase in equilibrium water content of the treated wood arising from the hygroscopicity of the retained salts. In contrast, the dicyandiamide-phosphoric acid process increases shear strength of yellow pine parallel to the grain by as much as 30% [146]. This increase is attributed to reaction with the cellulose (plus, perhaps, formation of less soluble nonhygroscopic products) on curing.

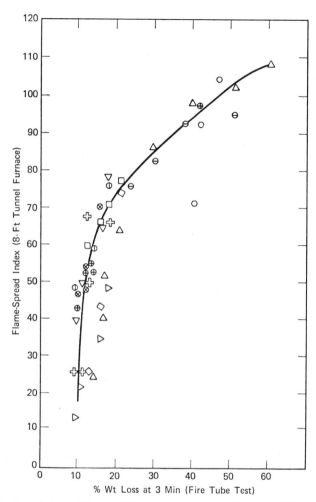

Figure 4-7b Correlation of flammability measurements made by the 8-ft tunnel furnace method and the fire tube tests on $\frac{3}{8}$-in. Douglas fir plywood treated to different retention levels with elementary fire-retardant chemicals [145]. *Key:* ○ NaCl; ⊖ $Na_2Cr_2O_7 \cdot 2H_2O$; ⦶ $Na_2O \cdot 2B_2O_3 \cdot 10H_2O$; ⊗ $Na_2O \cdot 4B_2O_3 \cdot 4H_2O$; ⊕ $Na_2O \cdot 5B_2O_3 \cdot 10H_2O$; □ H_3BO_3; △ $(NH_4)H_2PO_4$; ▽ $(NH_4)_2SO_4$; ◇ $ZnCl_2$; ▷ Ammonium polyphosphate fertilizer 11-37-0; ⊹ High phosphate fertilizer 18-46-0. (U.S. Forrest Products Laboratory Data reproduced by permission from *Fire Technology*, © National Fire Protection Association, Boston, Mass.)

Moisture retention is obviously important. Some of the commercial treatments are restricted as to location of use so as not to be exposed to high humidities. Very moist air will ultimately cause migration of the soluble salts to the wood surface and, finally, loss of the salts completely. For those salts that do not react on curing, predictions may be made (from basic properties)

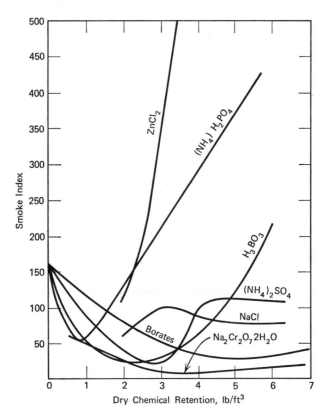

Figure 4-8a   Relationship of smoke density to level of chemical retention in ⅜-in. Douglas fir plywood evaluated by the 8-ft tunnel furnace method [145].

of their effect on the final equilibrium water content of treated wood and on its susceptibility to high humidities. Table 4-8 (p. 142) shows the relative humidities over saturated solutions of several of the fire retardants. The effect of mixing two or more of these salts together is not predictable *a priori*. There may be counteracting effects; for example, $NH_4H_2PO_4$ has a lower relative humidity over its saturated solution than does $(NH_4)_2HPO_4$, but is more acidic and will reduce strength of treated wood more. It is also less soluble. A mixture of the two salts may be the best compromise. One study of the practical result of this is shown in Figure 4-10, in which the dicyandiamide-phosphoric acid (D:P) treatment gives the best performance and zinc chloride the worst [146]. In Figure 4-11 various treatments are compared in actual outdoor weathering tests. The results are weight loss in the fire tube which is assumed to be proportional to leaching of the retardants [105]. The sulfate-phosphate-borate-boric acid formula leaches most readily; zinc chloride does not perform too

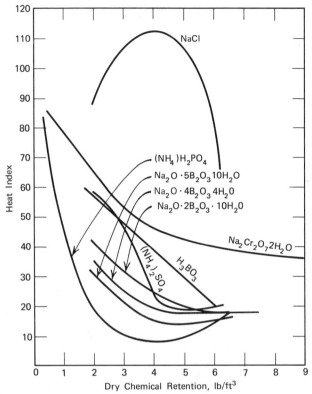

Figure 4-8b Relationship of heat evolved to level of chemical retention in ⅜-in. Douglas fir plywood evaluated by the 8-ft tunnel furnace method [145]. (U.S. Forrest Products Laboratory Data reproduced by permission from *Fire Technology*, © National Fire Protection Association, Boston, Mass.)

well either. However, application of a sealer greatly alleviates the problem. There has been one study of the effect of length of time of air drying on ultimate resistance to leaching. This dealt with preservatives, primarily, but there are limited data to show a reduction in leachability of zinc chloride with prolonged air drying prior to exposure (reduced from 45% leached at 1 day to 25% at 3 weeks) [149].

Leaching may also be reduced by adding large molecules such as polyethylene glycol [89] to entrap the fire-retardant salt physically or by adding dispersed insoluble retardants such as chloroparaffins which will plug the wood pores [149a].

Corrosion of metal fasteners is normally of concern for uses of wood members for structural purposes. Corrosion rates have been determined on 1 × 2 in. metal strips sandwiched between treated blocks of wood and exposed to high humidities at a high temperature. Table 4-9 shows the findings.

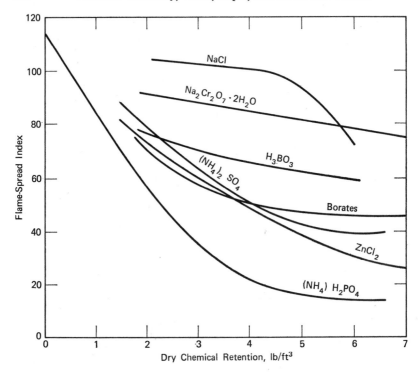

Figure 4-9 Relationship of flame spread to level of chemical retention in ⅜-in. Douglas fir plywood evaluated by the 8-ft tunnel furnace method.

Assuming an acceptable rate to be 10 mils/yr or less (this is a common target for industrial tanks, pipes, and equipment), then only the ammonium sulfate result on steel is of any real concern. Presumably, this can be reduced to a satisfactory rate by adding a corrosion inhibitor. It would appear that corrosion of fasteners should not be a problem. A combination of ammonium sulfate, phosphate, and sodium chromate (78:19:3) has been suggested to obtain corrosion resistance superior to either borax-boric acid or inhibited zinc chloride [150].

The effect of fire-retardant salts on the adhesives used in making plywood has also been studied [150a].

Studies of the decay resistance of wood treated with fire retardants show that the zinc chloride and boron compounds are excellent preservatives [132]. Ammonium phosphates and sulfate also serve as preservatives but are somewhat less effective than the zinc or boron compounds. In no case were any of these materials deleterious to the decay resistance of wood; all were beneficial [151].

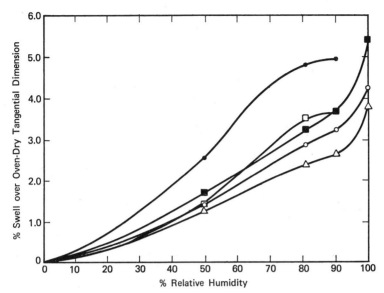

Figure 4-10 Swelling of untreated ponderosa pine and of similar wood treated with four fire-retardant salts when exposed to various relative humidities [146]. *Key:* ○ monoammonium phosphate; □ ammonium sulfate; △ dicyandiamide:phosphonic acid; ● zinc chloride; ▧ untreated.

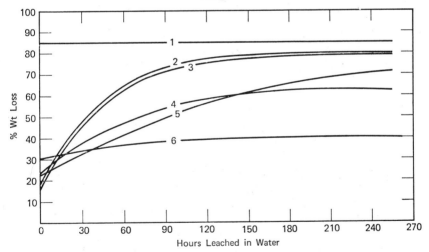

Figure 4-11 Small southern yellow pine sapwood sticks, $\frac{3}{8} \times 1\frac{5}{8} \times 28$ in., pressure-treated with solutions of fire-retardant salts in concentrations normally employed in commercial treatments. Treated sticks were dried and alternatively immersed in running water for 8 hr and air dried for 16 hr over a total period of 256-hr immersion and 512-hr air drying, or 32 days. Curves illustrate relationship between actual immersion time and fire tube weight loss [105]. Curve 1: Untreated. Curve 2: Federal formulation. Curve 3: Commercial fire retardant. Curve 4: Chromated zinc chloride (fire retardant). Curve 5: Chromated zinc chloride. Curve 6: Sealed chromated zinc chloride.

Table 4-7   Effect of Fire Retardants on the Modulus of
Rupture of Treated Wood [147]

Treatment	Kiln Dried	MOR, psi		
		200°	300°	400°
Control	10,208	11,519	10,954	11,256
$Na_2B_4O_7$	11,887	9,600	7,860	9,825
H_3BO_3	8,130	11,346	$\cdots$	5,580
$50Na_2B_4O_7:50H_3BO_3$	10,560	13,779	11,523	7,800
$(NH_4)_2HPO_4$	8,946	10,294	9,300	9,573
$(NH_4)_2SO_4$	7,739	10,575	6,861	10,300
AWPA Type D[a]	6,438	9,069	3,600	8,106

[a] Believed to be $ZnCl_2$, $(NH_4)_2SO_4$, H_3BO_3, $Na_2Cr_2O_7 \cdot 2H_2O$—35:35:25:5

Table 4-8   Vapor Pressure of Water over Solutions
Saturated with Several Fire-Retardant Salts

Salt	Concd Sat. Solution, Wt. %	Temp, °C	Rel. Humidity, % of sat.
$NH_4H_2PO_4$	27.2	20	93
$(NH_4)_2HPO_4$	37.3	20	88.2
$(NH_4)_2SO_4$	43.5	25	81
$ZnCl_2$	81	20	10
H_3BO_3	3.5	10	76
$Na_2B_4O_7$	2.5	20	99+

Table 4-9   Corrosion Rate on Three Metals [147]
Ten days, 95% Relative Humidity, 49°C (120°F)
(Mil-L-0019140B, Bureau of Ships Test)

Treatment	Corrosion Rate, mils/yr		
	Aluminum	Brass	Steel
None	0.14	0.42	13.35
$Na_2B_4O_7$	2.12	0.54	0.72
H_3BO_3	1.56	0.63	1.02
$50Na_2B_4O_7:50H_3BO_3$	1.59	1.23	3.42
$(NH_4)_2HPO_4$	2.08	4.25	1.96
$(NH_4)_2SO_4$	7.86	4.29	21.84

Future

Looking ahead, one can see a clear need for fire-retardant treatments which are permanent and which have the least detrimental effects possible at low cost. The dicyandiamide-phosphoric acid treatment seems to be a step in this direction. The result desired could perhaps be achieved with expensive chemicals like tris(aziridinyl) phosphorus oxide or with combinations of chemicals including tetrakis(hydroxymethylphosphonium) chloride and the like. These treatments are expensive under present price structures and are not widely used. It is the writer's view that the next generation of fire retardants for wood will not be developed for wood but will be adapted from new treatments for cotton textiles. In the textile market there is developing legislative pressure that is forcing rapid strides in low-cost, permanent treatments for cellulosics. Many of these will be adaptable to wood-treating conditions (indeed for paper, too).

It is rare that one sees publication of basic synthetic work directed to wood treatments. An exception is a recent synthesis of polymerizable phosphoramidate monomers for wood treating, for example [93]:

$$(\phi O)_2-\overset{\overset{\displaystyle O}{\|}}{P}-Cl + HN(CH_2CH{=}CH_2)_2 \xrightarrow[\frac{1}{2}\ hr,\ 5°]{benzene}$$

$$(\phi O)_2-\overset{\overset{\displaystyle O}{\|}}{P}-N(CH_2CH{=}CH_2)_2$$

Wood is impregnated with an alcohol solution of the monomer and polymerization is affected either by heating (130°, 1 hr) or by irradiation with a CO-60 source. The treated wood resisted 14 days of washing with cold water and remained fire resistant and exhibited no afterglow. One can visualize many variations on this theme, the obstacle being cost. Research efforts should be bent to obtaining such materials via low-cost processes. The raw materials are often inexpensive but the processing itself is inefficient and the yields are often poor. However, with larger volumes of sales in prospect, it is not unreasonable to expect suitable chemicals selling for $0.35–$0.50/lb and containing, say, up to 35% phosphorus or its equivalent.

PAPER

Early in this century substances such as $(NH_4)_2HPO_4$ and $(NH_4)_2SO_4$ were suggested as fire retardants for paper products [152]. In 1916 the possibility of using silicates and other insoluble metal salts was suggested [153]. For roofing paper, oil-soluble phosphate esters such as triphenyl or tricresyl phosphate were claimed in a patent in 1913 [154]. Boron compounds were used in this same period. Sometimes it would appear that there is nothing new under the sun; these same chemicals, 50 years later, are still the most commonly used fire retardants for paper products. That this is so is largely because the requirements in most cases have been for low-cost, temporary methods. In many uses of paper, this remains true today. The literature reflects this emphasis and deals almost entirely with simple salts, readily available, and easy to handle. The basic aspects of the chemistry of fire-retardant paper are essentially those of wood and will not be repeated here. Many of the applied papers and patents deal with both wood and paper; the preceding section on wood should be consulted by the reader concerned with paper in order to gain a fuller background. The literature has been reviewed uncritically in abstract form in two segments by the Institute of Paper Chemistry at Appleton, Wis. One covers the period up to 1940 [155]; the other from 1940 through 1958 [156]. In what follows selected references from reference 156 are used to supplement those found in the searches made for this book (and which cover the literature through 1968).

A host of treatments for paper to render it resistant to fire are given in Table 4-10. The treatments are classified roughly by the principal elements used, although this is not always easy to divine. Only enough information is presented to enable the reader to determine the extent of his interest. He is expected to seek the original references for details.

The most commonly used materials are ammonium sulfate and ammonium phosphates, with or without boron compounds. Most of the other treatments are higher in cost than these, and for large volume uses the high cost is not tolerable. In various special applications (certain molded and laminated products) the end result is a fairly expensive item with stringent performance requirements. In these areas higher-cost retardants may be used. An example would be molded electric insulators based on paper pulp.

Current uses for fire-retardant paper require principally temporary protection. Durable treatments resistant to repeated soaking with water or long-term exposure to high humidities are not needed in most cases. For such applications ammonium sulfate or phosphate with borax or boric acid is adequate and has the required low cost. A new area is developing, however, in which at least semipermanence is needed. This is the so-called disposables product line in which nonwoven fabrics based on paper play a key role. These are used in hospital gowns and bedding, in paper dresses, and the like, where, for example, a spilled liquid must not cause rapid leaching of the retardant. Indeed, some users are asking that such materials should retain fire-retardant properties through one or two mild launderings. These uses often are in products that come into contact with the skin and the retardants must therefore not cause irritation. A variety of approaches are being studied. Among these are the use of resinous binders to coat the fire-retardant particles and thus to give some resistance to leaching. Certain less soluble

Table 4-10   A List of Fire-Retardant Treatments Used for
Paper Products

Chemicals	Conditions	Reference
Phosphorus and phosphorus-nitrogen compositions		
$NH_4H_2PO_4$, $(NH_4)_2HPO_4$	For paper board: soak in conc solution of mixed salts 1–3 min, 110°–120°, dry	157
NH_4 phosphate or sulfate	Added to particle board with resins	158
NH_4 phosphate or sulfate	Added to sheet in dispersion of rosin in water	159
$NH_4H_2PO_4$, urea	Added as impregnant with resins	160
NH_4 phosphate or sulfate or sulfamate with H_3BO_3 or borax	10–50 $(NH_4)_2SO_4$, 10–50 NH_4 phosphate, 1–15 H_3BO_3, 1–5 $Na_2B_4O_7$ g/l. Impregnate paper with this solution	161
NH_4 or urea phosphates, polyols	4.4–8 moles urea, 1 mole H_3PO_4, 1.2 moles glycerol heated 13 min, 130–160°. 10% aq solution of this prod. applied to paper, dry at 60°	162
Urea or guanylurea phosphate	Used either in CH_2O resin or in alcohol solution (50–60°) for paper or board	163
Guanylmelamine pyrophosphate	Apply from separate solutions of melamine deriv. and pyrophosphate	164
Urea and pyrophosphoric acid	30$H_4P_2O_7$, 80 urea heated $\frac{1}{2}$ hr at 160°, add H_2O, chill, filter, use solution with resins dry and harden paper at 140°	165
Nitrogen resin salts of H_3PO_4	Insoluble phosphate for coating on paper	166
NH_4 salt of alkyl phosphate	Alkyl has 5C or less. Solution in H_2O	167
Ethyleneimine phosphate (or Br^- or $SO_4^=$)	50 polyethyleneimine, 50(75%)H_3PO_4. Use salt in hot water to add to Kraft pulp in beater	168
$POCl_3$, NH_3 reaction product	Add 3–25% of prod. to sol cellulosic solutions, then cast films	169
$\left(\begin{array}{c}CH_2\\ \mid \quad\quad N\\ CH_2\end{array}\right)_3 PO$	Add to Kraft pulp in beater	170
$Ca(H_2PO_3)_2$ or $MgNH_4PO_4$	Add with fibers and phenol-formaldehyde resin to make board at 165–200°	171
Phosphorus-halogen compositions		
$(ClCH_2CH_2O)_3PO$, $NH_4H_2PO_4$	Mixed with polyol and blowing agents to give intumescence on heating; used in paper laminates	172
$(HOCH_2)_4P^+Cl^-$, vinyl chloride resins	Used with or without Sb_2O_3; urea-formaldehyde resins, triethanolamine, impregnate and cure 2–15 min at 120–160°	173
$(ClCH_2CH_2O)_2\overset{\overset{O}{\mid}}{P}CH{=}CH_2$	Impregnate paper with at least 2% by weight	174

Table 4-10   (Continued)

Chemicals	Conditions	Reference
$(ClCH_2CH_2O)_2\overset{\overset{\displaystyle O}{\displaystyle \shortmid}}{P}CH_2CH_2Cl$	Impregnate from nonaqueous solvent	175
$(PNX_2)_x$, NH_3	Impregnate with H_2O solution; cure at 100–170°	176
$\begin{array}{c}CH_2\\ \diagdown\\ NH{\cdot}B \text{ where B is}\\ \diagup \quad H_3PO_4, HBr, \text{ or}\\ CH_2 \quad H_2SO_4\end{array}$	Impregnate with H_2O solution; e.g., with phosphate use 25% solution	177
$TiCl_4$, H_3PO_4, $SbCl_3$	Ppt insoluble salts by following impregnation with alkaline rinse	178
Miscellaneous nitrogen compositions		
$(NH_4)_2SO_4$	9–16% in crepe paper with asphalt and wax	179
$(NH_4)_2SO_4{\cdot}Al_2(SO_4)_3{\cdot}24H_2O$	Immerse Kraft board in molten bath of cpd of 180–189°	180
$NH_4SO_3NH_2$	10–40% add-on, amine salts give softening in addition to FR; used in nonwovens	181
Urea sulfamate	Used molten to treat paper, film	182
$NH_4SO_3NH_2$, with H_3BO_3, dicyandiamide	For Kraft paper, immersion process in 10% solution (based on $NH_4SO_3NH_2$)	183
Urea, $MgCl_2$	Used to impregnate paper with emulsions of high boilers from fatty acid distillation	184
Dicyandiamide	Used in coating paper	185
NH_4 glycol monosulfate	Used as FR agent for paper with softening effect	186
NH_4 or amine salts of sulfonic acid esters of glycolic amides	Immersion of paper in H_2O solution	187
Guanidine thiocyanate	Applied as NH_4CNS and guanidine carbonate	188
Boron compositions		
H_3BO_3	Surface treatment with thickened H_2O solution; cure at 115°	189
$Na_2B_4O_7$	6–9% add-on in fiber slurry for board making; cure at 160–200°	190
$Na_2B_4O_7$, H_3BO_3	Use as surface coating (or as integral part of board slurry at 8% add-on) with resinous binders	191
ZnB_4O_7	Used in phenolic resin varnishes for molding or laminating paper	192
$Na_2B_4O_7$, $NH_4SO_3NH_2$	Use equal parts each in 30% H_2O solution (35–50°) to coat paper, then seal with bitumen	193

Table 4-10   (Continued)

Chemicals	Conditions	Reference
4NaF·5B$_2$O$_3$·5H$_2$O	Use in H$_2$O solution as surface coating or impregnant for paper; 8% add-on to plywood makes it FR	194
NH$_4$BO$_2$, RCl R = benzyl-, benzylidene-, Clϕ-	Use in molding cellulosics	195
Halogen compositions		
Chloroparaffins	Use in waste paper pulping with Al(OH)$_3$ followed by methylol urea binder system	196
Chloroparaffin, pentachlorophenol, cresyl diphenyl phosphate	Impregnate cellulosic with 8–20% paraffin (40+% Cl), 3–8% "penta," 3–10% CDP	197
Poly(vinyl chloride)	Use in electric insulators with phenolic resins; in paper sheets as surface coating	198
Polyamine·HBr	Used as surface treatment for fibrous products	199
Chlorocarbons, metal oxides	Applied as coating as organosol or plastisol plus oxide powder	200
Tetrachlorophthalic anhydride	Used in making paper dielectrics with polyol, resins	201
Haloalkylcarboxylates	E.g., bis(dibromopropyl) phthalate is added to paper	202
Chloronaphthalene or chloroanthracene	Used in emulsion, at least 30% Cl in the organic	203
Antimony compositions		
Sb$_2$O$_3$, Chlorowax 70	Applied from solvent or from water-mineral spirits emulsion; add-on: up to 25%	204
Sb$_2$O$_3$, vinylidene chloride copolymers	Water emulsions used to treat paper	205
Sb$_2$O$_3$ (or As, Sn, Bi oxides) used with chloroorganic	Add to pulp in beater	206
Sb$_2$O$_3$, chlorocarbon (40+% Cl)	Covers parchment paper, added to pulp, sheet surface treated with vinyl chloride polymers	207
Sb$_2$O$_3$, poly(vinyl chloride)	Added to pulp in chest, followed by alum	208
SbCl$_3$, tertiary amines	Complexes with ethanolamines added to paper from H$_2$O solution, subsequent heating ppts Sb	209
Acid solution of Sb (Zr or Ti)	Add as soluble acid solution, neutralize with NH$_3$	210
Silicate compositions		
Na Silicate	Use with ZnO in H$_2$O spray for insulation and soundproof board on binder	211

Table 4-10   (Continued)

Chemicals	Conditions	Reference
Na or K silicates	Spray on paper, then spray H_2SO_4 or HCl to yield SiO_2	212
Na silicate, agar-agar or gums	Treat with 20–30% silicate	213
Na silicate, alum, with NH_4 sulfate, phosphate, chloride, $ZnCl_2$	Make board with best results at 15% Na silicate	214
Colloidal SiO_2, N bases	Use solution of SiO_2 with amines or quaternary N cpds to coat paper	215
Si oxyamide-formaldehyde polymer	Impregnate Kraft paper	216
Miscellaneous compositions		
Titanium chlorides	Treat paper with H_2O (colloidal) sol., make alkaline, dry	217
Ti sulfate, Al sulfate, Na borate, Na silicate	Impregnate with H_2O solution	218

retardants, among them the insoluble ammonium polyphosphate discussed in Chapter 2, p. 35, are being tried with some success. The market is not yet sophisticated enough to be able to use compounds of the type typified by tetrakis(hydroxymethyl) phosphonium chloride. It now appears that new materials capable of reacting with cellulose without impairing its physical properties must be developed. They will have to be priced in the range of $0.10–0.25/lb (assuming fire-retardant performance equal to, say, mono-ammonium phosphate).

There are several ways of adding fire retardants to paper products [219]. Very insoluble ones can be added in the wet end of the paper machine, that is, to the pulp. This is handled in the usual manner for pigments or filler additions. Normally, for this type of addition the retardant will be antimony trioxide usually with a chlorocarbon that is water-insoluble and perhaps an insoluble phosphorus compound. Retention aids such as alum, polyethylene-imine, and the like, will usually help to improve efficiency. Soluble fire-retardant salts can be added to the mat in boardmaking at the Fourdrinier machine after 90–95% of the free water has been removed. A method has been described [220] in which a concentrated aqueous solution of phosphate/sulfate/fluoride is flooded over the mat so as to leave 5–15% salt (dry board basis).

Some retardants can be added to paper on the size press if the sheet is reasonably porous and not too heavy. Either solutions or suspensions may be used (suspensions may require agitation). Losses via this method are small.

Saturation and coating may be carried out either on or off the machine. Saturation requires a porous sheet and, probably, separate off-machine equipment. The technique often permits greater add-on than can be achieved on the size press or by spraying. Control of addition is relatively easy (by control of solution or suspension concentration). Coating is done in the usual way and is eminently suited for medium weight paper that is to be coated in any event. Board products will not receive sufficient add-on from a simple coating but may be protected by an intumescent coating. Coating lightweight papers tends to make them too stiff for many purposes.

The effect of two types of fire-retardant systems on properties is shown in Table 4-11. It is clear that there is little detrimental effect and some benefit in some cases.

Table 4-11 Effect of Fire Retardants on Sheet Properties [219]. (Reprinted by permission of *Tappi.*)

Soln Bath Conc, % solids	Pickup, lb chem/ 100 lb paper	Tear Factor, tear/basis wt	Burst Factor, burst/basis wt
Inorganic Salts			
0	· · ·	0.51	0.68
5	11	0.49	0.64
10	24	0.46	0.54
15	39	0.49	0.44
20	53	0.53	0.44
Polymeric Retardant			
0	· · ·	0.51	0.68
5	11	0.70	0.86
10	23	0.80	0.98
15	36	0.83	1.04
20	48	0.91	1.10

It is worth pointing out here, too, that new compounds for paper will probably result from intensive synthesis and application efforts now directed toward cotton textiles. The reader is urged to consult the basic section at the beginning of this chapter as well as Chapter 5 on cellulosic textiles.

FOREST FIRE CONTROL

The use of chemicals to combat forest and brush fires was first studied in the 1930s at the same time that the voluminous researches on retardants for wood were carried out. Barrett [220] and Truax [221] published papers on forest fire control in this period. Truax found $NH_4H_2PO_4$ to be the best of the materials studied but suggested that the use of chemicals should be limited to areas where water was very scarce. In the 1950s the feasibility of using aircraft to drop fire-fighting liquids on or ahead of fires was demonstrated. Further studies of various chemical agents were stimulated by this development. There were four basic systems: water, thickened water, retardant solutions, and thickened retardant solutions. The first two are short-term or temporary systems and arrest or slow a fire by removing heat through evaporation of water. The third and fourth are longer-term systems and combine the effects of water evaporation and the effect of the retardant on pyrolysis and combustion. All four have found merit and have been used on actual forest fires.

Today plain water is used both from ground and air equipment. Ground tankers use virtually nothing else, although at this writing vigorous efforts are underway to develop efficient logistics for use of chemicals in ground tankers. Water is also used to a limited extent in aircraft. In the United States, Canada, and Europe some seaplanes are fitted with scoops. Refilling the plane's tank is a very fast operation, involving little more than a sea landing and take-off in rapid sequence. More efficient action is obtained with water containing either surfactants or thickeners (or both).

The surfactants promote penetration of the water into the fuel and thereby reduce runoff to the ground. The thickening agents reduce runoff and permit formation of fairly heavy layers of water on the fuel. Obviously, the more water per unit of fuel, the greater the cooling effect will be. Studies at Syracuse University [222] indicate that thickened water is up to four times as effective as plain water. Common thickeners [223] are gums and polymers—carboxymethylcellulose [224], guar gum, other polysaccharides [225], various acrylamides [226] and their copolymers—and clays (e.g., bentonite). Since these are not, strictly speaking, chemical fire retardants (they operate on a physical principle), they will not be discussed further here.

A variety of chemical retardants have been evaluated. Some less likely materials are thickened calcium silicate [227], a suspension of calcium phosphate [228] (slimes from phosphate rock mining), and potassium carbonate [220]. (The last named is currently finding use in fire extinguisher formulas for use on military and building fires; the other two are not in use.) Barrett [220] found 1 gal of K_2CO_3 solution equal to 2 gal of plain water.

Aerial studies were begun by the U.S. Forest Service in the 1930s and continued in the 1940s [229]. In 1954 Operation Firestop was undertaken [230]. This work culminated in the introduction of sodium calcium borate suspension for aerial bombing [230a]. Difficulties with this material—abrasiveness and soil sterilization primarily [231]—caused further studies to be conducted. Out of this work came the ammonium phosphate and ammonium sulfate systems, currently in use in the western United States, Canada,

Australia, and Europe. This development work is summarized well in several publications [223, 232].

Basic work in support of this effort has been carried out in other Forest Service laboratories, notably those at Macon, Ga., and Missoula, Mont. In work at Macon the following performance was found on wood crib fires [233]:

	Wt Units for Result Equal to $(NH_4)_2HPO_4$
$(NH_4)_2HPO_4$	1.0
$NH_4SO_3NH_2$	0.6
11-37-0 liquid ammonium polyphosphate	1.6
$(NH_4)_2SO_4$	2.0
Na, Ca borate	5.0

For use in aircraft, the weight factor may be critical; the data provide another clue as to the reasons for abandoning the borates. Ammonium sulfamate is apparently too expensive for this use. Thus, the ammonium phosphates and sulfate are the materials of choice.

In certain areas simple fertilizer solutions are used for this purpose. An ammonium polyphosphate solution [145] (11% N, 37% P_2O_5:11-37-0) has been used. Its viscosity at 20° is about 100 cp, much greater than that of water. Retention on the fuel will therefore be higher than with a solution of orthophosphate or sulfate. Another viscous fertilizer solution has been proposed for use on fires [233a]. It is obtained by adding calcium ion to ammoniated fertilizer-grade phosphoric acid. Precipitates of insoluble iron, aluminum, and calcium of colloidal size provide the necessary viscosity. Neither this nor the 11-37-0 solution makes provision for corrosion control, colorants, and other more sophisticated needs developed by extensive experience in the western states.

The most successful systems are proprietary mixtures based on $(NH_4)_2HPO_4$ or $(NH_4)_2SO_4$. These are provided as dry mixtures and contain, typically, a large proportion as the retardant and minor amounts of thickening agents, coloring matter, and corrosion inhibitors. In 1968 two products were predominant in the western United States: one based on $(NH_4)_2HPO_4$, the other based on $(NH_4)_2SO_4$. The $(NH_4)_2HPO_4$ system contained a polymeric thickener (carboxymethylcellulose) [234], the $(NH_4)_2SO_4$ retardant contained a clay thickener [235]. Both were colored red with iron oxide and both contained proprietary corrosion inhibitors. The thickener is required for two reasons: (a) to hold the liquid together during the air drop, and (b) to maintain a thick film of retardant on the fuel. The flow properties of the retardant liquid are important to its successful use. Figure 4-12 shows the

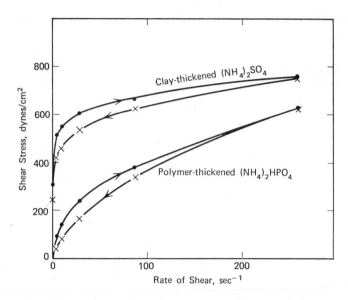

Figure 4-12 Flow curves for two commercial products for air attack on forest fires, showing the dependence of viscosity (stress/rate of shear) on rate of shear. The clay-thickened product exhibits a yield value or resting structure. Data obtained with Roto-visco viscometer. (Cup radius, 2.1 cm; gap, 0.13 cm; temperature, 22°C; data uncorrected— plotted rate of shear as calculated from Newtonian relations.) (Unpublished data from W. W. Morganthaler, Monsanto Co., St. Louis, Mo.)

shear stress versus rate of shear (flow curve) for the two systems. That thickened with clay exhibits a yield value (resting structure) which must be overcome by application of a finite amount of stress before flow begins. The system thickened with a polymer does not have a yield value. Both exhibit a viscosity which is markedly dependent on the rate of shear. Presumably one needs to know something of the viscosity at rates of shear which will prevail during the drop and stresses which govern once the fluid is on the fuel (e.g., gravity stress). The ideal has not yet been defined, but the foresters using the products have established some guidelines. The clay-thickened product is required to be typically 1400–3000 cp. when measured at low shear with a Brookfield viscometer; the system thickened with polymer is specified to give a value in the range of 1000 to 1500 cp.

The colorant is added for visibility to provide the pilot with guidance for succeeding drops. Usually the retardant is not dropped directly on a fire but, rather, ahead of the fire to form a line. Visibility is obviously critical. Iron oxide, pigment grade, is the preferred colorant and was used in both commercial products in 1968 and 1969.

The retardant fluid is prepared in metal equipment, pumped by metal pumps, and carried in metal aircraft. Substantial corrosion, especially of the aircraft, cannot be tolerated. Corrosion inhibitors are therefore added. The sulfate system requires an inhibitor for steel; the phosphate actually is an inhibitor for steel. (It has been claimed [235a] that mixtures of phosphate and sulfate are less corrosive than either one by itself.) Both require inhibitors to protect brass, copper, and aluminum. Various materials have been described in the patent literature [235]. For best results, a mixture of several different and specific inhibitors may be required.

It is common to prepare in advance many thousands of gallons of finished retardant fluid so that aircraft may be loaded rapidly on demand. If, at the end of the season, the storage tanks are full, then it is important to have a stable fluid, that is, one with no settling or degradation of the components. A "soft" settle is acceptable if the original properties can be restored by recirculation and mixing. The polymeric thickeners may be subject to biological degradation. Preservatives are commonly added to minimize this effect. The clay-thickened products generally exhibit some settling; the polymer-thickened products, some viscosity loss on storage over the winter months. For the two commercial products these effects have been controlled to a tolerable degree.

The addition of surfactants to such a system has been proposed [236] but has not been adopted in the United States. The reason is in part because the surface tension of such fluids cannot be lowered as dramatically as can that of plain water and retention of the retardant on the fuel by viscosity control appears adequate.

Figure 4-13 shows results of a test drop using the commercial phosphate product and indicates the kind of experimental work that has been carried out on this use of fire retardants. Determination of the variation of the drop pattern as the viscosity and shape of the flow curve are altered will ultimately allow these researchers to tailor a near-ideal product. And this product ideal will vary depending on whether fixed-wing craft or helicopters are used. Such studies are now active and are models of efficient and close cooperation among federal, state, municipal, and industry groups.

Such cooperation is required not only in developing chemical formulations but also in developing hardware for preparing, storing, and delivering the retardant to the fire. Equipment is being developed which will convert dry retardant powder in bulk silos to finished solution or suspension ready for loading into a plane at 500 gal/min. Units are being demonstrated which will allow the powder to be delivered by helicopter to remote mixing stations where finished retardant fluid may be prepared using water from a brook or stream; the pump and mixer (eductor) are also delivered by air. Some of the equipment and associated technology developed in the last few years are described in recent publications [223, 232, 237, 238].

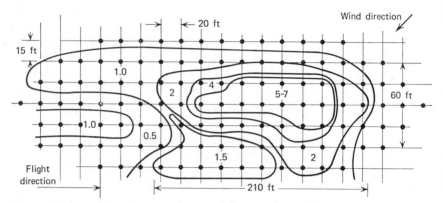

Figure 4-13 Drop-pattern/concentration relationship, location of test cups, and pattern. *Key:* Numbers represent gal/100 ft² coverage. *Conditions:* Solution viscosity, 1500 cps; drop speed, 120 mph; drop height, 88 ft; wind speed, 9 mph; temperature, 63°F.

The use of these same chemicals in fighting fires from the ground is becoming common practice as new logistical concepts are worked out. Since 1 volume of retardant solution is equal to 4 or more volumes of water on a fire, it makes sense to use retardant wherever water is hard to obtain. This advantage must be weighed against the additional cost. The goal is to have a tanker fitted with a reservoir of dry retardant powder and a system which can, on demand, be switched from plain water to retardant solution. This would require a foolproof mixer which, ideally, could work against the back pressure of a spray nozzle; the latter requirement is not essential but would add to the unit's flexibility. Such problems are currently under active study and more elaborate, yet dependable, systems may be anticipated.

An additional use for ground-based retardants is for control of deliberately set fires. These are most often encountered in disposal of slash from logging and in burning along certain rights of way. Spraying fire lines with retardant solution is an inexpensive substitute for bulldozing swaths around piles of slash. Special retardants have been developed for this use, which differ only in a somewhat lower cost because requirements for storage stability of the finished solution are less stringent than for retardants for wild-fires.

The viscosity of solutions of $(NH_4)_2HPO_4$ thickened with polymeric agents runs from 50 to 100 cp [238] for use for either controlled burning or wildfires. A solution at this viscosity is about as thick as can be sprayed through a nozzle and still give a suitably fine dispersal of droplets. Fairly thick layers are retained on the fuel though not as thick as with the air-drop products. This is partly overcome by raising the level of retardant salt in solution from about 10% for air drops to about 15% for ground use. Table 4-12 lists some properties of the two classes of commercial products.

Table 4-12 Product Data on Two Classes of Commercial Forest Fire Fighting Chemicals [223, 238, 239]. (Reproduced in part by permission from "Chemicals for Forest Fire Fighting", second edition, copyright 1967 National Fire Protection Association, Boston, Mass.)

| Solution Property | $(NH_4)_2HPO_4$ [a] | | $(NH_4)_2SO_4$ [b] |
	Air Drop	Ground	Air Drop
Salt content, %	10.5	15.0	16
Density, lb/gal	8.9	9.1	9.3
Solution pH	8.0	7.9	6.5–7.0
Solution viscosity, cp[c]			
Initial	1000–1500	50–100	1400–3000
After 24 hr	1200–1700	50–100	1400–3000
Corrosion, mils/yr			
Copper	<1	<1	1–2
Aluminum	<1	<10	<1
Steel	<1	<3	Not reported
Magnesium	<1	0	Not reported

[a] Solutions made from PHOS-CHEK® fire retardants, products of Monsanto Co.

[b] Solutions made from FIRETROL® fire retardants, products of Arizona Agrichemical Corp.

[c] Measured at low rate of shear with Brookfield viscometer

A problem which remains at least partially unsolved is that of preventive coatings for brush, tall grass, and so forth, along roadsides, around campsites, and in other regions of high fire risk. Current practice—and it is very limited—is to apply a solution of $(NH_4)_2HPO_4$ or similar retardant and let it dry. In California this is done as soon as the growing season is over and the grass has turned brown (early summer). The difficulty is that rainfall will wash the retardant off the fuel and it must be replaced. There is a clear need for what might be called a semipermanent retardant for spray application on such areas. The retardant would persist through normal rainfall for the entire fire season. One obvious solution would be to use suspensions of water-insoluble retardants, perhaps with latices to serve as binders in a manner analogous to current practice with various agricultural sprays. Certain organophosphorus compounds might serve the purpose, as might the "insoluble" ammonium polyphosphate discussed in Chapter 2, p. 35. Encapsulation of the low-cost salts with organic resins has been studied by Raff et al. [240], but the tiny pellets produced by the technique did not give adequate coverage and the fire burned around them. These investigators have suggested use of emulsions of polymers made from vinyl or allyl phosphorus monomers.

The cost of these alternative treatments will be higher per gallon than for the presently used water-soluble salts. This must be balanced against the need for repeated applications of the low-cost materials. The Forest Service is studying these concepts at this writing in several laboratories in the United States. It is probable that an economical solution will be available within a year or two as a result of such efforts.

REFERENCES

[1] Some good general references on cellulose chemistry are given in references 2–4.

[2] E. Ott, H. M. Spurlin, and M. W. Grafflin, eds., *Cellulose and Cellulose Derivatives*, 2nd ed., Wiley-Interscience, New York, Parts I, II, 1954, Part III, 1955.

[3] L. Roth and J. Weiner, Chemical Modification of Cellulose, *Bibliographic Series No. 228*, Inst. of Paper Chemistry, Appleton, Wis., Parts 1, 2, 1966, Part 3, 1967.

[4] L. F. Fieser and M. Fieser, *Organic Chemistry*, 2nd ed., Heath, Boston, 1950 Chapter 15.

[5] C. Quivoron and J. Neel, *J. Chim. Phys.*, **62**, 83 (1966).

[5a] C. P. Wade, E. J. Roberts, and S. P. Rowland, *Polym. Lett.*, **6**, 673 (1968).

[5b] M. Chene, *Papeterie*, **88** (11), 1434 (1966).

[6] Reference 2, Part I, pp. 296 ff.

[6a] J. O. Warwicker and A. C. Wright, *J. Appl. Polym. Sci.*, **11**, 659 (1967).

[7] Reference 2, Part I, p. 276.

[8] Reference 2, Part II, pp. 874 ff.

[9] K. Hamaguchi and E. P. Geiduschek, *J. Amer. Chem. Soc.*, **84**, 1329 (1962).

[10] Reference 2, Part II, pp. 825 ff.

[11] Reference 2, Part II, pp. 862 ff.

[11a] S. Okajima and A. Kai, *J. Polym. Sci., Part A-1*, **6**, 2801 (1968); T. L. Vigo, R. H. Wade, O. Mitcham, and C. M. Welch, *Text. Res. J.*, **39**, 305 (1969).

[12] Reference 2, Part I, pp. 99 ff.

[13] Reference 2, Part I, pp. 174 ff; P. K. Chatterjee, *J. Appl. Polym. Sci.*, **12**, 1859 (1968).

[14] Reference 2, Part I, pp. 140 ff; P. J. Kaugle, G. M. Nabar, *J. Appl. Polym. Sci.*, **12**, 2533 (1968).

[15] S. L. Madorsky, *Thermal Degradation of Organic Polymers*, Wiley-Interscience, New York, 1964, Chapter XII.

[16] Reference 2, Part II, pp. 713 ff.

[17] E. D. Klug, "Cellulose Derivatives," in *Encyclopedia of Chemical Technology*, 2nd ed., Vol. 4, Wiley-Interscience, New York, 1964.

[18] Reference 2, Part II, pp. 882 ff.

[18a] J. E. Heath and R. Jeffries, *J. Appl. Polym. Sci.*, **12**, 455 (1968).

[18b] G. V. Nikonovich, T. Saidaliyev, Yu. T. Tashpulatov, and Kh. V. Usinanov, *Vysokomol. Soedin.*, **A10**, (4), 960 (1968); J. G. Frick, Jr., *Amer. Dyestuff Reptr.*, **56** (18), 79 (1967).

[19] Reference 2, Part II, pp. 760 ff.

[20] W. L. Tanner, *U.S. 1,896,725* (to National Chemical and Manufacturing Co.) Feb. 7, 1933.

[21] G. P. Touey, *U.S. 2,759,924* (to Eastman Kodak Co.), Aug. 21, 1956.

[22] D. W. Gallagher, *Amer. Dyestuff Reptr.*, **53**, 23 (1964).

[23] C. A. Thomas and G. Kosolapoff, *U.S. 2,401,440* (to Monsanto Chemical Co.), June 4, 1946; J. D. Reid and L. W. Mazzeno, Jr., *Ind. Eng. Chem.*, **41**, 2828 (1949); J. D. Reid, L. W. Mazzeno, Jr., and E. M. Buras, Jr., *ibid.*, 2831.

[23a] C. J. Malm and C. E. Warin, *U.S. 1,962,827* (to Eastman Kodak Co.), June 12, 1934.

[24] A. C. Nuessle, F. M. Ford, W. P. Hall, and A. L. Lippert, *Text. Res. J.*, **26**, 32 (1956); R. Schiffner and G. Lange, *Faserforsch. Textiltech.*, **8**, 435 (1957) (in German); E. L. Donahue, *U.S. 3,253,881*, May 31, 1966; K. Katsuura and S. Nonaka, *Sen-i Gakkaishi*, **13**, 24 (1957); *ibid.*, 28; K. Katsuura and H. Mizumo, *ibid.*, **22**, 510 (1966) (in Japanese); R. A. Wells, A. J. Head, and N. F. Kember, *Brit. 899,284*, June 20, 1962. See also references 9 and 20. Chapter 5, for Little's work.

[25] Z. A. Rogovin, M. A. Tuganova, J. G. Krjazhev, and T. J. Zharova, *U.S. 3,391,096* (to Moscoswky Tekstiljny Institute, Moscow, USSR), July 2, 1968.

[26] I. N. Ermolenko and L. A. Churkina, *Dokl. Akad. Nauk Beloruss. SSR*, **3**, 11 (1959) (in Russian); I. N. Ermolenko and F. N. Kaputsky, *J. Polym. Sci. USSR*, **53**, 141 (1961).

[27] R. R. Dreisbach and J. L. Lang, *U.S. 2,900,279* (to Dow Chemical Co.), Aug. 18, 1959.

[28] R. F. Schwenker, Jr., and E. Pascu, *Text. Res. J.*, **27**, 173 (1957); *Ind. Eng. Chem.* **50**, 91 (1958); *U.S. 2,990,233* (to Textile Research Inst.), June 27, 1961; Textile Research Inst., *Brit. 873,555,873,556*, Appl. Dec. 17, 1957; Y. Yulashev, R. V. Perlina, M. M. Sadykov, and Kh. U. Usmanov, *Vysokomol. Soedin.*, **8**, 231 (1966) (Eng. trans.).

[29] R. G. Zhbankov, R. Marunov, U. Mei-yan′, M. A. Tyuganova, and Z. A. Rogovin, *Vysokomol. Soedin.*, **5**, 1292 (1963).

[30] M. L. Nielsen, *Text. Res. J.*, **27**, 603 (1957).

[31] G. L. Drake, Jr., L. H. Chance, J. V. Beninate, and J. D. Guthrie, *Amer. Dyestuff Reptr.*, **51**, 40 (1962); Dow Chemical Co. (by R. B. LeBlanc), *Belg. 666,043*, Dec. 28, 1965.

[32] G. L. Drake, Jr., W. A. Reeves, and J. D. Guthrie, *Text. Res. J.*, **29**, 270 (1959); S. R. Hobart, G. L. Drake, Jr., and J. D. Guthrie, *ibid.*, 884.

[33] S. R. Hobart, G. L. Drake, Jr., and J. D. Guthrie, *Amer. Dyestuff Reptr.*, **50**, 30 (1961).

[34] R. Schiffner and G. Lange, *Faserforsch. Textiltech.*, **9**, 417 (1958); *ibid.*, **11**, 276 (1960).

[35] U. Mei-Yan′ and Z. A. Rogovin, *Vysokomol. Soedin.*, **5**, 706 (1963).

[36] Mei-Yen Wu, M. A. Tyuganova, E. L. Gefter, and Z. A. Rogovin, *Tselliue. Proizvod., Sb. Statei*, **1963**, 37; N. M. Volgina, A. I. Meos, L. A. Vol′f, and E. E. Nifant′ev, *Zh. Prikl. Khim.*, **40**, 209 (1967).

[37] C. S. Marvel and B. R. Bluestein, *J. Polym. Sci.*, **6**, 351 (1951).

[38] Mei-Yen Wu, T. Zharova, and Z. A. Rogovin, *Zh. Prikl. Khim.*, **35**, 1820 (1962); Z. Yuldashev and G. A. Tsveshko, *Akad. Nauk Uzb. SSR Tashkent. Dokl.*, **22**, 32 (1965).

[39] K. A. Petrov, E. E. Nifant′ev, and I. I. Sopikova, *USSR 163,614*, filed November, 1962; K. A. Petrov, E. E. Nifant′ev, I. I. Sopikova, and M. Merkulova, *Tselliue. Proizvod., Sb. Statei*, **1963**, 86; K. A. Petrov and E. E. Nifant′ev, *USSR 136,347*, Mar. 14, 1961; *Vysokomol. Soedin.*, **4**, 242 (1962); D. A. Predvoditelev, M. A. Tyuganova, E. E. Nifant′ev, and Z. A. Rogovin, *Zh. Prikl. Khim.*, **40**, 171 (1967).

[40] K. A. Petrov, E. E. Nifant′ev, I. I. Sopikova, and M. A. Belavintsev, *Tselliue. Proizvod., Sb. Statei*, **1963**, 90.

[41] Z. A. Rogovin, M. A. Tyuganova, D. A. Prednoditelev, and M. V. Abramov, *USSR 183,322*, June 17, 1966.

[42] D. A. Predvoditelev, E. E. Nifant'ev, and Z. A. Rogovin, *Vysokomol. Soedin.*, **7**, 791 (1965); A. D. Kiselev and S. N. Danilov, *USSR 159,524*, Dec. 22, 1962.

[43] A. D. Koselev, N. A. Aksenova, and L. I. Kutsenka, *Zh. Prikl. Khim.*, **38**, 1355 (1965).

[44] K. A. Petrov, I. I. Sopikova, and E. E. Nifant'ev, *Vysokomol. Soedin.*, **7**, 1667 (1965).

[45] K. A. Petrov, E. E. Nifant'ev, L. V. Khorkhoyann, and V. F. Voblikov, *Vysokomol. Soedin.*, **5**, 348 (1963).

[46] N. I. Garbug, R. G. Zhbankov, D. A. Predvoditelev, E. E. Nifant'ev, and Z. A. Rogovin, *Vysokomol. Soedin.*, **8**, 613 (1966).

[47] A. D. Kiselev, L. I. Kutsenko, and N. A. Aksenova, *Vysokomol. Soedin.*, A9, 1052 (1967).

[48] Z. A. Rogovin, U. Mei-Yan', M. A. Tyuganova, T. Ya. Zharova, and E. L. Gefter, *Vysokomol. Soedin.*, **5**, 506 (1963).

[49] W. A. Reeves and J. D. Guthrie, *Ind. Eng. Chem.*, **48**, 64 (1956).

[50] J. D. Guthrie, G. L. Drake, Jr., and W. A. Reeves, *Amer. Dyestuff Reptr.*, **44**, 328 (1955).

[51] W. A. Reeves, O. J. McMillan, Jr., and J. D. Guthrie, *Text. Res. J.*, **23**, 527 (1953).

[52] D. A. Predvoditelev, M. A. Tyuganova, M. A. Korshunov, and Z. A. Rogovin, *Zh. Prikl., Khim.*, **39**, 1610 (1966).

[53] Reference 4, pp. 962 ff. A sample patent in this area is G. Faulhaber, D. Voges, E. Penning, H. Wilhelm, and N. Goetz, *U.S. 3,322,569* (to Badische Anilin- & Soda-Fabrik, A.-G.), May 30, 1967.

[54] E. Pacsu and R. F. Schwenker, Jr., *Text. Res. J.*, **27**, 173 (1957).

[55] A. I. Polyakov and Z. A. Rogovin, *Vysokomol. Soedin.*, **5**, 610 (1963).

[56] L. H. Chance and E. K. Leonard, *U.S. 3,350,164* (to U.S. Dept. Agr.), Oct. 31, 1967; J. P. Stevens & Co., Inc., *Neth. Appl. 6,600,539*, Oct. 24, 1966.

[57] The Fire Resisting Corp., Ltd., Eng., *U.S. 358,736*, Oct. 21, 1905.

[58] Salomon, Germany, *U.S. 352,960*, Apr. 4, 1905.

[59] Robine and Lenglen, *Rev. Gen. Chim.*, **11** (Jan. 26, 1907).

[60] J. A. Decew, *Brit. 5,411*, Mar. 4, 1912.

[61] R. E. Prince, *Proc. 10th Ann. Meeting Amer. Wood Preservers Assoc.*, **1914**, 158.

[62] G. M. Hunt, T. R. Truax, and C. A. Harrison, *Proc. Ann. Wood Preservers Assoc.*, **1930**, 130.

[63] *Ibid.*, **1931**, 104.

[64] *Ibid.*, **1932**, 71.

[65] T. R. Truax, C. A. Harrison, R. H. Baechler, *Proc. Ann. Wood Preservers Assoc.*, **1933**, 107.

[66] *Ibid.*, **1935**, 231.

[67] R. H. Mann et al., *Proc. Ann. Wood Preservers Assoc.*, **1944**, 261. For general references to wood chemistry and associated techniques see B. L. Browning, *Methods of Wood Chemistry*, Vols. I, II, Wiley-Interscience, New York, 1967; B. L. Browning, ed., *The Chemistry of Wood*, Wiley-Interscience, New York, 1963.

[67a] N. A. Richardson, *J. Soc. Chem. Ind.*, **56**, 202 (1937).

[68] W. L. Morgan, *Brit. 487,702*, June 24, 1938.

[69] Societe Française Albert, *Fr. 1,407,896*, Aug. 6, 1965.

[70] W. W. Bell, *Can. 384,554* (to Celluloid Corp.), Oct. 10, 1939; A. Winogradow, *U.S. 1,917,176*, July 4, 1932.

[71] Celanese Corp., *Belg. 699,355*, June 1, 1967.

[72] R. A. Raff, I. W. Herrick, and M. F. Adams, *Forest Products J.*, **16**, 43 (1966).

[73] Military Academy for Chemical Protection (by K. A. Petrov and I. I. Sopikova), *USSR 175,211*, Sept. 21, 1965.

[74] Omnium de Produits Chimiques pour Industrie et l'Agriculture (by G. Peyrue, R. Marquie, and J. J. Franck), *Fr. 1,162,292*, Sept. 10, 1958; O.P.C.I.A. and Societe Isorel (by J. J. Franck and B. Sandor), *Ger. 1,081,652*, May 12, 1960.

[75] *Engineering*, **119**, 11 (1925).

[76] W. P. Green, Jr., and W. N. Meek, Jr., *U.S. 2,766,139* (to Masonite Corp.), Oct. 9, 1956.

[77] J. M. Black, *U.S. Dept. Agr., Forest Serv. FPL Rept. No. R1427*, 8 pp. (March, 1943); L. Decker, *Chem. Ind.*, **65**, 23 (1942).

[78] G. K. Koeller, *Span. 281,233*, Nov. 3, 1962.

[79] A. C. Nuessle and W. P. Hall, *U.S. 2,606,115* (to Joseph Bancroft & Sons Co.), Aug. 5, 1952; Aktiebolaget Statens Skogsindustrier (by N. Strandberg), *Ger. 1,091,423*, Oct. 20, 1960; *Swed. 185,881; 201,415*, Oct. 29, 1963, Feb. 1, 1966; S. R. Oy, *Finn. 34,854*, Jan. 15, 1966; R. P. Langguth and H. L. Vandersall, *U.S. 3,398,019* (to Monsanto Co.), Aug. 20, 1968.

[80] S. Tamura, *Jap. 98,797*, Dec. 23, 1932; Bowater Research and Development Co., Ltd. (by G. F. Underhay, K. C. Saunders, and J. Ballantyne), *Ger. 1,061,503*, July 16, 1959; E. G. Hallonquist, J. M. Jaworsky, and V. G. Kassey, *U.S. 3,415,765* (to MacMillan, Bloedel, Ltd.), Dec. 10, 1968.

[81] Daika Development Co., Ltd. (by N. Tagachi, U. Oishi, and M. Fujimoto) *Jap. 12,750*, Sept. 5, 1960.

[82] Ruhrchemie A.-G., *Fr. 811,887*, Apr. 24, 1937; G. Shimoto, *Jap. 177,477*, Jan. 19, 1949; I. S. Goldstein and W. A. Dreher, *U.S. 2,917,408* (to Koppers Co., Inc.), Dec. 15, 1959; *Forest Products J.*, **11**, 235 (1961).

[83] I. S. Goldstein and W. A. Dreher, *U.S. 3,159,503* (to Koppers Co., Inc.), Dec. 1, 1964.

[84] Gebr. Giulini G.m.b.H. (by B. Garre), *Ger. 1,038,744*, Sept. 11, 1958.

[85] K. W. Müller and D. Delfs, *U.S. 2,872,355* (to Farb. Bayer A.-G.), Feb. 3, 1959.

[86] A/S Hunton Bruk., *Brit. 869,535*, May 31, 1961.

[87] Farb. Bayer A.-G. (by K. W. Mueller and D. Delfs), *Ger. 1,161,675*, Jan. 23, 1964

[88] Cooperative research at Tennessee Valley Authority and U.S. Dept. Agr., Forest Products Lab., at Madison, Wis. Write to either for details.

[89] Vedex DauskS kovindustri A/S, *Fr. 1,433,746*, Apr. 1, 1966; C. Christofferson and K-O. Sörenen, *U.S. 3,418,159* (to Vedex Dausk Skovindustri A/S), Dec. 24, 1968.

[90] H. R. Frisch, *Brit. 629,654*, Sept. 26, 1949.

[91] E. E. M. Payne, *Brit. 394,019*, June 22, 1933.

[92] S. Ishihara, *Zairyo*, **14**(143), 622 (1965) (in Japanese).

[93] P. C. Arni and E. Jones, *J. Appl. Chem.*, **14**, 221 (1964).

[94] Albright & Wilson, Ltd. (by W. A. Reeves and J. D. Guthrie), *Ger. 961,658*, Apr. 1957; *U.S. 2,927,050* (to U.S. Government), Mar. 1, 1960 (continuation in part of *U.S. 2,668,096*).

[95] S. S. Sakornbut, *U.S. 2,926,096* (to Monsanto Chemical Co.), Feb. 23, 1960; *U.S. 2,893,881* (to Monsanto Chemical Co.), July 7, 1959.

[96] C. F. Perizzolo, *U.S. 3,371,058* (to Stauffer Chemical Co.), Feb. 27, 1968.

[97] S. S. Sakornbut, *U.S. 2,769,730* (to Monsanto Chemical Co.), Nov. 6, 1956.

[98] Farb. Hoechst A.-G. (by V. Lorenz), *Ger. 1,013,863*, Aug. 14, 1957.

[99] I. S. Goldstein and W. J. Oberley, *U.S. 3,160,515* (to Koppers Co., Inc.), Dec. 8, 1964; *U.S. 3,285,774* (to Koppers Co., Inc.), Nov. 15, 1966.

[100] A. J. Erbel and D. L. Kenaga, *U.S. 2,803,562* (to Dow Chemical Co.), Aug. 20, 1957; *U.S. 2,725,311* (to Dow Chemical Co.), Nov. 29, 1955.

[101] A. Eichengrün, *Z. Angew. Chem.*, **42**, 214 (1929).

[102] M. Lewin, *U.S. 3,150,919* (to State of Israel), Sept. 29, 1964; State of Israel, *Neth. Appl. 6,608,865*, Dec. 27, 1966.

[103] S. Bonotto and M. C. Loni, *Energ. Nucl.* (*Milan*), **13**, 631 (1966) (in Italian).

[104] T. B. Monroe, *U.S. 1,338,322*, Apr. 27, 1919; D. E. Keeley, *Trans. Can. Inst. Mining Met.*, **52**, 606 (in *Can. Mining Met. Bull. No. 451*) (1949); E. P. Pershall, *U.S. 2,637,691* (to T. J. Moss Tie Co.), May 5, 1953.

[105] R. B. Hopkins, *Southern Power and Ind.*, **65**, 68 (1947); R. H. Bescher, W. T. Henry, and W. A. Dreher, *Proc. Amer. Wood Preservers Assoc.*, **1948**, 369.

[106] Vereinigte Chemische Fabriken Zu Leopoldshall, *Ger. 206,626*, Jan. 30, 1907; J. A. Clarke, *U.S. 3,245,867* (to Dow Chemical Co.), Apr. 12, 1966.

[107] A. Arent, *Brit. 132,813*, Sept. 20, 1919; *Brit. 138,641, 146,099*, Feb. 5, 1920, June 12 1920.

[107a] B. Lovell, *U.S. 3,031,374*, Apr. 24, 1962.

[108] M & T Chemicals, Inc., *Brit. 1,055,759*, Jan. 18, 1967.

[109] G. B. Fahlstrom, *U.S. 2,757,121* (to Osmose Wood Preserving Co. of America, Inc.), July 31, 1956.

[110] T. Holzer, *Ger. 1,008,907*, May 23, 1957.

[111] W. Mitkowski, W. Olpinski, and B. Zyska, *Prace Glownego Inst. Gornitwa Komun.* (*Katowice*) No. 223, 30 pp. (1958) (See *Chem. Abstr.*, **54**, 21589*g*).

[112] Hanriot et al., *Bull. Soc. Ind. Nat. Encour.*, **134**, 111 (1923); A. Van Kleek, *U.S. 2,387,865*, Oct. 30, 1945; J. C. Middleton, S. M. Draganov, and F. T. Winters, Jr, *Forest Products J.*, **15**, 463 (1965); United States Borax & Chemical Corp., *Brit. 1,023,690*, Mar. 23, 1966.

[113] British Plimber, Ltd., and CIBA (A.R.L.), Ltd. (by F. Bird and R. D. Warnes), *Ger. 1,205,694*, Nov. 25, 1965.

[114] Celcure & Chemical Co., Ltd. (by E. M. Wallace), *Ger. 1,056,356*, Apr. 30, 1959.

[115] Forina August Moralt (by M. Eder and R. Groessner), *Ger. 1,209,274*, Jan. 20, 1966.

[116] M. F. Weiss, *U.S. 1,339,488*, May 11, 1918.

[117] E. R. DuFresne and D. L. Campbell, *U.S. 3,306,765* (to General Dynamics Corp.), Feb. 28, 1967.

[118] A. Pataki, S. U. Hossain, and W. P. Johnson, *Fr. 1,407,478* (to Abitibi Power & Paper Co., Ltd.), July 30, 1965.

[119] R. B. Radkey et al., *Amer. Ry. Eng. Assoc. Bull.*, **63**, 315 (1961).

[120] G. Gunn, *Brit. 546,256*, July 2, 1942.

[121] Chemische Werke Albert, *Belg. 665,047*, Dec. 8, 1965.

[122] G. A. Martin, *U.S. 3,342,629* (to Callery Chemical Co.), Sept. 19, 1967.

[123] R. G. Quinn, *U.S. 2,030,653* (to International Paper Co.), Feb. 11, 1934.

[124] M. L. Nielsen, *U.S. 2,526,083* (to Monsanto Chemical Co.), Oct. 17, 1950.

[125] W. Herz, *Austrian 156,811*, Aug. 25, 1939.

[126] R. Slez, *Australian 229,198*, July 15, 1960.

[127] I. G. Farbenind, A.-G. (by K. Kaimler and M. Paquin), *Ger. 592,777*, Feb. 14, 1934.

[128] G. Gunn, *Brit. 425,781*, Mar. 21, 1935; Deutsche Solvay-Werke G.m.b.H., *Belg. 692,383*, Jan. 10, 1967.

[129] H. J. Thaler, *Fr. 1,429,099*, Feb. 18, 1966; Badische Anilin- & Soda-Fabrik A.-G. (by R. Gaeth, B. Schmitt, R. Breu, and H. Nebel) *Ger. 1,228,797*, Nov. 17, 1966.

[130] Kokoku Rayon and Pulp Co., Ltd. (by M. Osada, N. Tsubomoto, and S. Sasaoka), *Jap. 19,180*, Aug. 27, 1965.

[131] L. Schüler, *U.S. 502,867*, 1893 (cited by Mann et al. on p. 268 of reference 67).

[132] H. W. Angell, *Proc. Forest Products Res. Soc.*, **5**, 107 (1951).

[133] J. M. Scribner, *Scribner's Engineers' and Mechanics' Companion*, Huntington and Savage, New York, 1848, pp. 122–23.

[134] F. L. Browne and W. K. Tang, *U.S. Dept. Agr., Forest Serv., Res. Paper FPL 6*, 20 pp. (1963).

[135] W. K. Tang and H. W. Eickner, *U.S. Dept. Agr., Forest Serv., Res. Paper FPL 82*, 37 pp. (1967).

[136] H. W. Eickner, *Forest Prod. J.*, **12**, 194 (1962).

[137] F. L. Browne and J. J. Brenden, *U.S. Dept. Agr., Forest Serv., Res. Paper FPL 19*, 16 pp. (1964).

[138] H. W. Eickner, *J. Mater.*, **1**, 625 (1966).

[139] *ASTM Standard Method of Test E69-50*, Amer. Soc. for Testing and Materials, Philadelphia, Pa., 1950.

[140] *Method of Test No. 255*, Nat. Fire Protection Assoc., Boston, Mass., May, 1966.

[141] "Fire Test Methods Used in Research at the Forest Products Laboratory," *U.S. Dept. Agr., Forest Serv., FPL Rept. No. 1443* (1959).

[142] *Standard Method of Test C 16041T*, Amer. Soc. for Testing and Materials, Philadelphia, Pa., 1941.

[143] L. K. Andrews, *Proc. Amer. Wood Preservers Assoc.*, **38**, 462 (1942).

[144] H. D. Bruce, *Proc. Amer. Wood Preservers Assoc.*, **52**, 11 (1956).

[145] H. W. Eickner and E. L. Schaffer, *Fire Technol.*, **3**, 90 (1967).

[146] I. S. Goldstein and W. A. Dreher, *Forest Prod. J.*, **11**, 235 (1961).

[147] J. C. Middleton, S. M. Dragonov, and F. T. Winters, Jr., *Forest Prod. J.*, **15**, 463 (1965).

[148] F. E. Brink, *Tech. Rept. 485, U.S. Naval Civil Eng. Lab., Port Hueneme, Cal.*, 69 pp. (1966).

[149] J. C. Jain and A. Lagus, *J. Timber Dryers' Preserv. Assoc. India*, **6**, 10 (1960).

[149a] United States Borax & Chemical Corp. (by L. T. Arthur and R. Thompson), *Can. 803,409*, Jan. 7, 1969.

[150] H. Becker, *Seifen-Oele-Fette-Wachse*, **92**, 991 (1966).

[150a] R. E. Schaeffer, *U.S. Dept. Agr., Forest Serv., Res. Note FRL 0160*, 16 pp. (November, 1968).

[151] T. C. Scheffer and A. Van Kleek, *Proc. Amer. Wood Preservers Assoc.*, **1945**, 204.

[152] *Papierfabr.*, **10**, 263 (1910).

[153] R. G. Myers, *J. Ind. Eng. Chem.*, **8**, 888 (1916).

[154] A. Maschke, *Ger. 267,407*, Jan. 21, 1913.

[155] C. J. West, *Bibliographic Series No. 113*, Inst. of Paper Chemistry, Appleton, Wis., 1941.

[156] C. J. West, E. Stringham, L. Roth, and J. Weiner, *Bibliographic Series No. 185*, Inst. of Paper Chemistry, Appleton, Wis., 1959.

[157] W. P. Green, Jr., and W. N. Meek, Jr., *U.S. 2,766,139* (to Masonite Corp.), Oct. 9, 1956.

[158] J. N. Sears, *U.S. 2,658,878* (to L. J. Carr & Co.), Nov. 10, 1953.

[159] C. R. Outterson, *U.S. 2,867,549* (to Albemarle Paper Manufacturing Co.), Jan. 6, 1959; *U.S. 2,986,478* (to Albemarle Paper Manufacturing Co.), May 30, 1961.

[160] Badische Anilin- & Soda-Fabrik A.-G. (by H. Gerlach), *Ger. 1,004,036*, Mar. 7, 1957; G. Jones, W. Juda, and S. Soll, *U.S. 2,452,054* (to Albi Manufacturing Co., Inc.), Oct. 26, 1948.

[161] R. Aarons, W. H. Baumgartner, and D. R. English, *U.S. 2,935,471* (to E. I. duPont de Nemours & Co.), May 3, 1960; L. J. C. Van de Zande, *U.S. 2,769,729*, Nov. 6, 1956.

[162] D. X. Klein and M. N. Curgan, *U.S. 2,692,203* (to Heyden Chemical Corp.), Oct. 19, 1954.

[163] J. Leicester and C. S. Wright, *Brit. 587,366*, Sept. 14, 1944; R. A. Pingree and R. C. Ackerman, *U.S. 2,488,034* (to Sun Chemical Corp.), Nov. 15, 1949.

[164] A. M. Loukomsky, *U.S. 2,779,691* (to American Cyanamid Co.), Jan. 29, 1957.

[165] A. Berger, D. Gottfried, and A. Schürch, *Switz. 321,212*, June 15, 1957.

[166] W. Juda, G. Jones, and W. Altman, *U.S. 2,628,946* (to Albi Manufacturing Co., Inc.), Febr. 17, 1953; Goldschmidt A.-G., *Belg. 726,503*, Jan. 6, 1969.

[167] W. W. Cobbs, *U.S. 2,262,634* (to Monsanto Chemical Co.), Nov. 11, 1941.

[168] Cassella Farb. Manikur A.-G. (by E. Honold and O. Hansen), *Ger. 863,450*, Jan. 19, 1953.

[169] E. H. Rossin and M. J. Scott, *Can. 556,032* (to Monsanto Chemical Co.), Apr. 15, 1958.

[170] H. Osberg, J. W. Brook, and A. Goldstein, *Can. 787,295* (to Chemirad Corp.), June 11, 1968; D. L. Kenaga, *U.S. 3,312,520* (to Dow Chemical Co.), Apr. 4, 1967.

[171] A. R. McGarvey, *U.S. 2,690,100* (to Armstrong Cork Co.), Sept. 28, 1954; *U.S. 2,690,393* (to Armstrong Cork Co.), Sept. 28, 1954.

[172] Farb. Hoechst A.-G., *Belg. 630,422*, Oct. 21, 1962.

[173] R. J. Dearborn, *U.S. 3,087,836* (to Hooker Chemical Corp.), Apr. 30, 1963; Hooker Chemical Corp., *Belg. 629,820*, Oct. 21, 1963; Hooker Chemical Corp., *Neth. Appl. 6,413,907*, July 26, 1965.

[174] A. J. Erbel and D. L. Kenaga, *U.S. 2,803,562* (to Dow Chemical Co.), Aug. 20, 1957.

[175] D. L. Kenaga and A. J. Erbel, *U.S. 2,725,311* (to Dow Chemical Co.), Nov. 29, 1955.

[176] P. H. P. Vallette, *U.S. 2,782,133* (to Compagnie Française des Matières Colorantes), Feb. 19, 1957.

[177] E. Honold and O. Hansen, *Ger. 863,450*, Jan. 19, 1953.

[178] D. Duane, *Can. 531,072*, Oct. 2, 1956.

[179] E. H. Voigtman and J. C. Bletzinger, *Can. 448,590* (to Paper Patents Co.), May 18, 1948.

[180] I. Lichtenstein, *U.S. 2,710,264* (to Structural Paper Co.), June 7, 1955.

[181] E. T. Blakemore, *Amer. Paper Converter*, **23**, 11 (1949); K. Rosenlind, *U.S. 2,991,143* (to Kimberley-Clark Corp.), July 4, 1961.

[182] M. E. Cupery, *U.S. 2,212,152* (to E. I. duPont de Nemoura & Co.), Aug. 20, 1940.

[183] R. Aarons and D. Wilson, *U.S. 2,723,212* (to E. I. duPont Nemours & Co.), Nov. 8, 1955.

[184] Gebrüder Lohman G.m.b.H., *Ger. 909,663*, Apr. 12, 1954.

[185] D. D. Ritson and C. G. Landes, *Tappi*, **38**, 300 (1955).

[186] W. M. Fuchs and E. Gavatin, *U.S. 2,511,911*, June 20, 1950.

[187] Chemische Fabrik Kalk G.m.b.H. (by W. Seelinger), *Ger. 1,214,986*, Apr. 21, 1966.

[188] American Viscose Corp., *Brit. 647,582*, Dec. 20, 1950.

[189] L. R. Dunn and L. W. Eckert, *U.S. 2,875,044* (to Armstrong Cork Co.), Feb. 24, 1959.

[190] E. A. Lauring, *U.S. 2,849,316* (to Minnesota and Ontario Paper Co.), Aug. 26, 1958.

[191] E. A. Lauring, *U.S. 2,594,937* (to Minnesota and Ontario Paper Co.), Apr. 29, 1952; O. R. Videen, *Can. 549,782* Dec. 3, 1957; G. O. Orth, Jr., C. V. Pevey, and E. Reichman, *U.S. 3,245,870*, Apr. 12, 1966.

[192] J. J. Wachter, *U.S. 2,785,144* (to Westinghouse Electric Corp.), Mar. 12, 1957.

[193] H. C. O'Brien, Jr., *U.S. 2,648,615*, Aug. 11, 1953.

[194] J. C. Robertson, *U.S. 2,842,510* (to E. I. duPont de Nemours & Co.), July 8, 1958; N. D. Clare and A. J. Deyrup, *U.S. 2,823,145* (to E. I. duPont de Nemours & Co.), Feb. 11, 1958.

[195] C. G. F. Cavadino, *Brit. 608,668*, Sept. 20, 1948.

[196] H. L. Becker, *U.S. 2,611,694* (to Homasote Co., Inc.), Sept. 23, 1952.

[197] S. S. Sakornbut, *Can. 549,378*, Nov. 26, 1952.

[198] N. F. Arone, *U.S. 2,701,776* (to General Electric Co.), Feb. 8, 1955; G. A. Kilm, *Fr. 888,064*, Dec. 2, 1943.

[199] A. J. Wesson and H. C. Olpin, *U.S. 2,464,360* (to Celanese Corp. of America), Mar. 15, 1949.

[200] H. Hopkinson, *U.S. 2,610,920*, Sept. 16, 1952.

[201] G. J. Bohrer, *Can. 483,726*, June 3, 1952.

[202] J. C. Wygant, E. J. Prill, and R. M. Anderson, *U.S. 3,236,659* (to Monsanto Co.), Feb. 22, 1966.

[203] Badische Anilin- & Soda-Fabrik A.-G., *Brit. 731,130*, June 1, 1955.

[204] K. S. Campbell and J. E. Sands, *U.S. 2,462,803* (to the United States of America, as represented by the Secy. of Agr.), Feb. 22, 1949; J. F. McCarthy, *U.S. 2,518,241* (Treesdale Laboratories, Inc.), Aug. 8, 1950.

[205] G. H. Brown and E. D. Mazzarella, *U.S. 3,300,426* (to National Starch and Chemical Corp.), Jan. 24, 1967.

[206] D. D. Cameron, *Can. 454,507* (to Hercules Powder Co.), Feb. 8, 1949.

[207] E. R. Laughlin, J. L. Ayres, and P. J. Mitchell, Jr., *U.S. 2,416,447* (to E. I. duPont de Nemours & Co.), Feb. 25, 1947.

[208] Liselotte Quehl, *Ger. 1,222,785* (by K. Quehl and F. Lautenbach), Aug. 11, 1966.

[209] E. Tassel and P. Vallette, *Fr. 1,063,983*, May 10, 1954.

[210] W. W. Riches, *U.S. 2,785,041* (to E. I. duPont de Nemours & Co.), Mar. 12, 1957.

[211] W. Tak. *Neth. 68,567*, Aug. 15, 1951.

[212] T. Iha, *Jap. 2005*, June 4, 1952; H. J. Thaler, *Belg. 636,515*, Dec. 16, 1963.

[213] S. Akagi and S. Okada, *Jap. 558*, Feb. 2, 1954.

[214] Z. Pulikowski, *Prace Inst. Celuloz.-Papier*, **5**, 10 (1956).

[215] N. Higashi, *Jap. 6473*, Dec. 16, 1953.

[216] H. A. Walter, *U.S. 2,514,268* (to Monsanto Chemical Co.), July 4, 1950.

[217] H. H. Beacham and I. M. Panik, *U.S. 2,728,691* (to National Lead Co.), Dec. 27, 1955.

[218] O. Herting, *U.S. 1,804,417* (to Sani Paper Products Co.), May 12, 1931; *Can. 325,562*, Aug. 30, 1932.

[219] B. J. Sutker and E. D. Mazzarella, *Tappi*, **49**, 138A (1966).

[220] R. H. Bescher, *U.S. 3,271,238* (to Koppers Co.), Sept. 6, 1966; L. I. Barrett, *J. Forestry*, **29**, 214 (1931).

[221] T. R. Traux, *Fire Central Notes*, **3**, 1 (1939); *J. Forestry*, **37**, 677 (1939); see also the early Russian work: P. P. Serebrennikov, *USSR Central Forest Res. Inst. Bull.*, **2**, 43 (1934).

[222] A. R. Aidun, *Quart. Progr. Rept. No. 11-13*, Syracuse Univ. Res. Inst., Syracuse, N.Y., 1960, 57 pp.

[223] M. S. Lowden et al., *Chemicals for Forest Fire Fighting*, 2nd ed., Nat. Fire Protection Assoc., Boston, Mass., 1967, 112 pp.

[224] C. B. Phillips, *Fire Control Expt. Calif. Div. Forestry 2*, 5 pp. (1961).

[225] C. B. Phillips, *Fire Control Expt. Calif. Div. Forestry 1*, 10 pp. (1961).

[226] Dow Chemical Co. (by R. N. Bashow and B. G. Harper), *Ger. 1,201,180*, Sept. 16, 1965.

[227] S. V. Petertyl and D. W. Davis, *U.S. 3,080,316* (to Johns-Manville Corp.), Mar. 5, 1963.

[228] J. G. Gilchrist, Jr., *U.S. 3,247,107* (to International Minerals & Chemical Corp.), Apr. 19, 1966.

[229] Reviewed by S. J. Muraro, "A Laboratory Evaluation of Aerially Applied Forest Fire Retardants," Forest Serv. U.S. Dept. Agr., Washington, D.C., 1960, 43 pp.; F. W. Funk, "A Comprehensive Report for the Period July 1, 1936, to July 1, 1937," unpublished Forest Serv. Rept. dated July 15, 1937, U.S. Dept. Agr., Washington, D.C.; P. D. Hanson and C. L. Tebbe, "Aerial Bombing on Forest Fires," Forest Serv., Rocky Mountain Forest and Range Expt. Sta., unpublished, 1947; D. P. Godwin, "Fire Control Notes: Aerial and Chemical Aids," Forest Serv., U.S. Dept. Agr., Washington, D.C., 1936, pp. 5–10.

[230] "Operation Firestop," *Prog. Rept. No. 4: Fire Retardants—1*, Forest Serv., Pacific Southwest Forest and Range Expt. Sta., Berkeley, Calif., 1955, 12 pp.

[230a] G. A. Connell, *U.S. 2,858,895* (to United States Borax & Chemical Corp.), Nov. 4, 1958; G. A. Hesterberg, *Proc. World Forestry Congr. 5th*, Seattle, Wash., 2, 999 (1960).

[231] H. R. Miller and C. C. Wilson, *Tech. Paper 15*, Forest Serv., Pacific Southwest Forest and Range Expt. Sta., Berkeley, Calif., March, 1957, 20 pp.

[232] M. S. Lowden et al., *Chemicals for Forest Fire Fighting*, 1st ed., Nat. Fire Protection Assoc., Boston, Mass., 1963; see also J. B. Davis, D. L. Dibble, and C. B. Phillips, *Misc. Paper No. 57*, Forest Serv., Pacific Southwest Forest and Range Expt. Sta., Berkeley, Calif., April, 1967, 18 pp.; C. A. Connell and G. D. Holmes, *Forestry*, **36**, 91 (1963).

[233] R. W. Johansen, *Prog. Rept. No. 2102 (S5)*, Forest Serv., Southeastern Forest Fire Lab., Macon, Ga., 1965.

[233a] D. C. Young, *U.S. 3,245,904* (to Union Oil Co. of Calif.), Apr. 12, 1966.

[234] Other polymers than CMC have been tried. See, for example, M. Freifield and B. Churchill, *U.S. 3,345,289* (to General Aniline & Film Corp.), Oct. 3, 1967. No others had received widespread use as of 1968.

[235] See, for example, the use of bentonite: K. E. Nelson, *U.S. 3,334,045* (to Arizona Agrochemical Corp.), Aug. 1, 1967.

[235a] L. E. Gould, *U.S. 3,409,550* (to Shell Oil Co.), Nov. 5, 1968.

[236] R. P. Langguth, *U.S. 3,223,649; 3,275,566* (both to Monsanto Co.), Dec. 14, 1965, Sept. 27, 1966; R. P. Langguth and W. W. Morgenthaler, *U.S. 3,338,829; 3,350,305* (both to Monsanto Co.), Aug. 29, 1967.

[237] N. N. Krasavina and V. G. Lorberbaum, *Sb. Nauch.-Issled. Rabot po Lesu. Khoz., Leningr. Nauchn.-Issled. Inst. Lesn. Khoz.*, **1963**, 354. (See *Chem. Abstr.*, **64**, 11447f).

[238] "User Guide, PHOS-CHEK® Fire Retardant," Monsanto Co., St. Louis, Mo., 1968.

[239] "FIRE-TROL®," Bull. Ariz. Agrichemical Corp., Phoenix, Ariz., 1964, 12 pp.

[240] R. A. V. Raff, M. M. Mitchell, and M. F. Adams, *Fire Technology*, **3**, 33 (1967).

Chapter 5

CELLULOSE: TEXTILES

Of all the areas reviewed in preparation for writing this book, textiles made the greatest impact through sheer magnitude. Of the 6000-odd references initially recovered, a full quarter dealt with flameproofing fabrics. The history of sustained efforts to achieve fire-resistant textiles may be traced back over three centuries. No other aspect of fire safety has been studied so diligently for so long a time. Yet, at this writing, many problems remain and the need for better methods has never been more acute.

In the late 1960s the U.S. Congress acted to tighten control over flammable fabrics [1]. The Department of Commerce was asked to study the problems of death, injury, and damage created by flammable fabrics and to set new standards based on the results of such a study. At this writing the Department has begun its work and has already indicated that it is unhappy with the existing test methods [2]. A preliminary regulation has been issued on carpets; children's apparel will be regulated soon [2a]. These actions have catalyzed an upsurge of interest in academic, government, and industrial laboratories. The references which underpin the discussion in this chapter are therefore to be regarded as only the beginning of a vast new literature on this subject.

This chapter is entitled "Cellulose: Textiles," thereby reflecting the fact that almost all the references deal with cotton cloth. An occasional reference considers rayon. Sprinkled very lightly through the literature are reports on the flame retardance of silk or the nonflammability of wool. These are included in this chapter, as are certain new subjects, for example, reactions with fibers driven by ionizing radiation and production of nearly flameproof fabrics by controlled pyrolysis (carbonizing techniques). Discussion of flameproofing nylon, polyesters, and acrylics is reserved for succeeding chapters.

HISTORICAL

There is an interesting historical background to the search for flame retardants for fabrics. The record goes back over 300 years and includes a treatment for canvas used in Parisian theaters (1638) and a report at Oxford on a piece of "incombustible cloth" (1684) [3]. An early English patent—indeed, the 551st issued in England—was taken out by a man named Wyld (was he Obadiah [4] or Jonathan [3]?) in 1735. This patent describes the use of "alum, borax, vitriol, or copperas" for flameproofing paper pulp or textiles. The Montgolfier brothers, inventors of the lighter-than-air balloon in 1783, are said to have coated their early models with alum to reduce the fire hazard.

Conklin's comments [4] are particularly interesting on the role played by Gay Lussac and William Henry Perkin. King Louis XVIII commissioned Gay Lussac to look into means of protecting fabrics used in the theater (after some previous experiences in Europe with theater fires). After studying a variety of substances, Gay Lussac (1820) found that ammonium salts of sulfuric, hydrochloric, or phosphoric acid were very effective on kemp and linen fabric and noted improvement by using a mixture of ammonium chloride and ammonium phosphate. He further observed that borax did not prevent afterglow but was very effective when mixed with ammonium chloride. This work has withstood the scrutiny of subsequent workers for 150 years and remains valid and applicable today.

Flannelet was a cause of concern at the turn of the present century because of its extreme flammability. The renowned chemist William Henry Perkin became interested in the problem and conducted a series of researches. Perkin defined the requirements for a flameproofing process in words that apply as well today as in 1913:

A process, to be successful, must, in the first place, not damage the feel or durability of the cloth or cause it to go damp as so many chemicals do, and it must not make it dusty. It must not affect the colors or the design woven into the cloth or dyed or printed upon it; nothing (such as arsenic, antimony, or lead) of a poisonous nature or in any way deleterious to the skin may be used and the fireproofing must be permanent, that is to say, it must not be removed even in the case of a garment which may possibly be washed 50 times or more. Furthermore, in order that it may have a wide application, the process must be cheap [5].

Perkin's work confirmed the data obtained by Gay Lussac and others [6] and led to a commercial process called "Non-Flam." In this process the flannelet was first impregnated with sodium stannate and then ammonium sulfate, followed by washing and drying. The retardant is believed to be stannic oxide somehow reacted with the cellulose so as to be permanently resistant to laundering with soap. Perkin's results on the amount of a substance to render 100 parts of cellulose nonflammable are given in Table 5-1 [4].

Perkin's process did not win popular favor [3] and little more was done on flameproofing until World War II [7]. It was during the late 1930s and early

Table 5-1   Amount of Substance to Render Cellulose Nonflammable (Perkin)

Substance	Parts/100 Parts Cellulose	Substance	Parts/100 Parts Cellulose
Ammonium chloride	4.2	Silicate acid	30.0
Ammonium phosphate	4.5	Potassium chloride	45.0
Ammonium sulfate	4.5	Sodium phosphate	30.0
Zinc chloride	4.0	Aluminum borate	24.0
Calcium chloride	4.5	Aluminum phosphate	30.0
Magnesium chloride	4.5	Calcium phosphate	30.0
Aluminum hydroxide	3.8	Magnesium phosphate	30.0
Zinc sulfate	4.5	Zinc borate	20.0
Borax	8.5	Tungstic acid	>15.0
Boric acid	10.0	Sodium tungstate	>15.0
Magnesium sulfate	15.0	Ammonium tungstate	>10.0
Sodium chloride	35.0	Clay	>5.0
Sodium silicate	50.0		

1940s that waterproof and flameproof canvas was developed for outdoor use by the military. The treatment consisted of a chlorinated paraffin and an insoluble metal oxide (e.g., Sb_2O_3) along with a binder resin—all applied from an organic solvent [8]. During the war careful studies were conducted at Columbia University under the supervision of the U.S. Army Quartermaster and the National Research Council, National Academy of Science. Much of this work was aimed at gaining more understanding of the behavior of fire retardants on textiles. Temporary techniques (soluble phosphates, borates, halides, etc.) and more durable treatments were considered. Among the latter were urea-phosphate and other nitrogen-phosphorus combinations and Leatherman's modifications of Perkin's approach (which led to the chlorocarbon-metallic oxide treatment). These three means of achieving flame-resistant fabrics are presented in great detail in the record of that work published as a monograph in 1947 by R. W. Little and his associates [9]. This book has become a standard reference work and represents a significant point in the history of flameproofing research on textiles.

The British carried out extensive test work under the Royal Aircraft Establishment in the same period. Data from this period were published by Ramsbottom in 1947 [10]. The work was focused largely on physical and chemical properties of fabric treated with soluble salts. Data on strength after treatment and carbon char remaining after heating are particularly useful. A review of the processes available in the United States at this time was prepared by the National Bureau of Standards (NBS) [11] after the horror of the Coconut Grove fire in Boston and the Hartford circus-tent fire. The NBS report contains formulas and application conditions for 11 treatments and also discusses test methods in use at that time.

Over the years since World War II the search for better flame retardants has continued at an increasing pace so that the bulk of the total literature has appeared since these early reviews. (Workers in this field have been prolific in writing short reviews of their art but no comprehensive and authoritative monographs have appeared since Little's book.) In addition to *Chemical Abstracts*, the searcher can get a thumbnail sketch of developments by reading the sections on "Fireproofing" and "Flameproofing" in the editions of the *Review of Textile Progress* published annually since 1948 [12]. An hour or two spent with this series gives one a clear overview of the activity without clogging the mind with details. A noncritical compilation of very brief abstracts was published in 1959 by The Institute of Paper Chemistry. About half the Institute's 876 abstracts on flameproofing fall in the textile area [13]. (Note, however, that the search for the present book turned up some 6000 references.)

There are review articles in Polish [14], Spanish [15], German [16, 17], and French [18, 19]. Frieser's articles [17] in German are wide-ranging and fairly detailed. In the early 1950s Little and his colleagues published data on fire retardants for fabrics [20]. In 1957 McQuade reviewed developments to that time in a chapter for a book on textile chemicals [21]. He presents both basic and applied data with some useful comparisons between systems, including the urea-phosphate method, brominated triallyl phosphate, tetrakis(hydroxymethyl) phosphonium chloride (THPC), and modified THPC. (See subsequent sections of this chapter for a discussion of these systems.) In 1958 Perfect [22], in England, reviewed the same material, adding some further comments on phosphate-nitrogen systems (amines, amides, phosphoramides, etc.). Dorset subsequently added to this British review with articles from 1958 to 1962 [23]. Sandholzer at NBS presented results of comparative testing of some of the same systems carried out at the Bureau in 1958 [24]. The fabrics were treated in commercial finishing plants and sent in to the laboratory at NBS for evaluation. Data are given for flame resistance and for physical properties before and after laundering.

Beginning in the mid-1950s, reports began to appear on the use of THPC. Many progress reports were issued, especially by the Southern Regional Research Laboratory of the U.S. Department of Agriculture. A few of these are cited in references 25–27. Drake, in an encyclopedia article [28], reviewed briefly the status as of the mid-1960s. Recent symposiums [29–31] and abbreviated review articles have been devoted to this subject. They are generally devoted to discussions; relatively little hard data are given. References 9–31 constitute a reasonably extensive review of the field through 1968 and the references given in each of these papers in turn constitute very comprehensive coverage. In the remainder of this chapter most of the literature—all that is of real value—is discussed and intercompared.

PHOSPHATES

Ammonium Orthophosphates

The compounds $NH_4H_2PO_4$ and $(NH_4)_2HPO_4$ have been used by themselves for a very long time for "temporary" fire retardance. The effect persists
as long as exposure to water is avoided. In some cases, as in heavy draperies
in relatively dry climates, the treatment may last for many years. The
treatment can be traced back to Gay Lussac. In this century references appear
in the earliest citations [32]. Coppick, writing in the Little book [9], discusses
in detail the effect of ammonium phosphates on the thermal decomposition
of cotton fabric. The effects are very similar to those for wood described in
Chapter 4. Table 5-2 shows the influence of $(NH_4)_2HPO_4$ on various measures

Table 5-2 Properties of Cotton Fabric Treated with $(NH_4)_2HPO_4$ [33]

Add-on, %	No. of H_2O Leachings	Loss in Pyrolysis, mg/cm² of fabric	Loss via Afterglow, mg/cm² of fabric	Glow Time, sec
0	0	21.4	5.1	73
0.1	0	19.6	2.7	70
0.5	0	17.0	0.8	9
1.0	0	15.7	0.3	0
15.0	0	17.2	0.1	0
15.0	1	17.8	0.4	0
15.0	3	20.2	0.8	45
15.0	4	23.3	4.1	210
15.0	5	23.7	3.6	195

of thermal decomposition. It also gives an indication of the lack of resistance
of the treatment to water leaching; after four leachings (15 min each in plain,
static water) the effects of the retardant are removed. Whereas ammonium
phosphates, phosphoric acid, and other phosphorus compounds that, on
heating, generate moderately strong acid residues are effective, the metal
phosphate salts are not. For example, Little [34] shows the following results
in a microburner test at a 45° angle:

Compound	Add-on, %	Afterflaming, sec	Char Area, in²
$Na_3PO_4 \cdot 12H_2O$	77	3	Completely consumed
$Na_2HPO_4 \cdot 12H_2O$	57	60	Completely consumed
$NaH_2PO_4 \cdot H_2O$	17	1	2.4
H_3PO_4	15	0	2.4

Thus, only the free acid and the strongly acidic monobasic salt are flame retardants for cotton fabric.

> This is an important point. Many treatments confer wash-resistant flame retardance in the sense that the retardant will not be removed by laundering. But in the course of washing the sodium ions in the detergent solution will exchange for the ionizable volatile cations on the cellulose. If sufficient sodium ions exchange into the flame retardant on the fabric, the resistance to flaming will be much reduced. Thus, a requirement for a wash-fast treatment is not only chemical resistance to water but also immunity to ion exchange to some degree.

More data on the effect of ammonium phosphates on some physical and chemical properties of cotton are given on p. 227.

Methods of applying water-soluble fire retardants have been discussed by Little and his colleagues [9, 20]. Perhaps the easiest method is impregnation in a commercial laundry machine. A solution of the ammonium phosphate is prepared at a suitable concentration and the fabric immersed and then spun for a fixed period and the wet add-on is determined. By adjusting the solution strength to this measured value for wet add-on, the final dry add-on can be determined. If, for example, it should be desired to obtain 2% phosphorus from $(NH_4)_2HPO_4$ in the fabric and the wet add-on is found by trial to be 120% of the dry fabric weight, then the solution strength desired is:

$$X = \frac{\% \text{ P}}{(\% \text{ add-on})(\% \text{ P in } (NH_4)_2HPO_4)} (100)$$

$$= \frac{2}{(120)(24)} (100)$$

$$= 7\% \text{ } (NH_4)_2HPO_4$$

Drying should be carried out at as low a temperature as possible to minimize loss of ammonia through decomposition. $(NH_4)_2HPO_4$ loses ammonia and converts to $NH_4H_2PO_4$ at an ever-increasing rate as the temperature rises. The predicting equation for this is given in Chapter 2, p. 35. At $125°$, $NH_4H_2PO_4$ is stable but $(NH_4)_2HPO_4$ has an ammonia vapor pressure of 30 mm of Hg at $125°$ and 120 mm at $150°$. Thus, the final product on the fabric is more acidic than $(NH_4)_2HPO_4$; this may cause tendering of the cloth.

Phosphates with Other Nitrogen Compounds

Various combinations of less volatile bases that nonetheless decompose on heating at fire temperatures, but which can pass through the drying stage while maintaining neutral pH, have been evaluated. Some of these also appear to foster the otherwise sluggish direct esterification of cellulose by the phosphate group. Thus, the reaction

$$\text{ROH} + \underset{\underset{|}{\overset{\overset{O}{|}}{}}{\text{HO}-\text{P}-(\text{ONH}_4)_2} \rightarrow \underset{\underset{|}{\overset{\overset{O}{|}}{}}{\text{RO}-\text{P}-(\text{ONH}_4)_2} + \text{H}_2\text{O}$$

where R is the cellulose molecule, apparently goes very slowly if at all under normal curing conditions. But if urea or a similar compound is added, the esterification proceeds smoothly. In this case urea is probably serving as a dehydrating agent [35]:

$$\text{ROH} + (\text{NH}_2)_2\text{CO} + (\text{NH}_4)_2\text{HPO}_4 \rightarrow \underset{\underset{|}{\overset{\overset{O}{|}}{}}{\text{RO}-\text{P}-(\text{ONH}_4)_2} + 2\text{NH}_3 + \text{CO}_2$$

No free water is evolved. Under normal curing conditions it appears that the product is probably intermediate between that shown and $\text{ROP(O)(OH)(ONH}_4)$. Thus, the pH of the treated fabric is about 5 [36]. Presumably, the more urea and the quicker the drying, the higher the ammonia content of the finished goods. The Ban-Flame process was based on the idea that a large excess of urea was helpful in this regard [37]. The loss of strength with this urea-phosphate treatment is apparently acceptable and the flame retardance decreases only slowly with laundering. The loss is undoubtedly due to replacement of ammonium ions with sodium ions and not to leaching out of the phosphorus. Indeed, Little shows this ion exchange explicitly by first reducing flame resistance by leaching with a calcium chloride solution and then restoring it by leaching with ammonium chloride solution (followed by water rinsing to remove mobile salts) [38]. Excess urea increases resistance to ion exchange. For example, at a mole ratio of 4.4:1 [urea: $(\text{NH}_4)_2\text{HPO}_4$], immersion in 5% CaCl_2 solution had little effect on flammability of treated cloth [38]. This may be brought about by full esterification of the phosphate with attendant crosslinking.

The urea-phosphate treatment degrades cellulose to such an extent that some 35–45$\%$ of the tensile strength is lost [38, 39]. The higher the curing temperature, the lower the strength is likely to be despite the shorter times used at higher temperatures. After a great many experiments, the Columbia group selected 13 min at 160° as optimum for a treatment in which the 4.4:1 mole ratio (2:1 weight ratio) was used. The solution was adjusted so as to give a 46$\%$ dry add-on (3.6$\%$ P). Some improvement in strength has been noted by adding NH_4OH and formaldehyde [38] but the loss is still considerable.

Other nitrogen bases have been tried in place of urea. Guanidine, ammonium sulfamate, and dicyandiamide have been studied [38]. These bases react more slowly than urea and therefore require somewhat higher curing temperatures. This takes its toll of fabric strength. More recent work has shown cyanamide to be more reactive than urea [40, 41]. Curing temperatures as low as 105° (for 4 min) are reported to be effective. The preferred mole ratio of cyanamide to H_3PO_4 is about 3:1. A comparison of this method and a system based on THPC is given in Table 5-3. The cyanamide treatment gives a

Table 5-3   Comparison of Cyanamide-H_3PO_4 Method with THPC [40]

Retardant	Fabric	No. Washes[a] to Failure	Hand Relative to Untreated Fabric
Cyanamide-H_3PO_4	80 × 80	5	Very slightly firmer
	Sateen	6	Very slightly firmer
THPC	80 × 80	25	Much firmer (stiff)
	Sateen	35	Much firmer (stiff)

[a] Washed in hard water in hot detergent solution. Failure taken as ⩾6 in char length in a vertical test.

better hand, perhaps, but only partial resistance to laundering. The resistance to laundering is excellent in soft water (35–50 washings at 0 ppm hardness) but a serious reduction in flame retardance is found with increasing hardness (8–10 washings at 100 ppm hardness) [40]. This treatment might be effective for drapes, furniture fabrics, and the like. A recent patent claims improved performance by addition of haloalkylphosphates [42].

Variations on the urea-phosphate theme include addition of small amounts of pyridine [43], conversion of the H_3PO_4 to an ester (e.g., of ethylene glycol [44]) before adding urea to condense the ester with cellulose, and addition of various resins. The last technique has been widely studied as a means of protecting the phosphorus both from leaching and from ion exchange. These resins have invariably been aldehyde, usually urea-formaldehyde or melamine-formaldehyde systems [45]. The most recent report claims that using poly-phosphoric acid (average chain length of 4) with ammonia or other amine provides resistance to water leaching and washing in soap solution not attainable with orthophosphates [45a]. This same approach was tried two decades ago by the Columbia group [38] and there is no evidence that matters have progressed much commercially since that time. The urea-phosphate concept has drawbacks that apparently cannot be surmounted by subsequent modifications. These drawbacks are clearly associated with fabric tendering and lack of resistance to ion-exchange reactions.

Phosphorylamides

In the 1950s there was a flurry of research on reaction products of $POCl_3$ and NH_3 [46]. (See Chapter 2, pp. 37–39, for a discussion of the phosphorylamides.) Nielsen discusses the probable structure of his compounds and how they react with cellulose [47]. The molecule is a polymer and contains some

$$
\begin{array}{cc}
\text{O} & \text{H} \\
| & | \\
\end{array}
$$

—P—N— groups. It is very soluble in water and is neutral to slightly basic
|
NH$_2$

in aqueous solution. It is applied to fabric from a 10–25% solution; the fabric is dried and cured 5 min at 150°. Up to 12% losses in breaking strength are reported at 13% dry add-on. The treated cotton is semiresistant to water, that is, it withstood some 20 mild washings but only 5 hot washes. The phosphorus is firmly bound to the cellulose but suffers from salt formation and ion exchange with nonvolatile cations. From this standpoint the treatment is much like the urea-phosphate method, and because of its higher raw material cost it could not really compete with the latter. Later modifications, including use of a higher molecular weight [48], addition of resins [49], and substitution of ethylenediamine [50] for ammonia, are reminiscent of the relatively unsuccessful efforts to boost the permanence of the urea-phosphate method. Neither effort appears to have made significant headway. Mixed ester-amides of H$_3$PO$_4$ have also been tried but without commercial success [50a].

There are three other subcategories of P—N compounds that have been considered for cotton fabrics. These are reaction products of P$_2$O$_5$ with nitrogen bases (NH$_3$ [51], ethylenediamine [52]), phosphorus isocyanates [53], and a mixed bag of reaction products of amines with PCl$_3$ or POCl$_3$ [54]). (See Chapter 2 for chemical details on all these compounds.) The product of reacting anhydrous NH$_3$ with P$_2$O$_5$ contains hydrolytically unstable —P—O—P— and —P—N—P— linkages. These presumably react both with H$_2$O and with cellulose during fabric treatment. The final fabric is no doubt susceptible to ion exchange and therefore will lose flame resistance on repeated laundering with detergent. The ethylenediamine treatment of the NH$_3$—P$_2$O$_5$ product will provide more stability but cannot eliminate the ion-exchange problem.

The phosphorus isocyanates have not been reported on extensively. The patents indicate that strength is not sacrificed on treatment with OP(NCO)$_3$ using a curing cycle of 2–3 min at 130°. To the writer's knowledge this treatment has not had commercial success, suggesting that either the permanence is not as claimed or that the phosphoryl triisocyanate presents some problems in cost, stability, and the like. Presumably, the problem of ion exchange would only arise if the P—N bond were to hydrolyze to —P—O— groups.

Many different amines have been reacted with phosphorus halides in attempts to come up with new and better fire retardants for cotton. A few

Table 5-4   Three Types of P-N Compounds for Fire-Retardant Cotton

Compound	Use Conditions	Reference
$(CH_3)_2N\overset{\displaystyle O}{\overset{\displaystyle \|}{-}}P-(NH_2)_2$	With melamine-HCHO resin	54
$(RO)_2\overset{\displaystyle O}{\overset{\displaystyle \|}{-}}P-NH_2$, where R = lower alkyl or haloalkyl	Alone or with melamine-HCHO resin; in one case used with triallyl phosphate	55
$(CH_3NH)_2\overset{\displaystyle O}{\overset{\displaystyle \|}{-}}P-O-\overset{\displaystyle O}{\overset{\displaystyle \|}{}}P-O-\overset{\displaystyle O}{\overset{\displaystyle \|}{}}P-(NHCH_3)_2$ with N and $(CH_3)_2$ below central P	With melamine-HCHO resin	56

such structures are shown in Table 5-4. None have received serious attention because of cost, complexity, and the questionable stability of the P—N bond. However, there is one material that has received very serious attention. This is APO, the reaction product of ethyleneimine and phosphoryl trichloride:

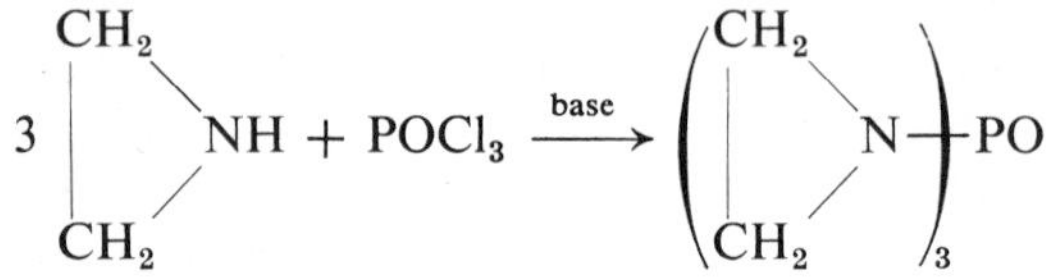

$$3 \underset{CH_2}{\overset{CH_2}{\diagdown}}NH + POCl_3 \xrightarrow{\text{base}} \left(\underset{CH_2}{\overset{CH_2}{\diagdown}}N \right)_3 PO$$

Let us repeat the admonition of Chapter 2: APO stands for the incorrect name "tris(aziridinyl) phosphine oxide." The compound is in no sense a phosphine oxide. The phosphorus oxidation state is $+5$, not -1. The product is a true phosphoramide (or phosphorylamide). Mention of APO first appeared in the *Review of Textile Progress* in 1956 in conjunction with THPC. The first references to APO by itself appeared in 1958–59 in reports from the Southern Regional Research Laboratory, U.S. Department of Agriculture (U.S.D.A.), New Orleans. Research since that time has been vigorously pursued on APO, THPC, and THPC-APO. The THPC systems are described later in this chapter; here we consider only APO taken alone.

Table 5-5 lists patents issued on the use of APO by itself for flame-resistant cotton. Some of the ideas in these patents have been amplified in articles from the U.S.D.A. laboratory and from Dow Chemical Co. The former noted in their work with APO-THPC fire retardants that APO seemed to confer crease resistance on treated cotton. The subject was pursued [74].

Table 5-5   Patents Issued on APO Alone as Fire Retardant for Cotton
(Reproduced by permission of The Dow Chemical Company)

Chemicals	Cure	Year	Reference
APO, urea/ethylenediamine	5 min, 140°	1958	57
APO, urea, H_3PO_4	3 min, 140°	1958	25
APO, carboxylic acid	· · ·	1959	58
APO, prepolymerized in NaOH	5 min, 145°	1959	59
APO, with carboxylated cotton	5 min, 90°	1959	60
APO, aminoplast resins	5 min, 90°	1959	61
APO, triethanolamine	10 min, 140°	1959	62
APO, polyol	10 min, 140°	1959	63
APO, phenols	· · ·	1959	64
APO, phosphonomethylated cotton	· · ·	1960	65
APO, NH_4 borate, urea	5 min, 130–150°	1962	65a
APO, NH_3—P_2O_5 product or ethylenediamine-P_2O_5 product or NH_3—$POCl_3$	5 min, 150°	1962	66
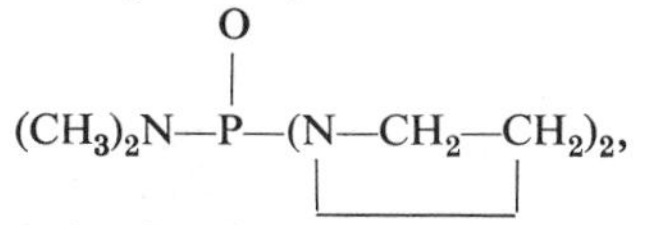 $(CH_3)_2N$—$\overset{\overset{O}{\|}}{P}$—$(N$—$CH_2$—$CH_2)_2$, simpler phosphoramides	10 min, 150°	1962	67
APO, $(NH_4)_2HPO_4$	5 min, 150°	1963	68
APO, guanidine phosphate, guanidine carbonate	5 min, 150°	1966	69
Substituted halophosphonic acid bis(alkyleneimides) 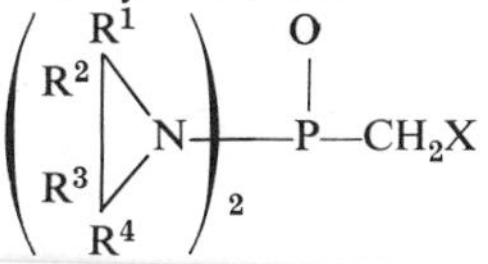 $\left(\begin{smallmatrix}R^1\\R^2\\R^3\\R^4\end{smallmatrix}\bigg\rangle N\text{—}\bigg)_2 \overset{\overset{O}{\|}}{P}\text{—}CH_2X$	7 min, 170°	1967	70
APO, epichlorohydrin (also THPC), NH_3	· · ·	1967	71
Mixed phosphoramides: 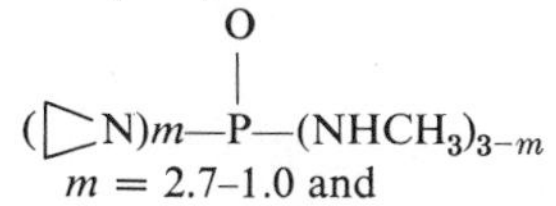 $(\square N)m$—$\overset{\overset{O}{\|}}{P}$—$(NHCH_3)_{3-m}$ $m = 2.7$–1.0 and 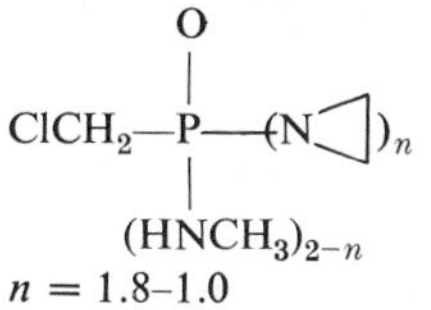 $ClCH_2$—$\overset{\overset{O}{\|}}{P}$—$(N\square)_n$ $(HNCH_3)_{2-n}$ $n = 1.8$–1.0	· · ·	1967, 1968	72
APO, thiourea	2 min, 160°	1968	73
APO, 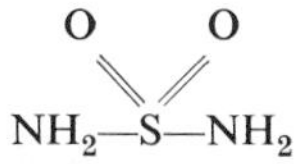 NH_2—$\overset{\overset{O\ \ O}{\diagdown\diagup}}{S}$—$NH_2$	1¾ min, 165°	1968	73a
APO, PVC, Sb_4O_6	2 min, 150°	1969	73b

The APO was polymerized in the presence of catalytic amounts of $Zn(BF_4)_2$ by curing 4 min at 140°. The treatment was found to resist laundering and chlorine bleaching satisfactorily. However, a considerable loss in strength was noted that could be counteracted only partially by use of a fabric softener. The hand was judged to be good. Good wrinkle recovery was obtained at 8–10% add-on of APO. (Later work by Dow led to the recommendation of about 5% add-on for good crease resistance [75].)

It is evident that APO can polymerize with itself [74]:

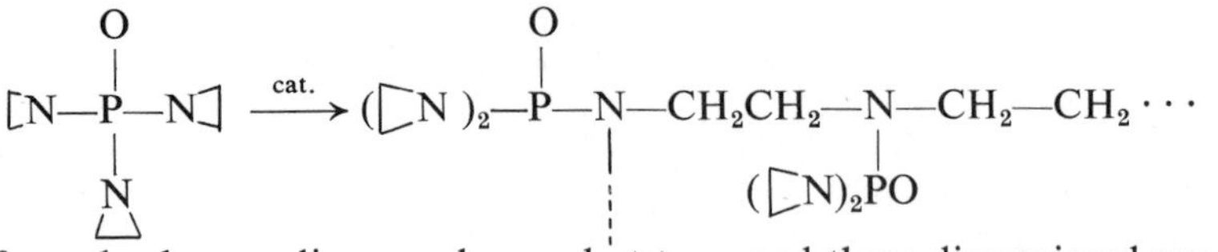

$$[N—P—N] \xrightarrow{\text{cat.}} ([N)_2—P—N—CH_2CH_2—N—CH_2—CH_2 \cdots$$

This formula shows a linear polymer, but two- and three-dimensional products can form just as readily. APO may also react with cellulose:

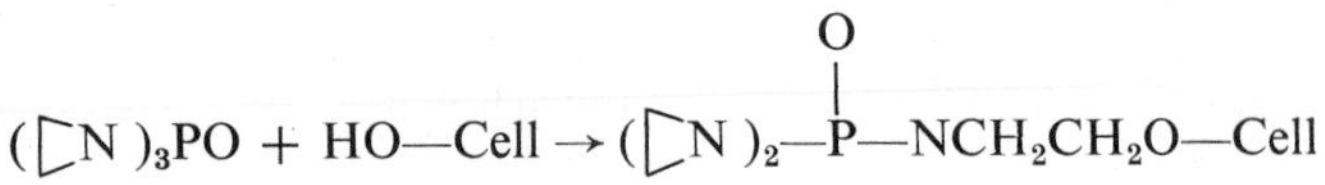

$$([N)_3PO + HO—Cell \rightarrow ([N)_2—P—NCH_2CH_2O—Cell$$

Presumably both types of reaction occur within the treated fabric.

There are two detailed reports, one from each laboratory, on the application of APO as a flame retardant [76, 77]. In the U.S.D.A. work the $Zn(BF_4)_2$

Table 5-6 Some Properties of APO [78]

Formula	$([N)_3—PO$
Molecular weight	173.2
Melting point	43.5°
Boiling point	Decomposes
Solubility	Sol H_2O, most common org. solvents limited soln in pet. ether
% P	17.9
% N	24.3
Commercial form	80% soln in anhydrous ethanol (stable to 0°)
Toxicity	Hazardous chemical. Handle strictly in accord with instructions. (Polymerized products are not considered toxic.)

catalyst was used; in the Dow work thiourea was added in stoichiometric quantities. Table 5-6 lists some properties of APO of interest. Note the warning on toxicity of monomeric APO. This has been reviewed by the U.S.D.A. and a bibliography has been published [77a]. APO is a very

Table 5-7 Two Treating Techniques Based on APO

	U.S.D.A. [76]	Dow [77]
APO, %	30	24.6
Catalyst [Zn(BF$_4$)$_2$], %	5	· · ·
Thiourea, %	· · ·	15.4
Surfactant, %	0.1	0.4
Softener, %	· · ·	3.9
H$_2$O, %	64.9	55.7
pH	5.7	?
Wet pickup, %	Ca. 60	Ca. 54
Drying	4 min at 80–90°	To 5% moisture [75]
Curing	3 min at 155°	1½ min at 165° [75]
Afterwash	Deterg. wash; hot and cold rinses	Home washer cycle
Resin add-on, %	16.9	21.6

reactive compound and must be handled with care. Tables 5-7 through 5-9 summarize the work done on APO and show the excellent durability of treated fabric to laundering; the fabric retains good flame resistance even after repeated exposure to industrial laundry practice. The treated fabric is even resistant to burial in the ground, retaining all of its strength and flame resistance after 15 days' burial whereas untreated cloth was mildewed and

Table 5-8 Results of Laboratory Studies on 8.5-Oz Sateen Cotton Fabric—U.S.D.A. Catalytic APO Method [76]

Conc of Zn(BF$_4$)$_2$	Cure Temp, °C	Resin Add-on, %	Char Length in.[a] Before Boil	Char Length in.[a] After Boil	Elmendorf Tear (Warp), g[b] Before Boil	Elmendorf Tear (Warp), g[b] After Boil	No. of Home Launderings, %P 0	No. of Home Launderings, %P 5	No. of Home Launderings, %P 15
5	140	14.5	2.7	2.9	1875	1925	1.95	2.03	2.02
5	160	16.8	2.7	3.2	1575	1675	2.16	2.26	2.16
5	180	17.4	3.4	2.7	1375	1575	2.33	2.45	2.37
7	140	15.3	2.6	2.7	1775	1800	2.34	2.11	2.09
7	160	16.9	3.0	2.8	1650	1750	2.25	2.33	2.28
7	180	17.8	2.5	2.7	1475	1600	2.37	2.41	2.40
10	140	17.6	3.7	2.6	1700	1675	2.22	2.30	2.26
10	160	18.4	3.1	2.4	1450	1600	2.31	2.35	2.39
10	180	17.9	2.5	2.7	1475	1475	2.35	2.37	2.39

[a] Boiled 3 hr in 0.5% soap, 0.2% Na$_2$CO$_3$ soln
[b] Elmendorf tear strength of control 5467 before laundering (drops substantially after laundering)

**Table 5-9   Properties of Cotton Fabrics Treated on Pilot or
Full Mill Scale**

U.S.D.A. pilot plant [76]	Control	Treated
Cotton sateen		
Weight, oz/yd^2	8.43	9.65
Thread count, warp/fill	85/53	86/53
Elmendorf tear, g[a]	5467	2300
Breaking strength, lb	121.4	105.0
Stiffness $\times 10^{-4}$, lb	11.85	28.20
Flame resistance, char length, in.	$\cdots$	2.9
Dow report [77]		
Cotton twill, 8.5 oz		
Initial		
Elmendorf tear, g	3825	$\cdots$
Breaking strength, lb	186	171
Stoll flex abrasion, cycles	337	957
Flame resistance, char length, in.	$\cdots$	$3\frac{1}{2}$
After 25 industrial launderings		
Elmendorf tear, g	1775	1800
Breaking strength, lb	152	143
Stoll flex abrasion, cycles	146	204
Flame resistance, char length, in.	$\cdots$	$3\frac{1}{4}$

[a] Warp filling

very weak [77]. The Dow report [76] shows that the strength of the finished goods compares favorably with untreated fabric *after* laundering. (Unlaundered, untreated cotton exhibits very high strengths which do not withstand the rigors of the washing process. The lower strength of APO-treated fabric perhaps reflects the effects of the laundering in the treating cycle.)

One study used $(NH_4)_2HPO_4$ as a catalyst for APO and found the system to be promising as competition for the THPC-APO combination. Little has developed from this [77b].

At this writing and despite these very encouraging results, one manufacturer has discontinued production of APO. The reasons appear to be toxicity, slow development of the market, and high cost per unit of phosphorus and hence high cost of finished goods. Nonetheless, the APO technique is clearly a major step in the right direction. We shall consider it further in conjunction with THPC in subsequent sections of this chapter. (The U.S.D.A. has published a literature review on APO as a textile chemical [78a].)

An interesting by-product of this research is a substance dubbed APN. This curious compound was looked at briefly by the U.S.D.A. group [79].

It is made from ethyleneimine and ($-PNCl_2t)_3$:

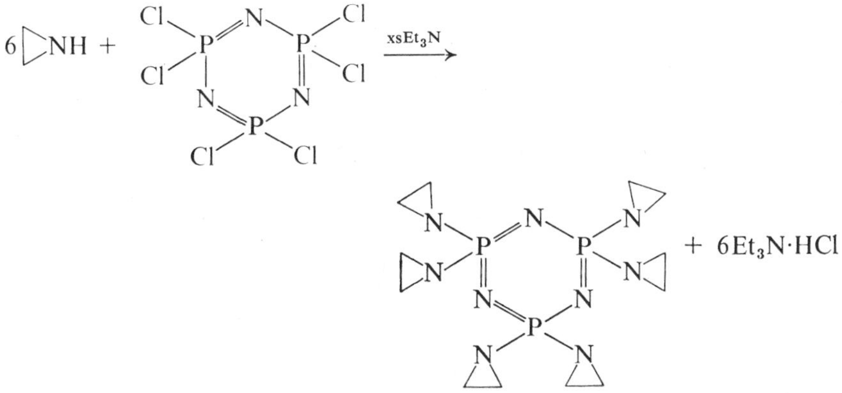

This is obviously a highly reactive polyfunctional molecule capable of establishing a large amount of cross-linking in a fabric. It was found that APN polymerization is catalyzed by $Zn(BF_4)_2$ just as APO is. Further, it reacts in such a way that curing may be effected at room temperature or by brief exposure at, say, 80°. The material is at present a laboratory curiosity but the development bodes well for future, analogous concepts.

The product of propyleneimine and $POCl_3$—$(CH_3—CH—CH_2N)_3PO$— has similar reactive properties and has recently been studied in various copolymer systems [79a].

Other Phosphates

Some work has been done with pentaerythritol-based phosphate esters [80]. The full ester of erythritol or dipentaerythritol was first prepared by reaction with polyphosphoric acid. Fabric was treated with polyethylenimine (7–12%) and then with the phosphate ester (4–10%), dried, and cured i in at 100°. The treatment is said to withstand soap and water laundering but was not brought to a commercial stage for reasons not altogether clear. (Guesses would be high cost, some ion-exchange problems, and losses in strength.)

Another rather remote idea was the treatment of cotton with poly(vinyl-pyridine) and then with phosphoric acid and ammonia, followed by heating to 100°. The fabric resists laundering to a degree but is reduced in strength by about 40% [81]. The idea has not made much of an impact to date and, again, one suspects that the reasons are poor strength and susceptibility to loss of flame resistance through ion exchange. Finally, the introduction of dialkyl phosphate esters of cellulose by reacting cellulose with methane sulfonylchloride, CH_3SO_2Cl, and subsequently replacing this substituent by treatment with dialkyl phosphoryl chloride has been mentioned in Chapter 4,

p. 117–21. In application to cotton fabric the treatment consists of (*a*) mercer-ization, (*b*) treatment with CH_3SO_2Cl in pyridine for 10 min at 65°, and (*c*) immersion for 4–5 min at 25° in a pyridine solution of the phosphoryl-chloride [82]. The process has not found favor, undoubtedly because of high costs.

Phosphate-Halogen Compositions

A few systems have been proposed for cotton, based on phosphate esters containing halogens either in the ester or mixed with it. One such system covered reacting a bromopropyl phosphorylchloride with polyvinyl alcohol to yield a polymer suitable for fabric impregnation [83]. Another method was a technique for esterifying cellulose with chloroethyl phosphate in basic solution by heating for nearly 1 hr at 110° [84]. Only marginal performance was obtained. Still another suggestion was to use mixtures of triesters of phosphoric acid and poly(vinylchloride) as an impregnating solution [85]. The solvent was gasoline. A recent patent [85a] describes the phosphorylation of atactic polypropylene so as to leave the group $-OPO(Cl)_2$ on the polymer. The finished polymer is then added to the cellulose for flameproofing. None of these notions has received commercial attention.

A reaction product of tri(haloalkyl) phosphate and a resin containing anhydride groups has been claimed as a flameproofing system for rugs and drapes [85b].

During the early 1950s, before THPC had been developed fully, the system receiving the most attention was a triallyl phosphate-bromine combination [85c]. The process has been described in a series of patents from the Southern Regional Research Laboratory of the U.S.D.A. [86] and in papers from that laboratory and the U.S. Army Laboratories at Natick, Mass. [87–89]. The U.S.D.A. process is a treatment with an emulsion of brominated polymeric allyl phosphate ester (BAP) prepared from triallyl phosphate and bromoform ($CHBr_3$). One recipe is:

$(CH_2{=}CHCH_2O)_3PO$	18.9
$CHBr_3$	13.2
$(CH_2CHOH)_x$	0.6
$NaHCO_3$	2.4
$K_2S_2O_8$	0.6
H_2O	64.3
	100.0

The emulsion polymerization is carried out at 80–85° for 2 hr (cooling is required to maintain this temperature through the first part of the reaction). The emulsion pH should be slightly basic (pH 7.2–7.5). The polymer is believed to have the following structure:

$$Br_3C \!-\!\!\left[\begin{array}{c} CH_2\!-\!CH \\ | \\ CH_2 \\ | \\ O \\ | \end{array}\right]_x\!\!-\!H \rightarrow \text{Cross-linked polymers via remaining allyl}$$

groups and $x = 1.0\text{--}2.5$

$(CH_2\!\!=\!\!CH\!-\!CH_2O)_2P\!-\!O$

The product typically contains about 10% phosphorus and 26% bromine. It does not, apparently, react with the cotton but provides instead a polymeric surface coating [89].

Fabric treatment consists of padding to obtain 18–22% dry add-on, followed by drying 4–10 min at 80–110° and curing 4–6 min at 140°. The authors note that some allyl alcohol may be driven off during curing and suggest venting to avoid eye irritation. The fabric is finally washed. The treatment does not affect breaking strength but does reduce tear strength. To offset the latter, a plasticizer may be added. Results of tests on fabrics from such treatments are presented in Table 5-10. The char lengths are not

Table 5-10   Results of Tests on 8.2-Oz Cotton Twill with Brominated Poly (Allylphosphate) [87]

	Tear Resistance, % of control				Flame Resistance, in. of char[a]			
	No. Laundry Cycles				No. Laundry Cycles			
	0	1	6	12	0	1	6	12
Unplasticized	70	82	84	89	5.0	4.8	5.3	5.7
2% tetrabutyl thiodisuccinate	84	97	92	92	5.3	5.4	5.3	6.1
4% tetrabutyl thiodisuccinate	91	97	100	100	5.4	4.9	5.2	5.8

[a] Vertical flame test. Assuming 20% add-on, the retention of flame-resistant agents would be about 2.0%P, 5.4%Br

particularly outstanding, 4.5 in. being considered by the authors as maximum for the fabric used. However, the wash resistance was excellent and the other fabric properties were very good. Presumably, more phosphorus would yield the necessary improvement in flame resistance.

An alternative use of a bromine-phosphate compound is the solvent application of the fully brominated triallyl phosphate monomer. A fair

The concept of applying a water-insoluble organic phosphate was proposed at least as early as 1941 [89a]. The suggestion was to use an aqueous dispersion of tris(chloroethyl)-phosphate for yarn or rugs at an add-on of 3–10%. Immersion in the dispersion for 3–5 hr at 75–78° was required. There is little evidence that this idea was pursued until the work described below was carried out.

amount of research has been carried out in the last decade on the use of bromopropyl phosphates as flame retardants.

A recent paper (1968) [90] described the use of tris(2,3-dibromopropyl) phosphate as a retardant for military fabrics. The ester is insoluble in water and, being monomeric, requires a polymeric binder for best adhesion to the fabric. A system was devised whereby the chemicals could be applied from a commercial dry-cleaning machine. The following recipe was suggested:

$Cl_2C{=}CCl_2$ (solvent)	80 parts
$\left(\begin{array}{c}CH_2{-}CH \\ \mid \\ O_2CCH_3\end{array}\right)_x$ (binder)	1.5 parts
$[BrCH_2CH(Br)CH_2O]_3PO$	20 parts

The flame retardant was applied without changes to the machine. No curing is required. Tear strength is reduced somewhat, but not as markedly as for most other treatments. Flame resistance is fair with no afterflaming.

	Char Length (Vertical), in.	
	Initial	After 15 Launderings
8.5-oz cotton sateen	5.1	6.3

Weather resistance and light-fastness was rated as very good by the U.S. Army investigators. The most recent work (1969) has been directed to radiation curing of trialkylphosphate and acrylic acid derivatives [90a]. Since radiation curing is already an accepted textile mill practice, there is much to be said for this approach. It would seem that more work is warranted in this direction to reduce flammability further. Although application from a solvent system is not going to be received as enthusiastically as would a water-based system, the advantages (no cure or other high-temperature treatments) may affect this. Certainly this avenue will be explored fully.

Phosphorus-Nitrogen-Halogen Compositions

In this category are the derivatives of phosphonitrilic halides—$(-PNX_2-)_x$—which have received attention off and on for the last 20 years. The $(PNX_2)_x$ polymers are intriguing to the chemist because of their all-inorganic nature. Research teams in several laboratories are still seeking the first large-scale commercial use for these compounds. In the mid-1950s some work was published by the U.S.D.A. group at New Orleans on allyl phosphonitrilates and their brominated derivatives. This group prepared the following products [91]:

$[PN(OCH_2CH{=}CH_2)_2]_x$,   brominated (A),
(A)

$[PN(OCH_2CH(Br)CH_2Br)_2]_x$,   adduct of (A) with $CHBr_3$

Table 5-11   Comparison of Bromine-Phosphonitrilic Compositions as Flame
Retardants for 8-Oz Cotton Twill [91]

Treatment	Add-on wt. %	Approx. Assay		Laundering	Elmendorf Tear Strength % of Control	Char Length, in.
		% P	% Br			
Partially brominated	12.7	0.8	8.4	None	56	4.3
[—PN(OR)$_2$—]$_x$;				Alk. soap[a]	70	Burned
R = allyl,						entire
cure 10 min, 140°						length
				Detergent[b]	54	5.3
[—PN(OR)$_2$—]$_x$;	14.8	1.0	9.7	None	79	5.5
R = 2,3-dibromopropyl;				Alk. soap	103	Burned
cure 30 min, 100°						entire
						length
				Detergent	82	7.0
[—PN(OR)$_2$—]$_x$—CHBr$_3$	14.7	1.2	8.9	None	77	3.9
adduct; R = allyl;				Alk. soap	107	Burned
cure 6 min, 140°						entire
						length
				Detergent	84	5.4

[a] Boiled 3 hr in 1% soap, 0.5% Na$_2$CO$_3$ soln
[b] After 15 launderings with a detergent (Fed. Spec. CCC-T-1916 Test No. 5556)

The three bromine-containing products were compared on cotton with the
results shown in Table 5-11. The adduct with CHBr$_3$ was prepared as an
emulsion with a persulfate initiator with the method described for triallyl-
phosphate and CHBr$_3$ in the previous section [92]. The ester made from 2,3-
dibromopropanol was made at room temperature in pyridine by letting a
mixture stand 24 hr [93]. Note the substantial loss in tear strength but the
generally excellent resistance to laundering. The resistance of the —P—N=
structure to water has its limits, as shown by its failure to withstand boiling
alkali. Presumably the linkages hydrolyzed to form sodium phosphate salts
of little use as flame retardants.

A series of patents have been issued on the use of (—PNX$_2$—)$_x$ derivatives
such as the amides, the methyl esters, and mixtures of the two:

[—PN(NH$_2$)$_2$—]$_x$, [—PN(OCH$_3$)$_2$—]$_x$, [—PN(OCH$_3$)(NH$_2$)—]$_x$, etc.

The full amide is cured on cotton 4 min at 160° [94] and flame resistance is reported to be retained after immersion in warm water. The mixed amide-methyl esters can be condensed by use of urea [95] or methylol melamine [96] with cellulose to give wash-fast flame resistance. Curing in the presence of either condensing agent is done for 5 minutes at 150°.

Other variations include a pyridinium additive to such a system to achieve water repellency [97]; use of fluoroalkyl esters of phosphonitrilics to add both oil and water repellent qualities [98]; and the use of aniline and chloro-aniline derivatives [99]. None of these treatments have attained any degree of commercial success. $(PNCl_2)_x$, the starting material, has not been

widely produced on commercial scale. Furthermore, the $-\overset{|}{\underset{|}{P}}-N=$ linkage leaves something to be desired in terms of stability.

Phosphites and Phosphonates

Esters of phosphorous acid have been claimed for treating fabrics as far back as 1927 [e.g., tris(chlorophenyl) phosphite] [100], but little was done with phosphorus in the $+3$ oxidation state for this purpose until the 1950s. In 1954 the use of phosphonates (and phosphates, phosphinates, and even phosphine oxides) in urethane coatings for fabrics was suggested as a means of increasing flame resistance [101]. The concept has since been modified and used in other urethane systems (see Chapter 8).

There are three, or perhaps four, well-studied systems based on phosphonates. They are polymers of alkenyl esters used as fiber coatings, reactive chloromethylphosphonic acids, and phosphonoalkanoicamides and variations of these. The polymers are derived from three related monomers [102]:

$$(CH_2{=}CHCH_2O)_2{-}\overset{\overset{\displaystyle O}{|}}{P}{-}CH_2Cl$$

$$(CH_2{=}CHCH_2O)_2{-}\overset{\overset{\displaystyle O}{|}}{P}{-}CH_2Cl, \text{ followed by partial bromination}$$

$$(CH_2{=}CHCH_2O)_2{-}\overset{\overset{\displaystyle O}{|}}{P}{-}CH_2CH_2CN$$

$$(CH_2{=}CHCH_2O)_2{-}\overset{\overset{\displaystyle O}{|}}{P}{-}CH_2{-}\overset{\overset{\displaystyle R}{|}}{C}H{-}\overset{\overset{\displaystyle O}{\|}}{C}{-}OR$$

All were polymerized in a solvent using benzoyl peroxide as a catalyst at 85–100°. Solutions in $CH_3OH/CHCl{=}CHCl$ or emulsions in water were padded on fabric and the product cured at 100° for 30 min. Good wash resistance is claimed. This work is very closely related to the studies with brominated allyl phosphate discussed earlier in this chapter. These particular systems apparently offer relatively little in improvement over the phosphates, and because of higher costs to produce these materials they have not caught on commercially. This is somewhat mystifying in view of some basic studies by Rogovin's group [103] which show that phosphonates (C—P) are superior to phosphates (C—O—P) and that

$$CH_3\overset{\overset{\displaystyle O}{\|}}{-}P\text{—}(OH)_2 > CH_3CH_2\overset{\overset{\displaystyle O}{\|}}{-}P\text{—}(OH)_2 > \phi\overset{\overset{\displaystyle O}{\|}}{-}P\text{—}(OH)_2$$

In the late 1950s techniques were devised to couple phosphonic acids directly to cellulose. The method employed a halomethyl derivative (see p. 118, Chapter 4):

$$Cell\text{—}OH + ClCH_2\overset{\overset{\displaystyle O}{\|}}{-}P\text{—}(OH)_2 \rightarrow Cell\text{—}OCH_2\overset{\overset{\displaystyle O}{\|}}{-}P\text{—}(OH)_2 + HCl$$

The method of Schiffner and Lange [104] was based on urea or other nitrogen bases as acceptors for the HCl released and also to serve as ammoniating agents so as to produce directly an ammonium salt of the phosphonomethyl ether of cellulose. The fabric is immersed in solutions of either the phosphonic acid mixed with the base or prereacted with it. The textile is cured 10–20 min at 150°. The U.S.D.A. technique [105] started with either chloromethyl-phosphonic acid or chloromethylphosphonyl dichloride. Both were converted to disodium chloromethylphosphonate, which was padded on from a caustic solution and cured 30 min at 110°. To obtain fire-retardant properties the sodium ions were replaced with ammonium ions by ion exchange:

$$Cell\text{—}OCH_2\overset{\overset{\displaystyle O}{\|}}{-}P\text{—}(ONa)_2 + xsNH_4^+ \rightarrow Cell\text{—}OCH_2\overset{\overset{\displaystyle O}{\|}}{-}P\text{—}(ONH_4)_2 + 2Na^+$$

Although tear strength was substantially reduced with the urea method, it was but little affected by the caustic method. This again demonstrates the marked stability of cellulose at elevated pH. The flame retardance of cotton so treated is shown in Table 5-12. Of course, the difficulty with such a treatment is the ionic form of the product. Laundering in detergent solutions will inevitably reconvert the ammonium form to the sodium form (or the calcium form in hard water). For this reason the treatment cannot be recommended for wash-fast applications. It would be suitable for such items as upholstery and draperies, where resistance to moisture but not to salts is required. No major commercial use for the method is known at this writing.

It is apparent from the foregoing that the acid groups on the phosphonyl group must be blocked if wash-fast flame resistance is to be achieved.

**Table 5-12   Flame Retardance of Cotton Phosphonomethylated
by the NaOH Method [105]** (Reprinted by permission of The
American Chemical Society.)
80 × 80 Cotton Print Cloth

Treatment	% P	Vertical Char Length, in.
Phosphonomethylated	0.72	Burned entire length
Cross-linked[a]		
phosphonomethylated	0.92	Burned entire length
Double phosphonomethylated	1.66	4.0
Cross-linked		
double phosphonomethylated	1.93	3.8

[a] Cross-linked with dichloropropanol and caustic

Apparently, little or no work has been reported with full esters of chloro-methylphosphonic acid—at least no literature on this has been uncovered in the textile area. For example, based on the stability of simple organic esters, one could certainly tailor a product for good stability to hydrolysis in base. In Chapter 2 the good stability of, for example, triphenyl phosphate was discussed. One would predict that diphenyl chloromethylphosphonate would produce a very stable system. (There might be other disadvantages to such a material; for example, low phosphorus assay.)

A related series of systems has been studied and one of these has been introduced commercially in the past year. These are the esters of phosphono-alkanoicamides and their derivatives [106]:

$$(RO)_2—\overset{\overset{\displaystyle O}{|}}{P}—CH_2CH_2—\overset{\overset{\displaystyle O}{\|}}{C}—NH_2 \qquad (RO)_2—\overset{\overset{\displaystyle O}{|}}{P}—CH_2CH_3—\overset{\overset{\displaystyle O}{\|}}{C}—NHCH_2OH$$

Free amide   Formaldehyde derivative

The preparation of these materials was discussed in Chapter 2 (Table 2-6); it begins with dialkyl phosphite, sodium, and acrylamide [106]. The ester group may be halogenated [107]. The methylol compound has been commercialized [108] and is recommended for use with a triazine resin [e.g., tris(hydroxymethyl) melamine], a surfactant, a small amount of urea, a polyethylene softener, and ammonium chloride. The product is padded on the fabric, dried, and cured 1–2 min at 175° (or 5 min at 160°). Presumably the molecule couples with cellulose:

$$(RO)_2—\overset{\overset{\displaystyle O}{|}}{P}—CH_2CH_2—\overset{\overset{\displaystyle O}{\|}}{C}—NHCH_2OH + HO—Cell$$

$$\rightarrow Cell—OCH_2NH—\overset{\overset{\displaystyle O}{\|}}{C}—CH_2CH_2—\overset{\overset{\displaystyle O}{|}}{P}—(OR)_2$$

The result is a wash-fast treatment. The hand of fabric treated with these compounds appears to be less affected than by any other durable flame-retardant system.

Tesoro and coworkers [109, 110] have published some details of their work on this system and a related one. They first looked at a two-step process:

$$\text{Cell—OH} + \text{HOCH}_2\text{NHCOCH}_2\text{X} \longrightarrow \underset{\text{(A)}}{\text{Cell—OCH}_2\text{NHCOCH}_2\text{X}}$$

$$\text{(A)} + \text{(RO)}_3\text{P} \xrightarrow{\text{Arbuzov}} \text{Cell—OCH}_2\text{NHCOCH}_2\overset{\displaystyle O}{\overset{\displaystyle \|}{—P}}\text{—(OR)}_2$$

The similarity to the product of the one-step reaction of the preceding paragraph is evident. Char lengths with the two-step process were 5–6 in.; strength was decreased appreciably, and stiffness was unaffected or only increased by a small amount. Char length was improved by addition of tris(hydroxymethyl) melamine. Similar results were found with the one-step process, using the ester of phosphonopropionamide hydroxymethylated with formaldehyde. Results of tests on this system are given in Table 5-13. The

Table 5-13   Test Data on Cotton Sheeting (4.3 Oz) Treated with $(EtO)_2P(O)CH_2CH_2C(O)NHCH_2OH$ (NMPA) and Aftertreated $(HOCH_2)_3N_6C_3H_3$ (TMM) [110]

Add-on, % NMPA	% TMM	% N	% P	Vertical Char Length in.	Strength, lb Tensile	Tear
16.5	8.2	5.0	2.0	5.0	53	1.5
16.9	1.2	1.8	2.1	6.0	50	1.7
17.0	0.0	5.3	1.6	4.7	54	1.6
12.6	4.5	3.1	1.6	5.2	49	1.6
6.4	16.6	7.2	0.8	4.5	51	1.5
6.3	10.4	2.9	0.9	5.3	51	1.6
0	0	0	0	Burned entire length	74	3.1

char length can be reduced below 5.0 in. with strength losses (tensile) of 25–30% and a loss of tear strength of about 50%. The fabric is stiffened somewhat. The total add-on is about 20% of the final fabric weight or 25–30% of the untreated, dry fabric weight. This high add-on is necessary because of the relatively low phosphorus content of the phosphonate. [The methylol derivative, $(EtO)_2P(O)CH_2CH_2C(O)NHCH_2OH$, contains only

13% P.] In addition, the cost per pound of phosphorus is relatively high at this writing [108, 111, 112]. The phosphorus in this product is selling at about \$7.50/lb (assuming no value for the rest of the molecule). Price at the moment is felt to be one of the major obstacles to the large-scale treatment of draperies, upholstery, blankets, and bedding. Certainly these substituted phosphonates appear to have something to offer technically. It will be interesting to observe the fate of the materials in the market place and see what the price trend will be. (The reaction of acrylamide using sodium is not likely to be a preferred, low-cost synthetic route.)

Tesoro has also experimented with derivatives of a commercial halogenated polyphosphonate [112a]. A polymeric derivative of uncertain structure from the following reactants

$$\text{H}_2\text{NCH}_2\text{CH}_2\text{NH}_2 + \text{ClCH}_2\text{CH}_2\text{O} - \overset{\overset{\displaystyle O}{\|}}{\underset{\underset{\displaystyle \text{ClCH}_2\text{CH}_2}{|}}{\text{P}}} - \text{O} - \overset{\overset{\displaystyle \text{CH}_3}{|}}{\underset{\displaystyle \text{H}}{\text{C}}} - \overset{\overset{\displaystyle O}{\|}}{\underset{\underset{\displaystyle \text{OCH}_2\text{CH}_2\text{Cl}}{|}}{\text{P}}} - \text{O} - \overset{\overset{\displaystyle \text{CH}_3}{|}}{\underset{\displaystyle \text{CH}}{}} - \overset{\overset{\displaystyle O}{\|}}{\text{P}} - (\text{OCH}_2\text{CH}_2\text{Cl})_2$$

was selected as the best candidate. From this work also came further data on the P—N combined effect. Figure 5-1 shows that this is a pronounced phenomenon and that this sum of P + N must be 4% or so for flame retardancy. The finishes readily resist up to 100 launderings in detergent solution. Like most phosphorus-nitrogen finishes, these are attacked by hypochlorite (at the nitrogen atom, presumably) and too much exposure to chlorine bleach will remove the finish.

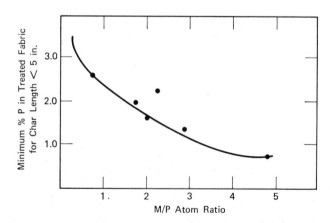

Figure 5-1 Minimum phosphorus content as a function of the N/P atom ratio (from Tesoro [112a]).

Another phosphonate receiving laboratory attention recently is the product of cyanuric chloride and trialkyl phosphite

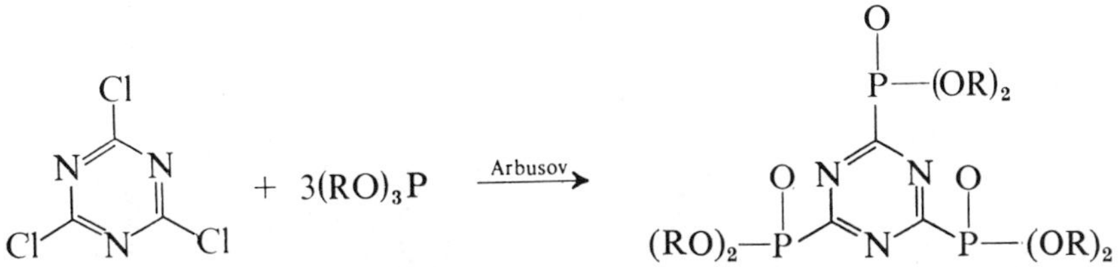

The mixed phosphonyl-amino compounds have been prepared and used

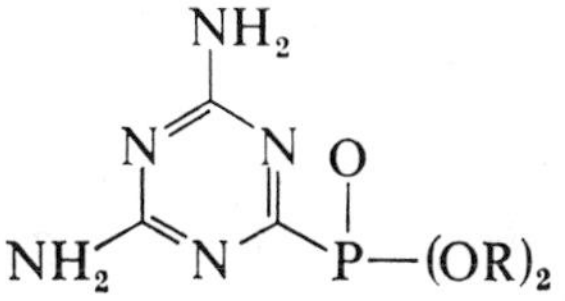

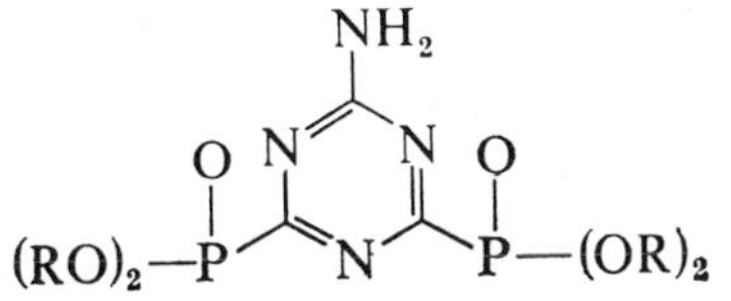

with formaldehyde and methylol melamine for durable cotton finishes. The development is still in its early stages [112b].

Phosphonium Salts, Phosphines, and Phosphine Oxides

More has been written about tetrakis(hydroxymethyl) phosphonium compounds for flame-retardant cotton than about any other flame-retardant topic. Not only are there a large number of review-like papers and many individual reports but also a sizable patent literature. All this centers on the compound known as "tetrakis" or THPC:

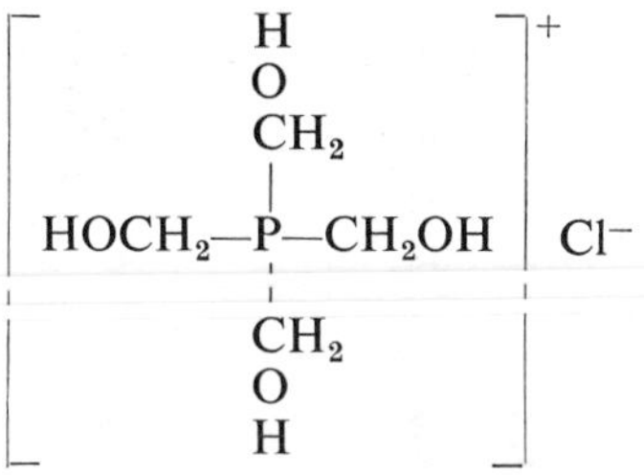

$$\left[\begin{array}{c} \overset{\displaystyle H}{\underset{\displaystyle CH_2}{\overset{\displaystyle O}{|}}} \\ HOCH_2-P-CH_2OH \\ \underset{\displaystyle H}{\overset{\displaystyle CH_2}{\underset{\displaystyle O}{|}}} \end{array} \right]^{+} Cl^{-}$$

The synthesis of this material has been discussed in Chapter 2. As of 1968, there are at least eight variants of the THPC process:

THPC, methylol melamine, urea, heat
THPC, methylol melamine, urea, partial heat cure, partial NH_3 cure
THPC-APO or APS with heat
THPC-brominated allyl phosphate (or phosphonitrilate)
THPC + NaOH, methylol melamine, urea, heat
THPC + NaOH, amide, partial heat cure, partial NH_3 cure
THPC + NaOH, methylol melamine, urea, copper, cure variously
THPC + NaOH, NH_3 cure

Table 5-14 A Chronological Patent History of THPC Treatments

Date	Patent No.	Inventors	Assignee	Invention
Feb. 2, 1954	U.S. 2,668,096	W. A. Reeves, J. D. Guthrie	U.S.A.	THPC + aminized cotton, cure 15 min at 120°; also add N base with THPC
Nov. 21, 1956	Brit. 761,985	J. G. Evans, G. Landells, J. R. W. Perfect, B. Topley, H. Coates	Bradford Dyers and Albright & Wilson	THPC [or $(HOCH_2)_3PO$], hydroxymethyl melamine, $(HOCH_2CH_2)_3N$, dicyandiamide; cure 5 min, 145°
Nov. 27, 1956	U.S. 2,722,188	W. A. Reeves, J. D. Guthrie	U.S.A.	THPC, $(HOCH_2NH)_3C_3N_3$, cure in aq NH_4OH no heat cure
June 11, 1957	U.S. 2,759,569	W. A. Reeves, J. D. Guthrie	U.S.A.	THPC, 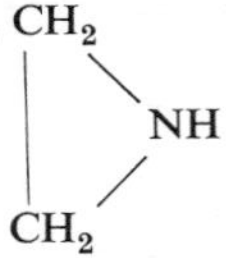 $CH_2\underset{CH_2}{\overset{\textstyle\diagdown}{\diagup}}NH$ copolymer, no heat cure, no loss in tear strength
Oct. 15, 1957	U.S. 2,809,941	W. A. Reeves, J. D. Guthrie	U.S.A.	THPC, various amines to make novel copolymers useful for treating cotton
Oct. 22, 1957	U.S. 2,810,701	W. A. Reeves, J. D. Guthrie	U.S.A.	Emulsion prepared from prepolymerized THPC$(HOCH_2NH)_3C_3N_3$; useful for emulsion coating of fabric
Nov. 5, 1957	U.S. 2,812,311	W. A. Reeves, J. D. Guthrie, J. Warren	U.S.A.	Water-sol prepolymer from THPC, melamine, or $(HOCH_2NH)_3C_3N_3$ prepared by heating soln 5 min, 87–96°; use on cotton
Nov. 21, 1957	Brit. 882,993		Albright & Wilson	THPC, triethanolamine, urea, cure in NH_4OH, no heat cure

Date	Patent	Inventors	Company	Description
Nov. 25, 1958	U.S. 2,861,901	W. A. Reeves, C. Hamalainen, J. D. Guthrie	U.S.A.	THPC, $(HOCH_2NH)_3C_3N_3$, urea plus bromoform polymer with diallylmelamine; cure 6 min, 140°
Nov. 27, 1958	Ger. 1,045,098	J. G. Evans, G. Landells, J. R. W. Perfect, B. Topley, H. Coates, W. A. Reeves, J. D. Guthrie	Bradford Dyers and Albright & Wilson	Nonflammable resin from THPC, $(HOCH_2CH_2)_3N$, $(HOCH_2NH)_3C_3N_3$ cyanamid, heat 1 hr, 130°
Jan. 14, 1959	Brit. 807,236		Albright & Wilson	Bromoform copolymer or brominated allylmelamine with THPC
June 30, 1959	U.S. 2,892,803	W. A. Reeves, J. D. Guthrie	U.S.A.	Add ammonium orthophosphate after standard THPC treatment
Oct. 7, 1959	Brit. 821,503		Albright & Wilson	THPC after treating with haloalkyl phosphonitrilate in 2 steps
Nov. 3, 1959	U.S. 2,911,322	E. Klein, J. W. Weaver	U.S.A.	Field method of setting THPC, methylol melamine, uses NH_4OH; addition of activated C gives resistance to lethal gas
May 9, 1961	U.S. 2,983,623	H. Coates	Albright & Wilson	THPC, urea; prepolymer; cure with gaseous NH_3; no heat cure
Dec. 20, 1961	Brit. 884,785	I. Christopher, W. Greenwood, J. R. W. Perfect, H. R. Richards	Bradford Dyers and Albright & Wilson	Pretreat with urea-formaldehyde resin, then use standard THPC treatment
Aug. 21, 1962	U.S. 3,050,522	H. Coates, J. J. Lawless	Albright & Wilson	Make THPC in soln in presence of urea, treat cotton with soln adding $(HOCH_2NH)_3C_3N_3$
Sept. 18, 1962	U.S. 3,054,698	G. M. Wagner	Hooker Chemical Corp.	Add poly(vinyl-chloride) resin to standard THPC bath, cure 10 sec in ir oven at 380°; tensile strength better than control

Table 5-14 (Continued)

Date	Patent No.	Inventors	Assignee	Invention
Mar. 21, 1963	Belg. 622,752	R. F. Zimmerman, P. E. Hoch, G. M. Wagner	Hooker Chemical Corp.	Use of sulfites to stabilize THPC system to prevent premature polymerization
Apr. 18, 1963	Belg. 623,764 (also see 675,066 for Jan. 13, 1966)	G. M. Wagner	Hooker Chemical Corp.	Use of $MgCl_2$, $Zn(NO_3)_2$, or $R_3N \cdot HCl$ to prevent premature polymerization
July 2, 1963	U.S. 3,096,201	H. Coates, B. Chalkey	Albright & Wilson	Partial polymerization in bath, finish with NH_4OH soln cure
Aug. 20, 1963	U.S. 3,101,278,- 3,101,279	G. M. Wagner, P. E. Hoch	Hooker Chemical Corp.	Covers bath stabilizers as in Belg. 622,752 atd 623,764
Oct. 9, 1963	Brit. 938,989	H. Coates, B. Chalkley	Albright & Wilson	Cure in gaseous NH_3; then finish in aq NH_4OH
Mar. 11, 1964	Brit. 951,988		Hooker Chemical Corp.	Polymer of THPC formed by blowing air through its melt $5\frac{1}{2}$ hr, 120°C; use pre-polymer for textiles
Aug. 25, 1964	U.S. 3,146,212	G. M. Wagner, P. E. Hoch, I. Gordon	Hooker Chemical Corp.	Same subject matter as in Belg. 622,752
Mar. 25, 1965	Neth. Appl. 6,404,149		Hooker Chemical Corp.	Partial heat curing, followed by cure in NH_4OH or $(NH_4)_2SO_4$, Na_2CO_3, $NaHCO_3$
Nov. 23, 1965	U.S. 3,219,478	G. M. Wagner	Hooker Chemical Corp.	THPC system plus poly(vinyl-chloride) and Sb_2O_3
Jan. 1, 1966	Neth. Appl. 6,508,901		CIBA, Ltd.	3 steps: (1) treat with $(HOCH_2NH)_3C_3N_3$; (2) THPC, $(HOCH_2CH_2)_3N$; (3) cure in gaseous NH_3
Jan. 13, 1966	Belg. 675,066		Hooker Chemical Corp.	See Belg. 623,764. Describes partial heat cure, NH_3 cure

Date	Patent No.	Inventors	Assignee/Country	Description
Feb. 22, 1966	U.S. 3,236,676	H. Coates, B. Chalkley	Albright & Wilson	THPC plus NaOH, NH₃ cure, then final cure in NH₄OH
July 28, 1966	Ger. 1,221,605	G. M. Wagner	Hooker Chemical Corp.	THPC standard bath with a polyethylene fabric softener; prevents chlorine retention
Aug. 19, 1966	Neth. Appl. 6,601,930		Hooker Chemical Corp.	Emulsion containing THPC system plus chloroparaffin, Sb₂O₃, and poly(vinyl-chloride); cure 2 min, 165°
Aug. 23, 1966	U.S. 3,268,360	I. V. Beninate, G. K. Drake, Jr., W. A. Reeves	U.S.A.	THPC (or the phosphine oxide) plus diisocyanate; heat cure
Oct. 4, 1966	U.S. 3,276,897	W. A. Reeves, R. M. Perkins, G. L. Drake, Jr.	U.S.A.	THPC, tris(carbamoylethyl)amine, cure in NH₃ gas, then NH₄OH
Nov. 16, 1966	Brit. 1,048,283		Hooker Chemical Corp.	Standard THPC bath plus polyethylene softener; claims no chlorine retention
Mar. 21, 1967	U.S. 3,310,419	G. M. Wagner	Hooker Chemical Corp.	Use NaOH or other alkali with THPC, methylol melamine, urea; cure partially with heat, then with NH₄OH
Mar. 21, 1967	U.S. 3,310,420	G. M. Wagner	Hooker Chemical Corp.	As in previous patent but use acid-releasing substances such as tertiary amine hydro-chlorides (made in situ from the amine and THPC)
Sept. 24, 1968	U.S. 3,403,044	H. Chance, G. L. Drake, Jr., W. A. Reeves	U.S.A.	THPC combined with phosphoramides, e.g. O=P(NH₂)₃
Oct. 1, 1968	U.S. 3,404,022	H. Chance, G. L. Drake, Jr., W. A. Reeves	U.S.A.	THPC polymerized with tris(carbamoyl-ethyl) phosphine or its oxide; NH₃ cure
Jan. 14, 1969	U.S. 3,421,923	G. Guth	Ciba, Ltd.	THPC melamine-CH₂O resin
Feb. 18, 1969	U.S. 3,428,480	G. M. Wagner, R. A. Schad	Hooker Chemical Corp.	THPC, amines, chlorocarbon, PVC

Table 5-14 (Continued)

Date	Patent No.	Inventors	Assignee	Invention
THPC-APO				
May 12, 1959	U.S. 2,886,539	G. L. Drake, Jr., W. A. Reeves, L. H. Chance	U.S.A.	THPC, APO, ethanolamine, methylol melamine, cure 3 min, 160°
June 15, 1960	Brit. 837,709		Albright & Wilson	THPC, ethanolamine, cure 10 min, 145°
Mar. 4, 1965	Neth. Appl. 6,410,278		Hooker Chemical Corp.	THPC, APO, poly(vinylchloride), Sb_2O_3, heat cure
Oct. 11, 1966	U.S. 3,273,497	J. B. Bullock, C. M. Welch	U.S.A.	THPC, APO, thiourea, triethanolamine; the S cpd improves weather resistance
May 9, 1967	U.S. 3,318,659	J. B. Bullock, C. M. Welch	U.S.A.	THPC, APO treatment with vinyl-chloride-acrylate copolymer and Zr acetate; gives water and rot resistance
Other Phosphine Compounds				
Mar. 1, 1961	Ger. 1,102,095	H. Malz, F. Kassack	Farb. Bayer A.-G.	$(NH_2\langle O\rangle)_3PO$, HOAc, $(HOCH_2NH)_3C_3N_3$ in dioxane; cure 10 min, 150°; halophenyl phosphine oxides may be used

Date	Patent	Author(s)	Company	Formula/Conditions
Mar. 24, 1966	Neth. Appl. 6,512,332		Farb. Hoechst A.-G.	$ClCH_2CH\overset{\displaystyle O}{\overbrace{-}}CH_2$, $(HOCH_2)_3P$, cure in NH_3 gas, then in NH_4OH
Aug. 30, 1966	U.S. 3,270,052	L. H. Chance, W. A. Reeves, G. L. Drake, Jr.	U.S.A.	$(NH_2-\overset{O}{\overset{\|}{C}}-CH_2CH_2)_3P$, CH_2O, $MgCl_2$, cure 3 min, $160°$
July 16, 1959	Ger. 1,061,513	M. Reuter, E. Wolf, L. Orthnier	Farb. Hoechst A.-G.	$(HOCH_2)_3P$, $ClCH_2CH\overset{\displaystyle O}{\overbrace{-}}CH_2$ with methylol melamine
May 15, 1963	Brit. 926,268		American Cyanamid Co.	$C_8H_{17}PH_2$, toluenediisocyanate copolymer
Jan. 14, 1969	U.S. 3,422,048	J. F. Cannelongo	American Cyanamid Co.	$[R'_3-\overset{+}{P}-R-\overset{+}{P}-R'_3]2X^-$
Jan. 14, 1969	U.S. 3,421,834	M. Grayson	American Cyanamid Co.	$Bu_3\overset{+}{P}CH_2CH_2OH$ or $Bu_3\overset{+}{P}CH_2CH_2O-\overset{O}{\overset{\|}{C}}-CH_3$ or $Bu_3\overset{+}{P}CH=CH_2$
Mar. 25, 1969	U.S. 3,434,981	C. F. Baranauckas, I. Gordon	Hooker Chemical Corp.	Mixed hydroxyalkyl-alkyl phosphonium salts
Mar. 25, 1969	Can. 809,123	V. J. Vullo	Hooker Chemical Corp.	$(CH_2=CH)_3PO$, Na_2CO_3, heat

Presently, almost all the permanent fire-retardant systems for cotton apart from military applications (for these, see the next section) are based on either THPC or the phosphonoalkanoicamides.

The patent history is tabulated in Table 5-14. The key advances in the art are clearly shown chronologically. The development owes its success to the research efforts of teams at U.S.D.A. and Hooker Chemical Corp. and the groups at Bradford Dyers Associated, Ltd., and Albright & Wilson, Ltd., in England. The nonpatent literature is, however, primarily from the U.S.D.A. group.

In 1953 the U.S.D.A. researchers first discussed their new method [113]. They found, in working with aminized cotton, that THPC would react with amino groups to give a composition stable to boiling caustic. The amino groups did not have to be prereacted with cotton but could participate as part of a comonomer. Thus, melamine and THPC could be copolymerized within and on the fibers to give a polymeric coating which could not be readily removed by laundering. The methylol groups on THPC undergo many of the reactions of typical formaldehyde derivatives [114], and it was quickly found that it would condense with methylol melamine to form a durable polymer [115, 116]:

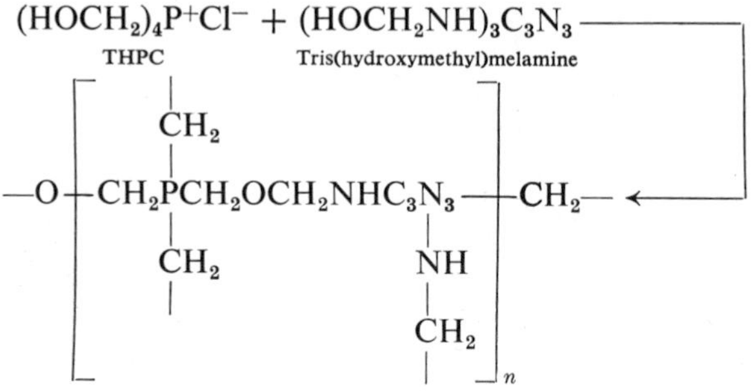

$$(HOCH_2)_4P^+Cl^- + (HOCH_2NH)_3C_3N_3$$

THPC   Tris(hydroxymethyl)melamine

$$-O\!-\!\left[\begin{array}{c} CH_2 \\ | \\ CH_2PCH_2OCH_2NHC_3N_3 \\ | \\ CH_2 \\ | \end{array} \right.\!-\!\left. \begin{array}{c} \\ \\ -CH_2- \\ NH \\ | \\ CH_2 \\ | \end{array} \right]_n$$

The quaternary phosphonium structure was first thought to be largely converted to the phosphine oxide during fabric processing, but results of later studies indicate [117] that the phosphonium structure is retained at least in part. Indeed, a study of the APO-THPC system has shown [118] that the THPC acts to a considerable extent as a tetrafunctional reagent and that much of the chlorine is retained by the polymer. However, there is some decomposition to the phosphine oxide (via the phosphine) with concomitant liberation of hydrochloric acid. Urea and/or triethanolamine are therefore added to act as acid acceptors. Since acid serves as a catalyst for the polymerization, the bases also serve as stabilizers of the bath to minimize premature reaction [119]. Thus, the first THPC bath contained four main ingredients [120]:

15–16% THPC
9–10% trimethylol melamine
10% urea
3% triethanolamine plus a wetting agent
61–63% H_2O

Cloth is padded with the solution, squeezed in with high pressure rolls, dried to about 15% moisture at about 85° (4½ min), and then cured 4½ min at 140°. Residual soluble salts are removed by washing. The solution (bath) should be kept cool and should be fresh to avoid undue prepolymerization before curing. Drying must be done at temperatures below 100° to avoid surface curing. Curing can be carried out at various temperatures: higher temperatures require shorter curing times but tend to reduce tearing strength; lower temperatures correspond to longer curing times.

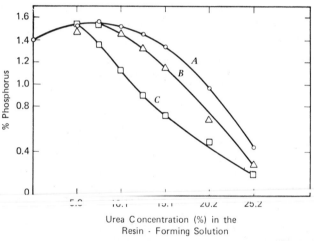

Figure 5-2 Effects of urea concentration on the durability of phosphorus in THPC-resin-treated fabric. The resin-forming solution contained 16% THPC, 9.7% Resloom HP, and 3% TEA 119. Curve A: Phosphorus introduced by the resin treatment. Curve B: Phosphorus remaining after 10 laundering cycles. Curve C: Phosphorus remaining after 3-hr boil in soap and sodium carbamate solution.

Figure 5-2 shows the effect of urea on the phosphorus retention of treated fabric; Figure 5-3 shows the tear strength as improved by the addition of softener (we have seen this effect of softeners several times earlier in this chapter). Table 5-15 gives some of the early data from the U.S.D.A. laboratories.

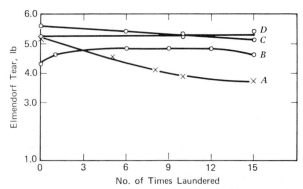

Figure 5-3 Effect of laundering on the tear strength of THPC-resin-treated 8-oz twill with permanent-type softener, compared with intreated control fabric 119. Curve *A*: Untreated control. Curve *B*: THPC-resin—treated. Curve *C*: THPC-resin—treated with 0.5% cetylamine as softener. Curve *D*: THPC-resin—treated with 0.5% stearamide as softener.

The treated fabric withstood 15 launderings satisfactorily and only lost 15%–20% of the resin by analysis after 3 hr in boiling soap/soda ash solution [113, 119]. The breaking strength of treated fabric is unaffected by the presence of the polymer, but the tear strength is reduced; addition of a

Table 5-15   Application of THPC Flame-Retardant Process to Various Fabrics [119]

| Kind of Fabric | Composition of Solution, % | | | | | Dry Add-on of Resin, % | Char length, in. |
	THPC	Methylol-melamine	Urea	TEA	Wetting Agent		
Cotton belting, 28 oz	15.5	9.3	9.8	3.9	1.0	11.6	0.3
Cotton blanket, 10.7 oz	11.1	6.7	5.2	2.8	· · ·	14.2	1.3
Cotton napped bathrobe fabric	16.5	10.0	7.8	4.1	1.0	11.5	2.3
Cut pile cotton rug	13.6	8.2	8.6	3.4	1.0	7.4	2.2
Cotton sateen, 8.5 oz	17.5	10.5	11.0	4.4	· · ·	18.0	3.2
Mercerized cotton drapery, 8.4 oz	16.9	10.2	10.7	3.5	· · ·	20.0	3.3
Upholstery fabric, cotton warp, wool filling	15.3	9.2	9.6	5.0	· · ·	16.0	1.9
Cotton whipcord	21.1	12.8	6.7	5.2	· · ·	21.6	2.4
Bleached cotton sheeting	16.8	10.2	7.9	4.2	· · ·	22.6	3.7
Viscose rayon, 8.3 oz	19.1	10.9	11.4	3.5	· · ·	20.1	2.1
Viscose napped flannel, cotton warp	15.4	9.3	9.7	4.1	· · ·	25.8	3.1
Cotton oxford, 6.5 oz	18.1	10.9	11.4	4.5	1.0	23.2	3.8
Cotton percale, 4 oz	16.4	9.9	7.8	4.1	· · ·	20.0	3.7
Cotton marquisette	15.7	9.4	9.9	6.2	· · ·	32.7	5.2
Low-count cheesecloth	15.3	9.2	9.6	3.4	· · ·	28.9	4.0

softener corrects this problem to some extent. The treated fabrics exhibit good crease resistance and mildew resistance and withstand prolonged burial in the ground—a test for outdoor textiles. The fabric withstands up to 3 months' weathering before its flame-retardant properties are lost.

The high-temperature curing step was considered undesirable because of the reduction in tearing strength, believed to be caused by this heating step. It was found early in the studies that addition of NH_4OH to the treating bath would cause polymerization [120]. This in turn led to the discovery that a cure could be effected by exposing fabric padded with THPC-methylol melamine to NH_3 gas [121]. The use of gaseous ammonia permits uniform distribution of the curing agent without undue migration of the still soluble prepolymer. The curing is completed by immersing the fabric in a solution of NH_4OH. The fabric properties are, as expected, somewhat better than with heat curing; however, there is little or no cross-linking of the resin and thus little or no crease resistance. The finish so produced is claimed to be very resistant to laundering [122].

The ingredients in the formula have been varied with partial success. Ethyl carbamate (A) and N,N',N''-tris(2-carbamoylethyl) amine (B) were

$$\underset{\text{(A)}}{C_2H_5O\overset{\overset{\displaystyle O}{\|}}{C}NH_2} \qquad \underset{\text{(B)}}{(H_2N\overset{\overset{\displaystyle O}{\|}}{C}CH_2CH_2)_3N}$$

evaluated as replacements for urea. It was concluded that (B) was the better of the two and superior to urea in that excellent hand and strength were obtained along with good flame retardance at relatively low add-on. A comparison is given in Table 5-16. Note the improvement in stiffness, tear strength, and resistance to flex abrasion. The char length is slightly poorer than is obtained with urea but the percentage of phosphorus is lower for the example shown. It appears that materials superior to urea would be used if

Table 5-16 Comparison of Urea and N, N', N''-Tris(2-Carbamoylethyl) Amine (TCEA) on O.D. Sateen: Pilot Plant Data [120]

	THPC-Urea	THPC-TCEA	Untreated
Char length, in.	3.0	3.8	· · ·
Glow time, sec	0.6	0.2	· · ·
Phosphorus, %	2.0	1.8	· · ·
Nitrogen, %	2.1	· · ·	· · ·
Stiffness $\times 10^{-4}$ (warp), in. lb	28.5	13.9	9.6
Breaking strength (warp), lb	134	118	116
Tearing strength (warp), g	4250	5280	5800
Flex abrasion (warp), cycles	834	1126	923

cost were no object. The economic advantage of urea will be very hard to offset.

It is apparent that the release of HCl in curing THPC is a problem and, to some extent, the potential for the release of irritating vapors of form-aldehyde may also be obnoxious. Clearly, a way around these difficulties would be desirable. One way is to exchange the chlorine for something less irritating, namely, hydroxide:

$$(HOCH_2)_4P^+Cl^- \qquad (HOCH_2)_4P^+OH^-$$
$$\text{THPC} \qquad\qquad\qquad \text{THPOH}$$

Treatment of THPC with base apparently does not normally yield the phosphonium hydroxide (THPOH) but instead yields a mixture of phosphines and phosphine oxides. A study [117] of the THPC—NaOH—H$_2$O system by P^{31} nmr has clearly shown that THPOH is not isolable and the results indicate that the following species are formed instead:

$$(HOCH_2)_3P \qquad (HOCH_2)_3PO \qquad HOCH_2OCH_2P(CH_2OH)_2 \qquad \text{etc.}$$

The methylol ether or hemiacetal shown is formed by the reaction of the formaldehyde liberated by the decomposition of THPC. In whatever way the reaction may proceed, the result is indeed to remove the HCl and to make less probable the release of formaldehyde on curing. In work published prior to late 1968, 1 mole of THPC was reacted with 1 mole of NaOH and the NaCl removed by filtration from a methanol-water solvent. To the resulting solution was added methylol melamine, urea, and so forth, and the fabric was then padded, dried, and cured by gaseous NH$_3$ [123] or by heating at 150° [124]. The treated fabric has excellent durability to laundering, gives char lengths in the range of 2.5–4.5 in., exhibits wash-wear and per-manent press properties, and is resistant to chlorine bleaching. The tear strength is only moderately reduced and could probably be brought back to that of the control by adding a softener. Some of the properties are shown in Table 5-17.

The recommended bath composition [124] is the first entry in the table, namely, 2:4:1 molar ratios. This, for a bath at 30% solids (not including the NaCl formed), produces the following formulation:

THPC	15.0 g/100 g solution
NaOH	3.2
Urea	9.5
Tris methylol melamine	7.0

Note that triethanolamine is not needed because of the absence of HCl.

A recent paper [124] recommends at least 15% resin add-on, drying 4 min at 85° and curing 4 min at 150°, followed by washing. As is the case for all

Table 5-17 Properties of Treated Fabrics Using THPC-NaOH Reaction Product [124]

Molar Ratio THPC-NaOH Reaction Prod/Urea/Methylol Melamine	Resin Add-on, %	Flex Abrasion, cycles	Breaking Strength, lb	Tear Strength, g	Char Length, in.			Phosphorus, %	
					Orig.	After Boil[a]	After 15 Washes[b]	Orig.	After 15 Washes
2:4:1	18.4	217	143	2130	2.1	2.3	2.2	3.5	2.2
1:1:1	21.7	72	109	1240	2.4	2.2	2.7	3.3	2.3
1:2:1	20.2	126	132	1660	2.5	2.2	2.7	3.1	1.9
1:1:2	21.5	43	107	1190	2.4	2.6	2.7	2.3	1.5
4:1:2	18.3	143	130	1800	3.1	2.6	2.8	4.8	3.1
2:2:1	19.1	181	132	1860	2.5	2.1	2.4	4.3	2.8
Control	...	307	142	2920	...	...	...	0.02	...

[a] 3 hr in boiling sodium soap and soda ash solution
[b] Laundering in home washer using synthetic detergent

THPC treatments, the fabric should be desized and well scoured before treatment to permit the maximum movement of the agents into the fibers. The most recent papers compare various curing techniques and indicate that the ammonia cure leaves the fabric in relatively good condition. Tear strength is reduced substantially but stiffness is affected relatively little [124a]. In an even more recent development [124b] a one-shot system has been devised wherein the ammonia and THPOH are used together. This is made possible by the presence of copper ions which form complexes with the ammonia until broken up by curing at 150–160°. This simplifies processing layout considerably by eliminating gaseous ammonia as a reactant.

The elimination of the HCl problem is a step forward in improving the THPC method and more such advances can be expected. For example, it now appears that there is no need to remove the by-product NaCl before treating the fabric. The THPC can be decomposed directly in the treating bath without losing efficacy [125]. The latest formula shows NaOH present in molar concentrations equal to that of THPC and no methylol melamine or urea. An ammonia cure is used. The drying step is not particularly critical with this formulation. Add-ons of 16–17% are recommended. Although it is likely that ways will be found to improve performance still further by use of related methylol phosphine oxides, it is unlikely that cost/performance will be markedly improved over a system which uses THPC as a key ingredient. This is because it is unlikely that any synthetic routes will be uncovered that will be more economical than that to THPC (raw materials: PH_3, CH_2O, HCl). Many of the alternatives that one can envisage will use THPC itself as a raw material.

Several systems have been developed which use other monomers containing phosphorus along with THPC. Two of these were promoted briefly in the mid-1950s and were essentially two-step processes using other systems already developed. In one, the bromoform-triallyl phosphate emulsion treatment was used, initially as a separate coating applied over a THPC treatment and then as a single combination application [126]. Flame resistance was good for most fabrics, but somewhat reduced compared to the regular THPC method for lighter fabrics. Tear strength was not particularly good.

Better results were found with a THPC solution mixed with an emulsion of a bromoform-allylphosphonitrilate (made as described earlier in this chapter) [127]. A one-step process was used. Treated fabrics had excellent flame resistance and good durability to washing, but they still had relatively poor tear strength. In terms of flame resistance, the treatment seems somewhat more efficient than the straight THPC method. Neither of these combinations has become commercially important, probably because the brominated compounds are not readily available at low cost and the increment in performance cannot justify their use.

Surely the best known of the combinations is the APO-THPC system, in use commercially in the United States in recent years. The chief advantage of the method is the high efficiency, that is, good flame resistance at relatively low add-on [128–130]. APO is a very reactive monomer and, no doubt, reacts with both the hydroxyl groups of cellulose and with THPC. The first stage of the latter may be illustrated as shown [118]:

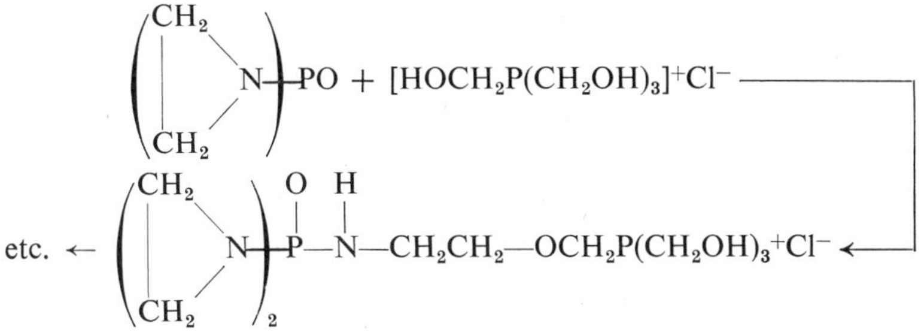

$$\left(\begin{array}{c} \diagup CH_2 \\ | \\ \diagdown CH_2 \end{array} N\right) PO + [HOCH_2P(CH_2OH)_3]^+Cl^- \longrightarrow$$

$$etc. \leftarrow \left(\begin{array}{c} \diagup CH_2 \\ | \\ \diagdown CH_2 \end{array} N\right)_2 \overset{O}{\underset{|}{P}}-\overset{H}{\underset{|}{N}}-CH_2CH_2-OCH_2P(CH_2OH)_3{}^+Cl^- \longleftarrow$$

That THPC may react to a considerable extent as a tetrafunctional monomer has been pointed out by Bullock and Welch [118]. They found that 81–96% of the chlorine in THPC remains bound in treated fabric and that mole ratios of APO to THPC greater than 1.0 were required for maximum durability. Neither fact necessarily proves that THPC remains as a phosphonium compound but the results are suggestive. The fact that the system is essentially neutral in pH, an environment in which the phosphonium structure is expected to be reasonably stable, further supports this idea.

The treating baths used in most of the research were at a 1:1 mole ratio; the following is the composition of one by weight:

17.3%	THPC
13.7%	APO
4.3%	Triethanolamine
0.2%	Surfactant
Balance	H_2O

The solution is kept at room temperature by cooling, if necessary. Wet add-ons of 20–70% are obtained; a total add-on of dry resin of as low as 8% is effective. (This is easily the lowest level of resin found effective of any of the commercial permanent treatments.) Proper attention to drying is important. Drying should be done with low tension on the fabric at 85° as soon after padding as possible. Care must be taken to insure uniform drying to minimize migration and stiffening. Temperatures above 85° will aggravate this. Heat curing can be 4 or 5 min at 140° or 3 min at 165°. The fabric is then washed to remove residual soluble salts.

Table 5-18 Properties of APO-THPC-Treated 8.5-Oz O.D. Sateen [129]

		Untreated Control[a]			APO-THPC-Treated Fabric					
					10% Add-on			15% Add-on		
Test		No. of Launderings			No. of Launderings			No. of Launderings		
		0	5	15	0	5	15	0	5	15
Thread count	W	87.0	86.6	87.1	86.4	88.4	88.4	86.2	88.4	89.6
	F	56.0	57.0	58.6	54.6	57.2	59.0	54.4	56.2	59.6
Width, in.		39.3	…	40.0	40.1	…	39.3	40.1	…	38.8
Weight, oz/yd^2		9.0	…	8.8	9.65	…	10.42	9.96	…	11.2
Thickness, in.		0.0248	0.0212	0.0211	0.0190	0.0211	0.0232	0.0190	0.0227	0.0250
Stiffness, bending moment $\times 10^4$ in. lb	W	13.5	6.8	13.8	14.2	11.6	14.2	36.6	9.0	15.5
	F	12.0	6.7	13.4	18.8	7.4	11.9	16.3	7.0	15.0
Char length, in.		…	…	…	2.97	3.68	4.12	2.65	2.77	2.87
% N		…	…	…	1.63	1.50	1.31	2.12	2.04	1.95
% P		…	…	…	1.40	1.20	0.91	1.76	1.87	1.39
Strip breaking strength, lb	W	134.0	120.8	125.5	123.2	115.0	107.4	113.6	96.2	92.6
	F	118.6	100.0	115.2	97.0	86.2	96.6	88.6	97.2	99.0
Elongation, %	W	11.6	13.2	13.5	7.9	11.3	13.5	6.3	7.8	19.9
	F	21.8	19.3	20.0	20.7	19.7	21.7	20.9	19.3	12.2
Abrasion resistance, flex	W	1856	443	228	1109	1263	1027	482	654	704
	F	2267	752	393	2252	1513	1416	1062	928	1048
Elmendorf tear, lb	W	14.8	8.3	6.2	10.6	8.4	8.6	8.8	7.1	6.9
	F	20.9	10.1	7.9	11.7	9.5	10.2	9.1	7.5	6.8
Tongue tear, lb	W	15.0	7.4	6.2	8.2	7.1	7.6	7.7	6.1	6.2
	F	15.9	9.5	8.6	10.8	8.1	8.9	8.8	7.8	8.4
Crease resistance (Monsanto W + F)		229	210	225	270	237	243	275	250	246
Breaking strength[b] lb, after chlorine bleach	W	118.5	134.5	…	102.4	123.4	…	97.4	100.6	…
chlorine bleach + scorch	W	127.0	127.9	…	101.8	112.4	…	101.2	105.8	…

[a] Fabric rinsed one time in hot water and frame-dried

[b] AATCC Tentative Test Method 92-1958

Table 5-18 provides data on O.D. sateen fabric so treated. Inspection of the table shows that on every count the performance of APO-THPC-treated fabric matches the control or exceeds it. Thus, there is little stiffening, good flame resistance, high strength (tear strength is especially good as compared to all other treatments discussed so far), and sufficient resistance to chlorine bleach. Table 5-19, taken from the same report, shows that army duck fabric again has very good properties and excellent flame retardance.

Table 5-19 Properties of Army Duck Fabric Commercially Treated with APO-THPC [129]

Test	Untreated	APO-THPC-Treated	
		Commercially	SURDD
Width, in.	32¼	31¼	· · ·
Thread count	53 × 43	56 × 40	· · ·
Weight, oz/yd²	10.1	10.9	· · ·
Elmendorf tear[a] (warp), lb	9.3	6.3	6.4
Afterflame, sec	· · ·	0	0
Afterglow, sec	· · ·	1	0
Char length, in.	· · ·	3.2	2.8
Char length, in. after 3-hr soap-soda boil	· · ·	3.3	2.7
Breaking strength (warp), lb	124.6	102.4	117.0

[a] Application of a softener to the treated sample would increase the tearing strength, making it about comparable to the untreated fabric

The key feature of this method is its effectiveness at low levels of treatment. At x% add-on the cost should be 30% less than for many other permanent treatments. This makes it very attractive.

However, it should be pointed out that in 1968 both APO and THPC were selling at prices such that the phosphorus cost was from $5.50 to $6.25 lb. (This is not a completely fair statement because the nitrogen in APO is worth something as a synergist.) Nonetheless, the cost of the chemicals is extremely high and is a limiting factor in the acceptance of these processes. Thus far, they are being used, but only for materials which must be treated to fulfill either fire codes or military specifications. With the new amendments to the Flammable Fabrics Act, the need may expand and the prices may come down because of increased volume.

Not only APO but its thio-analog, APS, has been studied [128]. APS is more stable than APO and therefore is of interest. APS is likely to be more expensive than APO. APO manufacture has recently been discontinued by its major supplier. The reasons have not been given, but one can guess that this was caused by (*a*) its toxicity as a monomer, presenting handling difficulties in the fabric finishing plants, and (*b*) the small volume of business for flame retardants at that time. This second problem, as indicated above, may be helped by new legislation. Nonetheless, it is a disquieting development in view of the very good performance of the APO-THPC system.

Two sidelights of some interest are the study of various THPC systems on stretch materials and on very light fabrics or fabrics of high surface area. Interestingly, the flame-retardant finishes in no way impair stretch properties [128a]; indeed some improvement in characteristics is noted, for example, in wash-wear performance. For high-surface-area fabrics, a surface treatment may be obtained by spraying on the solution rather than padding it to render only the surface flame resistant. With napped or shag fabrics, this method avoids difficulties associated with the compressive action of the padding process. Less effect on strength is encountered with this technique since the chemicals do not permeate the fabric. Good flame resistance is obtained [128b].

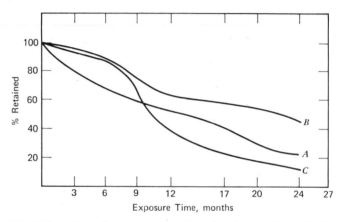

Figure 5-4 Effect of weathering on strength and resin retention of olive-drab sateen treated with various mole ratios of APO-THPC 118. Curve *A*: Phosphorus 1:1. Curve *B*: Phosphorus 2:1 and 3:1. Curve *C*: Strength all ratios.

A recent paper discusses the application of THPC alone or with polyvinyl chloride or 2,3-dibromopropyl phosphate for cotton-polyester fiber blends. These are more difficult to treat but success can be achieved with a THPC-NaOH, methylol melamine system using the brominated phosphate ester and a softener, for example [129]. This kind of study is just beginning to be undertaken. In connection with military fabrics, the U.S.D.A. group has conducted a number of weathering tests on treated fabrics [118, 130–132]. Despite the fact that treated fabric can even withstand boiling alkali, there is a gradual loss of phosphorus and flame retardance during prolonged outdoor exposure. This can be reduced for the THPC method by adding a finishing coating of a vinyl resin [130]. THPC systems are bacteriocidal [131], which helps in resisting rot and mildew. A very detailed study of the performance of fabrics treated with the APO-THPC system as a function of weathering has recently been reported [118]. To obtain best results, a higher ratio of APO to THPC is required, for example, 2:1 instead of the usual 1:1. The loss in phosphorus observed may be caused by photochemical processes; it is apparently not a bacteriological effect or simple water leaching. Figure 5-4 shows some of the results.

The foregoing data have been summarized for various segments of the industry [133]. Of interest to the potential finishing mill are several papers [134, 135] which discuss plant layout and plant costs, including, particularly, capital estimates for an entirely new operation. It is pertinent to point out here that all patents issued to U.S.D.A. are available royalty-free and licenses may be obtained on request [131].

As indicated in the table of patents (Table 5-14), some research has been done on other phosphines and their derivatives as fire retardants. Boyer and Vajda [136] reviewed some of these materials (along with many others) as of 1964. A Russian group adapted THPC to polyvinyl alcohol fibers using dimethylol urea [137]. No data were given on the properties of the treated fibers in fabric form.

In an attempt to combine features of THPC and APO in one molecule, the following scheme has been investigated [138]:

$$P_4 + 2KOH + 4H_2O + 9CH_2{=}CHCONH_2 \rightarrow$$
$$K_2HPO_3 + 3\ OP(CH_2CH_2CONH_2)_3$$
$$TCEP$$

TCEP is then methylolated with formaldehyde, $Zn(NO_3)$ or $MgCl_2$ catalyst added, and fabric padded, dried, and cured 5 min at 145°. (The phosphine was also prepared and studied). Only fair flame retardance was obtained and the treatment reduces fabric strength considerably The flame resistance is definitely inferior to that from the regular THPC method. High wrinkle resistance is obtained.

A recent patent [139] refers to the use of vinyl phosphonium compounds. The vinyl group reacts with the cellulosic hydroxyl group in the presence of alkali. For example, methyltrivinylphosphonium iodide is added to cotton with dimethylol ethylene urea and NaOH. Curing is carried out for 3 min at 150°. Crease resistance is obtained and, presumably, fire resistance, although that is not claimed in the patent. The table of patents indicates the beginnings of other sophisticated applied technologies based on phosphines or their oxides. Another example is the reaction product of epichlorohydrin with a methylol phosphine. We can expect much more of this approach in the years immediately ahead.

The various methods described are intercompared at the end of this chapter.

HALOGENS, ANTIMONY, AND ANTIMONY–HALOGEN COMBINATIONS

Halogens

The halogens by themselves are relatively ineffective as flame retardants for cellulose. However, there is some benefit from a halogen treatment, and the addition of secondary additives, especially certain metal oxides, produces excellent flame retardance. An early reference [140] claimed good results with halogenated aromatics: pentachlorophenol, trichlorophthalic acid, and the like. Chlorinated polyphenyls were suggested in the 1930s [141], and more recently, sophisticated polymers based on halogenated isocyanates [142] and halogenated polyesters [143]. Direct substitution on cellulose has been discussed in reference 28 of Chapter 4 [144]. In one basic paper the relative effectiveness of several inorganic halides was evaluated [145]. The results are shown in Table 5-20. The fluorides are ineffective; the authors imply that this

Table 5-20 Minimum Add-on of Inorganic Halides Required to Impart Flame Retardance to 8-oz Cotton Twill [145]

Compound	Add-on Required, $\%^a$			
	F	Cl	Br	I
NH_4^+	$\cdots$	>24	5–6	4–7
Li^+	>30	6–7	5–6	4–5
Mg^{2+}	>40	7–9	3–4	5–7
Zn^{2+}	14–17	5–6	7–8	$\cdots$

^{a} Add-on required to reduce char length below 10 in. in the vertical test.

is because there is no detectable evolution of fluoride radicals into the gas phase at burning temperatures, whereas, for example, chlorine radicals are given off during combustion of cotton treated with lithium chloride. Bromine is more effective than chlorine in lithium and magnesium salts, not in the zinc salt. The margin of superiority of bromine over chlorine is about the same factor of 2 cited in Chapter 3. Iodine is about as effective as bromine.

As far as is known, there are no commercial treatments of cellulosic textiles using halogens alone.

Antimony Compounds

The successful use of antimony compounds in the absence of either halogens or other transition metal oxides has not been reported in the literature. The most commonly available form of antimony is the oxide—Sb_4O_6—often referred to by the formula Sb_2O_3. This compound is essentially inert when combined with cellulose and exhibits no flame retardant properties. One patent claims effectiveness for Sb_4O_6 when applied to cotton with casein, glue, or a urea formaldehyde resin [146]. Performance of Sb_4O_6 is somewhat improved by adding Fe_2O_3 or SnO_2 [147].

When halogens are added the picture is remarkably changed.

Antimony-Halogen Compositions

The presence of halogens with antimony oxide serves to activate both components and yields fire retardant systems of high quality. On heating, the oxide is apparently converted into one or another of the chlorides:

$$R \cdot HCl \xrightarrow{\text{heat}} HCl + R$$

$$Sb_4O_6 + HCl \longrightarrow \underset{\text{dec } 170°}{SbOCl} \longrightarrow \underset{\text{bp } 223°}{SbCl_3}$$

The vapor of the latter compound is both a dehydrating agent and a free radical chain breaker. Cellulose treated with Sb_4O_6 and a chlorine-bearing compound capable of yielding HCl at a low temperature was heated in air and charring observed at 280°, a temperature at which untreated cotton remains unaffected. The effect is like that from phosphoric acid. In addition, analysis showed that most of the antimony had been converted to the chloride and finally had distilled from the residue. Only 20% of the antimony remained behind [148]. Thus, there are two distinct effects: one in the solid (or liquid) phase involving charring from dehydration, and the other in the gas phase apparently acting by interrupting free radical oxidation chains.

These properties have been used in developing several commercial treatments for cotton. Among these are the only presently accepted methods for producing high performance military fabrics (tentage, paulins, etc.) with the necessary durability to weather. The research leading to this began in the 1930s and built on the foundation laid 25 years before by W. H. Perkin. Snyder apparently was the first to combine Perkin's stannate treatment with a chlorinated vinyl compound [149]; Leatherman followed with an improved system and switched to chlorinated diphenyls [150]. Shortly thereafter he

Table 5-21 A Patent History of the Antimony Oxide-Chlorocarbon Treatment for Cotton

Date	Patent No.	Inventors	Assignee	Invention
1932	U.S. 1,885,870	C. Snyder		SnO_2, vinyl chloride
Aug. 20, 1935	U.S. 2,012,686	M. Leatherman		SnO_2, chlorodiphenyls
Oct. 20, 1942	U.S. 2,299,612	E. C. Clayton, L. L. Heffner	William E. Hooper & Sons, Inc.	Sb_4O_6, chloroparaffins
Feb. 6, 1945	Can. 425,400	C. F. Martin	Dominion Oilcloth & Linoleum Co., Ltd.	Sb_4O_6, chloroparaffins, plasticizers, $MgSiO_3$
Sept. 17, 1946	U.S. 2,407,668 (also see U.S. 2,439,395; 2,439,396)	M. Leatherman	U.S.A.	$ZnCO_3$, tricresyl phosphate, vinyl chloride-vinyl acetate or acrylate copolymers
Sept. 23, 1946	U.S. 2,427,997	C. B. White		Pretreat with Na_2CO_3, then with $SbCl_3$; ppts Sb salt, insol in H_2O
Feb. 17, 1948	U.S. 2,436,216	E. W. Leatherman		Add ZnO, $Fe(OH)_3$, and/or metal soap to Sb_4O_6, chloroparaffin system to reduce afterglow
Feb. 8, 1949	U.S. 2,461,302	J. Truhlar, A. A. Pantsios	Rudolf F. Hlavaty	Bath of $SbCl_3$, organic phosphite, steam-treated fabric to remove HCl. Produces finely dispersed Sb_4O_6
Aug. 8, 1950	U.S. 2,518,241	J. F. McCarthy	Treesdale Laboratories, Inc.	Sb_4O_6, chloroparaffins, phenolformaldehyde resin, and $CaCO_3$. Claims improved stability, some afterglow resistance

Date	Patent	Inventors	Assignee	Notes
Aug. 22, 1950	U.S. 2,520,103	A. M. Loukomsky, R. H. Kienle, T. F. Cooke	American Cyanamid Co.	Sb_4O_6 colloidal dispersion plus vinyl chloride copolymer, monoguanidine phosphate, cyanamid. The phosphate prevents afterglow
Aug. 22, 1950	U.S. 2,519,348	M. R. Burnell, W. J. van Loo	American Cyanamid Co.	Improves on previous patent by using NH_4OH or urea, biuret, and the like in the dispersion
Nov. 30, 1950	Jap. 4150	S. Kambara, S. Tokuma	Nippon Dyeing Co.	Chlorinated polyester, chlorin-ated rubber, chloroparaffin and Sb_4O_6 in trichloroethyl-ene medium
Jan. 2, 1951	U.S. 2,536,978	F. Fordemwalt	American Cyanamid Co.	Emulsion contg. Sb_4O_6, chloro-carbon, and heat-convertible alkyd resin
Apr. 17, 1951	U.S. 2,549,060	J. W. Creely	American Cyanamid Co.	Sb_4O_6, chlorocarbon, and low-melting phosphoramide, e.g., rxn prod. of urea-H_3PO_4 mixtures heated in air to 160°
Apr. 17, 1951	U.S. 2,549,059	J. W. Creely, T. F. Cooke	American Cyanamid Co	Like previous patent except uses direct amine salts of poly-phosphoric acid without heating
April 1, 1952	U.S. 2,591,368	J. F. McCarthy	Treesdale Laboratories, Inc.	Water dispersion of Sb_4O_6, poly(vinyl chloride), plasti-cizers, and carboxymethyl-cellulose as stabilizer
Sept. 16, 1952	U.S. 2,610,920	H. Hopkinson		Metal oxides, chloroparaffins, chlorinated rubber, and a glowproofing agent, e.g., rosin amine phosphate (or borate or silicate) or analo-gous octadecylamine salts

Table 5-21 (Continued)

Date	Patent No.	Inventors	Assignee	Invention
May 26, 1953	U.S. 2,640,000	L. P. Seyb, C. A. Neros	Diamond Alkali Co.	Sb_4O_6, chloroparaffins, and protective resin film, i.e. urea-formaldehyde. Resin is said to reduce acid tendering
Oct. 19, 1953	Jap. 5400	T. Tsuda		Sb_4O_6, chloroparaffin, and chlorinated residues from DDT. Applied from CCl_4
May 10, 1954	Fr. 1,063,983	E. Tassel, P. Vallette	Compagnie française des Matières Colorantes	$SbCl_3$ complexed with triethanol-amine forms neutral soln. Fabric is immersed; then, by washing, the Sb is hydrolyzed to finely dispersed insol Sb_4O_6
Jan. 24, 1956	Fr. 1,109,296	R. L. Morin	Establissements Dickson	Sb_4O_6, copolymer of vinyl chloride and vinyl acetate, trace Zn stearate are applied from ketone solvent to very light fabrics without undue stiffening
July 24, 1956	U.S. 2,755,534	I. Barnett	Johns-Manville Corp.	Standard Sb_4O_6-chlorocarbon treatment applied to asbestos-cotton fiber blends
Feb. 6, 1958	Ger. 1,023,744	K. Quehl		Sb_4O_6, chlorinated poly(vinyl chloride) (60% Cl), water-proofers from org. solvent

Date	Patent	Author	Company	Description
Sept. 16, 1958	U.S. 2,852,414 (see also India. 53,399 for Nov 7, 1956)	J. D. Broatch	British Jute Trade Research Assoc.	Sb phosphate and chlorine contg. vinyl polymer; phosphate gives glowproofing on decomp. to Sb chloride and acid phosphate
Apr. 7, 1959	U.S. 2,881,097	A. Giordano, W. J. Straka	Harshaw Chemical Co.	Cu-Sb mixed oxide plus chloroparaffin and a urea-CH_2O resin. Cu gives mildewicidal props
Mar. 9, 1961	Ger. 1,101,347	N. J. Read	Associated Lead Manufacturers, Ltd.	Sb_4O_6 in vinyl chloride latex, softeners such as dioctyl phthalate for lightweight fabrics
May 23, 1961	U.S. 2,985,540	L. J. Goldbeck	Kimberly-Clark Corp.	Cellulose-synthetic fiber blends are first treated with Sb_4O_6 in a non-chlorine-contg. resin system. Then a vinyl chloride resin is coated on top
May 27, 1961	E. Ger. 21,376	R. Scharf, K. A. Reif		Sb_4O_6 or TiO_2 in chlorocarbons with organic phosphate esters in emulsion
Jan. 10, 1964	Austrian 230,904	O. Zwoboda		Makes special fiber-bonded mats, then treat with Sb_4O_6 and chlorocarbons
Jan. 15, 1964	Czech 109,534	L. Ktema, M. Janko		Cellulose and synthetic fiber blends treated with Sb_4O_6-chlorocarbon from org. solvent. May add Cr_2O_3-stearic acid mixture or Al salts to improve H_2O resistance

described the preparation of chlorinated paraffins suitable for use with metal oxides [151]. In 1942 a patent was issued on the use of antimony oxide and chlorinated paraffins [152] and the stage was set for development of a finish in time for the military needs of World War II. During the war the Sb_4O_6-chloroparaffin treatment was intensively studied by the Army Quartermaster Corps at several locations and a satisfactory understanding of the technique was obtained. Immediately after the war a series of papers [153–155] and patents appeared which summarized this research. These discussed aspects such as the nature of ingredients which stabilize the chloroparaffin against premature decomposition (iron and zinc compounds are detrimental) [153], a typical formulation [154], and results of a variety of tests on treated fabrics [155]. The patent history is extensive and runs into the early 1960s; it is summarized in Table 5-21.

The literature in recent years has been sparse, indicating that the concept has matured; we may expect little more in the way of significant improvements. In the patent literature there are several instances of attempts to obtain very intimate dispersions of the Sb_4O_6 by adding aqueous solutions of $SbCl_3$ and subsequently hydrolyzing it to the oxide by washing. The acid halide solution tenders the fabric. In 1958 a group of Indian researchers suggested the use of antimonites [156]:

$$Sb_4O_6 + 12OH^- \rightarrow 4SbO_3^= + 6H_2O$$
$$\text{Antimonite ion}$$

The oxide is dissolved in alkali (excess) and the fabric impregnated. Subsequent washing hydrolyzes the anion back to the insoluble oxide. Phosphate is added for glowproofing and the whole coated with chlorinated rubber from benzene solution. Vertical char lengths of 2–3 in. have been obtained after washing with little effect on fabric strength. Some stiffening was encountered.

Recently a new vinylidene chloride copolymer—Sb_4O_6 emulsion was introduced primarily for paper saturation [157]. The composition is 1 part Sb_4O_6 to 1 part resin solids. The most recent work is contained in a Polish paper describing similar emulsions prepared with a colloid mill [158]. Add-ons in the 20–30 % range are used.

The bulk of the data on the method are to be found in Little's book [9] and are therefore 1941–45 vintage. Table 5-22 has data on screening tests designed to pick out promising chlorine carriers. It is clear that neither the Sb_4O_6 nor the chlorocarbon by themselves are worth much as flame retardants. It is equally clear that the combinations are very effective and that some chlorocarbons are definitely preferable to others. Some, such as the chlorinated polyphenyls, are too stable. Subsequent tests showed that only a fraction of the theoretical amount of HCl is liberated [159]:

Chlorocarbon	% Cl	% of Cl Liberated[a]
Diphenyl	54	3–6
Paraffin	42	17–41
Paraffin	70	25–66

[a] From pyrolysis on cotton fabric

The antimony-halogen system does not eliminate afterglow and it is necessary to add a phosphate or a borate to achieve this. Either zinc borate or a decomposable phosphate is acceptable. The ratio of the various ingredients is, of course, important and much of the research was aimed at defining this. Sufficient chlorine is required to convert this Sb_4O_6 to SbOCl. The amount of chlorocarbon required to effect this conversion cannot be calculated just from a knowledge of the chlorine content of the system since it is not all available. For paraffin containing 70% Cl a 1:1 mixture (wt basis) with Sb_4O_6 was found to be sufficient at all add-ons from 20 to 60% of the fabric

Table 5-22   Effect on Cotton of Various Antimony-Chlorine Systems [159]
Add-on ca. 40%; Equal Parts Sb_4O_6 and Chlorocarbon

Retardant	% Cl in Chlorocarbon	Afterflame Horizontal Test, sec	Flame Rate, in./sec
None	$\cdots$	>60	0.86
Sb_4O_6	$\cdots$	>60	0.57
Pentachlorodiphenyl (A)	54	>60	0.37
Sb_4O_6 + A		~60	0
Neoprene (B)	40	59	0.56
Sb_4O_6 + B		0	0
Chloroparaffin (C)	42	>60	0.81
Sb_4O_6 + C		21	0
Chloroparaffin (D)	70	>60	0.81
Sb_4O_6 + D		0	0
Vinylite (E)[a]	50	>60	0.65
Sb_4O_6 + E		0	0
Aniline·HCl (F)	27	>60	0.32
Sb_4O_6 + F		0	0

[a] Vinyl chloride-vinyl acetate copolymer

weight. For a whole group of chlorocarbons the weight ratio Sb_4O_6:chlorocarbon for good results varied from $4:1$ to $0.8:1$ [160]. However, other factors enter into the selection of chlorocarbon type and amount. The paraffin containing 70% Cl, for example, tends to cause a very stiff condition and is subject to crazing and cracking. A lower chlorine content in the paraffin prevents this; so do certain plasticizers. Stability of the emulsions or the solvent suspensions used as treating baths is also a consideration; there is much art on this last point (see Table 5-21). Laboratory work optimized both a water-in-oil emulsion and a solvent suspension, as shown in Table 5-23.

Table 5-23   Laboratory Formulas [160]

Emulsion		Solvent Suspension	
Water	50.0	$\cdots$	$\cdots$
Sb_4O_4	11.3	Sb_4O_6	8.8
Chloroparaffin		Vinyl chloride-vinyl	
(42% Cl)	5.7	acetate copolymer	
Chloroparaffin		(50% Cl)	8.8
(70% Cl)	3.8		
Zn borate	5.7	Zn borate	2.9
Rezyl	4.7	Triphenyl phosphate	8.8
Trichloroethylene	18.8	Solvent (ketone)	70.7

The triphenyl phosphate serves both as a plasticizer for the chlorocarbon copolymer and as a glowproofer along with the zinc borate. In various commercial formulas other additives may also be employed (see Table 5-21). Both the emulsion and the suspension shown in Table 5-23 are effective at 30% add-on (solids) or higher. Thus, a chlorine content of 3–4.5% with 5.6–9% Sb_4O_6 is required. (More would probably be needed were it not for the presence of the borate and/or phosphate.)

The treated fabric resists laundering reasonably well; that is, char length <5.0 in. after 6 cycles in alkaline soap solution. It withstands immersion in salt water without loss of flame resistance (unlike the systems prone to ion exchange). Eight weeks' exposure to the weather (fall, eastern United States) had little adverse effect; prolonged (180 hr) accelerated aging increased the char length from a value of 3.0 to 5.0 in. The results suggest that the glowproofing agent is the least durable to weather and is removed first. Little and his colleagues made exhaustive studies of the effect of laundering, wear, perspiration, rainfall, seawater, storage, and the like on the characteristics of the cloth used for Army uniforms. They also studied the effect of the flame retardants on physiological factors in the wearer. The only serious objections were to the extra weight of the treated fabric, its hardness, and its stiffness. Consult the monograph for details [161].

For tentage, the treating bath developed was a solvent suspension of the final composition shown below (1945) [162]:

Chlorinated paraffin (42%)	10–12%
Chlorinated paraffin (70%)	5–7
Binder resin	4–6
Sb_4O_6	5–7
Pigment	9–11
$CaCO_3$	5–7
Copper naphthenate	5–6
Solvent	45–55

The copper salt is used as a mildewicide. The finish is fire-, water-, weather-, and mildew-resistant, and known in the trade as an FWWMR system. The formula shown above evolved during World War II and is the result of several changes made as the war progressed. Data on its performance during up to 12 months' exposure at four sites in the United States are given in the monograph [162]. The tentage retained its FWWMR properties rather well, although considerable variation was noted in fabrics from different processing plants. Applications to camouflage netting and life jackets are also discussed.

A paper prepared in 1958, thirteen years after the work in Little's report was completed, discussed a more modern finish then in use in England. A poly(vinyl chloride) latex was applied with a chloroparaffin, Sb_4O_6, and a phthalate plasticizer [148]. No glowproofing agent was mentioned. Add-ons of 30% were recommended for 6-oz cotton twill; tensile strength was increased but tear strength was somewhat diminished. Clearly, the major advances were made at the time of World War II.

A brief word is due on a process which received attention in the early 1950s. This was the use of antimony and titanium compounds in the presence of chlorine [163]. Titanium tetrachloride and Sb_4O_6 were reacted

$$6\ TiCl_4 + Sb_4O_6 \rightarrow 6\ TiOCl_2 + 4\ SbCl_3$$

together to give an acid powder soluble in water (acidified with HCl). Fabric was immersed in this solution and then neutralized in a solution of Na_2CO_3. The product withstood 100 home launderings and was still fire-resistant. The chemical nature of the product might have been [164]

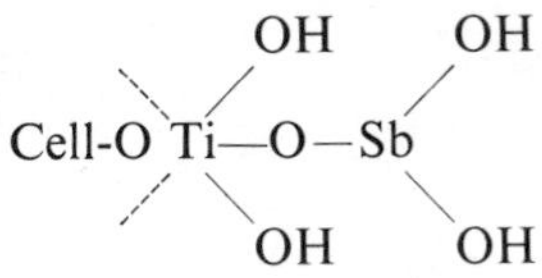

$$Cell\text{-}O\ Ti\underset{\diagdown OH}{\overset{\diagup OH}{\diagup}}O\text{-}Sb\underset{\diagdown OH}{\overset{\diagup OH}{}}$$

An add-on of 10–15% seemed to suffice; weather resistance was found to be adequate; breaking strength was at least 90% of original; hand was good.

There was no comment on tearing strength (the parameter most severely affected by most acidic treatments). The method is not known to be in use at this writing.

BORON AND NITROGEN COMPOUNDS; MIXED OXIDES

Boron Compounds

The long history of inorganic boron compounds as flameproofing agents for cellulose has been reviewed in Chapter 4 with respect to uses in wood. There are fewer references to the use of boron for cotton textiles. An early patent [165] disclosed $MgSO_4$ with borax for fabric; a paper in 1934 [166] discussed ammonium borates and boric acid and also mentioned vitreous boron compounds. In 1936 Clarence [167] suggested boric acid-borax mixtures at about 6% add-on, little loss in strength resulted if the drying temperature was held at 100° or less. The paper further states that less effect on strength of the fabric was observed with the acid-salt combination than with either acid or salt alone. However, Hartman and coworkers in 1939 [168] pointed out that such treatments are not durable and cautioned against their use where service for over 1 yr is anticipated. A simple impregnating solution comprising 10–20% solids in water (2 parts borax:1 part

Table 5-24   Effect of Borax-H_3BO_3 on Flammability of Cotton [9]

Retardant	Add-on, %	Afterglow, sec	Vertical Char Length, in.
Borax:H_3BO_3, (7:3)	5.4	775	6.0
	9.2	206	2.7
	13.4	161	2.5
Borax:H_3BO_3, (1:1)	5.0	265	4.1
	8.3	173	2.7
	12.1	25	2.5
Borax:H_3BO_3:$(NH_4)_2HPO_4$, (7:3:5)	3.1	· · ·	Burned entire length
	8.7	9	3.3
	12.4	7	3.1
Borax:H_3BO_3:$(NH_4)_2HPO_4$, (5:5:1)	7.0	112	3.7
	11.3	13	2.9
	16.9	6	2.5
Borax:$(NH_4)_2HPO_4$, (1:1)	7.9	2	3.3
	10.9	2	3.3
	16.7	2	2.7

boric acid) plus a wetting agent was described in a 1951 patent [169]. In recent years only two literature citations have been uncovered: one disclosed the use of a borotungstate complex in aqueous solution at pH 7 [170]; the other, a complex mixture of borates, bromide, urea, and phosphate from CH_3CCl_3 medium [171]. This literature is not sophisticated; little appears to have been done to attempt to develop an insoluble, wash-fast treatment based on boron compounds.

Data on temporary, water-soluble treatments may be found in various review articles (see comparisons in a later section of this chapter) and in Little's monograph [9]. Table 5-24 summarizes some of Little's information. The boron system that is best for flame retardance is not good in glow retardance. Addition of a little phosphate solves this problem; indeed, replacing H_3BO_3 with $(NH_4)_2HPO_4$ is the most promising result shown in the table. On a weight basis a borax-phosphate treatment is perhaps the most efficient overall of any flame-retardant system known, being effective at add-ons of less than 7% (note that borax-H_3BO_3 mixtures give char lengths below 5.0 at add-ons of only 5% but do not provide glowproofing). Other data from Little show the borax-based system to be even more sensitive to moisture than the soluble phosphates, losing all flame resistance after 2 weeks' exposure to 49°F, 85% relative humidity.

Nitrogen Compounds

No references are known to the author which treat nitrogen as a flame retardant for cotton in the absence of other flame-retardant elements. Thus ammonium salts of phosphoric, hydrochloric, sulfuric, sulfamic, and boric acids are the usual combinations. The ammonium ion is used because it is the least expensive volatile cation available. Yet there is some evidence that the nitrogen plays a greater role than this. If so, it should not be surprising since nitrogen is in the same family and has many of the chemical properties common to phosphorus, antimony, and other members of the Group V family.

Nitrogen can be introduced in a number of ways: simple salt addition [172], via aminization of the cellulose [173], and as part of reactive compounds, for example, the methylolamines and amides or APO. Table 5-25 shows the requirements for four ammonium salts. Sulfamate is more effective than sulfate; both are not good in terms of glowproofing. The two halides shown are at least as effective as the sulfur derivatives as flame retardants; the chloride would probably be less effective than the bromide.

In 1941 ammonium sulfamate was introduced as a flame retardant for cotton fabrics [174]. Glowproofers were added to the commercial product. Hexamethylene tetraamine was suggested in a 1945 patent [175], but sulfuric

Table 5-25   Effect of Four Ammonium Salts on Cotton Flammability

Retardant	Add-on, %	Afterglow, sec	Vertical Char Length, in.
Ammonium sulfamate	6.4	780	6.6
	9.8	642	5.1
	14.5	366	3.4
	26.8	276	2.6
Ammonium sulfate	9.5	· · ·	Burned entire length
	14.9	790	5.6
	17.3	549	4.0
	21.5	375	3.0
	26.5	258	2.9
Ammonium bromide	10	142	3.6
	20	11	2.7
Ammonium iodide	12	60	4.2
	16	9	3.3
	24	6	3.0

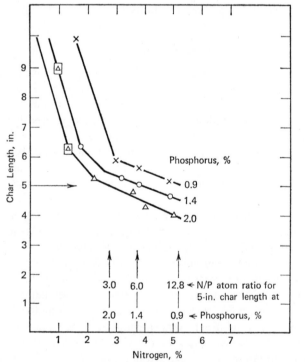

Figure 5-5   Effect of nitrogen content at various phosphorus levels on the vertical char length of cotton [109].

and phosphoric acids were a part of the treatment, thereby making it difficult to assess the role of the nitrogen in the treatment. The fact that nitrogen plays more than a fugitive's role is pointed up by studies on P—N systems [109, 110] for which a definite relationship has been shown. Figure 5-5 illustrates this clearly.

By increasing the N content from 2.8 up to 5.2%, the amount of P needed can be reduced from 2.0 down to 0.9% while the flammability is maintained constant at 5.0 in. This is very recent work; no doubt more will be heard on this point. For the present it remains true that nitrogen is used primarily as a convenient volatile base to provide acidic residues on heating.

Mixed Oxides

Perkin's method of flameproofing was to treat fabric with sodium stannate and then, in a second bath, to hydrolyze it to stannic oxide (hydrous) by use of an ammonium sulfate solution. In the 1930s and during World War II more research was done on such methods with some success. The oxide treatments are not efficient. A comparative listing often encountered in the literature is given in Table 5-26. Coppick and Hall have summarized the work done through 1945 [177]. The method is known as a double-bath precipitation technique simply because the insoluble hydrous oxides and hydroxides are placed on the fabric by (a) treatment with a soluble acid form (e.g., a halide) or an amphoteric basic salt (e.g., a caustic solution), and

Table 5-26 Relative Efficiencies of Various Insoluble Materials Versus Borax-Boric Acid (After Ramsbottom and Snoad [176])

Retardant	Min. Add-on for Fire Resistance, %
Borax-boric acid (1:1)	10
Iron(III) hydroxide	25
Antimony(III) oxychloride	30
Tin(IV) oxide (hydrous)	40
Titanium(IV) hydroxide	40
Bismuth trioxide (hydrous)	40
Zinc stannate	40
Aluminum borate	59
Aluminum hydroxide	70
Chromium(III) hydroxide	91
Silica (hydrous)	100
Aluminum silicate	100
Magnesium ammonium phosphate	125

then (b) hydrolysis is brought about by immersion in an aqueous solution of an acid or base. The hydrolysis causes precipitation [177]:

$$Na_2SnO_3 + (NH_4)_2SO_4 \rightarrow Na_2SO_4 + 2NH_3 + \underline{H_2SnO_3\downarrow}$$
(Perkin's method)

$$FeCl_3 + 3NH_4OH \rightarrow \underline{Fe(OH)_3\downarrow} + 3NH_4Cl$$

$$2Na_2WO_4 + SnCl_4 + 3H_2O \rightarrow 2NaCl + \underline{SnO(OH)_2\downarrow} + \underline{2WO_2(OH)_2\downarrow}$$

The tungstic-stannic acid system was judged to be the best of those studied by Coppick and even this was somewhat deficient in glowproofing properties. A phosphate addition was found which corrected this. The final formulation consisted of two baths. Bath No. 1 contained 8% $Na_2WO_4 \cdot 2H_2O$ and 3.0% $Na_4P_2O_7$; Bath No. 2, 8% $SnCl_4 \cdot 5H_2O$. By repeating the double-bath treatment twice, the recommended add-on of 30% could be obtained. By adding an overcoat of resin (5% add-on of a phenolformaldehyde type or about 2% of a melamine type) resistance to laundering (6 cycles in G.I. soap) was improved. There was an option of drying after the first bath, if, by so doing, the fabric was not damaged. This extra step was sometimes found helpful. On occasion, solvents other than water were employed, especially polar liquids such as alcohols and glycols.

There has been no new development in this technique since World War II, and it appears that Perkin's method, after half a century, is no longer of interest. However, inasmuch as his ideas were instrumental in the development of the antimony oxide system in use today, his efforts were clearly not in vain.

RAYON AND ACETATE FIBERS

Regenerated cellulosic fibers may be flame-retarded either by treatment of the finished fiber or fabric or by adding flame retardants to the spinning solution so as to disperse the retardant throughout the fiber. Each method has its advantages. Rayon responds to the same techniques as does cotton since the two are chemically essentially the same. Although such diverse notions as treatment with calcium oxalate [178] or polymerized silanes [179] have been offered, the old standard methods are still the soundest [179a]. Thus, phosphorus-halogen combinations [180] and antimony-chlorocarbon systems [181] are effective surface treatments for such fibers. $KSb(OH)_6$ has shown some utility as a spinning additive [181a].

In cellulose acetate some of the reactive sites are blocked by the acetate group and these will support combustion when released as acetic acid; the dehydration mechanism of the phosphorus acids is less effective. Addition of halogens to acetate fibers is almost mandatory to provide a means of blocking flaming reactions in the vapor phase. Thus, almost all references to acetates include chlorinated retardants.

Additions to the spinning solution are of interest because of the intimate dispersion of the retardant thereby obtained. Thus, to cellulose acetate solutions may be added chlorinated esters, for example, the haloacetate of 1,4-bis(hydroxymethyl)benzene [182]. Haloalkylphosphate esters may be added: the triesters [183] or the aluminum, tin, or titanium salts of diesters [184] have been proposed for cellulose ester fibers. In an interesting variation a reactive site was built into a viscose fiber as follows [185]:

1. Add to the viscose solution a soluble material containing a reactive site, for example, carboxylmethyl-, carboxyethyl-, or glycerolether cellulose.

2. Spin the fiber into acid solution.

3. Treat with a solution of a flameproofing metal, such as an antimony salt.

The effect is to build in sites tailored to the metal to be incorporated. In the case cited the derivatives were substituted for about 1% of the viscose cellulose.

One recent patent [186] claims addition of tris(2,3-dibromopropyl) phosphate to the spinning solution at a level of about 15% by weight of the cellulose followed by fiber preparation in the usual way. Good flame resistance is said to be obtained in this manner. Another patent [186a] by this company describes the addition of ethylene bromohydrin in a resin size to cellulose acetate. Still another [186b] claims dialkylphosphonitrilates as inert additives to rayon at levels up to 20%. In 1968 this same rayon producer introduced a new, commercial fire-retardant rayon said to have no halogen in it. No detailed descriptions that can be associated with this development have yet appeared (see, however, reference 186b) but undoubtedly they will in due time. The new rayon contains 2–3% phosphorus added as a nonreactive species to the spinning solutions [186c]. The fiber is being offered at premium prices.

COMPARATIVE DATA

It is not surprising, in view of the large number of treatments that have been proposed for cotton, that no one piece of research has included comparisons of all the treatments under exactly the same conditions. There are a few references in which several of the methods have been evaluated under the same conditions. These are cited here.

Figure 5-6 contains data from recent thermal analyses of several of the more recent phosphorus-based systems. Unfortunately for our purposes, a simple system like $(NH_4)_2HPO_4$ was not included for comparison. All the

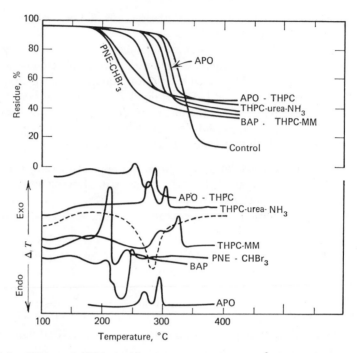

Figure 5-6a DTA and TGA in nitrogen atmosphere at 15°C/min (MM—methylol-melamine; PNE—phosphonitrilic ester; BAP—brominated allylphosphate) [187]. Dashed line is for $NH_4H_2PO_4$ from Figure 4-2 for cellulose.

treatments produce more charred residue than the control (35 to 50% versus 10%). Brominated allylphosphate (BAP) has the effect of facilitating the reaction the most; that is, the onset in TGA occurs more rapidly at lower temperatures. APO and THPC-methylolmelamine have the least effect by this criterion. Some of the DTA curves show two peaks. This may imply some molecular rearrangements before decomposition. (In all cases the first peak occurs before significant weight loss occurs.) In the case of the phos-phonitrilic ester a sharp exotherm occurs before the endotherm common to it and to BAP. The curves in oxygen are less easy to interpret and show less pronounced characteristics. The data in Figure 5-6b do show that fabrics treated with various phosphorus systems exhibit differing stabilities, the most stable materials being THPC and APO and the least stable, the phos-phate and phosphonitrilate esters. It appears that the ammonium ortho-phosphates confer intermediate stability.

In the monograph edited by Little [9] many data are presented on the decomposition products of fabrics treated with soluble flame retardants, for example, phosphates, borates, and halides. A sample of this is given in

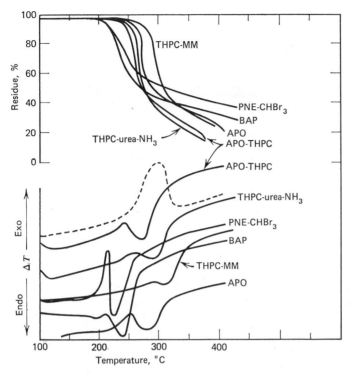

Figure 5-6b DTA and TGA in oxygen atmosphere at 15°C/min [187]. Dashed line is for $NH_4H_2PO_4$ from Figure 4-3 for cellulose.

Figure 5-7 [188] and in Table 5-27 [189]. This work was later pursued in even more detail by these workers in terms of studying the thermal response of fabrics at flaming temperatures [190].

Schuyten et al. [191] studied a group of flame retardants and compared measured properties against theoretical predictions of the extent of dehydration caused by each reagent. The results are shown in Table 5-28. The data show that the fabric was strongly dehydrated by each of the additives. Assay of water extracts of the samples treated with phosphorus compounds isolated species that behaved as polyphosphates, which implied that degradation of the additives was followed by combination of acidic fragments on further dehydration to form polyphosphoric acids This is consistent with current theories of the mechanism of the reaction of cellulosics with phosphorus compounds (see Chapter 1).

Little [192] offers a comparison of the effect of seven common soluble retardants studied by the vertical flame test. The data are given in Table 5-29. The difficulty with ammonium sulfamate is clearly shown. Afterglow

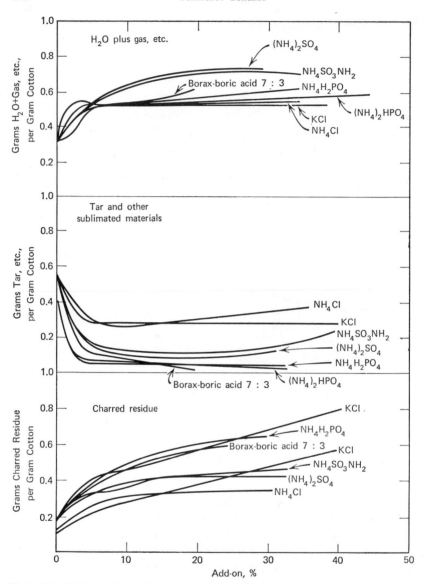

Figure 5-7 Influence of retardants on the quantity of the various phases resulting from the pyrolysis of cotton [188].

seems to be a problem with the boron compounds but they give a somewhat lower char length than the other retardants. The phosphates perform, as expected, as the best glowproofing agents. In a comparison of the nitrogen-phosphorus method and the antimony-halogen process, Little provides the data in Table 5-30. The superior glowproofing of the phosphate is again

Table 5-27 Effect of Large Amounts of Retardants on Extent and Duration of Isolated Glowing Reaction of Chars from Cotton Fabrics [189]

Retardant	Add-on, %	Initial Fabric Wt	Loss During Pyrolysis mg/cm² of fabric	Loss via Glow	Glowing time, sec
None	0	26.90	21.40	5.08	73
Sodium chloride	10	27.36	18.34	2.35	62
	20	30.66	18.51	2.64	75
Ammonium sulfamate	10	27.92	15.07	10.55	360
	20	29.52	16.70	9.89	300
	10	27.42	16.47	7.51	180
Borax	30	30.71	16.96	7.66	180
	40	34.12	17.00	9.39	240
Boric:acid	10	24.76	13.50	0.39	0
	20	25.48	13.80	0.35	1
Ammonium dihydrogen phosphate	10	27.64	15.26	0.07	0
Borax:Boric acid (3:7)	10	28.37	14.07	0.32	0
Borax:Boric acid (1:1)	10	26.30	13.20	1.22	130
	20	28.52	13.26	0.28	0
Borax:Boric acid (7:3)	5	25.42	13.58	7.86	195
	10	26.08	13.64	5.67	210
	20	29.22	15.21	0.48	13
Borax:boric acid:ammonium dihydrogen phosphate (1:1:2)	10	30.56	15.96	0.21	5
	20	30.73	13.15	0.20	0
Borax:boric acid:ammonium dihydrogen phosphate (7:3:5)	10	29.75	15.18	0.21	0
	20	33.42	15.32	0.22	0

Table 5-28   Dehydration of Cellulose by Flame-Retardant Agents [191]

		Ratio C/H in Residue	Moles H_2O Lost/Glucose
Theoretical: $C_6H_{10}O_5 - H_2O \rightarrow C_6H_8O_4$		0.75	1
$C_6H_8O_4 - H_2O \rightarrow C_6H_6O_3$		1.0	2
$C_6H_6O_3 - H_2O \rightarrow C_6H_4O_2$		1.5	3
$C_6H_4O_2 - H_2O \rightarrow C_6H_2O$		3.0	4
$C_6H_2O - H_2O \rightarrow C_6$		∞	5
Flame Retardant[a]	% Add-on		
LiCl	9.8	2.64	3.85
H_3PO_4	...	4.41	4.32
$(NH_4)_2HPO_4$	5.8	4.56	4.35
Brominated allylphosphate	19	3.48	4.15
THPC	16	2.78	3.92
Borax-boric acid	11	2.75	3.90

[a] Evaluated at 500°C

Table 5-29   Comparison of Soluble Retardants at a Constant
Add-on of 10% [192]

Retardant	Afterglow, sec	Char Length, in.
Borax:boric acid (7:3)	190	2.6
Borax:boric acid (1:1)	32	2.7
$(NH_4)_2HPO_4$	0	3.9
$NH_4H_2PO_4$	0	3.4
Borax:boric acid:$(NH_4)_2HPO_4$ (7:3:5)	8	3.2
Borax:boric acid:$(NH_4)_2HPO_4$ (5:5:1)	43	3.1
$NH_4SO_3NH_2$	550	5.0

Table 5-30   Urea-Phosphate Treatment versus Antimony Oxide-
Chloroparaffin [192]

Treatment	Add-on, %	Sea Water Soak	Afterglow, sec	Char Length, in.
Urea-phosphate	16	No	0	2.6
		Yes	0	3.7
Sb_2O_3-chlorocarbon	33	No	5	3.0
(Zn borate added)		Yes	6	3.2

Table 5-31   Effect of Various Treatments on Strength of Cotton
Sateen [193]

Treatment	Add-on, %	Strength, % of Untreated	
		0 Washes	14 Washes
THPC	15	91	75
THPC-BAP	14	66	82
Metal oxide-chloroparaffin	31	71	123
Urea-phosphate	17	73	···
Borax-boric acid	11	72	···
Resin treatment	25	68	93
Resin treatment	30	71	93

Note: Control (untreated) lost 25% of its strength after 14 washes;
strengths shown for washed fabrics are percentages of washed control.

evident but immersion in sea water (30 min) took some toll of the flame
resistance.

Freiser [193] collected data on the effect of various retardants on fabric
strength (Table 5-31). Most of the treatments are detrimental to strength, but
so are simple resin finishes. The reason is that the treatments involve immersion in various solutions. Washing of the control fabric in itself causes a

strength loss. Therefore it is unfair to judge strength losses too critically unless the control loss on washing is taken into account. Note that the metal oxide-chloroparaffin treatment is helpful in retaining, and indeed, in improving fabric strength.

Most recently Ciba workers [108] have published a comparison of their candidate (see pp 186) versus APO and THPC systems. Their results are shown in Table 5-32.

Table 5-32 Comparative Evaluation of Phosphonate-Compound 1, APO, and THPC Fire-Retardant Systems[a]

Property	Compound 1	APO Thiourea	THPC Resin	THPC Ammonia
Toxicity	Low	$--$	$-$	$-$
Application	Conventional	$=$	$=$	$-$
Flame-retardant effect:				
Afterburn	0 sec	$=$	$=$	$=$
Afterglow	0 sec	$=$	$=$	$=$
Char length	8–9 cm	$=$	$= \ldots +$	$= \ldots +$
Wash fastness 5 × SNV4	Unchanged	$-$	$=$	$=$
Dry cleaning	Unchanged	$=$	$=$	$=$
Hand	Soft to slightly full	$-$	$-$	$--$
Tensile strength loss	-25%	$=$	$--$	$+$
Abrasion resistance loss	-40%	$=$	$--$	$+$
Fabric appearance:[b]				
Dry	4	$-$	$+$	$--$
Wet	3	$=$	$=$	$-$
Key: = Almost equal				
+ Better				
− Inferior				
−− Distinctly inferior to finish with compound 1				

[a] From Aenishänslin, reference 108
[b] Overhead lighting

SUMMING UP AND A LOOK INTO THE FUTURE

Today there appear to be two or three main classes of flame-retardant treatments for cellulosic fibers for which there is significant commercial demand. These are the temporary water-soluble types for certain draperies, mattress stuffing, and the like where contact with water is deemed unlikely; permanent treatments for garments, mattress ticking, and interior fabrics where laundering may be required but severe weathering is unlikely; and treatments for outdoor fabrics, especially for the military. The first class of uses is well served by the soluble ammonium phosphates and the borates;

the second class consists at the moment primarily of the several THPC treatments and the phosphonoalkanoic amides; the third, the Sb_4O_6-chloroparaffin technique. Each serves its own purposes best. The greatest problem with the ammonium orthophosphates and the soluble borates is, of course, lack of resistance to laundering. The THPC systems and the phosphono-compounds suffer from the high cost of the chemicals and some deficiency in areas like stiffness and reduction in tearing strength. The Sb_4O_6-chlorocarbon treatments are heavy coatings which render the fabrics uncomfortable for use as apparel. There is still plenty of room for new treatments.

The ideal flame retardant for cotton textiles would have at least the following attributes:

1. React with surface hydroxyls of the fiber (to attain the wash-fast property).

2. Be high in content of flame-retardant element(s) (permits low add-on).

3. Be nonionic (avoids ion exchange with metals in washing solution or in sea water).

4. Be neutral or alkaline (avoids tendering of fibers).

5. Contain glowproofing elements as well as flame retardants.

6. Be nontoxic in the fiber and relatively nontoxic in the treating bath.

7. Be compatible with other finishes, such as, durable press, water- and soil-repellent resins, and dyeing processes.

8. Be low in cost per unit of effective element contained (somewhere between today's cost of ammonium orthophosphates and THPC, for example).

Much research is in progress in government, industrial, and institute laboratories in the United States and around the world. From the present activity we may look for an entire new generation of products stimulated by the rising desire for a safer environment and exemplified by such new legislation as the two federal acts of 1967 and 1968 [1, 2].

In addition to new products based on the kind of technology discussed in this chapter, entirely new approaches may be anticipated. One technique under study now and for the past several years is the grafting of various monomers and polymers onto cotton. Durable-press properties can be obtained by incorporation of vinyl monomers (e.g., methylolacrylamide) onto the fibers of cotton by high-energy radiation [194]. A kind of copolymer is formed via free-radical mechanisms. This has been done with acrylonitrile [195], chloroacrylonitrile [196], various acrylate monomers [197], and vinyl chloride [198]. Some work has been done with chemical initiators [199] in place of the radiation, but radiation is currently the method of choice. Commercial processes are now in operation for durable-press finishes. The method undoubtedly will be adapted to flame-retardant finishes by 1970. (However, no references have appeared on this aspect at this writing.)

A second and somewhat more dramatic new departure is the attainment of flame retardance by deliberately precharring or carbonizing the fabric. Carbonization processes have been studied largely for acrylic fibers; a fuller discussion of the technique is reserved for the section of Chapter 7 on acrylics. There are six patents on carbonizing cellulosic fabrics, all issued since 1966. Three concern rayon; the others are less restrictive. In the first patent [200] rayon is treated with fire-retardant salts and then heated to 290° or so to dehydrate the cellulose and produce a carbonized fiber. Some 40% of the original weight was lost and 40% of the strength was retained. The process is claimed only to work on rayon. The treated fabric is black, can withstand temperatures as high as 550° for short times, and can be further carbonized to true carbon fibers by heating in a reducing atmosphere. The second patent claims production of a carbonized rayon by heating the fabric or fiber in a reducing atmosphere to 325–390° at a rate of about 200°/hr. After working (or flexing) the fabric at cooler temperatures, it is finally heated above 500° to produce a fully carbonized fiber [201]. The working of the partially carbonized fiber improves the flexibility of the final product. The third patent [201a] on reconstituted cellulose fibers follows this sequence: (a) heat to 150° in helium; (b) add Cl_2 gas at 165° for 10 min; (c) heat to 650° in helium; and (d) finish at 1093° in helium.

The three patents on cotton fabrics as well as rayon cover three different processes. In the first one slow heating in an inert or reducing atmosphere is continued to 1200° [202]. A flame-resistant fabric results with good electric conductivity. In the second process treatment of the fiber with refractory metal salts, for example, $Zr(oAc)_4$, followed by heating under nitrogen to 800° produces a carbonized fabric with excellent flame resistance and good flexibility [203]. In the third process cellulose is coated with a polymer such as polyacrylonitrile or the cellulose is cyanoethylated or grafted with polyacrylonitrile. The cellulose thus treated is then subjected to a carbonizing treatment in a nonoxidizing atmosphere at temperatures above 300° [204].

These treatments produce in some cases fabrics which are as close to fireproof as one is likely to get with carbon in the molecule. Certain of the carbonized acrylic fabrics withstand a blowtorch for prolonged periods without visible effect. Such success has yet to be obtained with cotton, but there is no doubt in the writer's mind that this will be achieved in a year or two. The strength, flexibility, launderability, and the like will all be satisfactory. By partial carbonization—at the surface only—the original properties of the fibers can be retained while they are protected with carbon layers. The sticking point at present is the black color. It will take more work and considerable inventiveness to achieve sufficient color flexibility for the fabrics to come into widespread use. Cost is also a barrier at present but should be amenable to control on a large scale.

However, all of the treatments discussed in this chapter are for cotton or rayon cellulose used by themselves. Many of the newer treatments, and indeed some of the oldest, are adequate to the task. But today the trend is toward blends of cotton with synthetics—principally polyester fibers. This has brought about an entirely new situation. In a few years, say by 1973, most apparel now made of cotton will be made of blends of cotton with polyester. The same holds true for sheeting. The unfortunate fact is that these are the very areas likely to be affected first by the implementation of recent, restrictive legislation [1]. There are no good methods for flameproofing cotton/polyester. The THPC, THPC-APO, phosphonoalkanoicamide group of treatments is understood not to be the answer. Undue stiffening is one problem. Only very recently has much thought gone into this problem. Kruse [204a] has discussed the qualitative behavior of fiber blends and pointed out the problems that arise. Tesoro and Meiser [205] have indicated the direction new work must take. They have shown that blends of cotton and polyester might be more hazardous than either taken alone because the polyester melts and the cotton serves as a wick. In model studies using ammonium phosphate for the cotton and haloalkylphosphate ester for the polyester, much useful information was recorded by these workers. Since the fiber was more than half polyester, the results are presented under polyester in Chapter 8 (pp. 398–401). Suffice it to say that much remains to be done. It is not clear whether the answer will be in a new finish or in a combination of a new inherently fire-resistant polyester plus a standard finish on the fabric to treat the cotton component. The answer is a year or two ahead, at least, and its final form is hard to discern.

REFERENCES

[1] Amended FFA 1967: C. A. Baker, *Amer. Dyestuff Rept.*, **58** (6), 12 (1969).
[2] See *Federal Register*, *33*, 9963 (July 3, 1968); 14642 (Oct. 1, 1968); 15662 (Oct. 23, 1968); 17921 (Dec. 3, 1968).
[2a] *Federal Register*, *34*, 19812 (Dec. 18, 1969); 20434 (Dec. 31, 1969); *ibid*, 35, 1019 (Jan. 24, 1970); ibid. *35*, 6211 (Apr. 16, 1970).
[3] F. Ward, *J. Soc. Dyers Colour.*, **71**, 569 (1955).
[4] M. N. Conklin, *Color Trade J.*, **11**, 171 (1922).
[5] W. H. Perkin, *J. Ind. Eng. Chem.*, **5**, 57 (1913); *Text. Mfr.*, **39**, 423 (1913).
[6] W. H. Hunt, *The Chemical Engineer*, **11**, 22 (1910).
[7] For early reviews from Germany and France see G. Durst, *Kunststoffe*, **2**, 145 (1912), and R. Chesneau, *Bull. Soc. Ind. Rouen*, **60**, 265 (1932).
[8] W. W. Chase, *Text. World*, **93**, 90 (1943).
[9] R. W. Little, ed., Flameproofing Textile Fabrics, *Amer. Chem. Soc. Monograph Series No. 104*, Reinhold, New York, 1947, 410 pp.

[10] J. E. Ramsbottom, *The Fireproofing of Fabrics*, Royal Aircraft Establishment, His Majesty's Stationery Office, London, 1947.

[11] M. W. Sandholzer, Flameproofing of Textiles, *Nat. Bur. Std. Circ. 455*, U.S. Govt. Printing Office, Washington, D.C., 1946, 20 pp.

[12] Annual editions of the *Rev. Text. Progr.*, Text. Inst., Manchester, and Soc. Dyers Colour., Bradford, Eng., 1948 and annually thereafter.

[13] C. J. West, E. Stringham, L. Roth, and J. Weiner, Flameproofing, *Bibliographic Series No. 185*, Inst. Paper Chemistry, Appleton, Wis. 1959, 244 pp.

[14] K. Jack and W. Trzebny, *Prace Inst. Przemyslu Wlokien Lykowych*, **7**, 201 (1959).

[15] J. Gil Montero, *Ion*, **23**, 703 (1963).

[16] A. Schurch and A. Berger, *Textil-Rundschau*, **9**, 251 (1954); K. Quehl, *Melliand Textilber*, **35**, 434 (1954).

[17] E. Frieser, *SVF-Fachorgan Textilveredlung*, **15**, 143 (1959); *ibid.*, **16**, 460 (1960); *Spinner Weber Textilveredl*, **79**, 938, 1038, 1140 (1961); *ibid.*, **80**, 1202 (1962); *Melliand Textilber*, **39**, 795 (1958).

[18] K. Quehl, *Industrie Textile*, No. 812, 521 (1954).

[19] W. Stolbowsky, *Teintex*, **20**, 113 (1955); X. Bilger and G. Mangeney, *Teintex*, **29** 837 (1964). (See also *Melliand Textilber*, **46**, 294 (1965) (in German).

[20] J. M. Church, R. W. Little, and S. Coppick, *Ind. Eng. Chem.*, **42**, 418 (1950); R. W. Little, J. M. Church, and S. Coppick, *ibid.*, 432; J. M. Church, *Chem. Eng. News*, **31**, 325 (1953).

[21] A. J. McQuade, in *Textile Chemicals and Auxiliaries*, 2nd ed., H. C. Speel and E. W. K. Schwarz, eds., Reinhold, New York, 1957, Chapter 22.

[22] J. R. W. Perfect, *J. Soc. Dyers Colour.*, **74**, 829 (1958).

[23] B. C. M. Dorset, *Text. Mfr.*, **84**, 572 (1958); *ibid.*, **87**, 243 (1961); *ibid.*, **88**, 160, 482 (1962).

[24] M. W. Sandholzer, *Amer. Dyestuff Reptr.*, **48**, 37 (1959).

[25] J. G. Frick, Jr., R. L. Arcenaux, E. K. Leonard, and J. D. Reid, *WADC Tech. Rept. 58-130*, *ASTIA Doc. No. 206894*, OTS-U.S. Dept. Commerce, November, 1958.

[26] G. L. Drake, Jr., W. A. Reeves, and R. M. Perkins, *Amer. Dyestuff Reptr.*, **52**, 41 (1963).

[27] J. D. Reid, *Mod. Text. Mag.*, **1968**, 87 (September).

[28] G. L. Drake, Jr., *Kirk-Othmer Encyclopedia of Chemical Technology*, 2nd ed., Vol. 9, Wiley-Interscience, New York, 1966, p. 300.

[29] J. M. Church, A. Spiegelman, H. J. Franklyn, and G. F. Crikelair, *Amer. Dyestuff Reptr.*, **52**, 48 (1963).

[30] *Amer. Dyestuff Reptr.*, **56**, 49 (1967).

[31] *Textilveredl.; Schweiz. Z. Textilchem., Textilveredl. & Deren Randgeb.*, 3, 142 (1968); P. Lennox-Kerr, *Text. Ind.*, **133**, 156 (1969); J. E. Hendrix and R. H. Barker, *Clemson Univ. Rev. Ind.-Management and Text. Sci.*, **8**, 63 (1969); R. Aenishänslin, *Melliand Textilber.*, **10**, 1210 (1968); R. B. LeBlanc, *Text. Ind.*, **132**, 274 (1968); R. B. LeBlanc, *Amer. Dyestuff Reptr.*, **57** (27), 35 (1968); C. Reichman, *Knitted Outerwear Times*, **1969**, 23 (Mar. 10).

[32] G. W. Wilkens, *Australian 9,300*, Apr. 22, 1907.

[33] Reference 9, p. 67.

[34] Reference 9, p. 172.

[35] F. V. Davis, J. Findlay, and E. Rogers, *J. Text. Inst.*, **40**, T839 (1949).

[36] A. C. Nuessle, *J. Soc. Dyers Colour.*, **64**, 342 (1948).

[37] R. J. Taylor and M. Dishman, *Text. World*, **1947**, 113 (April).

[38] Reference 9, pp. 192 ff.; R. Aarons, W. H. Baumgartner, and D. R. English, *U.S. 2,935,471* (to E. I. duPont de Nemours & Co.), May 3, 1960.

[39] R. B. Seymour, J. A. Crumley, W. G. Agnew, and A. J. Kelley, *Amer. Dyestuff Reptr.*, **37**, 10 (1948).

[40] S. J. O'Brien, *Text. Res. J.*, **38**, 256 (1968); Pyroset® Flame Retardant CP, *Text. Finishing Bull. No. 240*, American Cyanamid Co., Bound Brook, N.J.

[41] T. H. Morton and F. Ward, *U.S. 2,530,261* (to Courtaulds, Ltd.), Nov. 14, 1950.

[42] D. Blackburn and E. Haelam, *Can. 788,894* (to Courtaulds, Ltd.), July 2, 1968.

[43] E. L. Donahue, *Brit. 856,360*, Dec. 14, 1960.

[44] Bozel-Maletra, Societe ind. prod. chim., Pierrefitte-Kalaa Djerda, Soc. gen. d'engraiset prod. chim. (by R. Lehmann, J. Lintner, and P. Williame), *Fr. 1,100,929*, Sept. 26, 1955.

[45] Chemische Fabrik Pfersee G.m.b.H. (by H. Enders, R. Zoerkendorfer, and W. Deckelmann), *Ger. 1,151,486*, July 18, 1963; L. Krema, *Czech. 100,291*, July 15, 1961; Chemische Werke Albert (by K. Stumpp and H. Metschke), *Ger. 969,419*, May 29, 1958; CIBA, Ltd. (by A. Berger), *Ger. 1,011,393*, July 4, 1957; J. Dabrowski, *U.S. 3,304,194* (to United Merchants and Manufacturers, Inc.), Feb. 14, 1967; A. Berger, *U.S. 2,781,281* (to CIBA, Ltd.), Feb. 13, 1957; CIBA, Ltd. (by A. Berger, G. During, and A. Schurch), *Swiss 321,212*, June 15, 1957; M. R. Burnell and J. E. Lynn, *U.S. 2,582,691* (to American Cyanamid Co.), Jan. 22, 1952; E. L. Donahue (M. S. Donahue, executrix), *U.S. 3,253,881*, May 31, 1966; Deutsche Gold-und Silber-Scheideanstalt, vorm. Roessler (by W. Schulenburg), *Ger. 922,482*, Jan. 17, 1955; F. F. Pollak and J. Fassel, *U.S. 2,464,342*, Mar. 15, 1949;

[45a] H. Yamagushi and M. Takada, *U.S. 3,436,250* (to Asahi Kogyo Kabushiki Kaisha), Apr. 1, 1969.

[46] M. L. Nielsen, *U.S. 2,648,597* (to Monsanto Chemical Co.), Aug. 11, 1953.

[47] M. L. Nielsen, *Text. Res. J.*, **27**, 603 (1957).

[48] J. L. Iannazzi, *U.S. 3,032,440* (to General Tire & Rubber Co.), May 1, 1962.

[49] Badische Anilin- & Soda-Fabrik A.-G. (by W. Krause), *Ger. 1,148,968*, May 22, 1963; J. E. Malowan and M. L. Nielsen, *U.S. 2,949,385* (to Monsanto Chemical Co.), Aug. 16, 1962.

[50] R. C. Steinhauer, *U.S. 3,029,283* (to Victor Chemical Works), Apr. 10, 1962.

[50a] J. R. Geigy A.-G., *Brit. 790,663*, Feb. 12, 1958.

[51] M. E. Cupery, *U.S. 2,163,085* (to E. I. duPont de Nemours & Co.), June 20, 1939; J. R. Costello, *U.S. 2,964,377* (to Victor Chemical Works), Dec. 13, 1960; J. Greenblatt, *U.S. 3,391,079*, July 2, 1968.

[52] R. C. Steinhauer, *U.S. 3,112,154* (to Stauffer Chemical Co.), Nov. 26, 1963.

[53] R. L. Holbrook, *U.S. 2,898,180* (to Olin Mathieson Chemical Corp.), Aug. 4, 1959; Imperial Chemical Industries, Ltd. (to H. C. Fielding and F. Nyman), *Brit. 966,525*, Aug. 12, 1964.

[54] Farb. Bayer A.-G. (by F. Kassack, H. Malz, and F. Lober), *Ger. 1,087,109*, Appl. May 11, 1956: H. Tolkmith and E. C. Britton, *U.S. 2,805,256* (to Dow Chemical Co.), Sept. 3, 1957.

[55] N. J. Glade, *U.S. 2,971,929* (to American Cyanamid Co.), Feb. 14, 1961; N. J. Glade, I. Hechenbleikner, and D. W. Kaiser, *U.S. 2,828,228* (to American Cyanamid Co.), Mar. 25, 1958; Courtaulds, Ltd. (by F. Ward), *Brit. 835,581*, May 25, 1960; I. Hechenbleikner, *U.S. 2,832,745* (to Shea Chemical Corp.), Apr. 29, 1958; Courtaulds, Ltd. (by F. Ward), *Brit. 858,582*, Jan. 11, 1961.

[56] J. R. Geigy A.-G. (by J. Bindler, E. Model, and R. Keller), *Ger. 1,040,498*, Oct. 9, 1958.

[57] W. A. Reeves, L. H. Chance, and G. L. Drake, Jr., *U.S. 2,859,134* (to U.S. Dept. Agri.), Nov. 4, 1958.

[58] W. A. Reeves, G. L. Drake, Jr., and J. D. Guthrie, *U.S. 2,915,480* (to U.S. Dept. Agr.), Dec. 1, 1959.

[59] W. A. Reeves, L. H. Chance, and G. L. Drake, Jr., *U.S. 2,917,492* (to U.S. Dept. Agr.), Dec. 15, 1959; L. H. Chance, G. L. Drake, Jr., and W. A. Reeves, *U.S. 2,901,444* (to U.S. Dept. Agr.), Aug. 25, 1959.

[60] W. A. Reeves, J. D. Guthrie, and G. L. Drake, Jr., *U.S. 2,906,592* (to U.S. Dept. Agr.), Sept. 29, 1959.

[61] W. A. Reeves, L. H. Chance, and G. L. Drake, Jr., *U.S. 2,889,289* (to U.S. Dept. Agr.), June 2, 1959.

[62] L. H. Chance, G. L. Drake, Jr., and W. A. Reeves, *U.S. 2,891,877* (to U.S. Dept. Agr.), June 23, 1959.

[63] L. H. Chance, G. L. Drake, Jr., and W. A. Reeves, *U.S. 2,870,042* (to U.S. Dept. Agr.), Jan. 20, 1959; L. H. Chance, G. L. Drake, Jr., and W. A. Reeves, *U.S. 2,886,538* (to U.S. Dept. Agr.), May 12, 1959.

[64] W. A. Reeves, J. D. Guthrie, and L. H. Chance, *U.S. 2,912,412* (to U.S. Dept. Agr.), Nov. 10, 1959.

[65] W. A. Reeves, G. L. Drake, Jr., and J. D. Guthrie, *U.S. 2,933,367* (to U.S. Dept. Agr.), Apr. 19, 1960.

[65a] Reggiana S.p.A., *Brit. 903,820*, Aug. 22, 1962.

[66] R. C. Steinhauer, *U.S. 3,034,919* (to Stauffer Chemical Co.), May 15, 1962.

[67] Farb. Bayer A.-G., *Brit. 892,131*, Mar. 21, 1962.

[68] T. O. Miles, F. A. Hoffman, and A. Merola, *U.S. 3,085,029* (to U.S. Dept. of the Army), Apr. 9, 1963.

[69] Chemische Fabrik Pfersce G.m.b.H., *Neth. Appl. 6,602,865*, Sept. 7, 1966.

[70] Badische Anilin- & Soda-Fabrik A.-G. (by H. Brandeis, R. Fikentscher, H. Petersen, and H. Bille), *Ger. 1,234,671*, Feb. 23, 1967.

[71] Farb. Hoechst A.-G., *Belg. 701,987*, June 28, 1967.

[72] Badische Anilin- & Soda-Fabrik A.-G., *Belg. 704,847, 717,595*, Oct. 9, 1967, July 4, 1968.

[73] R. B. LeBlanc, *U.S. 3,376,160* (to Dow Chemical Co.), Apr. 2, 1968.

[73a] R. B. LeBlanc and R. H. Symm, *U.S. 3,409,462* (to Dow Chemical Co.), Nov. 5, 1968.

[73b] Hooker Chemical Corp. (by G. M. Wagner), *Can. 809,255*, Mar. 25, 1969.

[74] G. L. Drake, Jr., and J. D. Guthrie, *Text. Res. J.*, 29, 155 (1959).

[75] R. B. LeBlanc, *Amer. Dyestuff Reptr.*, **53**, 42 (1964).

[76] G. L. Drake, Jr., L. H. Chance, J. V. Beninate, and J. D. Guthrie, *Amer. Dyestuff Reptr.*, **51**, 40 (1962).

[77] R. B. LeBlanc, *Text. Res. J.*, **35**, 341 (1965).

[77a] G. L. Drake, Jr., B. M. Kopacz, and F. S. Perkerson, *Agr. Res. Serv. Document 72-14*, U.S. Dept. Agr., September, 1958.

[77b] T. D. Miles, F. A. Hoffman, and A. Merola, *Amer. Dyestuff Reptr.*, **49** (17), 596 (1960).

[78] Tris(1-Aziridinyl) Phosphine Oxide, *Tech. Data Sheet*, Dow Chemical Co., Midland, Mich., 1963.

[78a] R. M. Perkins, G. L. Drake, Jr., and W. A. Reeves, *Agr. Res. Serv. Document 72-32*, U.S. Dept. Agr., December, 1964.

[79] S. R. Hobart, G. L. Drake, Jr., and J. D. Guthrie, *Amer. Dyestuff Reptr.*, **51**, 21 (1962).

[79a] R. A. Strecker and A. S. Tompa, *J. Polym. Sci., Part A-1*, **6**, 1233 (1968).

[80] A. McLean, W. Kilbride, and S. F. Marrian, *U.S. 2,472,335* (to Imperial Chemical

Industries, Ltd.), June 7, 1949; A. McLean, S. F. Marrian, and Imperial Chemical Industries, Ltd., *Brit. 604,490*, July 5, 1948; A. McLean and S. F. Marrian, *U.S. 2,470,042* (to Imperial Chemical Industries, Ltd.), May 10, 1949.

[81] R. R. Dreisbach and J. L. Lang, *U.S. 2,992,942* (to Dow Chemical Co.), July 18, 1961.

[82] E. Pacso and R. F. Schwenker, Jr., *U.S. 2,990,233* (to Textile Research Institute), June 27, 1961.

[83] N. O. Brace, *U.S. 2,733,229* (to E. I. duPont de Nemours & Co.), Jan. 31, 1956.

[84] L. H. Chance, J. Warren, and J. D. Guthrie, *U.S. 2,743,232* (to U.S. Dept. Agr.), Apr. 24, 1956.

[85] L. Horvath, F. Eder, J. Sprung, Z. Bartha, and T. Hegedus, *Hung. 147,527*, Oct. 1, 1960.

[85a] B. Bohmer, Z. Manasek, and D. Bellus, *Czech. 110,140*, Mar. 15, 1964.

[85b] Farb. Bayer A.-G., *Belg. 699,765*, June 12, 1967.

[85c] G. E. Walter and I. Hornstein, *U.S. 2,574,515* (to Glenn L. Martin Co.), Nov. 13, 1951.

[86] J. G. Frick, Jr., J. W. Weaver, R. L. Arceneaux, and M. F. Stansbury, *J. Polym. Sci.*, **20**, 307 (1956); J. G. Frick, Jr., and J. W. Weaver, *U.S. 2,686,768* (to U.S. Dept. Agr.), Aug. 17, 1954; J. G. Frick, Jr., and J. W. Weaver, *U.S. 2,686,769* (to U.S. Dept. Agr.), Aug. 17, 1954; J. W. Weaver, J. G. Frick, Jr., and J. D. Reid, *U.S. 2,711,998* (to U. S. Dept. Agr.), June 28, 1955; J. W. Weaver, *U.S. 2,778,747* (to U.S. Dept. Agr.), Jan. 22, 1957.

[87] J. G. Frick, Jr., J. W. Weaver, and J. D. Reid, *Text. Res. J.*, **25**, 100 (January, 1955).

[88] J. G. Frick, Jr., J. W. Weaver, R. L. Arceneaux, and M. F. Stansbury, *J. Polym. Sci.*, **20**, 307 (1956).

[89] A. J. McQuade, *Amer. Dyestuff Reptr.*, **44**, 749 (1955).

[89a] W. J. Taylor, H. C. Olpin, and K. R. House, *Brit. 531,651*, Jan. 8, 1941.

[90] T. D. Miles and A. C. Delasanta, *Text. Res. J.*, **38**, 273 (1968).

[90a] *Ibid.*, **39**, 357 (1969).

[91] C. Hamalainen and J. D. Guthrie, *Text. Res. J.*, **26**, 1261 (1956).

[92] C. Hamalainen, *U.S. 2,825,718* (to U.S. Dept. Agr.), Mar. 4, 1958.

[93] C. Hamalainen, *U.S. 2,681,295* (to U.S. Dept. Agr.), June 15, 1954; Albright & Wilson, Ltd., *Brit. 821,505*, Oct. 7, 1959; W. A. Reeves, C. Hamalainen, and J. D. Guthrie, *U.S. 2,814,573* (to U.S. Dept. Agr.), Nov. 26, 1957.

[94] Compagnie Française des Matières Colorantes (by P. H. P. Vallette), *Ger. 1,067,770*, Oct. 29, 1959; P. H. P. Vallette, *U.S. 2,782,133* (to Compagnie Française des Matières Colorantes), Feb. 19, 1957.

[95] Etablissements Kuhlmann (by C. Senez), *Fr. 1,357,831*, Apr. 10, 1964.

[96] Compagnie Française des Matières Colorantes (by C. Senez and M. Bilger), *Fr. 1,322,330*, Mar. 29, 1963.

[97] Etablissements Kuhlmann (by C. Senez), *Fr. 1,388,083*, Feb. 5, 1965.

[98] M. T. Beachem, *Can. 758,039* (to American Cyanamid Co.), May 2, 1967; C. J. M. Senez, *U.S. 3,441,433* (to Ugine Kuhlmann), Apr. 29, 1969.

[99] F. R. Hurley, *U.S. 2,637,704* (to Monsanto Chemical Co.), May 5, 1953; W. R. Grace & Co., *Brit. 961,912*, June 24, 1964; C. J. M. Senez, *U.S. 3.437,518* (to Ugine Kuhlmann), Apr. 8, 1969.

[100] I. G. Farbenind A.-G. (by R. Engelhardt), *Ger. 507,104*, May 3, 1927.

[101] D. E. Kvalnes and N. O. Brace, *U.S. 2,691,566* (to E. I. duPont de Nemours & Co.), Oct. 12, 1954.

[102] A. D. F. Toy and R. S. Cooper, *U.S. 2,867,547* (to Victor Chemical Works), Jan. 6, 1959; J. R. Costello, Jr., and T. P. Traise, *U.S. 2,841,507* (to Victor Chemical Works), July 1, 1958; A. D. F. Toy and K. H. Rattenbury, *U.S. 2,735,789* (to Victor Chemical Works), Feb. 21, 1956; J. R. Costello, *U.S. 2,811,469* (to Victor Chemical Works), Oct. 29, 1957.

[103] L. A. Rogovin, Mei-Yan Wu, M. A. Tyuganova, T. Ya. Zharoua, and E. L. Gefter, *Vysokomol. Soedin.*, **5**, 506 (1963).

[104] R. Schiffner and G. Lange, *Faserforsch. Textiltech.*, **14**, 375 (1963); R. Schiffner and G. Lange, *Ger. (East) 15,357, 18,253*, Sept. 22, 1958, Feb. 25, 1960; R. Schiffner and G. Lange, *Ger. 1,150,651* (to Deutsche Akademie der Wissenschaften zu Berlin), June 27, 1963.

[105] S. R. Hobart, G. L. Drake, Jr., and J. D. Guthrie, *Amer. Dyestuff Reptr.*, **50**, 30 (1961); G. L. Drake, Jr., W. A. Reeves, and J. D. Guthrie, *Text. Res. J.*, **29**, 270 (1959).

[106] H. Nachbur, A. Berger, C. Guth, and A. Maeder, *U.S. 3,423,369* (to Ciba, Ltd.), Jan. 21, 1969; Ciba S.A., *Belg. 718,060; 718,696*, July 12, 1968; July 26, 1968.

[106a] CIBA, Ltd., *Fr. 1,395,178*, Apr. 9, 1965.

[107] CIBA, S.A., *Belg. 699,289*, May 31, 1967.

[108] (Pyrovatex® CP), CIBA Chemical Dye Co., Fairlawn, N.J., 1958; *Amer. Dyestuff Reptr.*, **57**, 54 (1968); R. Aenishänslin and N. Bigler, *Textilveredlung*, **3**, 467 (1968); R. Aenishänslin, *Text. Bull.*, **94**, 48, (December, 1968); R. Aenishänslin, C. Guth, P. Hofmann, A. Maeder, and H. Nachbur, *Text. Res. J.*, **39**, 375 (1969).

[109] G. C. Tesoro, S. B. Sello, and J. J. Willard, *Text. Res. J.*, **38**, 245 (1968).

[110] G. C. Tesoro, S. B. Sello, and J. J. Willard, Paper presented at the 155th Annual Meeting of the Amer. Chem. Soc., San Francisco, April, 1968; *Text. Res. J.*, **39**, 180 (1969). A recent thesis includes this subject: see J. E. Hendrix, J. E. Bostic, Jr., E. S. Olson, and R. H. Barker, Paper presented at the 157th Annual Meeting of the Amer. Chem. Soc., Minneapolis, April, 1969.

[111] *Chem. Eng. News*, **35**, 15 (1957) (Oct. 2).

[112] *Chem. Week*, **82**, 58 (1958) (Apr. 6).

[112a] G. C. Tesoro, *Textilveredlung*, **2**, 435 (1967).

[112b] L. H. Chance and J. P. Morean, Paper presented at the 9th Cotton Utilization Res. Conf., New Orleans, La., Apr. 30–May 2, 1969.

[113] W. A. Reeves and J. D. Guthrie, *U.S. Bur. Agr. and Ind. Chem., Mimeographed Circ. Ser. AIC-364, 1955, 11 pp.*, M. F. Stansbury and C. L. Hofpamir, *Amer. Dyestuff Reptr.*, **44**, 19 (1955).

[114] W. A. Reeves and J. D. Guthrie, *Ind. Eng. Chem.*, **48**, 64 (1956).

[115] J. D. Reid, *Text. Res. J.*, **26**, 136 (1956).

[116] D. M. Jones and T. M. Noone, *J. Appl. Chem. (London)*, **12**, 397 (1962).

[117] W. J. Vullo, *J. Org. Chem.*, **33**, 3665 (1968).

[118] J. B. Bullock and C. M. Welch, *Text. Res. J.*, **36**, 441 (1966).

[119] J. D. Guthrie, G. L. Drake, Jr., and W. A. Reeves, *Amer. Dyestuff Reptr.*, **44**, 328 (1955).

[120] W. A. Reeves and J. D. Guthrie, *Dyer*, **111**, 567 (1954); W. A. Reeves and J. D. Guthrie, *Text. World*, **104**, 101 (1954); G. L. Drake, Jr., W. A. Reeves, and R. M. Perkins, *Amer. Dyestuff Reptr.*, **52**, 608 (1963); Albright & Wilson, Ltd., *Brit. 882,993*, Appl. Nov. 21, 1956.

[121] W. A. Reeves and J. D. Guthrie, *U.S. 2,722,188* (to United States), Nov. 27, 1956.

[122] W. A. Reeves and V. K. Bourdette, *Text. Ind.*, **128**, 105 (1964).

[123] J. V. Beninate, E. K. Boylston, G. L. Drake, Jr., and W. A. Reeves, *Text. Ind.*, **131**, 110 (1967).

[124] *Ibid.*, **38**, 267 (1968).

[124a] J. V. Beninate R. M. Perkins, G. L. Drake, Jr., and W. A. Reeves, *Text. Res. J.*, **368** (1969); J. V. Beninate E. K. Boyleston, G. L. Drake, Jr., and W. A. Reeves, *Amer. Dyestuff Reptr.*, **57** (25), 74 (1968).

[124b] D. J. Donaldson and D. J. Daigle, *Text. Res. J.*, **39**, 363 (1969); *Text. Chem. Colorist*, *1* 534 (1969).

[125] J. V. Beninate, E. K. Boylston, G. L. Drake, Jr., and W. A. Reeves, Paper presented to the 1967 AATCC Nat. Tech. Conf., New Orleans, Oct. 20, 1967.

[126] J. D. Reid, J. G. Frick, Jr., and R. L. Arceneaux, *Text. Res. J.*, **26**, 137 (1956).

[127] C. Hamalainen, W. A. Reeves, and J. D. Guthrie, *Text. Res. J.*, **26**, 145 (1956).

[128] W. A. Reeves, G. L. Drake, Jr., L. H. Chance, and J. D. Guthrie, *Text. Res. J.*, **27**, 259 (1957).

[128a] R. M. Perkins, G. L. Drake, Jr., and W. A. Reeves, *Text. Ind.*, **130**, 3093 (1966).

[128b] R. M. Perkins, G. L. Drake, Jr., and W. A. Reeves, *Amer. Dyestuff Reptr.*, **54**, 17 (1965).

[129] R. B. LeBlanc, *Text. Ind.*, **132**, 274 (1968); G. L. Drake, Jr., J. V. Beninate, and J. D. Guthrie, *Amer. Dyestuff Reptr.*, **50**, 27 (1961); J. R. W. Perfect, *Textil-Rundschau*, **13**, 464 (1958); J. R. W. Perfect, *J. Soc. Dyers Colour.*, **74**, 829 (1958); R. P. Barber et al., *Amer. Dyestuff Reptr.*, **57**, 40 (1968).

[130] G. L. Drake, Jr., E. K. Leonard, and W. A. Reeves, *Text. Ind.* **130**, 145 (1966).

[131] P. E. Hoch, G. M. Wagner, and W. J. Vullo, *Text. Res. J.*, **36**, 757 (1966).

[132] J. B. Bullock, C. M. Welch, and J. D. Guthrie, *Text. Res. J.*, **34**, 691 (1964).

[133] G. L. Drake, Jr., *Amer. Dyestuff Reptr.*, **56**, 560 (1967); also Paper presented at the New York Academy of Medicine, Dec. 2, 1966.

[134] O. J. McMillan, Jr., *Text. Ind.*, **123**, 94 (1959); K. M. Decossas, S. P. Koltun, and E. L. Patton, *ibid.*, **125**, 101 (1961); K. M. Decossas, H. L. E. Vix, and E. L. Patton, *Amer. Dyestuff Reptr.*, **51**, 23 (1962); H. K. Gardner, Jr., G. L. Drake, Jr., and N. B. Knoepfler, *Hospitals*, **37**, (1963); R. G. Charlesworth, *Text. Ind.*, **131**, 119 (1967); K. M. Decossas, *Text. Ind.*, **125**, (1961); New Durable Flame Retardant Cotton Fabric with APO-THPC, *CA-S-22* (*Revised*), U.S. Dept., Agr., Southern Utility Research and Development Division, New Orleans, March, 1961.

[135] K. M. Decossas, B. H. Wojcik, A. de B. Kleppinger, W. A. Reeves, and H. L. E. Vix, *Text. Ind.*, **130**, 128 (1966).

[136] N. E. Boyer and A. E. Vajda, *SPE Trans.*, **4**, 45 (1964).

[137] G. A. Kiselev, L. A. Vol'f, and A. I. Meas, *Zh. Prikl. Khim.*, **39**, 388 (1966).

[138] L. H. Chance, W. A. Reeves, and G. L. Drake, Jr., *Text. Res. J.*, **35**, 291 (1965).

[139] L. M. Valentine, H. Fletcher, R. Cowling, J. A. Moyse, and J. H. Lupton, *Can. 789,443* (to I.C.I., Ltd., and Tootal, Ltd.), July 9, 1968.

[140] J. W. Aylsworth, *U.S. 1,085,783*, Feb. 3, 1913 (*Chem. Abstr.*, **8**).

[141] R. Engelhardt, *U.S. 2,073,004* (to I. G. Farbenind), Mar. 9, 1936.

[142] L. W. Georges and C. Hamalainen, *U.S. 2,428,843* (to U.S. Dept. Agr.), Oct. 14, 1947.

[143] Hooker Chemical Corp., *Belg. 676,221*, Feb. 8, 1966.

[144] E. Pascu and R. I. Schwenker, Jr., *U.S. 2,990,232* (to Textile Research Institute), June 27, 1961.

[145] J. G. Frick, Jr., J. D. Reid, and H. B. Moore, *Text. Res. J.*, **26**, 525 (1956).

[146] L. A. Jordan and L. A. O'Neill, *Brit. 573,471*, Nov. 22, 1945.

[147] L. A. Jordan and L. A. O'Neill, *Brit. 573,472*, Nov. 22, 1945.

[148] N. J. Read and E. G. Heighway-Bury, *J. Soc. Dyers Colour.*, **74**, 823 (1958).

[149] C. Snyder, *U.S. 1,885,870*, Nov. 1, 1932.

[150] M. Leatherman, *U.S. 2,012,686* (to United States), Aug. 27, 1935.

[151] M. Leatherman, *Circ. No. 466*, U.S. Dept. Agr., Washington, D.C., March, 1938, 17 pp.

[152] E. C. Clayton and L. L. Heffner, *U.S. 2,299,612* (to William E. Hooper & Sons, Inc.), Oct. 20, 1942.

[153] J. R. Redmond, *Amer. Dyestuff Reptr.*, **32**, 375 (1943); R. Van Tuyle, *ibid.*, 297.

[154] K. S. Campbell and J. E. Sands, *Text. World*, **96**, 118 (1946).

[155] S. A. Rulon, M. J. Sostmann, and I. L. Phillips, *Amer. Dyestuff Reptr.*, **35**, 489 (1946).

[156] M. P. Khera, R. M. Desai, and M. G. Bhargava, *J. Sci. Ind. Res.*, **17A**, 452 (1958).

[157] *Mod. Text.*, **45**, 44 (1964).

[158] M. Orzel and D. Zyzka, *Prace Inst. Wlokiennietwa*, **14**, 179 (1964).

[159] S. Coppick, in reference 9, pp. 239–48.

[160] R. W. Little, in reference 9, pp. 248–66.

[161] R. W. Little, E. F. Taylor, and W. R. Christensen, in reference 9, pp. 293–357.

[162] R. B. Finch, A. J. McQuade, and H. A. Rutherford, in reference 9, pp. 360–79.

[163] W. L. Dills, *U.S. 2,607,729* (to E. I. duPont de Nemours & Co.), Aug. 19, 1952; I. M. Panik and W. F. Sullivan, *U.S. 2,668,780* (to National Lead Co.), Feb. 9, 1954; D. Duane, *U.S. 2,728,680* (to National Lead Co.), Dec. 27, 1955.

[164] H. C. Gulledge and G. R. Seidel, *Ind. Eng. Chem.*, **42**, 440 (1950).

[165] J. W. Bannister and C. A. Rittel, *Brit. 1,744*, Feb. 3, 1915.

[166] W. F. Brosnan, *Text. World*, **84**, 2399 (1934).

[167] R. Clarence, *Tiba*, **14**, 405 (1936).

[168] H. Hadert, *Farbe Lack*, **55**, 324 (1949); E. F. Hartman, R. Hicks, and F. Hartman, *Fire Eng.*, **92**, 336 (1939).

[169] R. R. Oliver, *U.S. 2,553,781* (to Lockport Cotton Batting Co.), May 22, 1951.

[170] J. P. Seren and C. M. Armand d'Aix de Tour Sanite, *Fr. 1,418,288*, Nov. 19, 1965.

[171] A. L. Benarous, *Fr. 462,192*, Dec. 16, 1966.

[172] R. Aarons and D. Wilson, *U.S. 2,823,212* (to E. I. duPont de Nemours & Co.), Nov. 8, 1955; Ni Ho Boseki Co., Ltd., *Belg. 700,351*, June 22, 1967.

[173] O. J. McMillan, Jr., and J. D. Guthrie, *U.S. 2,695,833* (to U.S. Dept. Agr.), Nov. 30, 1954.

[174] W. E. Gordon, *Rayon Text. Monthly*, **22**, 98 (1941); W. E. Gordon, Paper presented at 40th Annual Convention, National Retail Dry Goods Assoc., New York, Jan. 14, 1941.

[175] F. W. Hochstetler, *U.S. 2,368,660* (to Hochstetler Res. Lab., Inc.), Feb. 6, 1945.

[176] J. E. Ramsbottom and A. W. Snoad, "Second Dept. of the Fabrics Coordinating Res. Committee," Dept. of Scientific and Industrial Research (Great Britain), 1930.

[177] S. Coppick and W. P. Hall in reference 9, pp. 217–39.

[178] J. Downing, J. W. Fisher, and E. W. Wheatley (deceased—by M. F. Wheatley administratrix), *U.S. 2,780,571* (to British Celanese Ltd. & Corp. of Great Britain), Feb. 5, 1957.

[179] K. Nakao, *Jap. 7099* (to Seiten Sato), Aug. 20, 1956.

[179a] C. J. Mahn, N. G. Baumer, and G. D. Hiatt, *Ind. Eng. Chem.*, **47**, 2521 (1955); M. Bentivaglio and B. E. Cash, *Mod. Plast. (Overseas Ed.)*, **1945**, 10 (June).

[180] Nelsons Silk, Ltd., *Belg. 610,903*, May 28, 1962.

[181] R. Schabert, *Ger. 1,189,516* (to Firma Carl Freudenberg), Mar. 25, 1965; H. E. Freudenberg, K. Kraft, O. Schildhauer, P. Wentuler, H. Freudenberg, and D. Freudenberg, *Brit. 1,066,132* (to Helmut Fabricius), Apr. 19, 1967; H. Boe, *Ger. 1,149,688* (to Carl Freudenberg, K-G.), June 6, 1963.

[181a] O. H. Sindl, *U.S. 2,805,176* (to R. S. Robe), Sept. 3, 1957.
[182] Societe Rhodiaceta, *Fr. 1,407,354*, July, 1965: Deutsche Rhodiaceta A.-G. (by A. Sippel and F. Engelbrecht), *Ger. 1,222,203*, Aug. 4, 1966.
[183] Nelsons Silk, Ltd. (by C. A. Redfarn), *Ger. 1,187,344*, Feb. 18, 1965; C. A. Redfarn, *U.S. 3,471, 318* (to Nelsons Silk, Ltd.) Oct. 7, 1969. W. K. Mohney, *U.S. 3,321,330* (to FMC Corp.), May 23, 1967; W. D. Paist and N. Van Gorder, *U.S. 2,662,834* (to Celanese Corp. of America), Dec. 15, 1953.
[184] R. C. Harrington, Jr., and J. L. Smith, *U.S. 2,933,402* (to Eastman Kodak Co.), Apr. 19, 1960.
[185] B. C. M. Dorset, *Text. Mfr.*, **84**, 626 (1959); R. S. Robe, *Brit. 795,133*, May 14, 1958.
[186] J. W. Schappel and A. I. Bates, *U.S. 3,266,918* (to FMC Corp.), Aug. 16, 1966.
[186a] M. R. Lytton, *U.S. 3,454,588* (to FMC Corp.), July 8, 1969.
[186b] L. E. A. Godfrey, *U.S. 3,455,713* (to FMC Corp.), July 15, 1969.
[186c] J. W. Schappel, *Mod. Text.*, 54 (July, 1968).
[187] R. M. Perkins, G. L. Drake, Jr., and W. A. Reeves, *Appl. Polym. Sci.*, **10**, 1041 (1966).
[188] S. Coppick, in reference 9, p. 44.
[189] S. Coppick, in reference 9, p. 68.
[190] S. Coppick, J. M. Church, and R. W. Little, *Ind. Eng. Chem.*, **42**, 415 (1950); R. W. Little, *Text. Res. J.*, **21**, 901 (1951).
[191] H. A. Schuyten, J. W. Weaver, and D. J. Reid, *Ind. Eng. Chem.*, **47**, 1433 (1955).
[192] R. W. Little, in reference 9, pp. 293–320.
[193] E. P. Frieser, *Spinner Weber Textilvered*, **79**, 1140 (1961).
[194] *Text. Bull.*, **1968**, 36 (August); *Chem. Eng. News*, **46**, 41 (July 22, 1968).
[195] J. C. Arthur, Jr., J. A. Harris, and T. Mares, *Text. Ind.*, **132** (1968).
[196] *Chem. Eng. News*, **46**, 71 (Oct. 7, 1968).
[197] F. A. Blouin, A. M. Cannlzzaro, J. C. Arthur, Jr., and M. L. Rollins, *Text. Res. J.*, **38**, 811 (1968).
[198] A. M. Feibush and H. P. Kieltyka, *U.S. 3,407,088* (to U.S. Atomic Energy Comm.), Oct. 22, 1968.
[199] Textile Institute, Moscow (by Z. A. Rogovin, A. M. Tyuganova, Yu. G. Kryazhev, and T. Y. Zharoua), *Brit. 1,022,083*, Mar. 9, 1966.
[200] E. M. Peters, *U.S. 3,235,323* (to Minnesota Mining & Manufacturing Co.), Feb. 15, 1966.
[201] Great Lakes Carbon Corp., *Fr. 1,519,765*, Feb. 21, 1967.
[201a] *Ger. 1,272,801*, July 13, 1966.
[202] Societe le Carbone-Lorraine, *Brit. 1,016,351*, Jan. 12, 1966.
[203] J. A. Lynch, *U.S. 3,242,000* (to Deering Milliken Res. Corp.), Mar. 22, 1966.
[204] G. Machell, *U.S. 3,395,970* (to Deering Milliken Res. Corp.), July 3, 1968.
[204a] W. Kruse, Paper presented to the Conference Future Flammability, Zurich, Switzerland, January, 1969.
[205] G. C. Tesoro and C. H. Meiser, Jr., Paper presented at the 39th Annual Meeting of the Textile Res. Inst., New York, Apr. 11, 1969.

Chapter 6

COATINGS

There are two kinds of fire-safe coatings. Fire-resistant coatings are themselves not likely to contribute fuel to a fire or to promote the spread of flame by rapid combustion. But many coatings do more than this; they also protect flammable substrates from the effects of fire. The problem in the first category is to minimize the flammability of the coating itself through a proper choice of ingredients. In the second case the coating must be formulated both to eliminate its own fire hazards and to protect the substrate. The problem is affected by the nature of the substrate; the approach may differ in the cases of wood and steel, for example.

Many of the earliest references to fire retardation by chemical treatments refer to coating wood, textiles, and so on. In most of these references the treatment did not consist of application of a paint or mastic as we know these terms today. In the present chapter we are dealing with complete formulations that fulfill all the requirements of normal protective coatings and, additionally, that provide fire safety. If a treatment does not have all these properties, it is discussed elsewhere in the book, for example, in the chapters on wood and textiles.

In 1948 Ware and Westgate [1] published a review of the U.S. patent literature to that time (covering 339 patents). The earliest citation is to the use of various inorganic salts as fire-retardant additives for paints (L. Paimboeuf, *U.S. 449*, Nov. 11, 1837). By 1900 many of today's concepts had been suggested in the patent literature.

The patent review is a convenient means of studying the development of such coatings through the mid-1940s. The review covers eight categories— chlorocarbons, bituminous systems, borates, phosphates, silicates, antimony compounds, nitrogenous organics, and miscellaneous—and not all entries are true coatings concepts. Thus, the earliest dates on patents covering paints as opposed to simpler treatments are (each date represents the corresponding entry in the foregoing categories) 1940, 1883, 1898, 1919, 1933, 1921, —, and 1837. (There are no entries under nitrogen compounds regarded as pertaining to paint.)

There have been a few scattered reviews of the status of the art. Late in World War II the New York Club of the Federation of Paint and Varnish Production Clubs presented in Cleveland a lengthy and detailed look at retardant paints for steel surfaces [2]. This was followed in 1945 by a second report [3] in which suggestions of the U.S. Navy were followed up and new comparative test data were presented. We shall come back to this work later.

Black and Sanders, writing in 1948 [4], listed a number of formulations then current and commented on each briefly. Sternberg [5] reviewed patents to 1952, including some foreign ones, and offered brief commentary and some formulas. A panel chaired by Westgate reviewed the need for fire-retardant coatings and discussed approaches to the problem [6]. And in 1954 the American Chemical Society published the proceedings of a 1953 symposium on this same subject. The publication is the well-known Monograph No. 9, "Fire Retardant Paints" [7]; contained therein are comments by Westgate, Cook, Yochers, Well, Mod, Chapman, Murray, Liberti, and Allen which are of a general nature and concern the need for and value of various types of fire-retardant paints. We shall discuss some of these articles subsequently. Since this landmark publication there have been only a few articles that are reviews of a general nature.

Holderried discussed paints containing bromine compounds in 1958 [8] and Schofield authored a very brief but general résumé in 1964 [9]. Vandersall's discussion [10] of intumescent systems contains some review material also.

Discussions of methods of testing have been a prominent part of the literature and rightly so. The cabinet test was discussed in two papers [11], five different methods were intercompared in another publication [12] (the cabinet test was preferred), and some Canadian research was presented in two papers in 1952 [13]. In Monograph No. 9 [7] Grubb and Cranmer presented the results of extensive simulated service tests run by the U.S. Navy. This was one of the earliest full-scale tests on which good data are to be found in the available literature. Recent emphasis has been on tunnel testing and on obtaining correlations between small tunnels and the 25-ft horizontal tunnel used by Underwriters' Laboratories. One such small tunnel was discussed in 1967 by Vandersall [14]; it seems to be gaining in popularity.

NON-INTUMESCENT COATINGS

Classifying coatings as to whether they are merely fire-resistant or whether they can also protect a flammable substrate (and how) is a risky business when only published information is used as the basis. It is not always possible to tell whether intumescence will occur just by examining a formula. If data from testing are given, the case may be easier to decide—but not always. Thus, in the following discussion there are undoubtedly errors in classification.

The attempt is made to separate the literature into those systems which intumesce on heating (i.e., the coating puffs up and forms a cellular structure which remains attached to the substrate and which provides insulation and protection when all goes well) and those which act through other mechanisms.

The patent literature contains almost all of the original research data in this field. Journal articles are by and large restricted to comparative studies carried out by trade associations and by consumers; for example, government agencies and the like. The development of a given fire-retardant coating is not often described by the company doing the work. In the succeeding paragraphs the patent story of non-intumescing coatings is presented, this is followed by comparisons from journal articles.

Phosphorus-Nitrogen, Phosphorus-Halogen Systems

In the simplest case a substance like NH_4MgPO_4 is added to a coating in place of some of the pigment normally required. This idea was patented in the first decade of this century by two Frenchmen [15]. There are many citations in which the addition of ammonium orthophosphates is disclosed. One formula called for about 40% $NH_4H_2PO_4$ in a water-base paint containing aldehyde condensation resins, zinc oxide, and clay [16]. Another used $(NH_4)_2HPO_4$ with borax and thickeners all emulsified with chlorocarbons and resins [17]. In another recipe poly(vinylacetate) emulsion was the base and aminoplast resins and $NH_4H_2PO_4$ were added [18]. In this case more than half the total wet coating weight was phosphate.

Slightly more complex systems containing nitrogen and phosphorus have been evaluated. A recent reference claims the reaction product of ϵ-aminocaproic acid [19] and $(NH_4)_2HPO_4$, presumably

$$NH_4(HPO_4) NH_2(CH_2)_5 COONH_4$$

added to a methanolic solution of an aminoplast resin for use on lumber. The amino acid may be assumed to reduce the solubility of the phosphate to improve resistance of the film to leaching. Complex compounds obtained from heating P_4O_{10} and dry NH_3 (Chapter 2) have been formulated using dicyandiamide in a water system [20]. Polyphosphorylamides—products from the reaction of $POCl_3$ and NH_3 (Chapter 2)—have been suggested in a chlorinated rubber base [21] or in a water-base system using aminoplast resins and alkyds [22]. A solution of basic zinc phosphate in aqueous ammonia is said to dry at 105° to yield a water- and fire-resistant coating [23]. Other transition metals may be used along with fillers. Further attempts to reduce water attack by altering the solubility of the phosphate included the use of less soluble organic bases, for example, $NH_2(CH_2)_6NH_2$, dicyandiamide, aniline [24], and hexamethylene tetramine [25]. In a departure from the ordinary, a compound like phenylphosphorodiamidate [26] has been tried as an ingredient in a flame-resistant wood varnish. Recently [27] a phosphonitrilic has been claimed as a comonomer for producing fire-resistant urethane coatings:

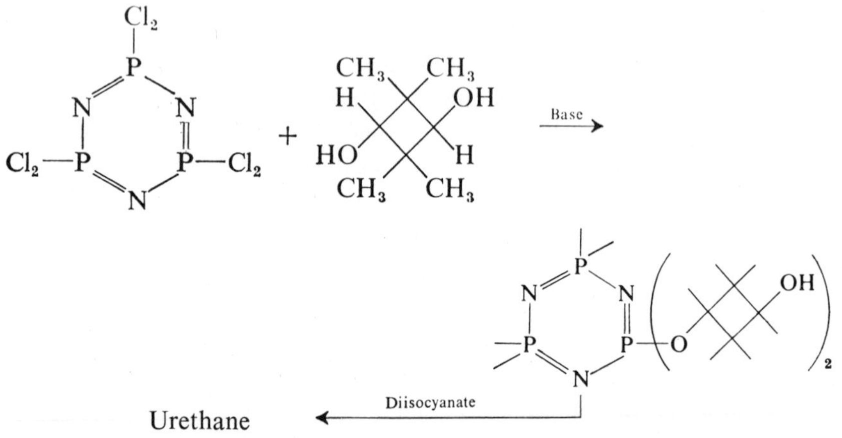

Of all of these possibilities, very few have attained commercial significance. The ammonium orthophosphates, because they are inexpensive, find some use but more in intumescent systems than in the types being considered here. The same is true for other salts of phosphoric acid. The reaction products with $POCl_3$ have not found favor; the phosphonitrilics are still being studied. The reduction of the solubility of the phosphate by using amines in place of ammonia succeeds technically but is costly and therefore has not been very popular. We shall return to this discussion in the intumescent category.

There are but three citations to phosphorus-halogen combinations: the use of chlorinated phosphate esters in lacquers [28] and in various resinous coatings [29] and phosphorus derivatives obtained through reactions with hexachlorocyclopentadiene [30]. The phosphorus-halogen synergism has evidently not been studied overly much for coatings.

Halogens and Antimony-Halogen Compositions

Chlorinated organics have long been used in coatings either as chlorinated aromatics [31] or as aliphatics. The chlorinated paraffins were developed for coatings for wood, textiles, and the like (see Chapter 5). This development was reviewed by the manufacturer in an article in 1948 [32]. The paper is really a discussion of Chlorowax 70 in combination with metal oxides and especially antimony oxide (see below). Chloroparaffins are not particularly effective by themselves in coatings, just as they are not in textiles. They require metal oxides to facilitate decomposition. Leatherman suggested zinc oxide as the agent [33]; and Craig found $BaSO_4$ to be helpful [34].

Chlorinated rubber [35] and chlorinated alkyd resins [36] are claimed to improve the fire resistance of coatings.

The addition of metal oxides to chlorocarbons makes a dramatic difference in performance, apparently by facilitating the decomposition of the former so as to give halogen radicals in the flame. These radicals break the oxidation chain in the flame and thereby achieve flame suppression (see Chapters 5 and 1). Of the oxides studied the majority of references are to antimony oxides—Sb_4O_6—often referred to as the trioxide. [There are scattered references to other forms, for example, $SbCl_3$ [37] and pyroantimonic acid ($H_4Sb_2O_7$) [38].] In 1934 McCulloch received a patent [39] on a paint for electric machinery based on chlorinated rubber and Sb_4O_6. And by the early 1940s numerous citations appeared in the literature [40]. House paints (exterior) were formulated using chlorinated paraffins and Sb_4O_6 with or without $CaCO_3$ [32, 41]. Using a paraffin containing 70% chlorine as a replacement for linseed oil, a formula was constructed. When some 35–50% of the linseed oil solids was replaced by the chloroparaffin and an equal weight of $CaCO_3$ and 1 lb of Sb_4O_6 was added per finished gallon of paint, a fire-retardant finish was obtained. The paint was judged to be good in coating properties, becoming more enamel-like with increasing chloroparaffin content. Some difficulty was encountered with drier life in storage in the container and some brushability problem at the higher chloroparaffin levels.

In a study of interior house paints having pigment/binder ratios of 1.5:1.0 or less, at least 30% of the vehicle solids should be derived from the chloroparaffin (70% Cl) to obtain fire-retardant coatings (with 1 lb of Sb_4O_6 per gallon) [32]. Two-coat application was superior to one coat. Iron oxide pigment seemed to help, perhaps because it accelerates decomposition of the chlorocarbon. The stability of this type of formula was claimed to be entirely satisfactory.

In a lengthy discussion of coatings for steel (for use by the Navy) antimony and chlorocarbon combinations performed well, especially with regard to flammability behavior [2, 3]. The challenge was to achieve rust inhibition and compatibility with primers. A study directed specifically at antimony and chloroparaffins in coatings for metals was reported by Birnbaum and

Table 6-1

	Lb/100 Gal	Vol gal
Pigment	853.5	26.96
Alkyd resin (70%)	229	29.00
Petroleum spirits	281	43.05
Lead naphthenate drier	7.8	0.81
Cobalt naphthenate drier	1.0	0.12
Manganese naphthenate drier	0.5	0.06
	1372.8	100.00

Markowitz [42]. The flame tester was an electrically heated metal substrate capable of reaching 930° (C) in 10 sec and 1266° in 30 sec. Several series were run, starting with the control formula given in Table 6-1. This formula has 60% pigment by volume on a dried basis. It was varied systematically by adding Sb_4O_6 and chloroparaffin (70% Cl). A flame evaluation was devised by rating the behavior at the periphery of the test piece, somewhat removed from the hot spot. In the absence of chlorine, flaming could be eliminated as follows:

Pigment Vol, %	Sb_4O_6 for Fire Resistance (arb.)	
	Vol, %	Wt. % Sb_4O_6 in Dried Film
40	8.0	~21.0
50	3.85	8.9

Addition of chlorinated paraffin (70%) for the alkyd resin vehicle enables one to reduce the antimony level sharply at a given pigment volume concentration (PVC), as shown in Table 6-2. The values in the paper are given in one

Table 6-2

	20% PVC	30% PVC	40% PVC
Vol % Sb_4O_6 in dry film	1.56	2.34	3.12
Approx. wt % Sb_4O_6 in dry film	4.5	6.1	7.4
Approx. vol % Cl in dry film	~46	~35	~15
Wt % Cl in dry film	36.8	24.0	9.3

case in vol %, in another, in wt % without precise density data. Thus, the figures in the Table 6-2 are somewhat imprecise [42]. But it is clear that a fire-resistant coating can be produced in an oil-based system more or less as given in Table 6-3.

Table 6-3

	Wt %
Sb_4O_6	3.51
Other pigments	26.4
Chloroparaffin (70% Cl)	19.5
Alkyd resin (70% solids)	11.3
Petroleum spirits	37.7
Driers	1.6
	100.0

Another way of introducing chlorine into such a coating is to use a chlorinated alkyd resin. For example, tetrachlorophthalic anhydride may be used in preparing the alkyd. One means of doing this has been described: linseed oil and tetrachlorophthalic anhydride were cooked in xylene and the product formulated with an antimony-containing pigment system. Two possible formulas are given in Table 6-4 [43].

Table 6-4

	Wt %		Wt %
Antimony oxide	10	Antimony oxide	7.75
Titanium dioxide	5	Titanium and other pigments	45
Barium sulfate	10	Alkyd resin solution (60%)[b]	33.4
Magnesium silicate	15		
Chlorinated alkyd resin solution (60% solids)[a]	60	Petroleum spirits	11.1
Driers and tints to suit		Driers, etc.	2.8
	100		100.0

[a] 41% linseed oil, 19% tetrachlorophthalic anhydride, 40% xylene
[b] Made from Het-acid (chlorendic acid)

Similarly, work has been done on oil-base [44] and water-base [45] systems using chlorinated rubber or chloroparaffins. In water-base systems acetates, acrylics, and styrene-butadiene emulsions have been evaluated. The chlorinated rubber was a neoprene latex emulsion.

Pentachlorophenol has been suggested for use along with chloroparaffins and antimony [46], and recently siliceous antimony oxide pigments have been evaluated. In these pigments the antimony oxide is present as a coating on a silica base in grains of pigment size. Improvements in performance have been observed with many active pigments prepared in this way. Fire retardance was improved by as much as one-third in this case [47].

Boron Compounds

Although there are very early references to the application of boron compounds to the problem of fire resistance [1, 48], most of the recipes published are mixed systems. It is not unusual to encounter formulas with three or more retardants, for example, sodium tungstate, NH_4Cl, boric acid, borax, and various silicates [49]. One approach has been simply to add borax to an existing oil-base paint formula; up to 50% borax may be needed [50]. Typical formulations are given in Table 6-5.

Table 6-5

	Wt %
Borax paint [4]	
Basic carbonate of white lead	41.0
Borax	32.0
Raw linseed oil	22.8
Turpentine	3.6
Japan drier	0.6
	100.0
Exterior zinc borate paint (*PVC* 34%) [7]	
Zinc borate	18.3
Titanium dioxide	
Blanc fixe	
Lead sulphate	42.6
Lead carbonate	
Zinc oxide	
XX Refined linseed oil	10.6
Isano oil	10.6
Chloroparaffin (70% Cl)	4.4
Polyamide resin solution	10.0
Driers	2.2
Mineral spirits	1.2
	100.0
Latex formula [50a]	
Sodium borate (FR-28®)	4.6
Whiting (Purecal M)	25
Clay ASP-200	5
Titanium dioxide	2.5
Mica	2.5
Polyvinyl acetate emulsion (58%)	13.3
Tricresyl phosphate	1.2
Emulsifiers	0.3
Water	47.6
	100.0

Zinc borate has been suggested in several patents (references 29 and 30 of Chapter 3). Recently, more sophisticated types of formulas have been developed. One example claims effectiveness for a glycol monoborate in paints, varnishes, and lacquers [50b].

Silicates

The use of water glass dates from 1901 [1]. The early formulas were exceedingly simple: one was a mixture of coal tar and water glass [51]; another, a paint with water glass, ammonium sulfate, and ammonium carbonate admixed [52]; a third contained water glass, asbestos, and lithopone [53]; yet another contained water glass, ground mica, titanium dioxide, mineral oil, and a drying oil [54]. Often the concept was simply to add water glass to an otherwise standard formula [55]. A half dozen or so formulations have been developed especially to achieve fire protection from sodium silicate, these are shown in Table 6-6. Some proposed recipes are complex [58] and depend on several components that are either applied separately or mixed just prior to use. In Table 6-6 paint B is really a ceramic coating which

Table 6-6   Some Formulations Containing Sodium Silicate

Paint A [4]		Paint B [56]	
	Wt %		
Sodium silicate	27 ("S")	Sodium silicate	(add to make
	8.8 ("C")		normal
			consistency)
Zinc oxide	24.1	Zinc orthosilicate	10 parts
Lithopone	14.7	Fieldspar	10 parts
Mica	14.7	Aluminum powder	3 parts
Water	10.7	Alumina hydrate	3 parts
	100.0	Asbestos	10 parts
		Lead carbonate	50 parts
		Wetting agents	8 parts
Paint C [57]			
Sodium silicate	38		
BaSO$_4$	57.3		
Graphite	4.7		
	100.0		
Plus binder resin			

confers protection by melting and forming a layer of glassy armor over the substrate. In addition to the formulas shown for water glass, an occasional reference appears to silica aerogel [59]. There is one reference to zirconium silicate used as a pigment in fire-resistant coatings [60].

Miscellaneous

This catchall section in this instance is very thin. Early references include the use of carbontetrachloride in coatings [61] (why there would be any lasting effect from a volatile solvent is a mystery), a mixture of aluminum sulfate and zinc chloride with resins to form a resistive coating [62], and the following intriguing recipe from a 1904 patent [63]:

Crocidolite fibers	6 parts
Mica	15 parts
Kaolin	10 parts
Defibered and bleached aloe scrap	5 parts
Pure wheat starch	5 parts
Pure white gelatin	5 parts
Plus vehicle	?

20

Coatings

Table 6-7 Flame Spread and Smoke Data
(Gross and Loftus [67])

	Flame-Spread Index	Smoke, mg
Red oak	99	0.3
Gypsum board	22	0.1
Plywood	195	0.8
Polystyrene	590	13.7
Linoleum	129	10.1
Paint on cement asbestos board	2.0	0.0
Paint on steel	7.4	0.0
Fire-resistant paint on plywood	33.0	1.2
Latex paint		
On plywood	93.0	1.3
On gypsum board	8.9	0.3
Oleoresinous paint		
On plywood	58.0	1.0
On gypsum board	3.5	0.0
Alkyd paint, gloss		
On plywood	97.0	1.7
On gypsum board	8.0	0.6
Alkyd paint, flat		
On plywood	69	0.4
On gypsum board	0.8	0.0
Varnish		
On plywood	162	0.6
On gypsum board	...	...
Shellac		
On plywood	832	0.2
On gypsum board	...	...

More recently there has been mention of arsenic compounds [64], potassium or sodium carbonate in paint [65], and the use of magnesium hydroxide as a smoke retardant [66]. This last concept appeared in a 1967 patent and is especially interesting in that smoke generation from paints is a critical property and is evaluated as closely today as the flame spread value. In the reference, use of $Mg(OH)_2$ as part of the filler for a polyester sheet reduced the smoke factor to 13 versus 99 for the control. One expects more research on specific additives to reduce smoke hazards.

Comparative Test Data

There are several studies from which to draw comparative data on non-intumescing coatings [2–4, 7, 67]. Unfortunately, each group used different

Table 6-8 Some Comparisons of Silicate, Borate, and Phosphate Coatings [4]

					Blistering		Discoloration		Blackened Area		Charred Area	
Paint	Amount of Alcohol, cc	Alcohol Boils, sec	Time Alcohol Burned, sec	Panel Starts Burning, sec	Front, sec	Back, sec	Front, sec	Back, sec	Front, in.	Back, in.	Length of Char, in.	Depth of Char, %
Commercial silicate paint	$1\frac{1}{2}$	61	216	190	12	115	47	. . .	$1\frac{1}{8} \times 1\frac{3}{8}$	None	$1\frac{1}{8}$	30
Commercial silicate paint	3	45	255	52	19	81	48	. . .	$1\frac{1}{8} \times 1\frac{1}{2}$	None	$1\frac{1}{4}$	55
Borax paint	$1\frac{1}{2}$	62	217	115	11	84	14	164	$1\frac{1}{4} \times 2\frac{3}{8}$	None	$1\frac{5}{8}$	25
Zinc borate paint	$1\frac{1}{2}$	60	215	138	11	95	20	195	$1\frac{3}{8} \times 2\frac{1}{2}$	None	$1\frac{1}{2}$	20
52-P-22 (solvent type)	$1\frac{1}{2}$	61	216	64	50	. . .	20	188	1×3	None	$2\frac{1}{2}$	50
House paint	$1\frac{1}{2}$	61	215	59	20	180	47	190	$1 \times 4\frac{1}{4}$	None	$2\frac{3}{4}$	35
House paint	3	45	266	57	10	30	25	150	$1\frac{3}{4} \times 4\frac{1}{2}$	None	3	50
52-P-22 (emulsion type)	$1\frac{1}{2}$	61	215	66	11	95	18	160	$1\frac{1}{4} \times 2$	None	2	30
52-P-22 (emulsion type)	3	48	265	57	32	81	36	146	$1\frac{3}{4} \times 4\frac{3}{8}$	None	$3\frac{1}{4}$	60
Emulsion flat wall paint	$1\frac{1}{2}$	61	230	65	21	110	14	155	$1\frac{1}{4} \times 2\frac{1}{4}$	None	$1\frac{3}{4}$	25
Commercial phosphate type paint	$1\frac{1}{2}$	60	210	None	30	None	25	None	$1\frac{1}{8} \times 2$	None	1	5
Commercial phosphate type	3	43	263	None	37	None	35	None	$1\frac{3}{4} \times 3\frac{3}{4}$	None	2	15
								. . .				

test methods, so a word of explanation is necessary before presenting results from each paper. To set the framework, let us examine the flammability of ordinary paints. Gross and Loftus of the National Bureau of Standards [67] studied a wide variety of building materials using a radiant panel test. They devised a flame-spread index composed of a factor relating the time to burn a given distance along the sample placed at an angle to the panel and a factor relating to the heat evolved from the sample. The results for a few systems selected from their work are shown in Table 6-7. Several points need to be made. Most paints do not burn vigorously (note, however, that varnish and shellac are very flammable, especially shellac). Indeed, application of ordinary paints to plywood reduces the fire hazard markedly. The paints applied to steel, cement board, and gypsum board all received flame-spread indices less than 10. This suggests very little hazard as measured by this particular test. Thus, the problem is not with the paints *per se* but with the substrate. Good fire-resistant paints not only do not assist in spreading flames but they also contribute to snuffing flames feeding on decomposition products from the substrate.

Some of the tests run in the past have not been meaningful in terms of the ability to stop a fire. Thus studies in which the amount of charring or the area of char is measured are not helpful in making a judgment as to whether or not a fire would have been helped or hindered in its further spread to other areas. Data have been given on length and depth of char for borax, silicate, and phosphate paints (see Table 6-8), but the results are not especially helpful, [4].

Westgate, in studying the inclined panel test, evaluated some paints available commercially in 1950 [68]. In this work the duration of afterflame was recorded and we can get some guidance from these data in terms of flame propagation. Table 6-9 shows some of the results. Here again we see the

Table 6-9   Ratings of Some Interior Coatings by an Inclined
Panel Test [68]

Coating	Char Area, in.^{2}	Afterflame, sec
Undercoated wood	43.6	92
Oleoresinous varnish	23.6	128
Synthetic resin emulsion paint	18.0	29
Rubber latex emulsion paint	17.0	36
Oil-base semigloss	15.7	46
Oil-base gloss	15.4	57
Modified caesin emulsion paint	14.7	22
Oil-base flat paint	11.4	14
Moderately intumescing fire-resistant paint	8.9	0

reduction in flammability of wood even by a varnish. Note that oil-base flat paints are very effective in protecting the wood.

A very thorough study was reported by Murray, Liberti, and Allen in 1954 [69]. They used the cabinet test (alcohol burner against a panel inclined at 45°). Ten pigments were studied and 10 vehicles. The pigments were evaluated in an alkyd vehicle with and without chlorinated paraffin at both 40 and 60% PVCs. The vehicles were studied with a fixed pigment containing 42% TiO_2, 38% Zn borate, 12.5% $CaCO_3$, and 7.5% asbestine. The results are given in Tables 6-10 and 6-11. Again, it is important to point out that these results do not give us an indication of the tendency of a flame to spread or be propagated over the treated area but only tell us if it chars and loses weight in some way (with or without actual flaming). The entries have been arranged in order of increasing weight loss. This trend, however, has little or no real significance since the nature of the chemicals volatilized governs whether or not they propagate the fire. (Obviously, volatilization of carbon as CO_2 is desirable, loss as an aldehyde or ketone is not.)

Table 6-10 Pigment Study [69]

Pigment	Av Wt Loss, %		Av Char Area, in.^{2}		Av Char Vol, in.^{3}	
	60% PVC	40% PVC	60% PVC	40% PVC	60% PVC	40% PVC
Zinc borate, plus Cl[a]	8.96	8.4	25.1	36.3	3.8	6.82
Calcium carbonate, plus Cl	9.9	15.3	22.5	38.1	4.22	6.65
Lead silicate, plus Cl	10.4	11.5	39.1	39.3	7.4	7.4
Antimony oxide, plus Cl	11.1	10.54	39.0	31.8	7.34	6.8
Zinc borate, no Cl	11.4	9.1	24.3	25.5	4.27	4.56
Antimony oxide, no Cl	11.45	11.53	41.9	41.3	7.8	7.78
Magnesium silicate, plus Cl	12.1	10.8	36.6	31.5	6.91	5.95
Blanc fixe, plus Cl	12.5	17.6	39.8	45.0	8.16	9.15
Zinc oxide, plus Cl	13.5	11.5	46.2	36.3	8.0	6.82
Lead silicate, no Cl	14.0	13.67	45.7	39.3	7.33	6.83
Lithopone, plus Cl	14.6	16.5	38.0	58.0	7.11	8.31
Zinc oxide, no Cl	17.1	15.5	42.7	34.5	7.98	6.53
Lead carbonate, plus Cl	19.6	16.2	50.8	50.4	10.2	8.31
Calcium pyrophosphate, plus Cl	21.2	22.7	55.8	49.0	10.9	9.23

[a] Plus Cl means chloroparaffin was added to the vehicle

Table 6-11   Burn Test Data on Vehicles [69]

Vehicle	Av Wt Loss, %	Av Char Area, in.^{2}	Av Char Vol, in.^{3}
Alkyd resin plus chlorinated paraffin	9.63	32.6	6.2
Alkyd resin plus silicone resin	9.8	28.1	5.44
Refined linseed oil plus chlorinated paraffin	12.2	39.0	7.36
Alkyd resin plus ethyl silicate	14.4	42.8	7.96
Alkyd resin plus chlorinated rubber	14.6	47.5	8.94
Vinyl chloride acetate resin	14.93	48.3	9.0
Oil-modified epichlorhydrin resin	16.54	45	9.04
Acrylic resin plus chlorinated rubber	18.9	52	9.7
Urea-formaldehyde resin	19.98	53	12.18
Polystyrene resin emulsion	22	53	12.18

Table 6-12   Test Data on Interior and Exterior Coatings

Coating	Before Leaching		After Leaching		
	Wt Loss, %	Char Area, in.^{2}	Wt Loss, %	Char Area, in.^{2}	Scrub Strokes[a] to Failure
Interior paints					
Ammonium phosphate/ chlorocarbon/alkyd					
A	6.5	16.22	15.6	38.2	600
B	5.2	15.50	· · ·	· · ·	1436
C	6.0	15.36	11.72	32.6	· · ·
Ammonium phosphate/ alkyd (no Cl)	6.6	17.3	· · ·	· · ·	2275
Control coating (no fire-resisting agents)	21.38	44.65	· · ·	· · ·	· · ·
Exterior paints					
Zinc borate/chloro- carbon/linseed oil/ alkyd resin/mineral spirits	8.19	23.0	10.3	29.1	· · ·
Zinc borate/chloro- carbon/linseed-isano oil/polyamide resin/ mineral spirits	5.33	13.75	5.68	13.9	· · ·

[a] Gardner scrub instrument

Table 6-12 lists some data on the water resistance of these paints. The interior paints do not resist exposure to water whereas the exterior paints are, of course, formulated for water resistance. There is a clear need for more research on the durability of fire-resistant paints not only as to the integrity of the film but also as to the retention of the fire retardant in an effective form.

There are two fairly detailed reports of investigations of paints on steel or other metal substrates; the studies were carried out largely for the U.S. Navy. In this instance the entire aim is that the paint does not propagate a flame originating elsewhere, perhaps even on the reverse side of the metal. In

Table 6-13

	Wt %
Primer	
Zinc chromate	31.23
Iron oxide	10.33
Magnesium silicate	10.83
Congo linseed	40.81
Mineral thinner	6.80
	100.00
Top coat	
Titanium dioxide	30.75
Lithopone	16.87
China clay	5.37
Methyl cellulose (5% soln)	1.25
Ester gum	4.68
Linseed oil	9.36
Driers	0.15
Emulsifier (Dupanol ME)	0.14
Water	31.43
	100.00

the work done by the New York Club of the Federation of Paint and Varnish Production Clubs, a great many coatings were reviewed. The original papers should be consulted for details [2, 3]. Silicate paints were found to be poor in adhesion to imprimed metal and were dropped for a while from the study. Emulsion paints were better than organic solvent coatings from the standpoint of flashing and overall flammability and because of their better adhesion to metal. High PVCs favor good fire resistance. The preferred primer-top coat system was as shown in Table 6-13. Much detail on such problems as emulsification, prevention of rust spotting, adhesion, washability, and brushability is given in the papers.

This group later returned to the silicate paints and found excellent fire resistance and good adhesion when the paints were applied over a primer. The paint used was paint A of Table 6-6. Problems were cited with can stability, brittleness of film, and the need for heat to set to a nonleachable coating.

Finally, some field testing was carried out in simulated ship structures by the U.S. Navy on certain of these coatings [70]. Panels coated with several kinds of fire-resistant paint were placed over portholes in a simulated hangar deck with the paint side out. Fires were set by flooding the interior deck with oil and igniting it. The paint was therefore heated indirectly through the steel plate. The maximum temperature behind the plates was 920–960°C; on the test panels, 285–570°C. The stock enamel paint flamed and burned entirely in a short time, whereas the fire-retardant coatings in thicknesses less than 5 mils did not flame or flash at all. In thicker films the control again burned entirely and some flashing and intermittent flaming was observed with some of the fire-resistant formulas. The contents of these commercial paints were not disclosed but the results are nonetheless useful. Large-scale tests of systems rather than of components will always be more meaningful in the long run and one can only hope that more such work will be reported in the future. The cost of such tests is, of course, high.

INTUMESCENT COATINGS

The Fourth of July "snake" is the most widely known example of intumescence. When heat is applied, combustion produces a residue which is puffed up by escaping gases. In the case of the snake, the light char forming the curls of the snake's body has little adhesion or integrity and it quickly disintegrates and blows away. It is, however, possible to puff up combustion residues in such a way as to produce a tough insulating foam which will protect materials over which it is formed in a very efficient manner. Such techniques have been refined for use in fire-retardant paints and mastics.

An extensive discussion of intumescent coating technology has been given by Vandersall [10]. Paints are formulated using materials especially selected to provide the best in intumescence and at the same time to produce the necessary coating properties.

It is important to emphasize the latter requirement. Too many coatings have been proposed which have good fire-resistant properties but which are decidedly inferior as paints.

The intumescent paint must have ingredients which will react on heating to form large amounts of an incombustible, or nearly so, residue. At the same time this residue must be expanded to a cellular foam with good insulating properties. And the foam must be tough and adherent so as to resist violent drafts and other forces arising from the fire. In most cases the residue is a carbon char, and it is formed by the dehydration of a polyhydric substance such as a polyalcohol (polyol). The reactions all take place within the coating—the polyol is added along with the dehydrating agent—and in no way does the system depend on the substrate. Thus, there is no requirement that the polyol be, for example cellulose derived from a wood substrate.

The polyol is dehydrated by a substance which is a Lewis acid. There is some question as to whether this acid must be strongly dehydrating in the sense that it has a very low vapor pressure of water in equilibrium with it or whether it merely must promote formation of an intermediate complex which will decompose in the desired direction. Thus, we have two possibilities:

(a)
$$R(OH)_2 + Acid \rightarrow RO + Acid \cdot H_2O$$
Simple dehydration

(b)
$$R(OH)_2 + Acid \rightarrow [R(OH)_2 \cdot Acid] \rightarrow RO + Acid + H_2O(g)$$
(or Acid·H$_2$O)

The two are in fact quite different. The first could, in theory, occur without direct contact between reagents, although actually the time required would be excessive. It is generally agreed that the second case is the more probable.

The most commonly used acid is phosphoric acid. It is introduced as a salt or ester in a form that decomposes on heating to produce acidic residues. In the case of ammonium phosphates and simple polyols the following is a possible reaction sequence on heating:

$$x(NH_4)_2HPO_4 \xrightarrow{\text{heat}} HO \left(\begin{array}{c} O \\ \| \\ P-O \\ | \\ O \\ | \\ H \end{array} \right)_x H + 2xNH_3 + (x-1)H_2O \qquad (1)$$

(2)

$$HO \left(\begin{array}{c} O \\ \| \\ P-O \\ | \\ O \\ | \\ H \end{array} \right)_x H + RCH_2CH_2OH \rightarrow RCH_2CH_2OPO_3H_2 + HO \left(\begin{array}{c} O \\ \| \\ P-O \\ | \\ O \\ | \\ H \end{array} \right)_{x-1} H$$

$$RCH_2CH_2OPO_3H_2 \xrightarrow{\text{more heat}} RCH=CH_2 + H_3PO_4 \qquad (3)$$

$$xH_3PO_4 \xrightarrow{\text{heat}} HO \left(\begin{array}{c} O \\ \| \\ P-O \\ | \\ O \\ | \\ H \end{array} \right)_x H + (x-1)H_2O \qquad (4)$$

Of this series of reactions all of the stages are known for simple alcohols and for the phosphates and phosphoric acids [71]. Producing carbon in a lower state of oxidation is the result (olefin from an alcohol). It is not so clear how one proceeds from the olefin down to carbon char. But it does occur, and voluminous amounts of char are produced in contrast to the result of combustion in the absence of the acid. In the literature the acid has been termed a catalyst because it is regenerated in the reaction and is available for reuse. However, reuse many times does not, in fact, appear to occur. This is because these are heterogeneous systems where mixing does not occur during the reactions. Therefore one normally has to add enough acid to react with all the possible sites of char formation.

Vandersall has summarized the carbon content of a variety of polyols that might be used (see Table 6-14) for protecting wood. These materials vary in

Table 6-14   Possible Sources of Carbon for Char Formation in an Intumescent Paint for Wood Protection [10]

	Formula	% Carbon	Reactivity, sites/100 g
Sugars			
Glucose	$C_6H_{12}O_6$	40	2.8
Maltose	$C_{12}H_{22}O_{11}$	42	2.3
Arabinose	$C_5H_6O_4$	45	3.0
Polyhydric alcohols			
Erythritol	$C_4H_6(OH)_4$	39	3.3
Pentaerythritol	$C_5H_8(OH)_4$	44	2.9
dimer	$C_{10}H_{16}(OH)_6$	50	2.5
trimer	$C_{15}H_{24}(OH)_8$	53	2.4
Arabitol	$C_5H_7(OH)_5$	39	3.3
Sorbitol	$C_6H_8(OH)_6$	40	3.0
Inositol	$C_6H_6(OH)_6$	40	3.0
Polyhydric phenols			
Resorcinol	$C_6H_8(OH)_2$	63	1.8
Starches	$(C_6H_{10}O_5)_n$	44	2.1

decomposition temperature but all are expected to decompose in the presence of the dehydrating agent in the paint film at a temperature lower than the combustion temperature of the underlying wood cellulose. Therefore the cellulose will not begin to degrade before the reaction in the paint film has occurred and insulation is provided. The erythritols have received much attention in recent years as the source of the carbon char.

The dehydrating agents that have received most study are the various ammonium and amine phosphates and phosphate esters and the borates. There is almost no mention of any others.

The materials which actually cause the char to expand are known as blowing agents. These are selected for their decomposition to volatile products at the desired temperatures. If this temperature is too low, the gases will escape before the char is formed; if too high, the char will merely be lifted off the substrate or be blown apart. Some common blowing agents are shown in Table 6-15.

Table 6-15   Some Blowing Agents [10]

Material	Gaseous By-Products	Decomposition Temp, °C
Dicyandiamide	NH_3, CO_2, H_2O	~210
Melamine	NH_3, CO_2, H_2O	250
Guanidine	NH_3, CO_2, H_2O	160
Glycine	NH_3, CO_2, H_2O	~233
Urea	NH_3, CO_2, H_2O	~130
Chlorinated paraffin (70% Cl)	HCl, CO_2, H_2O	190

Occasionally, other materials are added to improve the toughness of the carbon foam. These may be aminoplasts (urea- or melamine-formaldehyde resins), which are believed to form a tough skin over the cellular mass. Vandersall rightly points out that most of the intumescent ingredients fulfill more than one function and in this sense must be selected with all factors in mind. Most of the ingredients, for example, release some gas on heating and are therefore to be considered as blowing agents. Similarly, many of the ingredients bring with them some carbon which they can contribute to the char. And so on.

Every coating must be formulated as a special case. The combination of blowing agents will depend on the polyol and the dehydrating agent. The combination will be different for ammonium orthophosphate (dec 147°) than for ammonium polyphosphate (dec 325°). Considerable effort is normally required to optimize the intumescence; formulation work is time-consuming and tedious, and is made more so by the requirement that the final paint not only have excellent fire-protective properties but also be a first-class coating. This means that the effect of changing each intumescent ingredient on many factors—stability in the can, brushing characteristics, leveling, wet edge, final surface appearance, resistance to moisture and to scrubbing, and so on—must be carefully taken into account. Will the coating take tints well? Will it hide satisfactorily? Can it be made in current equipment? In the early days one could offer an inferior coating if the fire protection was good. Today every property must come up to strict standards and the job is much more difficult and demanding. To meet these demands new ingredients and new techniques have been advanced. Let us now review the developments in intumescent paints using as a means of classification the dehydrating agent.

Boron Compounds

An early intumescent paint based on borax and boric acid was simple indeed [72]. These two ingredients were added to an ordinary thermoplastic paint vehicle along with $CaCO_3$ and other pigments. By 1956 the art was more sophisticated and we find intumescence claimed in a formula such as that shown in Table 6-12 using zinc borate [73]. In another coating borax was used as a glass-former. A cross-linked resin system was proposed (a tung oil-phenolic or an epoxy resin), an exfoliated mica, and a blowing agent for the range 200–350° [74]. The most recent patent (1965) [75] reads more like the classical intumescent system described above. In this is claimed penta-erythritol, boric acid, and zinc acetate. The combination is intumescent and is water-soluble. (No journal articles on this coating have been recovered as yet.)

Phosphorus Compounds

There are only two instances in the literature where phosphorus compounds were used in forms other than esters or nitrogen derivatives. In one, phosphoric acid itself was used [76] in a two-part coating system; in the other, a water-insoluble metal metaphosphate (Zn, Ca, Na, or K salts) was suggested [77] for use with an amine resin, a polyol, a chlorocarbon, and an antimony compound in an oil-based coating. But these are unusual formulas. More typical are those based on ammonia and amine compounds of phosphorus.

Paints Containing $NH_4H_2PO_4$ or $(NH_4)_2HPO_4$

The early intumescent coatings were based on ammonium orthophosphate as the dehydrating agent. A formula from the 1940s calls for paraldehyde, $NH_4H_2PO_4$, urea, and starch (polyol) as a coating for wood [77]. This was improved by using a urea-formaldehyde resin and some protein to toughen the foam [78]; a pigment system was suggested soon afterward [79]. A latex paint containing basically the above ingredients (aminoplast resin, $NH_4H_2PO_4$, starch, etc.) was claimed in a 1954 patent where the latex was a vinyl chloride-vinylidene chloride copolymer [80]. The problem was to find a latex which would not be thoroughly coagulated by the high soluble salt content of the formula—the salt, of course, being the ammonium ortho-phosphate. In 1956 the outlines of today's formulations were taking shape

Table 6-16   Typical Water-Based
Intumescent   Paint   Containing
Monoammonium Phosphate [87]

Ingredient	%
Water	20.4
Surfactants	0.5
Titanium dioxide	8.5
Monoammonium phosphate	25.6
Melamine-formaldehyde resin	2.1
Dicyandiamide	8.5
Pentaerythritol	4.3
Chlorinated paraffin (70% Cl)	5.1
Carboxymethyl cellulose (2%)	6.4
Polyvinyl latex (60% solids)	17.6
Dibutyl phthalate	1.0
	100.0

[81]. The latex was the above-mentioned chlorine-containing copolymer, the polyol was pentaerythritol, and dicyandiamide was added as a blowing agent. This coating was said to withstand 550 scrubbing cycles. A similar formula was developed around a new, compatible, poly(vinylacetate) latex [82]. Techniques were given to improve plasticization of the films [83] and to provide other polyols (sorbitol [84], dextrin [85]) and other resins [84] and blowing agents [85]). An oil-based coating was built around these same intumescing ingredients with chlorinated rubber in naphtha to make a scrub-resistant system [86]. A modern formula for a paint based on $NH_4H_2PO_4$ is given in Table 6-16.

A formula that is claimed to be more water-resistant by virtue of the added silicone is given in Table 6-17 [87].

Table 6-17

	Wt %
Titanium dioxide	12.8
Starch	5.1
Ammonium phosphate	29.5
Aminoacetic acid	5.4
Chlorinated rubber	10.6
Alkyd resin solution (50%)	10.2
High-flash naphtha	25.7
Silicone resin solution (60%)	0.7
	100.0

Table 6-18 Patents on P-N Compounds other than Ammonium Orthophosphate in Paints

Date	Patent No	Inventor	Assignee	Invention
May 20, 1952	Jap. 1786	T. Sakurai, T. Izumi		$NH_2(CH_2)_6NH_2 \cdot H_3PO_4$, sucrose, aminoplast resin
May 13, 1952	U.S. 2, 596, 939	M. L. Nielsen, P. E. Marling	Monsanto Chemical Co.	Polyphosphorylamide, aminoplast, polyol, water
June 16, 1952	U.S. 2, 642, 405	M. L. Nielsen	Monsanto Chemical Co.	Polyphosphorylamide, aminoplasts, phenol-formaldehyde resins, water
June 17, 1952	U.S. 2, 600, 455	I. V. Wilson, R. Marotta	Monsanto Chemical Co.	NH_3-P_4O_{10} reaction product, aminoplast resin, starch, water
Feb. 17, 1953	U.S. 2, 628, 946	W. Juda, G. Jones, N. Altman	Albi Mfg. Co., Inc.	Aminoplast resin phosphate (guanidine melamine resin)
Apr. 20, 1954	U.S. 2, 676, 162	R. Marotta	Monsanto Chemical Co.	Polyphosphorylamide, epoxy resin, other resins in organic solvent
Nov. 1, 1955	U.S. 2, 722, 523	A. E. Gilchrist, L. D. Harrup, R. C. Hendrickson, D. T. Rohor	Glidden Co.	Guanylurea phosphate, starch, aminoplast resin, bake the coating at 150°C
Oct. 11, 1960	U.S. 2, 956, 037	J. M. Venable	Vimasco Corp.	Polyphosphorylamide, plasticized aminoplast resin system, dicyandiamide, pentaerythritol, organic solvent.
Mar. 15, 1961	Brit. 862, 569	G. Quelle, C. A. Redfarn, R. Thompson	Alim Chemical Corp.	Butylphosphate, urea resin, alcohols, organic solvent
Feb. 10, 1967	Neth. Appl. 6, 611, 210		Hooker Chemical Corp.	Phosphotrianilide, chlorocarbon, alkyd resin, organic solvent
July 30, 1968	U.S. 3, 396, 129	D. A. Yeadon, E. T. Rayner, G. B. Verburg, L. L. Hooper, Jr., F. G. Dollear, H. P. Dupuy	U.S.A.	Polyaminotriazine phosphates and/or poly(urea-urethane-phosphate-phosphonates in polyurethanes) and/or polycarbonate coatings
Jan. 14, 1969	U.S. 3, 422, 046	F. H. Thomas, H. M. Headrick, E. L. Schulz	Sherwin-Williams Co.	Phosphate ether polyol, melamine compound, chlorocarbon

Paints Containing Phosphorus Compounds of Reduced Water Solubility

Much of the difficulty with paints based on ammonium orthophosphates has centered on their high solubility in water. This has caused problems with crystal formation in the paint can on temperature cycling in storage. It often has led to trouble in terms of leaching of paint films or of migration of the phosphate through the coating (efflorescence). To avoid these problems phosphorus compounds of reduced water solubility have been sought. The requirements are low water solubility, ability to decompose on heating to an acidic residue, compatibility with other ingredients, and reasonable cost per unit of phosphorus. A dozen patent references on such approaches are summarized in Table 6-18. One of the more important materials, ammonium polyphosphate, is not mentioned in the table. Melamine phosphate is noted but once (see, however, Chapter 2). References in the patent literature on the use of these latter two compounds in coatings are not numerous. However, they have been thoroughly discussed in other literature.

In a series of four papers, workers at the Southern Regional Research Laboratory of the U.S.D.A. at New Orleans have described a series of oil-based intumescent coatings containing melamine phosphate and halogenated phosphate esters as the sources of acidic dehydrating agents [89]. These researches provide some relatively rare comparative data on different tests and different formulations. Six of the formulas are shown in Tables 6-19 and 6-20. They have in common the use of pentaerythritol (as a urethane

Table 6-19   Some Oil-Based Formulas from Work at U.S.D.A. [89]

	Formulation, wt %			
Components	B	C	G	H
Oil-based vehicle				
alkyd (tung/linseed)	· · ·	30	30	30
2 parts alkyd (from above)				
plus 1 part air blown tung	30	· · ·	· · ·	· · ·
Pentaerythritol polyurethane[a]	25	25	25	25
Tris(1-chloro-3-bromo-				
isopropyl) phosphate	· · ·	10	· · ·	· · ·
Tris(2,3-dibromopropyl)				
phosphate	10	· · ·	10	10
Chlorinated paraffin (70% Cl)	5.5	5.5	5.75	5.65
Polyamide resin	4.54	4.54	4.54	4.54
Pigments and driers, etc.	24.96	24.96	24.71	24.81
	100.00	100.00	100.00	100.00

[a] From 2 moles of pentaerythritol, 1 mole of TDI, 5 moles phenylisocyanate

Table 6-21(A) Performance Data on U.S.D.A. Oil-Based Coatings Containing Phosphate Esters [89](a)

| | Dry Paint on Panel | | | Fire-Retardant Performance | | | | | | |
| | | | | Wt Loss in Cabinet | | | | Performance in 16-Ft Furnace | | |
Formulation	No. of Coats	Weight, g	Thickness, mils	Nonleached, g	Leached, g	Weathered, g	mo	Flame Spread	Blistering Rupturing	Carbonaceous Buildup
B	2	33.1	8.3	9.0	...	...	...	...	...	...
	2	32.3	8.1	...	6.9	...	...	...	...	...
	3	45.2	11.3	6.8	...	...	...	...	...	...
	3	44.6	11.2	...	6.0	...	...	...	...	...
	3	46.3	11.6	...	...	4.1	18	...	...	...
	2	36.8	9.2	...	...	...	...	55	Negligible	Moderate
	3	48.4	12.1	...	...	...	...	45	Slight	Moderate
	2	32.2	8.1	...	...	...	...	40	Negligible	Moderate
	3	41.1	11.0	...	...	...	...	40	Slight	Moderate
C	2	26.0	6.5	6.4	...	...	...	...	...	...
	2	28.1	7.0	...	5.4	...	...	...	...	...
	3	39.0	9.8	4.9	...	...	...	...	...	...
	3	38.1	9.5	...	4.4	...	...	...	...	...
	3	41.0	10.2	...	...	3.6	15	...	...	...
G	2	30.5	7.6	...	...	...	...	>60	Negligible	Moderate
	3	36.6	9.2	...	...	...	...	30	Negligible	Moderate
	2	30.4	7.6	3.8	...	...	...	...	...	...
	2	30.9	7.7	...	5.0	...	...	...	...	...
	3	40.2	10.1	3.5	...	...	...	...	...	...
	3	39.7	9.9	...	3.9	...	...	...	...	...
	3	49.9	12.5	...	...	3.4	16	...	...	...
H	2	33.9	8.5	...	...	...	...	40	Negligible	Moderate
	3	47.1	11.8	...	...	...	...	>60	Considerable	Slight
	3	35.1	8.7	6.4	...	...	...	...	...	...
	3	36.2	9.1	...	6.5	...	...	...	...	...
	3	36.3	9.1	...	...	6.5	9	...	...	...
	2	28.2	7.0	...	...	...	...	30	Negligible	Good

(a) Material in Tables 6-20, 21A, 21B and 22 is reprinted, by permission, of the Federation of Societies for Paint Technology, Phila., Pa. (See Ref. 89 for journals and dates.)

Table 6-20 Two Formulas from U.S.D.A. with Both Melamine Phosphate and a Phosphate Ester [89]

Components	Formulation, wt % C'	A'
Oil-based vehicle[a]	30	25
Pentaerythritol polyurethane	20	20
Melamine phosphate	10	12
Tris(2,3-dibromopropyl) phosphate	5	8
Chlorinated paraffin (70% Cl)	5.6	5.6
Pigments, driers, etc.	29.4	29.4
	100.0	100.0

[a] From 4 parts tung oil/1 part castor oil

derivative) as the polyol and primary carbon source in the charring reaction and chlorocarbon as a primary source of blowing gas. All these formulas had comparable performance and none was outstanding. As shown in the performance tests in Table 6-21, ratings of flame spread below 30 were not

Table 6-21(B) Performance Data from the Cabinet Test on U.S.D.A. Oil-Based Coatings Containing Melamine Phosphate [89]

Formulation	Dry Paint on Panel Coats, g 1st	2nd	3rd	Total, g	Thickness, mils	Panel Wt Loss after Burning Nonleached, g	Leached, g
C	17.1	14.3	...	31.4	7.9	5.8	...
	16.2	13.9	...	30.7	7.7	...	6.2
	15.6	12.5	11.6	39.5	9.9	5.2	...
	16.4	13.9	12.4	42.7	10.7	...	4.6
E	10.9	11.7	10.5	33.1	8.3	9.0	...
	12.3	12.0	8.0	32.3	8.1	...	6.9
	14.8	15.9	14.5	45.2	10.8	6.8	...
	15.7	16.5	12.4	44.6	11.2	...	6.0

achieved with these materials. (A flame spread of 25 or less is required by Underwriters' Laboratories in the 25-ft tunnel test to achieve a Class A rating.) This work points the way to good coatings but falls just short of the mark. There are commercial coatings on the market today based on melamine phosphate that are undoubtedly very similar to these formulas. Table 6-22 shows how melamine phosphate performs in furnace tests.

Table 6-22 Performance of Melamine Phosphate Paint in Furnace Tests [89]
Coating A' from Table 6-20

| | Dry Paint on Panel | | Fire-Retardant Performance | | | | | | Remarks | |
| No. of Coats | Weight, g/ft² | Thickness, mils | Wt Loss in Cabinet | | Performance in Tunnel Furnaces | | | | | |
			Nonleached, g	Leached, g	2 ft g	2 ft fs[a]	16 ft, fs[a]	25 ft, fs[a]	Blistering and Rupturing	Carbonaceous Buildup
2	29.5	7.4	10.9	...	...	...	...	...	...	...
2	27.1	6.8	...	12.1	...	...	...	...	...	...
3	40.7	10.2	8.6	...	...	...	...	...	...	...
3	40.5	10.1	...	8.9	...	...	...	...	...	...
3	41.8	10.4	...	7.4(18)	...	...	...	...	...	...
2	27.9	7.0	...	...	13.4	58	...	...	Negligible	Slight
3	45.9	11.4	...	...	11.5	49	...	...	Slight	Moderate
2	35.8	8.9	...	...	13.6	54	...	...	Negligible	Slight
3	48.0	12.0	...	...	12.5	51	...	...	Slight	Moderate
3	41.8	10.4	...	...	14.7	59(18)	...	...	Slight	Moderate
2	30.8	7.7	...	...	...	...	>60	...	Negligible	Slight
3	42.4	10.6	...	...	...	...	45	...	Negligible	Slight
2	30.2	7.5	...	...	...	...	50	...	Negligible	Slight
3	43.5	10.9	...	...	...	...	40	...	Negligible	Slight
2	21.9	5.5	...	...	...	...	...	55	Negligible	Slight
3	37.8	9.1	...	...	...	...	...	45	Localized	Slight

[a] fs ≡ flame spread

The other major gap in the table of patents is the lack of data on paints containing ammonium polyphosphate. This material has reduced water solubility compared with ammonium orthophosphate and has been formulated into some very water-resistant, water-based and oil-based paints by Vandersall and others. The water-based systems suggested by these workers are shown in Table 6-23. In these paints all the ingredients have been chosen

Table 6-23 Two Water-Based Paints Based
on Ammonium Polyphosphate

	Wt %	
	A [10]	B [90]
Water	28.9	27.3
Diethylene glycol monoethyletheracetate	2.0	· · ·
Melamine-formaldehyde resin	1.9	· · ·
Melamine	7.7	7.4
Dipentaerythritol	3.8	7.0
Ammonium polyphosphate	22.9	24.2
Chloroparaffin (70% Cl)	4.6	3.7
Titanium dioxide	7.7	5.6
Polyvinyl acetate (60%)	15.7	20.0
Dispersants, wetting agents, plasticizers, etc.	4.0	4.8
	100.0	100.0

with an eye to reducing water solubility and therefore to improving water and scrub resistance. Thus dicyandiamide is replaced by melamine and pentaerythritol by its dimer or, occasionally, its trimer. The ether-acetate of the glycol is present as a "coalescing" agent to improve film integrity and scrub resistance. The finer points of the formulation work are in reference 90. Vandersall has compared various ingredients and their effect on performance [10]. Melamine phosphate forms a gel in water and is not deemed suitable for water-based paints. Monoammonium phosphate generally does not resist water.

Figures 6-1 to 6-6 show various properties as a function of the ingredients. Figure 6-1 shows the effect of coverage on flame spread as determined in the 2-ft tunnel. Figure 6-2 shows the improved scrub resistance achieved with the polyphosphate in terms of solution of this ingredient; Figures 6-3 and 6-4 show this in terms of flame spread as measured in the 2-ft tunnel. Figure 6-5 presents the effect of pigment volume concentration on fire retardancy and

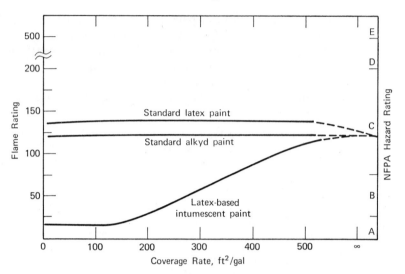

Figure 6-1   Effect of coverage rate on flame rating of ¼-in. Douglas fir plywood coated with various types of paint [90]. Panels primed with alkyd undercoat.

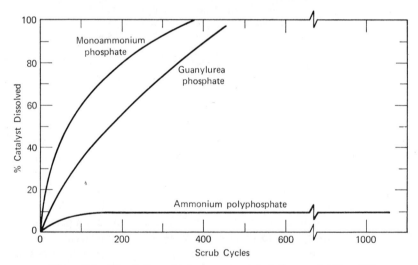

Figure 6-2   Dissolution of intumescent catalysts from paint films [10].

shows a very strong dependence on this. Finally, Figure 6-6 shows the effect of the coalescing agent on scrub resistance and flame spread.

    In 1968 Vandersall published his findings [91] on oil-based paints containing ammonium polyphosphate. The key finding in this work is the conclusion that convertible resins are not, in general, suitable. A convertible

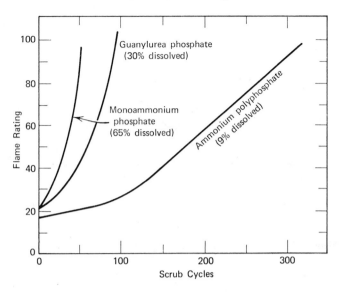

Figure 6-3 Effect of scrubbing on the flame of intumescent paints containing various catalysts (soluble carbonific, blowing agent, and foam stabilizer) [10].

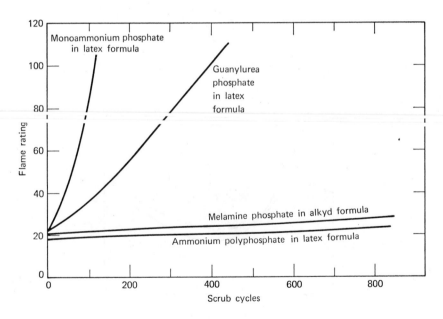

Figure 6-4 Scrub resistance of intumescent paints catalyzed with various reagents [10].

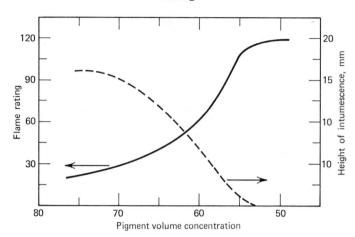

Figure 6–5 Effect of PVC on flame rating and intumescence [10].

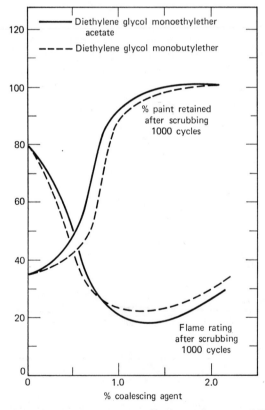

Figure 6-6 Effect of coalescing agents on abrasion resistance and flame rating [10].

resin is one that continues to polymerize on aging in air through oxidation, primarily. Nonconvertible resins set through solvent evaporation; many can be softened readily with the original solvent and many are thermoplastic. Vandersall found that certain convertible resins give paint films that are quite satisfactory just after drying. However, after aging a few weeks the film becomes tough and unyielding. This blocks the formation of good cellular foam on heating, with the result that the flame-spread ratings become unsatisfactory. Table 6-24 shows this for one of the least troublesome of the

Table 6-24   Effect of Film^a Oxidation on Intumescence [91]

Storage^b Time, days	Estimated^c Oxidation, %	Flame Rating	Intumescent Foam Height, mm	Insulation Value (ΔT),°C
3	0^d	9	28	81
14	33	14	25	95
25	66	15	15	105
33	90	35	7	118
48	>90	33	6	126

^{a} Ammonium polyphosphate "catalyzed" intumescent paint utilizing a chlorendic anhydride-glycerol medium soya-modified oxidizing alkyd
^{b} 72°F and 50% relative humidity
^{c} Manufacturer estimates 90% oxidation in 30 days
^{d} Initial 3 days required for solvent evaporation

convertible resins evaluated. With these results in mind, a variety of resin systems were evaluated for coating properties and flame-spread rating after suitable aging. Reference 91 contains 16 preliminary formulations based on nonconvertible resins. Three of the more promising ones are shown in Table 6-25 along with the flame-spread ratings and comments on other points. Note that Class A ratings are readily attained. The report should be consulted for detailed comments on the important variables.

The question of aging effects may explain why the alkyds used by the U.S.D.A. did not produce Class A ratings. Those paints were aged 30 days and therefore there was sufficient time for oxidation and film toughening to occur.

There are few other references to intumescent paints. Some work on intumescence from proteins in coatings was reported in 1956–57 [92], but nothing further has apparently been reported. There is a report [78] of intumescence obtained from a system in which the phosphorus is incorporated in the polyol (sugar-phosphate). Such a combination could only be justified if some final property were improved, such as intumescence, water resistance,

Table 6-25   Three Preliminary Formulas for Oil-Based
Intumescent Paints Containing Ammonium Polyphosphate
and Nonconvertible Resins [91]

Ingredient	Wt %		
	A	B	C
Resin			
Vinyltoluene butadiene	7.13	· · ·	· · ·
Ethylhydroxyethyl cellulose	· · ·	5.25	· · ·
Ethyl cellulose	· · ·	· · ·	5.49
Dipentaerythritol	8.18	6.03	6.29
Melamine	8.75	6.45	6.73
Ammonium polyphosphate	28.40	20.91	21.85
Chloroparaffin (70% Cl)	8.74	6.44	6.72
Titanium dioxide	6.40	4.72	4.92
Mineral spirits	32.40	46.18	· · ·
Isopropanol	· · ·	4.02	· · ·
Toluene	· · ·	· · ·	48.0
	100.00	100.00	100.00
Flame spread	14	10	10
Height of intumescence, mm	10	10	16
Coverage rate, ft²/gal	137	92	107
Brushing	Good	Good	Good

or serviceability. There is apparently no evidence that such is the case for phosphated polyols unless they are further reacted before use, for example, with isocyanates. An intumescent clear coating has been reported [93] based on a proprietary polyurethane and makes use of halophosphate esters. The coating is heat-cured or air-cured and gives a flame spread in the 25-ft tunnel of 45. Concomitantly, data on pyrolysis of several polyurethane coatings were given. Another recent patent discusses urethane coatings containing $NH_4H_2PO_4$ [94]. A family of diyne polyols (polyesters made from dibasic acids and conjugated diacetylene diols) have been studied as additives to long-oil alkyds to improve the intumescence [95]. Optimization studies such as this are under way in many laboratories and will constitute a major part of the literature on intumescence for the next decade.

MASTICS FOR METAL PROTECTION

In industrial, institutional, and multifamily dwellings structural steel is controlled as to fire rating by building codes. Of course, steel beams and columns do not burn; they do lose strength when heated, however, and may collapse early in a fire if they are unprotected. This collapse could occur when wooden members would be far from burning through, thus endangering the occupants earlier. Fire ratings on structural steel are given in terms of the length of time a given coating will maintain the temperature of the steel below roughly 540° (1000°F) when exposed to a prescribed thermal gradient. The temperature is selected to be representative of the maximum allowable temperature for continued load bearing for a variety of steels. Ratings are commonly 1 hr, 2 hr, and 4 hr, although intermediate times are sometimes used, for example, 45 min or $1\frac{1}{2}$ hr (all by ASTM E-119 or a variant). Some of these aspects, as they apply to chemical plants, have recently been reviewed by Waldman [96]; his paper is recommended.

There are currently at least three ways of attaining such ratings: concrete encasement, layers of gypsum board or plaster, and intumescent and other insulative mastics. Concrete and gypsum casings are generally rather heavy for a given rating and impose severe weight penalties on foundation design. Nonetheless, they are often the methods of choice for they are well-known and reliable techniques. Despite the massiveness of 8 or 10 in. of concrete about a post of steel, concrete is the most frequently used technique for the lower stories of high-rise buildings (where ratings up to 4 hr are needed). Similar arguments apply to the gypsum techniques. Where weight reduction is important, as in ceiling trusses and protection of upper stories, the insulative, lightweight coatings are useful. In the past 15 years or so several such systems have been devised; they are either intumescent or contain low-density insulative fillers. The present discussion is directed toward these light-weight systems. The concrete and gypsum systems are not really dependent on chemical fire retardants but rather on the heat-conducting, -absorbing, and -reflecting properties of these materials.

Some early concepts on steel were outlined by Zola [97] and on aluminum alloys for aircraft by Lasch and Jukkola [56]. In the case of aluminum, protection at 177° (350°F) is needed. In these early discussions (1954) solutions were largely just the fire-retardant paints of the period. Thus, paints based on monoammonium phosphate are described as are ceramic coatings (silicates) and antimony-halogen combinations. An early patent [98] claimed asbestos, silica, dolomite, and fluorspar in water glass as a ceramic barrier. Another [99] claimed magnesia, fiberglass, and foamed glass in a chlorocarbon resin as a troweling mastic. Way and Hilado have tested a number of insulative coatings of this general type [99a].

In recent years the trend has been toward new intumescent mastics. The intumescent paints described in earlier sections of this chapter have served as starting points for the formulation of such mastics. For water-based mastics an ammonium polyphosphate system serves well; for oil-based mastics either melamine phosphate or ammonium polyphosphate can be used. These systems are currently being used commercially.

To the paint formula is added a fibrous material to give more toughness to the mastic. This is necessary in thick coatings (up to $\frac{1}{2}$ in.) to prevent cracking and shrinking. Coated glass fibers have been used; so have several forms of asbestos fibers [100, 100a]. Adhesion to the metal surface must be obtained and corrosion prevented. Some formulas that show promise are given in Table 6-26.

Table 6-26   Three Developmental Mastic Formulas Based on Ammonium Polyphosphate

	A[a]	B[b]	C[b]
Water	27.5	26.7	· · ·
Toluene	· · ·	· · ·	27.9
Melamine	5.7	8.9	6.5
Dipentaerythritol	5.4	8.1	7.1
Ammonium polyphosphate	18.9	22.2	25.0
Chlorocarbon (70% Cl)	3.0	4.5	4.0
Asbestos fibers	16.2	4.8	4.8
Resin			
Polyvinylacetate (modified) (60%)	18.7	24.4	· · ·
Acrylic	· · ·	· · ·	22.7
Surfactants, thickeners, etc.	4.6	0.4	2.0
	100.0	100.0	100.0
Approximate rating at $\frac{3}{16}$-in. thickness	1 hr	1 hr	1–1½ hr

[a] Formulation by the Resin Division, National Starch and Chemical Corp., Plainfield, N.J. Formula R822-1376

[b] Formulations from H. L. Vandersall, A. J. Burns, and G. F. Snow, Inorganic Chemicals Division, Monsanto Co., St. Louis, Mo.

The evaluation of mastics is done in a variety of ways. The method in all cases is to record the temperature of the steel, detected by an embedded thermocouple, as a function of the temperature of the gases impinging on the coating. In the early days this was done by coating one side of the metal, directing hot gases to the other side, and recording both temperatures (see pp. 245–246). Today a piece of steel is fitted with thermocouples, coated completely with mastic, and then placed in a furnace. Results of such a test are shown in simplified form in Figure 6-7. Ratings greater than 2 hr appear attainable with such coating systems, though "listed" ratings are not yet greater than 2 hr (for listings: see Table 6-27). There is much activity in progress at this writing and there is no doubt that more 2-hr ratings will be "listed" soon. Whether 4-hr ratings can be achieved via these relatively thin coatings is not clear yet.

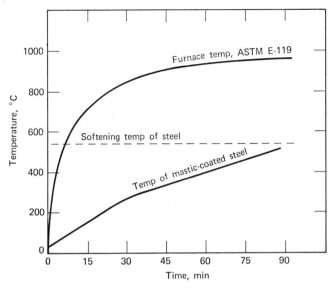

Figure 6-7 Time-temperature relation for a piece of angle iron coated with a water-based intumescent mastic (see Table 6-26) $\frac{3}{16}$-in. thick.

Table 6-27 Current Listings on Intumescent Mastics for Structural Steel (Underwriters' Laboratories, 1969[101])

Mastic Type	Structural Form	Mastic Thickness, in.	Rating, hr
Interior, intumescent, solvent-based	Beam (I)	$\frac{7}{16}$	2
Interior, intumescent, solvent-based	Beam (I) with ~~$\frac{5}{8}$-in. gypsum~~ on two faces)	$\frac{3}{8}$	2
Interior, intumescent, solvent-based	Beam (I)	$\frac{3}{16}$[a]	1
Interior, intumescent, solvent-based	Beam (I)	$\frac{1}{8}$	$\frac{3}{4}$
Interior, intumescent, soluent-based	Column (I) with $\frac{5}{8}$-in. gypsum on two faces	$\frac{3}{8}$[b]	$1\frac{1}{2}$
Interior, intumescent, solvent-based	Column (I)	$\frac{9}{16}$[c]	$1\frac{1}{2}$

[a] Applied in two coats to this total thickness

[b] Applied in two coats to this total thickness with glass fiber gauze between layers

[c] Applied in two coats to this total thickness; steel to be scratch-coated where necessary

The requirements for mastics for structural steel are much more severe than for fire-retardant paints. In the 25-ft flame tunnel used for paints over wood, for example, the test duration is only 10 min. Wood can readily be protected by 10 mm of intumescence from a correctly formulated paint for this length of time. A time of several hours is another matter. The difference is in the critical nature of the structural steel. Because this steel is used to support high-rise buildings, as much time as possible must be provided for evacuating personnel and for giving firefighters ample time to extinguish a fire. It is not only a question of a few minutes needed to save lives; rather, hours may be required to protect entire city blocks. Thus have evolved the long times in the tests for mastics. Coatings a few mils thick have become $\frac{3}{16}$ to $\frac{1}{2}$ in. in thickness to provide the added time.

GENERAL CONSIDERATIONS

Very serviceable fire-safe coatings are on the market. They cost more than regular paints and mastics, partly because of the moderately expensive ingredients but in some cases primarily because of the extra-thick coatings required to achieve top performance. Thus, to obtain Class A ratings with a paint, coverages as low as 150 ft²/gal are required versus the usual 400 ft²/gal. This factor of 2.7 confronts the user before he even considers the cost per gallon. And it is not reasonable to expect the cost per gallon to be any less than for an ordinary coating. Indeed, it is likely to be somewhat higher, though not as much as one might think. There are many situations where the additional cost is not a factor because Class A performance is required by codes. Then the significant factor is relative cost against performance between types of coatings. Here the pricing of the ingredients is important. Today certain of these ingredients are manufactured in small volume for this market only. As consumption rises the unit prices will fall for these products as volume leverage in the producing plants begins to make itself felt. So we may expect some economies in the various special ingredients as more and more codes are modified to require performance that can only be attained by the most modern, sophisticated coatings. In this regard, incidentally, it is likely that not only will improved flame retardance be needed but also improved washability or scrub resistance. All the other paint properties will have to be as good as the best regular paints on the market. This performance is either completely or nearly matched by the best fire-retardant coatings available today. In the case of mastics, the cost is not prohibitive but the performance is in some instances marginal. Where 1-hr ratings suffice the mastics now available are adequate, but few, if any, can give 2 hr. Here improvements in technology are needed, though not in a major way. On the other hand, 4 hr mastics will require a breakthrough of considerable magnitude. A 4-hr rating with a coating thickness of less than 1 in. would yield substantial savings in building costs and provides a compelling incentive for research efforts. Among the developments that we may anticipate are shop-coated structural members using the new mastics and leaving exposed only the areas where the joints will be. After assembly, only touching up will be required. This will provide reductions in the waste incurred when gunning mastics in the field.

REFERENCES

[1] R. P. Ware and M. W. Westgate, *National Paint, Varnish, Lacquer Assoc., Sci. Sect., Circ. No. 727*, 17–39 (1948).

[2] *Official Digest*, **231**, 435 (1943); *ibid.*, **250**, 408 (1945).

[3] *Paint, Oil, Chem. Rev.*, **106**, 10 (Dec. 2, 1943); **106**, 7 (Dec. 30, 1943); **108**, 129 (Nov. 1, 1945).

[4] G. Black and H. Sanders, *Product Eng.*, 140 (August, 1948).

[5] L. Sternberg, *Amer. Paint J.*, **36**, 69 (May 26, 1952).

[6] *Official Digest*, **27**, 982 (December, 1953).

[7] M. W. Westgate et al., "Fire Retardant Paints," *Advan. Chem. Series No. 9*, American Chemical Society, Washington, D.C., 1954.

[8] J. A. Holderried, *Paint, Oil, Chem. Rev.*, **121**, 8 (July 24, 1958).

[9] F. Schofield, *Ind. Eng. Chem.*, **56**, 49 (1964).

[10] H. L. Vandersall, Paper presented at the Wayne State Univ. Polymers Conf. Series, Program 5, Detroit, Mich., June 13–17, 1966.

[11] *Official Digest*, **275**, 756 (1947); A. W. Van Heuckeroth, G. S. Cook, and R. W. Hill, *CADO, Tech. Data Digest*, **15**, 27 (1950).

[12] A. W. Van Heuckeroth, R. W. Hill, and G. S. Cook, *N.Y. Univ. Coll. Eng. 2nd Symposium on Varnish and Paint Chem.*, **1949**, 67.

[13] R. C. Hubbard, *Chemistry in Can.*, **4**, 29 (1952); J. W. Suggitt, *Can. Chem. Processing*, **36**, 90 (1952).

[14] H. L. Vandersall, *J. Paint. Technol.*, **39**, 494 (1967).

[15] H. Terrisse and C. Coffignier, *Fr. 398,956*, Jan. 30, 1909; *Ger. 247,372*, Dec. 13, 1910.

[16] C. J. Christianson, *U.S. 2,681,326* (to Sherwin-Williams Co.), June 15, 1954.

[17] K. Okoshi, *Jap. 4770*, May 9, 1960.

[18] J. Urbanek (by P. R. Maser), *Brit. 864,099*, Mar. 29, 1961.

[19] R. R. Mueller, E. Wacker, and H. Krisch, *Ger. (East) 33,185*, May 15, 1965.

[20] R. E. Ellis, *U.S. 3,102,821*, Sept. 3, 1963.

[21] P. E. Marling and M. L. Nielsen, *U.S. 2,596,496* (to Monsanto Chemical Co.), May 13, 1952.

[22] M. L. Nielsen, R. W. Arnold, and E. C. Chapin, *U.S. 2,596,938* (to Monsanto Chemical Co.), May 13, 1952.

[23] H. R. Frisch, *U.S. 2,530,458*, Nov. 21, 1950.

[24] T. Sakurai and T. Izumi, *Jap. 685*, Feb. 18, 1953.

[25] N. V. Albi Chemie, *Dutch 71,463*, Dec. 1952.

[26] O.P.C.I.A. (by G. L. Quesnel), *Fr. 1,145,836*, Oct. 30, 1957.

[27] H. W. Coover, Jr., and R. L. McConnell, *U.S. 3,415,789* (to Eastman Kodak Co.), Dec. 10, 1968.

[28] R. J. Polack and E. H. Holmstrom, *Paint & Varnish Production*, **45**, 35 (1955).

[29] Farb. Bayer A.-G. (by H. Grahmann and N. Pusch), *Ger. 1,065,116*, Sept. 10, 1959.

[30] Hooker Chemical Corp., *Neth. Appl. 6,410,622*, Mar. 15, 1965.

[31] S. Cabot, *U.S. 1,308,575*, July 1, 1924; H. Dodd, *Can. 364,798* (to Canadian Industries, Ltd.), Mar. 16, 1937.

[32] K. S. Wade, *Paint & Varnish Production Mgr.*, 238 (August, 1948).

[33] M. Leatherman, *U.S. 2,472,112*, June 7, 1949.

[34] R. W. Craig, *U.S. 2,627,475* (to Diamond Alkali Co.), Feb. 3, 1953.

[35] B. B. Kaplan, *U.S. 2,984,640* (to Albi Manufacturing Co., Inc.), May 16, 1961.

[36] K. Suzuki, *Nippon Suisangaka Kaishi*, **24**, 381 (1958).

[37] W. K. Hearn, *U.S. 2,247,633*, July 1, 1939.

[38] D. Cascio, *U.S. 2,254,471*, Sept. 2, 1939.

[39] L. McCulloch, *U.S. 2,044,176* (to Westinghouse Electric & Manufacturing Co.), June 16, 1934.

[40] Siemens-Schuckertwerke A.-G., *Ger. 716,219*, Dec. 11, 1941; Herbig-Haarhaus A.-G. Laekfabrik Koln-Bickendorf, *Belg. 444,811*, Apr. 30, 1942.

[41] F. Petke, *Paint, Oil, Chem. Rev.*, **109**, 28 (1946); *Official Digest*, **262**, 512 (1946); *Paint Ind. Mag.*, **62**, 25 (1947); R. C. Crippen et al., *Paint, Oil, Chem. Rev.*, **117**, 40 (1954).

[42] L. S. Birnbaum and M. Markowitz, *Ind. Eng. Chem.*, **40**, 400 (1948); also see T. R. Walton, U.S. Naval Res. Lab. Rept. 6304, Wash., D.C., Aug. 5, 1965.

[43] N. Levit and E. S. Gurevich, *Lakokresock. Mater. Ikh. Primen.*, *1967* (4), 55 (in Russian); *Official Digest*, **28**, 942 (1956); W. M. Ewalt and A. W. Hopton, *Paint & Varnish Production*, **51**, 53 (November, 1961).

[44] D. Lurie, *Fr. 1,145,584*, Oct. 28, 1957; G. V. Thornberg, *Swed. 153,715*, Mar. 13, 1956.

[45] *Official Digest*, **27**, 779 (1953).

[46] Akira Tanaka, *Jap. 9585*, Nov. 9, 1956.

[47] E. K. Zimmermann and W. A. Ingram, *Ind. Vernice* (*Milan*), **19**, (4), 8 (1965) (in Italian); *Paint & Varnish Production*, **51** (12), 71 (1961).

[48] A. T. Hall, *Brit. 8513*, April, 1906.

[49] F. S. Vivas, *U.S. 1,369,857*, Mar. 1, 1920; *U.S. 1,707,587* (to International Fireproof Products Corp.), Apr. 2, 1928; *Can. 284,576* (to International Fireproof Products Corp.), Nov. 6, 1928; Y. Nagase and K. Nagano, *Jap. 3741*, Oct. 27, 1950; C. E. Wilkinson, *U.S. 2,939,794* (to Texaco, Inc.), June 7, 1960; W. P. Fairchild, *U.S. 3,281,252* (to Kelco Co.), Oct. 25, 1966; G. O. Orth, Jr., and J. Spinelli, *U.S. 2,840,478* (to American Potash & Chemical Corp.), June 24, 1958.

[50] *Forest Products Lab., Tech. Note No. 249*, 1941, 4 pp.; *Rev. Current Lit. Paint Colour Varnish & Allied Ind.*, **14**, 318 (1941); O. Bowles, *Mining and Met.*, **24**, 85 (1943).

[50a] *Paint & Varnish Production*, **51**, 65 (November, 1961).

[50b] D. L. Hunter, K. Kitasaki, and G. W. Willcockson, *U.S. 3,131,071* (to U.S. Borax & Chemical Corp.), Apr. 28, 1964.

[51] G. Kelly, *U.S. 882,891*, Mar. 24, 1907.

[52] L. Vleek, *Australia 19,455/34*, Sept. 19, 1935.

[53] P. C. Oswald, *Brit. 489,464*, July 27, 1938.

[54] M. Jacobs, *U.S. 2,438,339* (to Albert W. Clurman), Mar. 23, 1948.

[55] I. Holesckak, *Ital. 457,746*, June 6, 1950.

[56] H. W. Lasch and E. E. Jukkola, in reference 7, pp. 67 ff.

[57] V. Bumba, *Indian 44,674*, Oct. 31, 1951.

[58] O. A. Valenciano, *Rev. Facultad Cienc. Quim.* (Univ. Nacl. La Plata), **18**, 129 (1943) (pub. 1945); J. P. Seren and C. M. Armand d'Aux de Tour Sainte, *Fr. 1,427,038*, Feb. 4, 1966; G. Raynor and W. H. Baxter, *Brit. 1,075,909*, July 19, 1967.

[59] R. M. Hooks, *U.S. 2,734,827* (to Southwestern Petroleum Co., Inc.), Feb. 14, 1956.

[60] G. Gebrüder G.m.b.H., *Ger. 872,766*, Apr. 2, 1953.

[61] Boucheron, *Fr. 366,564*, May 25, 1906.

[62] S. Gausseff, *Brit. 3,368*, Feb. 14, 1908.

[63] Andrieu, *Fr. 349,982*, June 13, 1904.

[64] C. S. Thompson, *U.S. 1,996,089* (to Thompson Paint Manufacturing Co.), Apr. 2, 1934.

[65] R. J. White and W. R. Dauncey, *Brit. 433,645*, Aug. 19, 1935.

[66] Burns & Russell Co., *Brit. 1,080,468*, Aug. 23, 1967.

[67] D. Gross and J. J. Loftus, *ASTM Bull. (TP/35)*, p. 57 (May, 1958).

[68] M. W. Westgate, *National Paint, Varnish, Lacquer Assoc., Sci. Sect., Circ. No. 747*, 460 (1951).

[69] T. M. Murray, F. Liberti, and A. O. Allen, in reference 7, pp. 35 ff.

[70] R. Grubb and W. W. Cranmer, in reference 7, pp. 48 ff.

[71] F. B. Clarke and J. W. Lyons, *J. Amer. Chem. Soc.*, **88**, 4401 (1966).

[72] E. A. Lauring, *U.S. 2,594,937* (to Minnesota and Ontario Paper Co.), Apr. 29, 1952.

[73] A. O. Allen, T. M. Murray, and F. P. Liberti, *U.S. 2,754,217* (to Vita-Var Corp.), July 10, 1956.

[74] United Kingdom Atomic Energy Authority (by D. G. Stevenson and P. Walker), *Brit. 968,336*, Sept. 2, 1964.

[75] J. H. Bosto, M. M. Capp, C. R. Davison, and G. H. Garbaden, *U.S. 3,037,951* (to Benjamin Moore & Co.), June 5, 1962; R. P. Silver, *U.S. 3,221,035* (to Hercules Powder Co.), Nov. 30, 1965.

[76] Chemische Werke Albert, *Belg. 669,341*, Mar. 8, 1966.

[77] G. Jones and S. Soll, *U.S. 2,452,054* (to Albi Manufacturing Co.), Oct. 26, 1948; *U.S. 2,452,055*, Oct. 26, 1948.

[78] T. Sakkurai, *Jap. 8290*, Dec. 15, 1954; T. Sakkurai and T. Izumi, *J. Chem. Soc. Japan, Ind. Chem. Sect.*, **56**, 156 (1953); G. Jones, W. Juda, and S. Soll, *U.S. 2,523,626* (to Albi Manufacturing Co.), Sept. 26, 1950.

[79] H. A. Scholz and E. E. Saville, *U.S. 2,566,964* (to U.S. Gypsum Co.), Sept. 4, 1951.

[80] E. K. Stilbert, Jr., I. J. Cummings, and W. B. Guerrant, *U.S. 2,684,953* (to Dow Chemical Co.), July 27, 1954.

[81] E. K. Stilbert, Jr., I. J. Cummings, and J. P. Talley, *U.S. 2,755,260* (to Dow Chemical Co.), July 17, 1956.

[82] Farb. Hoechst A.-G., *Brit. 874,762*, Appl. Oct. 28, 1957.

[83] N. K. Peterson and I. J. Cummings, *U.S. 2,917,476* (to Dow Chemical Co.), Dec. 15, 1959.

[84] H. Sano, *Jap. 5278* (to Oshika Shinko Co.), June 23, 1959; Intubloc, Ltd. (by K. H. G. Fenner and R. Thompson), *Brit. 978,623*, Dec. 23, 1964.

[85] H. Goellert and H. Kleinert, *Ger. (East) 40,957*, Sept. 15, 1965.

[86] P. Simon, *U.S. 3,021,293* (to Sherwin-Williams Co.), Feb. 13, 1962.

[87] F. P. Liberti, *Paint & Varnish Production*, **51**, 57 (November, 1961).

[88] *Paint & Varnish Production*, **51**, 79 (November, 1961).

[89] G. B. Verburg, E. T. Rayner, D. A. Yeadon, L. L. Hopper, Jr., L. A. Goldblatt, F. G. Dollear, and H. P. Dupuy, *J. Amer. Oil Chem. Soc.*, **41**, 670 (1964); D. A. Yeadon, E. T. Rayner, G. B. Verburg, L. L. Hopper, Jr., F. G. Dollear, H. P. Dupuy, and H. Miller, *Official Digest*, **37**, 1095 (1965); E. T. Rayner, D. A. Yeadon, G. B. Verburg, F. G. Dollear, H. P. Dupuy, L. L. Hopper, Jr., and H. Miller, *J. Paint Technol.*, **38**, 1951 (1966); G. B. Verburg, D. A. Yeadon, E. T. Rayner, F. G. Dollear, H. P. Dupuy, L. L. Hopper, Jr., and E. York, *ibid.*, 407.

[90] H. L. Vandersall, "PHOS-CHEK® P/30 Brand Fire Retardant—Its Use in Intumescent Paint," *Rept. No. 6512*, Inorganic Chemicals Division, Monsanto Co., St. Louis, Mo., Oct. 20, 1965.

[91] H. L. Vandersall, "PHOS-CHEK® P/30 Brand Fire Retardant—Its Use in Intumescent Organic-Solvent Paints," *Rept. No. 7088*, Inorganic Chemicals Division, Monsanto Co., St. Louis, Mo., Apr. 23, 1968.

[92] N. Shikazono and T. Kanayama, *Repts. Govt. Chem. Ind. Research Inst. Tokyo*, **51**, 275 (1956); N. Shikazono, H. Kamata, and T. Kanayama, *Tokyo Kogyo Shikensho Hokoku*, **52**, 229 (1957); N. Shikazono and T. Kanayama, *ibid.*, 235; N. Shikazono, H. Kamata, T. Kanayama, and K. Maeda, *ibid.*, 242.

[93] C. C. Clark, A. J. Krawczyk, G. R. Reid, and E. V. Lind, *Amer. Chem. Soc., Div. Org. Coatings Plast. Chem. Preprints*, **26**, 233 (1966); *U.S. 3,365,420; 3,448,075* (to Textron, Inc.), Jan. 23, 1968, June 3, 1969.

[94] Intubloc, Ltd. (by R. H. G. Fenner), *Brit. 1,029,963*, May 18, 1966.

[95] E. T. Rayner, D. A. Yeadon, L. L. Hopper, Jr., H. P. Dupuy, and F. G. Dollear, *U.S. 3,269,989* (to United States), Aug. 30, 1966.

[96] S. Waldman, *Chem. Eng. Prog.*, **63** (8), 71 (1967).

[97] J. C. Zola, in reference 7, pp. 82 ff.

[98] E. Roach, *U.S. 2,429,946*, Oct. 28, 1947.

[99] W. P. Ellis, L. I. Smith, and I. J. Steltz, *U.S. 2,861,967* (to Benjamin Foster Co.), Nov. 25, 1958.

[99a] D. H. Way and C. J. Hilado, *J. Cell. Plast.*, **4**, 222 (1968).

[100] B. Kaplan, *U.S. 3,284,216* (to Albi Manufacturing Co.), Nov. 8, 1966; *Can. 808,233*, Mar. 11, 1969.

[100a] Chemische Werke Albert, *Belg. 719,938*, Aug. 26, 1968.

[101] *Building Materials List*, Underwriters' Laboratories, Inc., Chicago, Ill., January, 1969, p. 510.

Chapter 7

SYNTHETIC POLYMERS WITH ALL-CARBON BACKBONES

Chapters 7 and 8 deal with flammability problems in synthetic polymers, including both plastic products and synthetic fibers. Fire-safe properties in plastics have been the subject of intensive investigation since the mid-1950s and there is much literature on the subject. Two recent books have dealt solely with this question: Vogel's book [1] (in German) is a rather full recitation of the literature with emphasis on patents but relatively little discussion and interpretation, Hilado's book [2] tabulates the properties of plastics and lists approaches that have been taken to reduce flammability. The two books complement one another and should be part of the serious student's library. These books and the review literature are concerned mostly with plastics rather than fibers. The emphasis was almost entirely on plastics until the last 4 or 5 years, therefore the fiber situation is less clear. Fibers will be treated here in segments, with each category of fiber discussed as a separate subsection of the appropriate polymer. Thus, polyester fibers are found under the section on "Polyesters" in Chapter 8.

In addition to the two books cited, several useful review papers have been published in recent years. There are four in Japanese [3], two in German [4], and one in Hungarian [5], which will not be discussed since they offer primarily the advantage of the particular language in which they are written. Both Boyer and Jacques [6]* discuss aspects of flameproofing polymers and emphasize the usefulness and mode of effect of phosphorus compounds. Sherwood comments briefly on the situation as seen in England in 1963 [7]. Four lengthy but relatively nontechnical reviews were issued in 1966 [8], 1967 [9], 1968 [10], and 1969 [10a]; these are helpful for the reader who wants an overall view without too many data. Schmidt [11] prepared a lengthy patent review in 1965 (289 patents cited) and included in his paper an

281

Table 7-1 Some Properties of Synthetic Polymers (as Listed by Hilado) [2]

	Glass Trans. Temp, °C	Decomp. Temp. Range, °C	Ignition Temp. °C		Limiting Oxygen Conc (Fenimore-Martin Test), mole fraction oxygen to support combustion	Smoke Density, max value spec. opt. density (nonflaming)	Surface Flammability (Horiz.Position) Burning rate, in./min
			Flash	Self			
Cellulose	...	...	230–266	250–260	...	...	...
Various woods	...	...	...	...	...	260–660	...
Polyethylene	−125	335–450	341	349	0.174	470	0.3–1.2
Polypropylene	−20	328–410	...	570	0.174	456	0.7–1.6
Polybutylene	−25	...	...	...	...	...	1.1
Polybutadiene	−85	...	...	...	...	...	...
Polytetrafluoroethylene	−113, 127	508–538	...	530	0.95	0	Nonigniting
Polyvinyl chloride	80	200–300	391	454	0.45–0.49	300–470	SE
Polyvinylidene chloride	−18	225–275	532	532	0.60	...	SE
Polystyrene	100	300–400	345–360	488–496	0.181	345–372	0.5–2.5

Acrylonitrile-butadiene-styrene	...	...	...	...	...	...	1.0–2.0
Styrene-butadiene copolymer	...	327–430	366	...	...	71	...
Styrene-acrylonitrile	...	...	346–399	454	...	...	0.4–1.6
Polyester	...	...	...	483–488	...	350–360	...
Polyethylene terephthalate	...	283–306	...	...	...	...	...
Polyamide (nylon 6)	75	310–380	421	424	0.29	320	SE
Polyacrylonitrile	...	...	280–300	560	...	319	...
Polymethylmethacrylate	50	170–300	...	450–462	0.173	...	0.6–1.6
Polyurethane (rigid foam)	...	...	310	416	...	117	...
Polyformaldehyde	−85	222	...	...	...	...	...
Polycarbonate of bisphenol A	149	...	...	...	0.26–0.28	12	SE
Melamine-formaldehyde	...	...	...	...	...	...	SE
Urea-formaldehyde	...	...	...	...	...	...	SE
Rubber	...	...	...	...	...	236	...

informative discussion of the mechanism of the action of halogens as fire retardants acting as flame suppressants.

Several additional reviews, as well as Hilado's book, offer tabulated entries that will be useful as we proceed with the polymer-by-polymer discussion. These tabulations are given in this chapter but they also apply to much of Chapter 8 and, indeed, to parts of earlier chapters. Hildo [2] discusses test methods and lists, for surface flammability alone, some 30 different procedures. Hauck [12] discusses some of these in his paper. Tunnel testing is more and more the method of choice especially as the 2-ft tunnel improves in reliability in correlating to the 25-ft tunnel. Some properties of common plastics are given in Table 7-1. Polystyrene, polyolefins, and the acrylates are among the more flammable polymers as measured either by determining the limiting oxygen concentration or by testing for surface flammability. The flammable properties listed in Table 7-2 are taken from one of Hauck's reviews [12]; Table 7-3 is a list of less flammable polymers [13]. Table 7-4 summarizes the gaseous products of pyrolysis and combustion; the data are much condensed and give only a rough indication of what happens on heating. Madorsky's book [15] should be consulted for details on the thermal degradation of polymers. (In the tables in Chapters 7 and 8 SE stands for self-extinguishing and NB for nonburning.)

It should be clear from these tables that (a) there is a very great variation in properties among the family of synthetic polymers, (b) some polymers are much more flammable than others, (c) some polymers really need no additional fire-retardant treatment, and (d) the mechanism of degradation varies greatly [some unzip to monomers (see below); others degrade by random bond rupture]. The means of conferring flame-resistant characteristics, when needed, varies with the polymer type. In this chapter on polymers with an all-carbon backbone the emphasis is largely on the halogens and combinations of halogens with antimony, phosphorus, and so on. In Chapter 8 the emphasis shifts to some extent to phosphorus compounds. The subject matter was divided to reflect this difference.

Finally, the addition of fire-retardant additives invariably alters important physical properties of the finished polymer system. Rockey [16] has summarized some of the effects on color, strength, moduli, electrical aspects, and the like. Nonreactive additives usually act as plasticizers if they are compatible with the polymer and as fillers if they are not. The reactive fire retardants (comonomers) have less predictable effects. The trend seems to be toward reactive types in many areas to achieve permanence and controllable physical properties. This control can often be exercised by proper tailoring of the comonomer.

Let us now consider polymers with all-carbon backbones, beginning with the simplest—polyolefins—and proceeding through various substituted vinyls, acrylics, polystyrene, and on to rubber, asphalt, and the like.

Table 7-2 Comparison of Burning and Thermal Characteristics
of Plastics (after Hauck [12])

Material	Burning Rate (ASTM D-635), in./min	Heat Resist. (Continuous), °C	Ignition Temp (ASTM D-1929),°C	
			Flash	Self
Acetals	1.1	85	· · ·	· · ·
Acrylics:				
Cast sheet, rods	0.5–2.2	60–93	388	· · ·
Moldings	0.8–1.2	165–190	· · ·	· · ·
Cellulose acetate butyrate	0.5–1.5	68–88	· · ·	· · ·
Cellulose acetate moldings	0.5–2.0	· · ·	327	· · ·
Cellulose acetate sheet:				
Thickness 0.031–0.060 in.	6	· · ·	· · ·	· · ·
Thickness 0.061–0.250 in.	4	· · ·	· · ·	· · ·
Cellulose propionate	1.0–1.5	· · ·	· · ·	· · ·
Chlorinated polyethers	SE	121	· · ·	· · ·
Diallyl phthalates (ignition time, sec):				
Orlon filled (68)	· · ·	149	· · ·	· · ·
Dacron filled (84–90)	· · ·	149–188	· · ·	· · ·
Asbestos filled (70)	· · ·	177–232	· · ·	· · ·
Glass Fiber filled (70–400)	· · ·	204–232	· · ·	· · ·
Epoxies, cast, moldings	0.3 (SE)	79–204	· · ·	· · ·
Ethyl cellulosics	0.5–1.5	· · ·	291	296
Fluorocarbons	SE	193–260	· · ·	· · ·
Melamines	SE	99–204	602	· · ·
Phenolics	SE	121–260	482	· · ·
Phenoxies	Group II	77–79	· · ·	· · ·
Polyallomers	· · ·	82–110	· · ·	· · ·
Polyamides (nylons)	SE	93–121	421	424
Polycarbonate	SE	121	482	· · ·
Polyesters:				
Cast, unfilled	1.5	121–149	· · ·	· · ·
Reinforced with chopped glass	2.0	· · ·	· · ·	· · ·
Reinforced with glass fabric	0.5	· · ·	399	485
Polyethylenes, general purpose extrusion grade	1.0	· · ·	341	349
Polypropylenes	0.7	121	343	388
Polystyrenes:				
Heat, chem res type (SAN)	0.4–1.0	79–88	366	454
Shock res type	0.5–2.0	52–74	· · ·	· · ·
General purpose type	1.0–1.5	60–71	360	496
Lubricated type 2	0.5–2.0	· · ·	· · ·	· · ·
Improved heat res type	0.7–2.0	49–71	· · ·	· · ·
ABS Materials	1.3	· · ·	404	· · ·
Polyvinyl chlorides	SE	60–104	391	454
Polyvinylidene chlorides	SE	77–100	532	532
Silicones	SE	316–371	· · ·	· · ·
Ureas	SE	· · ·	· · ·	· · ·

Table 7-3　Some Less Flammable Polymers (after Hauck [13])

Material	ASTM D-635 Rating	Underwriters' Laboratories Fire Hazard Classification[a]	Remarks and Other Ratings
Thermoplastics			
Chlorinated polyether	SE	· · ·	· · ·
CTFE fluorocarbon	NB	SE, Group I	· · ·
FEP fluorocarbon	NB	SE, Group I	· · ·
PTFE fluorocarbon	NB	SE, Group I	· · ·
PVF$_2$ fluorocarbon	SE	SE, Group I	· · ·
Type 6 nylon	SE	· · ·	· · ·
Type 6/6 nylon	SE	SE, Group II	· · ·
Type 610 nylon	SE	· · ·	· · ·
Polycarbonate	SE	SE, Groups I II	· · ·
Polyphenylene oxide	SE	SE, Group I	· · ·
Polyimide	NB	· · ·	· · ·
Polysulfone	SE	SE, Group II	FAA flame-spread index: 23 at $\frac{1}{16}$ in., 19 at $\frac{1}{32}$ in.
Rigid vinyl	SE	· · ·	ASTM E-84 rated at 0–25 (Noncomb)
Vinyl/acrylic alloy	SE	SE, Group I	· · ·
Thermosets			
Alkyd	SE	· · ·	· · ·
Melamine	SE	Meet req. of UL Bull No. 484	MIL-14F
Phenolic	SE, NB	SE, Group I	MIL-14F; ratings depend on reinforcements
Silicone	NB	· · ·	· · ·
Urea	SE	· · ·	· · ·

[a] UL Bull. No. 94

POLYOLEFINS

The polyolefins here are taken to include polyethylene, polypropylene, and, to a lesser extent, polybutenes. These are made by the catalytic polymerization of the monomers:

$$x\text{RCH}{=}\text{CH}_2 \xrightarrow{\text{catalyst}} \left(\begin{array}{c} \text{CH}-\text{CH}_2 \\ | \\ \text{R} \end{array}\right)_x$$

**Table 7-4   Products of Pyrolysis and of Burning of Plastics
(After Schmidt and Reichherzer [14])**

Substance	Pyrolysis Products	Combustion Products
Polyolefins	Olefins, paraffins, cyclic hydrocarbons	CO, CO_2
Polystyrene	Styrene monomer, dimer, trimer	CO, CO_2
Polyvinyl chloride	HCl, aromatic, polynuclear hydrocarbons	HCl, CO, CO_2
Fluoroplastics	Perfluoroethylene, octafluoroisobutene	. . .
Polyacrylonitrile	Acrylonitrile monomer, HCN	CO, CO_2, NO_2
Polymethacrylate	Acrylate monomer	CO, CO_2
Polyvinyl alcohol	Acetaldehyde, acetic acid	CO, CO_2, acetic acid
Nylon 6	Caprolactam	CO, CO_2, NH_3
Nylon 66	Amine, CO, CO_2	CO, CO_2, NH_3, amine
Phenol resins	Phenol, formaldehyde	CO, CO_2, formic acid
Urea resins	Ammonia, methylamine, coal-like residue	CO, CO_2, NH_3
Epoxy resins	Phenol, formaldehyde	CO, CO_2, formic acid
Terephthalate resins	Olefins, benzoic acid	CO, CO_2
Silicone	SiO_2, CO, formic acid	SiO_2, CO, CO_2, formic acid
Cellulose	CO, CO_2, acetic acid	CO, CO_2, acetic acid
Cellulose nitrate	CO, nitrogen oxides	CO, CO_2, nitrogen oxides
Natural rubber	Dipentene, isoprene	CO, CO_2
Chlorinated rubber	HCl, dipentene, isoprene	HCl, CO, CO_2

The situation is not quite as simple as shown. There is a certain amount of cross-linking and branching in ordinary polyethylene, for example. For those members of the family that have at least one hydrogen on every carbon atom in the chain, degradation proceeds by a more or less random scission process to yield a mixture of fragments in the range C_2 up to C_6–C_{10} in the volatile fraction as measured in pyrolysis (no oxygen present) [15, 15a]. In the case of polyethylene less than 1% of the total volatiles is ethylene; for polypropylene, again less than 1% is propylene. Only a few percent of the volatiles from polyethylene are volatile at room temperature, the remainder being a mixture of high-boiling residues. For polypropylene the fraction of high-boilers drops to about three-fourths of the total volatiles.

Polyisobutylene presents a different picture:

$$\left(\!\!\begin{array}{c} CH_3 \\ | \\ C\!-\!CH_2 \\ | \\ CH_3 \end{array}\!\!\right)_{x}$$

Polyisobutylene

Table 7-5 Use of Phosphorus Compounds as Flame Retardants for Polyolefins

Date	Patent No.	Inventors	Assignee	Invention
Mar. 20, 1964	Fr. 1,355,604	P. S. Blatz	E. I. duPont de Nemours & Co.	Adds H_3PO_4, H_3PO_3, $ClCH_2POCl_2$, $MePh_2PO_4$, allyl-$P(O)(OEt)_2$ to, e.g., polypropylene and mills in at 200°, 10 min; gives SE prod.
Mar. 14, 1967	Can. 754,638	P. S. Blatz	E. I. duPont de Nemours & Co.	
May 30, 1967	U.S. 3,322,859	A. E. Sherr, H. C. Gillham	American Cyanamid Co.	20% of aromatic ester of $(PNX_2)_3$ milled in at ~170°, prod. passes ASTM D-635 $(PNX_2)_x$
Aug. 30, 1961 Nov. 16, 1965	Brit. 876,035 U.S. 3,218,290	E. Hofmann A. E. Sherr, H. C. Gillham	American Cyanamid Co.	R_3P=CHCOCH=PR_3, R = phenyl; add 15% to polyethylene milling in at 170°, prod. passes ASTM D-635
July 9, 1966	USSR 183,938	E. V. Kuznetsov, V. I. Gusev, T. N. Zhidkova, I. N. Andreeva, L. S. Semenova,	USSR	Unsaturated mixed esters of H_3PO_3 copolymerized with olefins
Apr. 22, 1966	Jap. 7448	H. Hatakeyama, O. Fukumoto	Toyo Rayon Co., Ltd.	Diesters of H_3PO_3 added at 5% level to polyolefins; claims fibers with flame resistance
June 3, 1952	U.S. 2,599,501	R. W. Upson	E. I. duPont de Nemours & Co.	Copolymers of general structure: 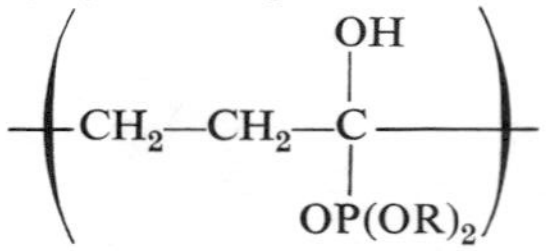 films and fibers are said to be flame-resistant

Date	Patent	Inventors	Assignee	Description
Sept. 15, 1964	Czech. 111,995	D. Bílbes, J. Pavlínec, Z. Manasek, M. Lazar, B. Bohmer	Czech.	Treat polyolefin with PCl_3, $POCl_3$, after partial oxidation. SE reached at 4.7% P
July 30, 1965	Fr. 1,407,334	H. C. Gillham, H. C. Klein	American Cyanamid Co.	20% of dialkylphosphinic acid in polyethylene milled at 170°, prod. passes ASTM D-635
May 4, 1966	Brit. 1,028,158		American Cyanamid Co.	15% of $(\phi CH_2)_3PO$ in polyethylene milled at 170°, prod. passes ASTM D-635
Nov. 8, 1966	U.S. 3,284,543	H. C. Gillham, A. E. Sherr	American Cyanamid Co.	$(\phi)_2\!-\!P\!-\!CH_2CH_2\!-\!P\!-\!(\phi)_2$ (with =O on each P) and variations used in equal parts with various polymers
Sept. 12, 1967	U.S. 3,341,625	H. C. Gillham, A. E. Sherr	American Cyanamid Co.	15% of $(\phi CH_2)_3PO$ milled as above in polyethylene, prod. passes D-635
Mar. 14, 1967; Mar. 4, 1964	U.S. 3,309,425; U.S. 3,431,324	H. C. Gillham, A. E. Sherr	American Cyanamid Co.	20% of 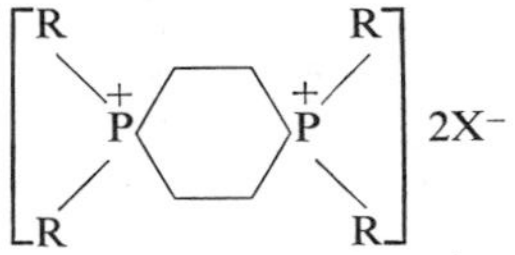 ⎡R—[cyclohexane ring with two P⁺ bearing R and R′]—⎤ 2X⁻ milled in polyethylene at 170°, prod. passes ASTM D-635
May 30, 1967	U.S. 3,322,861	H. C. Gillham, A. E. Sherr	American Cyanamid Co.	20% of $[(CNCH_2CH_2)_3P\!-\!CH_2P(CH_2CH_2CN)_3]Br_2$ in polyethylene milled and tested as above
Feb. 20, 1968	U.S. 3,370,030	J. F. Cannelongo	American Cyanamid Co.	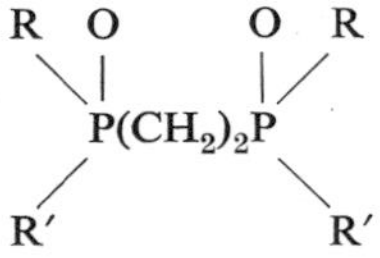 20% of $\dfrac{R}{R'}P(CH_2)_2P\dfrac{R}{R'}$ (with O between R groups)
Jan. 14, 1969	U.S. 3,422,048	J. F. Cannelongo	American Cyanamid Co.	See previous entry.
Jan. 14, 1969	U.S. 3,422,047	J. F. Cannelongo	American Cyanamid Co.	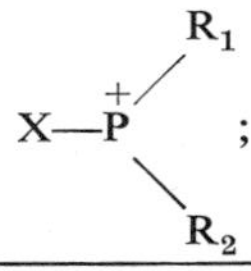 $X\!-\!\overset{+}{P}\big(R_1\big)\big(R_2\big)(R')$; X = Cl, BR, I

Here one carbon is fully substituted and decomposition is directed to formation of radicals which eventually become isobutylene monomer: a process known as unzipping. Some 18% of monomer is recovered from volatiles from pyrolysis for this substance. Thus, substitution may encourage unzipping. Unzipping in turn often causes flammability problems because the monomers are often highly volatile and very combustible. Random degradation gives heavier, less volatile, and thus less flammable products. (17a)

Phosphorus Compounds

A review of the literature on phosphorus used alone for polyolefins shows two striking facts: there are essentially no references to ammonium orthophosphates and there are almost no articles. One very recent exception is the pair of review articles by Sherr, Gillham and Klein [16a] on phosphorus compounds in thermoplastics. These papers contain some tabulated data on phosphonic and phosphinic acids, phosphine oxides and phosphonium compounds. Otherwise the literature is all patents, as shown in Table 7-5. Experience with ammonium polyphosphate in polyethylene, for example, has shown phosphorus in this form to be very inefficient [17]. Some 15% phosphorus was needed, or 50% of ammonium polyphosphate, to achieve reasonable flame resistance. This level of additive converts the polymer to a highly filled composite and is clearly outside the normally accepted level for an additive. That the form of the phosphorus is important can be seen from the entries in Table 7-5. Fire-resistant properties are achieved in several cases at 5% P or so (20% additive) when the agent is an organophosphorus compound. Presumably, the products studied by Gillham, Sherr, and coworkers were thoroughly wetted by the polyolefin so that when the polymer was heated in a flame the phosphorus remained in intimate molecular dispersion in the molten polymer, unlike the case for the inorganic phosphate. Some success has been achieved with copolymerization of phosphorus compounds and with treatment of the polymer so as to introduce phosphorus as a substituent [18].

A self-extinguishing polymer has been reported based on the following:

$$\left(\!\!\begin{array}{c}\text{CH}_2\!-\!\text{CH} \\ | \\ \text{CH}_3\end{array}\!\!\right)_{\!x} + \text{PCl}_3 + \tfrac{1}{2}\text{O}_2 \ \rightarrow$$

polypropylene

$$\left(\!\!\begin{array}{c}\text{P(O)Cl}_2 \\ | \\ \text{CH}_2\!-\!\text{CH}\!-\! \\ | \\ \text{CH}_3\end{array}\!\!\right)_{\!x} \xrightarrow{\text{H}_2\text{O}} \left(\!\!\begin{array}{c}\text{P(O)(OH)}_2 \\ | \\ \text{CH}_2\!-\!\text{CH}\!-\! \\ | \\ \text{CH}_3\end{array}\!\!\right)_{\!x}$$

At 4.68% P, the polymer was immediately self-extinguishing no matter how long a flame was held to it [19]. Of course, the polymer so produced is hydrophilic, ionic, and more brittle than polypropylene. However, the result is another confirmation that 5% of phosphorus molecularly dispersed in a polyolefin produces fire-safe properties.

Phosphorus-Halogen Compositions

A few examples of the P-X combination follow. Halogenated phosphates or phosphonates may be milled into the polyolefin at about 200°. For example, 13% of the following compound in polypropylene provides flame retardancy:

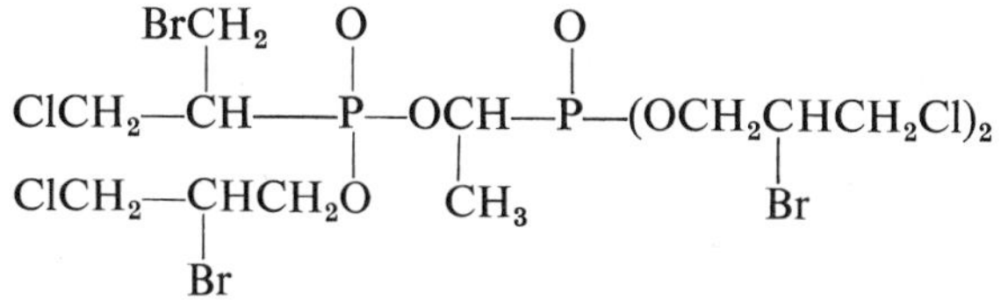

$$
\begin{array}{l}
\text{BrCH}_2 \quad \text{O} \qquad\quad \text{O} \\
\quad | \qquad\quad | \qquad\qquad\quad | \\
\text{ClCH}_2\text{—CH——P—OCH—P—(OCH}_2\text{CHCH}_2\text{Cl)}_2 \\
\qquad\qquad\quad | \qquad | \qquad\qquad\quad | \\
\text{ClCH}_2\text{—CHCH}_2\text{O} \quad \text{CH}_3 \qquad\quad \text{Br} \\
\qquad | \\
\qquad \text{Br}
\end{array}
\qquad
\left(
\begin{array}{l}
40\%\ \text{Br} \\
17.8\%\ \text{Cl} \\
7.8\%\ \text{P}
\end{array}
\right)
$$

(The final product contained 5.7% Br, 2.3% Cl, and 1% P.) Or tris(2,3-dibromopropyl) phosphate may be added at a 10% level to give a flame-retardant polypropylene containing 6.9% Br and 0.4% P [19a]. In a very

Table 7-6 Results with Polypropylene and Polybutene, ASTM D-635-56T

Additive	%	Flammability	Distance burned time to SE
Polypropylene			
Bis(bromochloropropyl)bromo-	1.95	Burns	· · ·
chloropropylphosphonate	2.85	SE	0.79/47
(44% Br, 19.3% Cl, 5.6% P)	3.70	SE	0.52/33
	4.00	SE	0.18/16
Tris(2,3-dibromopropyl)phosphate	3.80	Burns	· · ·
(69% Br, 4.4% P)	4.60	SE	0.34/26
	5.80	SE	0.22/15
	6.90	SE	0.09/15
Tris(bromochloropropyl)phosphate	4.90	Burns	· · ·
(42.5% Br, 18.8% Cl, 5.5% P)	5.90	SE	0.36/28
	7.40	SE	0.33/16
Polybutene			
Bis(bromochloropropyl)bromo-	0	Burns	· · ·
chloropropylphosphonate	3.37	SE	· · ·
	5.74	SE	· · ·
	7.98	SE	· · ·

recent patent [20] fairly low concentrations of bromohalophosphonates were found effective in polypropylene or polybutene, but whereas these two polymers can be rendered self-extinguishing at 4% of the fire retardant, polyethylene at 4% additive actually burns faster than an untreated control. The greater effectiveness for the higher polyolefins has been attributed to compatibility of the resins and the additives. Some of the data are shown in Table 7-6. Notice the unusually good results with the mixed halophosphonate in polypropylene. These systems are assembled by hot mixing but are based on additives, not on copolymerization.

Twenty percent of the following phosphorane produces flame-retardant polyethylene [21]:

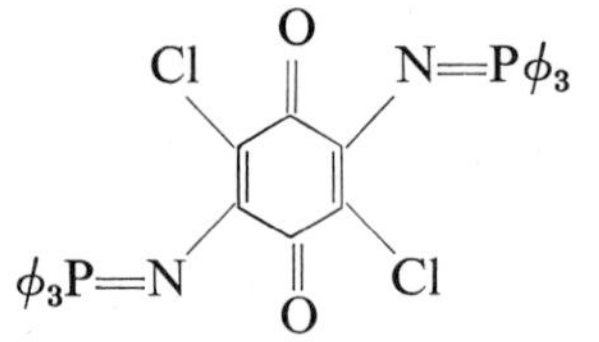

$$\left(\begin{array}{c} 9.8\% \ Cl \\ 8.5\% \ P \end{array}\right)$$

Therefore, only 1.7% P is required in this case when about 2% Cl is added. The effectiveness must again be attributed to good dispersion of the additive, made possible, undoubtedly, by the six phenyl groups. (The cost would probably be prohibitive.)

The use of two additives to obtain this same sort of effect has been discussed. For example, a variety of phosphonium compounds have been added to polypropylene along with a chlorocarbon. A possible synergistic effect is suggested by the following results [22]:

Phosphonium Salt	%	Chlorocarbon	%	Flame Test Result
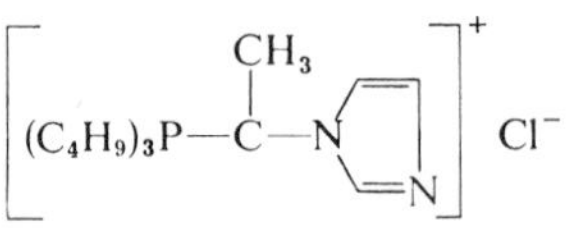	2.5	Chloroparaffin (70% Cl)	2.5	Passed
	5.0	Chloroparaffin (70% Cl)	0.0	Failed
	0.0	Chloroparaffin (70% Cl)	5.0	Failed

The reference contains similar data on a number of phosphonium compounds in which nitrogen was sometimes present, sometimes not; the halogen was varied; and the chlorine-bearing additive was also varied.

Halogens

Polyvinyl chloride may be regarded as a heavily chlorinated polyethylene (57% Cl). It burns slowly and is self-extinguishing when held horizontally. Further chlorination makes a very difficultly flammable polymer; for example, polyvinylidene chloride [2, 23]. There is some evidence that part of the effect is in the solid or liquid phase—simply adding HCl to the vapor space above the polymer does not suppress the flame [24]. On the other hand, bromine appears to act on the flame itself.

Polyethylene may be chlorinated under pressure at 90° with Cl_2. In one method [25] a polymer containing 29% Cl was obtained, and further treatment at 110° with Cl_2 and a peroxide raised the level to 36% Cl. This is about the chlorine content of an ethylenevinyl chloride copolymer which can also be synthesized directly from the two monomers [26]. The flammability of such a product would be far superior to polyethylene but it would not be rated as self-extinguishing. (A recent patent claims "flameproof" properties for a copolymer containing 90 mole % ethylene and chlorinated to 42% Cl [27].)

Polyvinylbromide is not described in the literature as a commercial polymer, probably because of its poor chemical stability. (The iodide would be even less stable.) For good flame resistance a great deal less bromine is required than that in a vinylbromide homopolymer. For example, copolymers have been prepared from ethylene and tribromoallylacrylate or 5-bromo-1-pentene [28]. The latter copolymer was flame-resistant at about 20% Br. If one grants marginal fire retardance for the chlorinated product in reference 27 at 42% Cl and for this material, then the weight ratio of effectiveness of Br to Cl is again confirmed to be about 2 to 1; that is, 1 part of Br is equal in fire retardation to 2 parts of Cl (by weight).

Foams are more difficult to retard than are more dense shapes with lower surface-to-volume ratios. Thus, a polyethylene foam was considered non-flammable only at a chlorine content greater than 66% Cl [29]. Reduced flammability can be attained with as little as 35% Cl with 4–5% sulfur added [30, 31a]. This combination with sulfur has also been suggested for an acrylonitrile-styrene system. For the latter system a chlorosulfonated polyethylene has been suggested [31] at the 5% level to improve flame resistance (not necessarily to render the polymer self-extinguishing).

Halogen-Metal Oxide Combinations

The literature in this category is far more extensive than for the other combinations. Unfortunately, it is essentially all patent literature, but there are enough citations so that one can obtain a fairly clear picture of the requirements by reviewing them in groups. Table 7-7 lists the significant

Table 7-7 Halogen-Metal Oxide Flame Retardants for Polyethylene

Halogen Compound	Metal Oxide	% X	% M_2O_3	Comments	Reference
Aliphatic chlorine					
NH_4Cl	Sb_4O_6	4.6	5.2	NB	32
Chlorinated polyethylene (28% Cl)	Sb_4O_6	5	10.0	"Flame-resistant"	33
Chloroparaffin (65–75% Cl)	Sb_4O_6	9.2	18.8	SE as partially filled composite	34, 35
Chloroparaffin (46.5% Cl)	Sb_4O_6	8.6		Passes ASTM D-635	36
Chloroparaffin (70% Cl)	Sb_4O_6	13.0	18.6	SE	37
Chloroparaffin (72% Cl)	Sb_4O_6	4.0	5.0	"Flame-retardant"	38
Chloroparaffin (55–80% Cl)	Sb_4O_6	10.5	28.0	SE	39
Chloroparaffin (70% Cl)	Sb_4O_6	9.8	24.0	Flame-resistant	40
Chloroparaffin (70% Cl)	Sb_4O_6	12.0	19.5	Flame-resistant	41
Neoprene (39% Cl) Chloroparaffin (70% Cl)	SbOCl	5.8	6.5	Flame-resistant	41a
Aliphatic with phosphorus					
Tris(chloroethyl) phosphate	Sb_4O_6	8.7	7.7	Contains 2.5% P—NB	42
Chloroparaffin (70% Cl) plus tetraester of pyrophosphoric acid[a]	Sb_4O_6	9.8	7.0	Contains 0.6% P—SE	43
Aromatics					
Cl_2—C—COϕCl$_3$ $\mid$ O Cl_2—C—COϕCl$_3$ O	Sb_4O_6	5.75	3+	SE	44
Octachlorodiphenylether	Sb_4O_6	5.6?	5?	SE	45
Chlorinated $(CH_2)_6$ (72% Cl)	Sb_4O_6	14.4	5	SE (<2 sec)	45a
Aromatic bromine					
Brominated dinuclear compounds (64% Br)	Sb_4O_6	6.4	5	SE	46
Br—⟨◯⟩—NH_2 (73% Br) Br	Sb_4O_6	5.8	6	SE (<5 sec)	47
Pentabromodiphenylether (71% Br)	Sb_4O_6	6.4	5	SE	48
Brominated phenoxysilane (70% Br)	Sb_4O_6	9.2	9.3	SE (<4 sec)	49
Aromatic bromine with chlorine					
Poly(tribromostyrene) plus chloroparaffin (70% Cl)	Sb_4O_6	4.1 Br 1.1 Cl	3.5	SE (<1 sec)	50
Pentabromochlorodiphenylether (67% Br, 6% Cl) plus chloroparaffin (70% Cl)	Sb_4O_6	4.1 Br 2.2 Cl	2.6	NB, drippings SE	51
Tribromophenyl caproate (56% Br) plus chloroparaffin (70% Cl)	Sb_4O_6	2.5 Br 2.5 Cl	4.4	Does not ignite in ASTM D-635	52
Fluorine					
NH_4BF_4(72% F)	Sb_4O_6	?	?	SE	53
NH_4BF_4 (72% F)	Sb_4O_6	4	20	SE	54

Table 7-8 Halogen-Metal Oxide Flame Retardants for Polypropylene

Halogen Compound	Metal Oxide	% X	% M_2O_3	Comments	Reference
Chlorine					
Chlorinated polypropylene (40% Cl)	Sb_4O_6 SnO	13.6	6 Sb_4O_6 1 SnO	Nonflammable	55
Chlorinated phthalimide derivative (57% Cl)	Sb_4O_6	11.5	10	SE (5 sec)	56
Dimer of hexachlorocyclo-pentadiene (78% Cl)	Sb_4O_6	19.5	8.4	SE (<2 sec)	57
Bromine-aliphatic					
Hexabromocyclododecane (75% Br)	Sb_4O_6	2.2	1	SE	58, 59
1,2-Dibromo-4-(α,β-di-bromoethylcyclohexane) (75% Br)	Sb_4O_6	3.0	2	SE	58, 59
Bromine-aromatic					
Bis(pentabromophenyl)-adipate	Sb_4O_6	4.7	2.7	SE	60
Bromine-chlorine					
Tris(tribromophenyl) phosphate plus chloroparaffin (70% Cl)	Sb_4O_6	3.1 Br 2.5 Cl	4.5 (0.13 P)	Nonflammable	61
Fluorine					
$Zn(NH_3)_4(BF_4)_2$ (49% F)	Sb_4O_6	5	Up to 40	SE	62

entries for polyethylene; Table 7-8, for polypropylene. The claim that NH_4Cl and Sb_4O_6 render polyethylene nonburning is intriguing since more compatible sources of chlorine do not appear, from Table 7-8, to be as effective. This may be because of the nitrogen in the NH_4Cl. Self-extinguishing polyethylene is produced from chloroparaffin and Sb_4O_6 only at rather high levels. Something like 10% chlorine and 20% Sb_4O_6 is required. Addition of phosphorus sharply reduces the requirement for Sb_4O_6 so that less than 10% of Sb_4O_6 is needed with 0.5% phosphorus and 8–10% chlorine to provide self-extinguishing properties.

Aromatic chlorine compounds seem to be more effective than the aliphatics, but there are too few examples from which to judge this. (Some say the aliphatics are to be preferred [2].)

Bromine is more effective than chlorine, though the only sound comparison is between aromatic bromine derivatives and aliphatic chlorine compounds. It would appear that 5 or 6% bromine with 5 or 6% Sb_4O_6 will render polyethylene self-extinguishing. Some further reduction in percentages may be obtained with mixed bromo-chloro compounds but not much: 5 or 6% halogen is still required.

Fluorine is not impressive when the inorganic compound is compared with NH_4Cl.

Antimony oxide in combination with alumina barely confers self-extinguishing properties when 14% Sb_4O_6 and 17% Al_2O_3 $3H_2O$ are added [63].

Polypropylene is somewhat easier to treat. In the aliphatic chlorine compounds, for example, self-extinguishing properties are achieved with far less Sb_4O_6 and only slightly more chlorine.

If one increases the halogen level the required level of Sb_4O_6 decreases and vice versa.

The requirements for bromine and Sb_4O_6 are very low indeed. Reference 59 is worthy of special note. The bromine derivative works best if its boiling point is 200° or greater. Thus, at this same level of additive, lower-boiling bromine compounds such as certain brominated aromatics are not effective. The bromine-chlorine mixture shown is no improvement over the other examples despite the presence of a very small amount of phosphorus. Again, fluorine is relatively ineffective.

There is one reference [64] to the use of titanium dioxide pigment at a 10% level in polypropylene. The treated polymer (actually a filled composite) would not burn. The pigment probably serves as a heat sink.

To sum up, the method of choice for polyethylene and polypropylene is to combine halogens with antimony trioxide (Sb_4O_6). Though bromine is far more effective than chlorine, its use is limited by its higher cost and poorer stability in the polymer. On occasion, performance may be further improved by use of some phosphorus, added as a halogenated phosphate or phosphonate. A more difficult problem is to render a solid shape fire-resistant and also prevent molten drippings from flaming. The avoidance of flaming drippings can be accomplished with the better systems. Polypropylene is somewhat easier to protect than polyethylene—probably reflecting effects of crosslinking, branching, and so forth, on the decomposition route. Fluorine is ineffective in these systems because it is not part of the polymer but merely an additive. Were it substituted for hydrogen on the polymer, many changes in physical and chemical properties would be expected. The stability of the C—F bond would confer flame resistance on that part of the molecule. Added fluorine in an inorganic compound probably leads only to volatilization without much interaction with radicals present in the flame.

No references directed exclusively to polyolefin fibers have been uncovered, though some of those cited above relate to both fibers and shapes. Solid additives like Sb_4O_6 will cause difficulty in spinnerets and therefore one would look for easier handling with halogenated organophosphorus compounds. To minimize the effect of the retardants on physical properties a bromine-phosphorus combination would perhaps be most useful and require the lowest additive level. Reactive comonomers containing high levels of bromine and phosphorus may be the long-range solution.

VINYLS (CHLORIDE, ACETATE, ALCOHOL, MISCELLANEOUS)

Polyvinyl Chloride

This widely used, versatile resin, known as PVC, is made by polymerizing vinyl chloride:

$$x CH_2{=}CHCl \xrightarrow[\text{processes}]{\text{variety of}} \left(CH_2{-}\underset{\underset{Cl}{|}}{CH} \right)$$

As stated earlier, PVC is flammable and burns with a slow, smoky flame [2]. On combustion it yields HCl, CO_2, and CO plus some aromatics and traces of aliphatics. Very little monomer is recovered in the gases and essentially no phosgene [65]. The generation of HCl is itself a serious problem reported to lead to secondary damage from corrosion [65a]. The discussion of halogens in polyethylene and polypropylene applies to PVC, which is a special case of chlorinated polyethylene. Much of what was presented under polyolefins applies to PVC and should be reviewed as part of the present discussion. PVC is not usually a serious flame hazard; indeed, vinyl chloride has been used as a comonomer to reduce the flammability of styrene-acrylonitrile copolymers [66] and of acryolonitrile fibers (see the section on "Acrylic Fibers"). However, there are many circumstances in which lower flammability is needed than PVC can provide. There are three ways of achieving the necessary improvement: additives, comonomers high in flame-retardant elements, and protective coatings. Protective coatings are often used for appliance housings where the fire hazard is primarily localized near electric junctions and heating elements. Intumescent coatings are frequently used on the inside of such housings. These coatings have the advantage of protecting without affecting the physical properties of the PVC.

The use of comonomers as fire retardants for PVC is not a widespread practice. Two vinyl compounds containing phosphates have been discussed as

Table 7-9 Some Phosphate Esters Used as Flame Retardants for PVC

Ester	Formula	Recommended Level	% P in PVC	Flame Rating	Reference
Tributyl phosphate	$(C_4H_9O)_3PO$	20	2.2	Flame-resistant	69
Diisopropyl phosphate-Ba salt	$[((CH_3)_2CHO)_2PO_2]_2Ba$	3.4	0.45	SE	70
Tritolyl phosphate	$(CH_3\phi O)_3PO$	15	1.3	Flame-resistant (resin compounded 44% of nitrocellulose)	71
Tris(trichlorophenyl) phosphate	$(Cl_3\phi O)_3PO$	33	1.6	...	72
Bis(pentachlorophenyl) phosphate	$(Cl_5\phi O)_2PO(OH)$	31	1.6	...	72
Brominated tritolyl phosphate	$(Br_{4.6}CH_3\phi O)_3PO$	50	1.1	Flame-resistant film	73

possibilities [67, 68]. One of these is a diphosphonate containing a high percentage of phosphorus [68]:

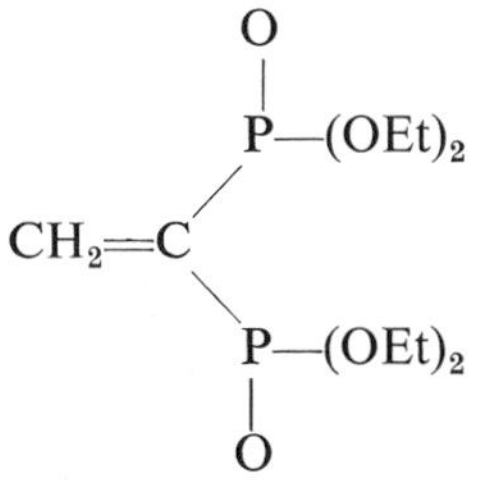

$$CH_2{=}C \begin{matrix} \overset{O}{\underset{|}{P}}{-}(OEt)_2 \\[1em] \underset{\overset{|}{O}}{P}{-}(OEt)_2 \end{matrix} \qquad (20.8\% \ P)$$

This is recommended for PVC at the 20% level ($\sim$4% P). The compound is not yet available commercially.

Additive retardants are the most common and are primarily phosphate esters or antimony oxide compositions. Table 7-9 shows some of the phosphate esters that have been claimed as flame retardants; Table 7-10 contains

Table 7-10   Some Results with Mixed Esters of Phosphoric
Acid in PVC [74]

Ester	% Ester in PVC	% P in PVC	SE Time, sec
Tricresyl phosphate	33	2.8	0
	50	4.2	0
Dibutylphthalate	33	0	Burns up to 30 sec
	50	0	Burns up to 30 sec
Dioctylphenyl phosphate:			
From 2-ethylhexanol	33	2.6	3–7
	50	3.8	0–3
From C_7–C_9 normal	33	2.6	3–7
alcohol blend	50	3.8	0–3
Diheptylphenyl phosphate			
from C_6–C_8 iso-alcohol	33	2.8	3–8
	50	4.2	0–3
Diphenyl 2-ethyl-hexyl			
phosphate	33	2.8	0–2
	50	4.2	0

comparative data on a series of one class of these esters. It is clear from the two tables that (a) 1–2% phosphorus added as a phosphate ester will produce a self-extinguishing PVC, (b) large amounts of ester are required to obtain even this low level of phosphorus, and (c) addition of bromine may further reduce the phosphorus requirement. Use of such large amounts of additive will obviously affect the physical properties of the polymer; these esters are used as plasticizers and are substituted for more flammable varieties when fire retardance is needed.

A few highly halogenated additives have been offered as flame retardants for PVC: halogenated adipate [75], hexachlorocyclopentadiene-adducts [76], and ring-chlorinated alkoxybenzenes [77], for example. All of these materials serve as plasticizers for PVC without adding to its flammability. They would not be expected to reduce flame ratings materially and are probably not as effective as the halogenated phosphorus plasticizers.

Addition of antimony oxide, Sb_4O_6, has an effect similar to that of phosphorus on flammability, as pointed out in the section on "Polyolefins." Various ranges have been claimed [78–80], but 5–15% appears most frequently. Two formulas using different plasticizers are:

	A [79]	B [80]
PVC	46	75
Plasticizers	28 DOP	. . .
Other	18 Carbon black	10
Sb_4O_6	5	15
	100	100

Zinc borate·$3\frac{1}{2}$ H_2O at a level of 2–10% has also been claimed [80a].

Table 7-11　Methods of Adding Phosphorus as a Comonomer for PVAc

Comonomer	Wt % Comonomer	% P	Flammability	Reference
H_3PO_4 (yields vinyl phosphate on milling)	5	1.8	SE (ASTM D-635)	81
$(CH_2=CHCH_2O)_3PO$	5–40	0.7–5.7	Resists ignition	82
$(CH_2=CHCH_2O)_2-\overset{\overset{O}{\|}}{P}-CH_2-\overset{\overset{CH_3}{\|}}{C}=CH_2$ $(CH_2=CHCH_2O)_2-\overset{\overset{O}{\|}}{P}-\phi$	25} 26}	3.6} 3.4} 7.0	Flame-resistant	83
$ClCH_2-\overset{\overset{O}{\|}}{P}-(OCH_2CH=CH_2)_2$	. . .	. . .	. . .	84
$CH_2=CHPO_3H_2$ or $CH_2=CHPO_3Et_2$	4.6	0.87	. . .	85
$CH_2=CH-\overset{\overset{O}{\|}}{\underset{\underset{H}{\overset{\|}{O}}}{P}}-OC_8H_{17}$	10	1.4	. . .	86

In summary, the principal means of achieving flame resistance in PVC is via the phosphate ester plasticizers; the effect may be boosted by adding Sb_4O_6.

Vinyl Acetate

All the references on polyvinyl acetate (PVAc) refer to copolymerization as the means of incorporating the fire-retardant element. The element invariably is phosphorus.

Other approaches are cited by Hilado [2], but in the references the application in vinyl acetate apparently has always been secondary to other polymers so that the references were placed in other categories. Thus, phosphate esters are useful and so are halogenated paraffins and the like.

A list of the comonomers that have found utility in PVAc is shown in Table 7-11. A phosphorus level of from about 1 to 7% is suggested by these entries, but only one actual numerical result is given. That result indicates that a minimum of 2% P should be considered for self-extinguishing behavior.

Polyvinyl Alcohol

Here, too, phosphorus is the preferred retardant, though relatively little work has been reported. Kosolapoff [86] in 1950 prepared a variety of phosphate esters of polyvinyl alcohol. An example is:

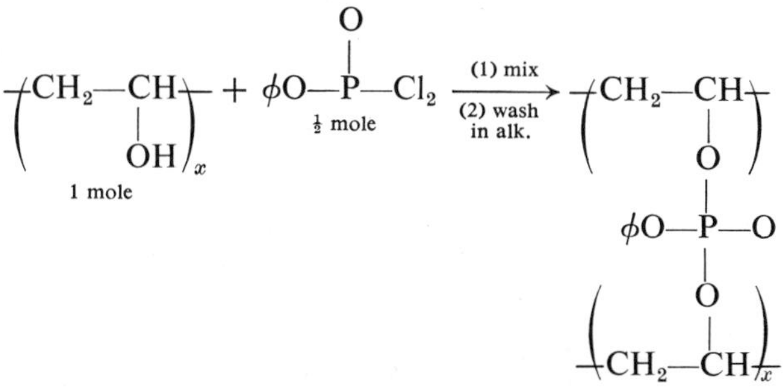

The product is insoluble in water, contains 13.7% P, and is, as one would expect, nonflammable. An ionic ester is prepared in a manner reminiscent of that for phosphorylating cellulose:

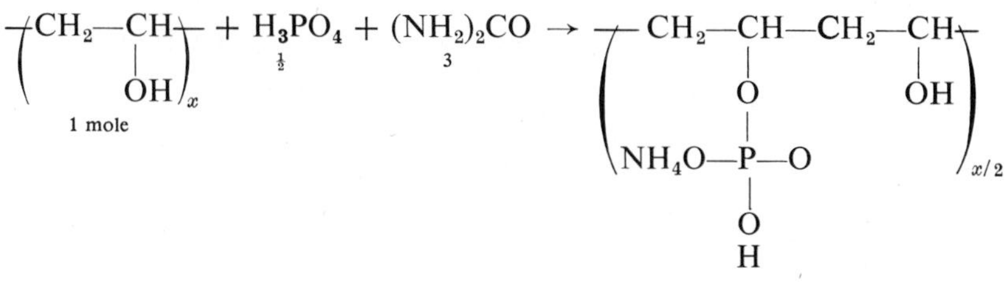

The product was spun from water into alcohol to produce a self-extinguishing fiber [87]. However, such a fiber should be very water-sensitive. It was therefore suggested as a surface treatment for insolubilized PVOH fibers. Such fibers treated with the ammonium polyvinyl phosphate were self-extinguishing [88]. The treatment would, presumably, not resist laundering but might be useful in draperies and other areas not requiring such water resistance. $POCl_3$ has also been employed to the same end, with pyridine used in the bath to avoid acid conditions [88a].

A more water-resistant surface treatment for PVOH fibers has been developed [89] from THPC—$(HOCH_2)_4PCl$—and polyfunctional organic nitrogen compounds. These surface copolymers are similar to those used so successfully for cotton.

Polyvinyl alcohol can be plasticized and fire-retarded with phosphate esters. One such ester [90], an ethoxylate of butyl acid phosphate, has been claimed specifically for this use.

PVOH fibers have also been flameproofed by immersion in acid baths of antimony-tin halides followed by immersion in ammonia solution [91]. A borate ester has also been found effective [91a].

Miscellaneous Vinyl Phosphorus and Vinyl Halogen Polymers

Several polymers have been reported for which the basic monomer contained the fire-retardant element. Thus, polymers of

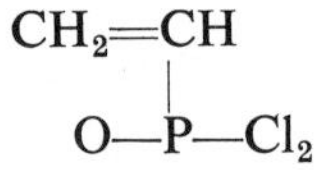

$$CH_2{=}CH$$
$$|$$
$$O{-}P{-}Cl_2$$
Vinyl phosphonyldichloride [92]

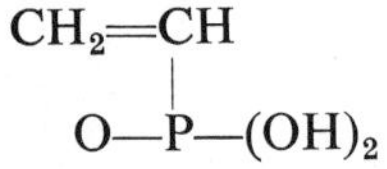

$$CH_2{=}CH$$
$$|$$
$$O{-}P{-}(OH)_2$$
Vinyl phosphonic acid [93]

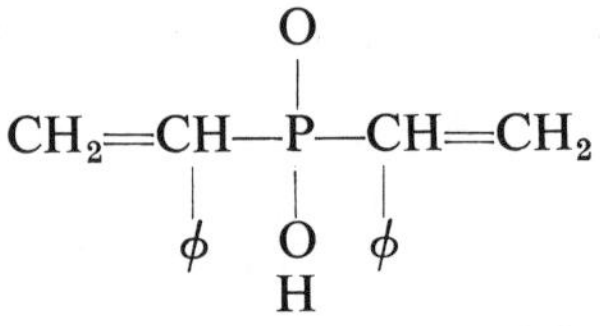

$$O$$
$$||$$
$$CH_2{=}CH{-}P{-}CH{=}CH_2$$
$$| \quad | \quad |$$
$$\phi \quad O \quad \phi$$
$$H$$
α-Phenylvinyl phosphinic acid [94]

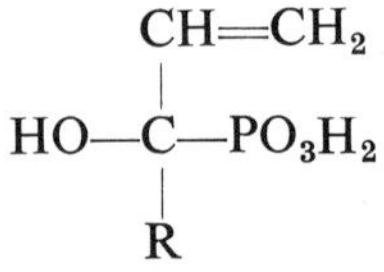

$$CH{=}CH_2$$
$$|$$
$$HO{-}C{-}PO_3H_2$$
$$|$$
$$R$$
Vinyl-ketone or -aldehyde
adduct with PCl_3 [95]

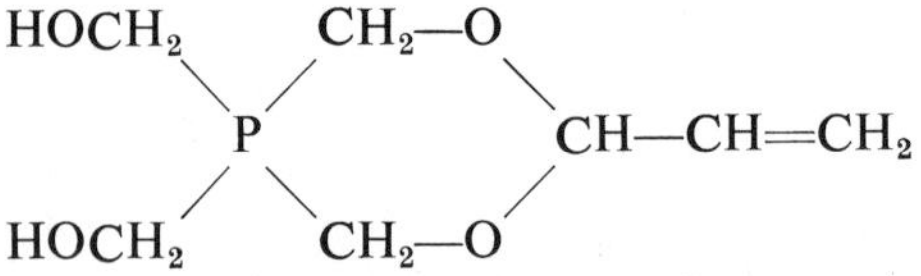

$$HOCH_2 \quad\quad CH_2{-}O$$
$$\diagdown \quad \diagup \quad\quad \diagdown$$
$$P \quad\quad\quad CH{-}CH{=}CH_2$$
$$\diagup \quad \diagdown \quad\quad \diagup$$
$$HOCH_2 \quad\quad CH_2{-}O$$
Acetal from THPC and acrolein [96]

$$(CH_2{=}CHCH_2)_2PH$$
Diallyl phosphine [97]

are all self-extinguishing. The paper by Marvel and Wright [95] is interesting in the variety of materials covered. Table 7-12 summarizes some of their

Table 7-12 Phosphorus-Containing Polymers from the Reaction of Phosphorus Trichloride and Carbonyl-Containing Polymers (Marvel and Wright [95])

No.	Copolymer Used	Inherent Viscosity	Solvent	% Carbonyl Containing Unit	% P in Product	% Extent of Reaction	% Solubility in Benzene	Inherent Viscosity[a]	Solvent[a]
1	Styrene-methyl vinyl ketone	3.40	Benzene	13	0.96	40	68	1.75	Chloroform
2	Styrene-methyl vinyl ketone	3.72	Benzene	24	1.81	37	36	1.61	Chloroform
3[b]	Styrene-methyl vinyl ketone	3.72	Benzene	24	0.27	5[b]	100	2.64	Benzene
4[b]	Styrene-methyl vinyl ketone	3.72	Benzene	24	0.24	5[b]	100	1.49	Benzene
5	Methyl vinyl ketone (100%)	0.60	Chloroform	100	18.10	89	0	Insoluble	...
6	Butadiene-methyl vinyl ketone	...	...	36	0.28	4	...	Insoluble	...
7	Methyl acrylate-methyl vinyl ketone	0.82	Chloroform	19	2.67	69	...	Insoluble	...
8	Methyl methacrylate-methyl vinyl ketone	4.78	Chloroform	32	4.09	62	...	Insoluble	...
9	Acrylonitrile-methyl vinyl ketone	...	...	24	0.07	1	...	...	...
10	Styrene-methylacrolein	0.45	Chloroform	29	7.33	100	14	0.38	Dimethyl-formamide
11	Styrene-cinnamaldehyde	0.81	Benzene	12	2.24	100	100	0.23	Benzene
12	Styrene-methyl isopropenyl ketone	1.46	Benzene	28	0.97	19	78	1.10	Chloroform
13	Styrene-benzalacetone	1.47	Benzene	11	0.01	1	100	1.48	Benzene
14	Styrene-benzalacetophenone	1.30	Benzene	29	0.07	2	100	1.50	Chloroform

[a] The inherent viscosities reported were measured on solutions containing 0.1 g of polymer in 25 ml of the solvent noted at 30°

[b] Water was used directly to hydrolyze the phosphorus trichloride addition complex without the intermediate action of acetic acid

preparations and gives some idea of the applicability of the reaction of PCl_3 with carbonyl groups (see p. 57, Chapter 2).

Two rather curious chlorine-containing monomers are diallyl chlorendate [98] and vinyl pentachlorophenyl sulfide: $CH_2=CHS\phi Cl_5$ [99]. Five percent

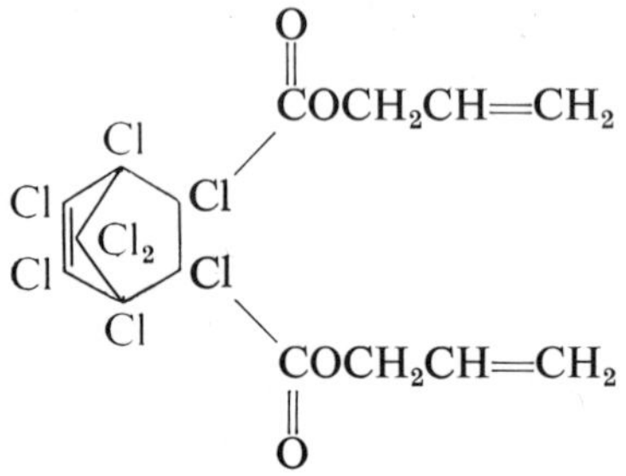

of the chlorendate copolymerized with a diallyl phthalate monomer gave a self-extinguishing copolymer by ASTM D-635. A polymer of the vinyl sulfide was found to be nonburning and formed fibers, sheets, and so on.

Acrylics

The acrylic polymers include, for our purposes, both acrylic acid derivatives and the acrylonitrile polymers. These are produced by a vinyl-type polymerization:

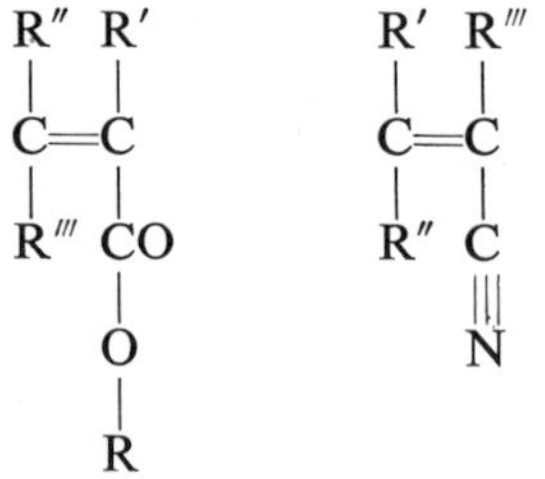

Acrylic acid derivative Acrylonitrile derivative

where the R's may be hydrogen, alkyl groups such as methyl, and so on.

Madorsky's discussion of the degradation of these materials is interesting in that substitution on the carbon atoms in the backbone has a marked effect on the nature of the decomposition. With no substitution, little monomer is recovered; with substitution, the product is almost entirely monomer, that is, the polymer unzips [15] (Table 7-13).

Addition of an acid-forming flame retardant alters the decomposition of methylmethacrylate so that large amounts of cokelike carbon [100] are formed in a manner apparently like the decomposition of cellulose. A recent

Table 7-13

Polymer	Monomer in Volatiles, wt %
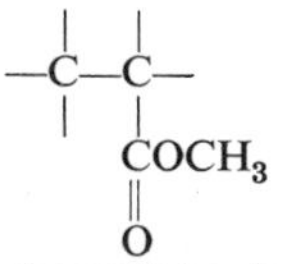 $-\overset{\textstyle\mid}{C}-\overset{\textstyle\mid}{C}-$ $\quad\quad\overset{\textstyle\mid}{C}OCH_3$ $\quad\quad\overset{\Vert}{O}$ Polymethylacrylate	0.7
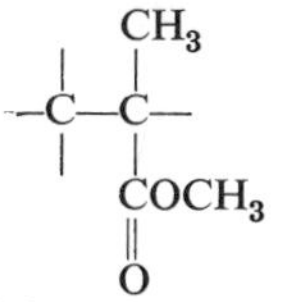 $\overset{\textstyle CH_3}{\underset{\textstyle COCH_3}{-\overset{\textstyle\mid}{C}-\overset{\textstyle\mid}{C}-}}$ $\quad\quad\overset{\Vert}{O}$ Polymethylmethacrylate	91.4
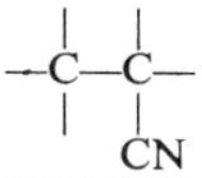 $-\overset{\textstyle\mid}{C}-\overset{\textstyle\mid}{C}-$ $\quad\quad\overset{\textstyle\mid}{C}N$ Polyacrylonitrile	5.2
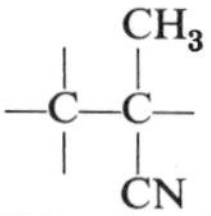 $\overset{\textstyle CH_3}{\underset{\textstyle CN}{-\overset{\textstyle\mid}{C}-\overset{\textstyle\mid}{C}-}}$ Polymethacrylonitrile	Up to 100

paper [101] discusses the role of anhydride formation and decarboxylation reactions in such materials but the exact mechanisms of such decompositions remain obscure.

In oxygen the acrylates and acrylonitrile polymers burn readily. Hilado [2] gives burning rates in the horizontal position as 0.6–1.6 in./min for polymethylmethacrylate and 1–2 in./min for acrylonitrile-butadiene-styrene terpolymer. (The flammability of the latter has been a center of controversy in the building industry for many years as soil or drain pipe made of this polymer finds use in more and more types of construction.)

The following discussion is divided into two sections: the first on acrylic polymers for shapes and fibers (almost all acrylic acid derivatives) and the second on fibers (almost all acrylonitrile polymers).

Acrylates

These are flame-retarded principally with phosphorus compounds to produce the charring cited above. Such compounds run the gamut from phosphates to phosphonium compounds. Self-extinguishing polymethyl-methacrylate (PMMA) has been produced by simply adding about 8%

Table 7-14 Phosphonates Used in Acrylates^a

Phosphonates	Polymer	Phosphonate Added, wt %	% P in Product	Flame Rating	Reference
$CH_2{=}\underset{\phi}{C}{-}PO_3H_2$	AA, MA	...	...	Resistant	106
$MAcrOCH_2{-}\overset{O}{\underset{\|}{P}}{-}(OMe)_2$	MMA	30	4.5	SE	107
$MAcrOCH_2CH_2{-}\overset{O}{\underset{\|}{P}}{-}(OMe)_2$	MMA	40	5.6	Resistant	108
$MAcrOCH_2CH_2{-}\overset{O}{\underset{\|}{P}}{-}(OEt)_2$	MMA	20	2.5	Resistant	109
$MAcrO\underset{CH_3}{C}H{-}\overset{O}{\underset{\|}{P}}{-}(OEt)_2$	MMA	40–50	5–6	SE	110
$MAcrO{-}\underset{CH_3}{C}\left[\overset{O}{\underset{\|}{P}}{-}(OMe)_2\right]_2$	MMA	25	4.7	SE	111

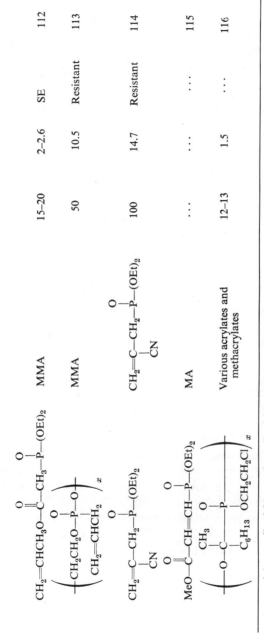

Structure	Comonomer				Ref.
CH$_2$=CHCH$_3$O—C(O)—C—CH$_3$—P(O)—(OEt)$_2$	MMA	15–20	2–2.6	SE	112
$\left(\text{CH}_2\text{CH}_2\text{O—P—O—}\atop\text{CH}_2=\text{CHCH}_2\right)_x$	MMA	50	10.5	Resistant	113
CH$_2$=C(CN)—CH$_2$—P(O)—(OEt)$_2$	CH$_2$=C(CN)—CH$_2$—P(O)—(OEt)$_2$	100	14.7	Resistant	114
MeO—C(O)—CH=CH—P(O)—(OEt)$_2$	MA	...	...	...	115
$\left(\text{O—C(O)—C(CH}_3)\text{—P—}\atop\text{C}_6\text{H}_{13}\ \ \text{OCH}_2\text{CH}_2\text{Cl}\right)_x$	Various acrylates and methacrylates	12–13	1.5	...	116

[a] AA : acrylic; MA : methacrylic; MAcr : CH$_2$=C(CH$_3$)CO—; MMA : methylmethacrylate

H_3PO_4 to the monomer and then polymerizing with an initiator [102] to give a polymer containing about 2.5% P. In another case, 4.2% of P_2O_5 was added along with 7.8% of 2-chloroethanol to give a nonburning PMMA containing about 1.8% P and about 3.5% Cl [103]. This presumably produced some chloroethyl phosphate esters *in situ*. Such esters can be added directly. Thus, a flame spread of 39 has been obtained with the following in PMMA [104]:

$$
\begin{array}{lll}
CH_2\!\!=\!\!CHCO_2Me & 78 & \left(\begin{array}{l}1.6\%\ P\end{array}\right. \\
ClCH_2CH_2(Br)CH_2O_3PO & 18 & \left(\begin{array}{l}3.4\%\ Cl\end{array}\right) \\
BrCH_2CH_2OP(O)(OH)_2 & \underline{4} & \left.\begin{array}{l}9.2\%\ Br\end{array}\right) \\
& 100 &
\end{array}
$$

An interesting nitrile has been proposed for flameproofing shaped polyacrylonitrile. The product is made from phosphoramide dichloride and 2-cyanoethanol and is added at about the 9–10% level to yield a flame-resistant polymer with 2% P. Both films and fibers were prepared [105].

Adding 1.5% P from compounds of the type $\begin{array}{c}R\\ \diagdown\\ P\!-\!NH_2\\ \diagup\\ R'\end{array}$, where R and R' are alkyl, alkoxy, on aryl, produces self-extinguishing behavior in ASTM D-635 [105a].

The phosphonates have been much studied for acrylates. Table 7-14 presents some of these compounds. (The synthesis of these materials—from alkyl phosphites—has been discussed in Chapter 2.) It would appear that from 2 to 10% phosphorus is required with good performance to be expected at about 5%. Most of the phosphonates contain 12–14% phosphorus and therefore as much as 40% of the phosphonate is required in the finished polymer. Physical properties may not be seriously affected by this large quantity when the phosphonate becomes a part of the polymer itself, but costs will be markedly increased. Thus, we have here a potentially valid technical solution that is not yet very attractive economically.

There is one reference to phosphinic acids; it is indicated that about 2% phosphorus in the form of an additive—diphenylphosphinic acid—is sufficient to pass ASTM D-635 [117]. A polymer based on THPC has been made that gives a flame-resistant product [118]:

$$
THPC + KOOCC(CH_3)\!\!=\!\!CH_2 \;\rightarrow\; [(HOCH_2)_4P]^+\left[\begin{array}{c}OOCC\!\!=\!\!CH_2\\ |\\ CH_3\end{array}\right]^-
$$

THP methacrylate

Methacrylate polymer $\xleftarrow{\quad Bz_2O_2 \quad}$

And PMMA mixed with 10% of $[(CNCH_2CH_2)_3PCH_2^-]_2^{2+}Br_2^{2-}$ gives a flameproof product containing 1.1% P, 1.5% N, and 2.9% Br [119]. A series of phosphines have also been claimed as fire retardants for acrylic polymers; for example, efficacy is cited for 15% triphenyl phosphine in polyethylacrylate (1.8% P) [120]. These uses of phosphorus in the -3 oxidation state remain relatively expensive curiosities but such developing technology suggests possible future directions.

Phosphorus-halogen compositions are potentially lower in cost per unit of flame retardance conferred and have found favor commercially. Despite a reference to an all-inorganic mixture of calcium phosphate and zinc chloride [121], the commonly used compositions are halogenated organic phosphate esters. Thus 0.5–3% P and 1.8–10% Cl from tris(chloroethyl) phosphate or bis(chloroethyl) phosphate provide self-extinguishing acrylates [122]. A methacrylate that self-extinguishes within 1 sec has been made from tris(bromoethyl) phosphate added so as to give 2% P and 7.6% Br [123]. These are of course nonpolymerizable additives and will have marked effects on the physical properties of the product—serving primarily as plasticizers. The effect of the chlorine is to reduce the phosphorus requirement from about 5% down to about 2% and the amount of fire-retardant compound from 30 to 40% down to 12 to 15% or so. The latter point is important in that the smaller amount of inert additives one puts in a resin system the easier it will be to maintain control over important properties. Bromine here seems not much more effective than chlorine.

Table 7-15

	Level in Methacrylate Polymer for Flameproofing		
	% P	% X	Reference
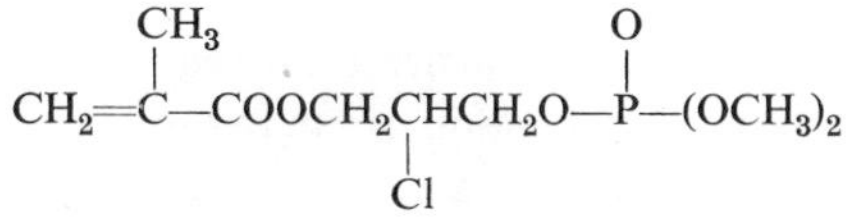 CH₂=C(CH₃)—COOCH₂CHCH₂O—P(=O)—(OCH₃)₂ with Cl	2.3	3.1	124
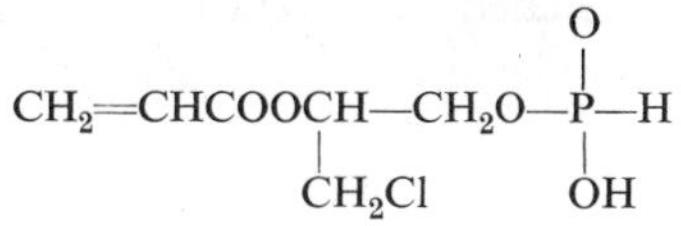 CH₂=CHCOOCH(CH₂Cl)—CH₂O—P(=O)(OH)—H	1.8–2.5	2.2–3.0	125
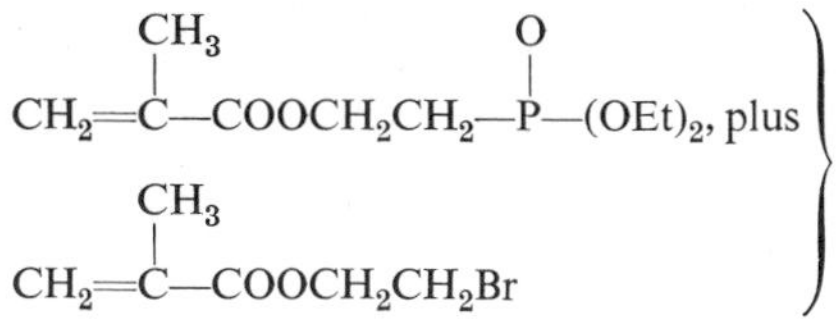 CH₂=C(CH₃)—COOCH₂CH₂—P(=O)—(OEt)₂, plus CH₂=C(CH₃)—COOCH₂CH₂Br	2.5	2	126

Reactive or copolymerizable phosphonates similar to those in Table 7-14 have been studied with halogens in the molecule. Essentially the same phosphorus and halogen requirements are encountered as with phosphate-halogen systems (Table 7-15).

Cass and Raether have reported in detail on the use of halogenated poly-phosphonates of the additive type in PMMA [127]. They compared the performance of the polyphosphonates with three halophosphate esters to provide the only good comparative data published in this area. Table 7-16 shows some of their results. About 2–2.5% P and 5% halogen are effective. The paper gives details on the effect of the retardants on heat distortion, light transmission, water absorption, strength, and so on. The higher molecular weight of the polyphosphonate apparently makes it less of a plasticizer for the polymer and therein lies some advantage.

There is but one reference to a chlorophosphinate and that is to a vinyl compound [128]:

$$CH_2{=}CH{-}O{-}\underset{\underset{\displaystyle}{|}}{\overset{\overset{\displaystyle O}{|}}{P}}{-}(CH_2CH_2Cl)_2 \quad \text{or} \quad CH_2{=}\underset{\underset{\displaystyle R}{|}}{\overset{}{C}}{-}O{-}\underset{}{\overset{\overset{\displaystyle O}{|}}{P}}{-}(CH_2CH_2Cl)_2$$

There would seem to be no particular advantage for this material nor to the use of phenyl dichlorophosphine [129] nor to the recently proposed phos-phorane [130]:

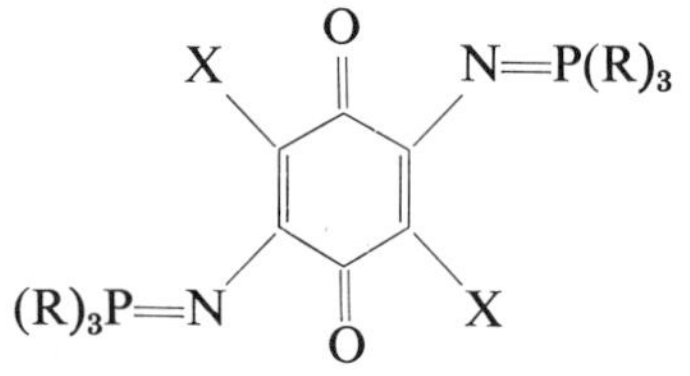

The phosphorane is certainly novel in structure as a fire retardant and is a stable, high-melting substance, but it will be a long time before such a material is economically justifiable.

Acrylics and acrylonitriles treated with halogenated flame retardants without phosphorus are less often encountered in the literature. Polymers of

$$\text{trichloroethylmethacrylate} {-} Cl_3CH_2CH_2O\overset{\overset{\displaystyle O}{\|}}{C}C(CH_3){=}CH_2 {-}$$

at 53% Cl [131] and of an acrylate made from a "hex"-cyclopentadiene condensate at 56% Cl [132] were found to be self-extinguishing. Another "hex" derivative rendered PMMA self-extinguishing when added to give 15% Cl in the product [133]. A pentachlorophenylmethacrylate when added as a comonomer yields flame-retarded PMMA at 21% Cl [134]. It appears that about 20% chlorine is

Table 7-16 Comparative Data for Several Halogen-Phosphorus Compositions in PMMA (after Cass and Raether [127])

Retardant	%P	%Cl	%Br	10%	17.5%	25%
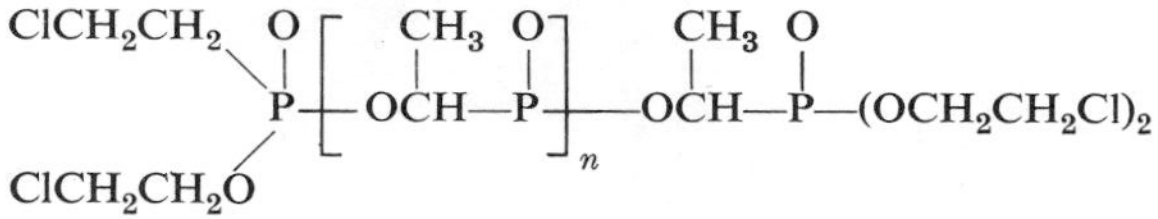 $(ClCH_2CH_2O)_2P(O)-[OCH(CH_3)-P(O)]_n-OCH(CH_3)-P(O)(OCH_2CH_2Cl)_2$ $n=2$	15.1	27.9	...	Burns (0.28 in./min)	SE	SE 3 sec
$n=6$	16.9	25.0	...		150 sec	SE 35 sec
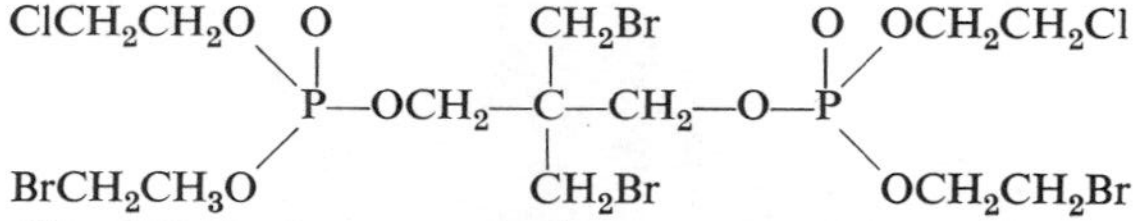 $(ClCH_2CH_2O)(BrCH_2CH_3O)P(O)-OCH_2-C(CH_2Br)_2-CH_2-O-P(O)(OCH_2CH_2Cl)(OCH_2CH_2Br)$	7.8	9.0	41.2	Burns (0.66 in./min)	Burns (0.33)	SE 75 sec
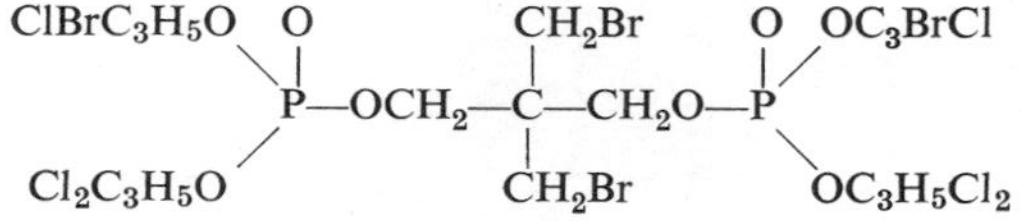 $(ClBrC_3H_5O)(Cl_2C_3H_5O)P(O)-OCH_2-C(CH_2Br)_2-CH_2O-P(O)(OC_3BrCl)(OC_3H_5Cl_2)$	6.6	20.8	33.6	...	...	Burns (0.67 in./min)
$(BrCH_2CH(Br)CH_2O)_3PO$	4.5	...	69.0	Burns (0.53 in./min)	Burns (0.43)	SE 350 sec
$(ClCH_2CH(Cl)CH_2O)_3PO$	7.2	48.8	...	Burns (0.53 in./min)	Burns (0.54)	Burns (0.43 in./min)
$(ClCH_2CH_2O)_3PO$	10.9	36.8	...	Burns (0.57 in./min)	SE 280 sec	SE 200 sec

Flammability ASTM D-635 (sec for SE) — Retardant Added: 10%, 17.5%, 25%

required. Bromine compounds appear to be effective at a few percentage units less than this: at about 16% (as bromoethylmethacrylate [135], dibromo-propylmethacrylate [136], or even the aromatic tribromophenoxyethyl-acrylate [137]) based on the total polymer weight. It is worthy of note that 1–2% of NH_4Br has been found to produce a self-extinguishing acrylonitrile-butadiene-styrene (ABS) terpolymer [138]. (It was earlier noted that NH_4Cl is unusually effective as used with Sb_4O_6 in polyethylene.)

Antimony oxide reduces the amount of halogen required, as expected; 7.5% Sb_4O_6 reduces the amount of bromine needed in acrylonitrile to 4.5–7.5% when supplied from epibromohydrin [139]. ABS can be made self-extinguishing by adding PVC [140]:

ABS resin	85
PVC	10
Sb_4O_6	5

Tin salts as stannous or stannic halides mixed with metal carboxylates have also been found useful in acrylates [141]. Boron esters also have shown some promise [141a].

"Acrylic" Fibers

These are polymers and copolymers of acrylonitrile. The fibers are spun from solution at moderate temperatures. For example, 22% acrylonitrile in dimethylacetamide is spun into a coagulation bath composed of solvent and water and the fibers are then washed, stretched, and dried. It is common practice to add small amounts of comonomers of the vinyl type to improve solubility for spinning, to enhance dye receptivity, and so on. It is therefore not surprising to find that most of the fire-retardant approaches use this comonomer technique. (Contrast this with PVC, for example.)

Polyacrylonitrile, on heating at temperatures up to 280° or so, tends to form what could be called a ladder polymer by internal cyclization [15, 141a]:

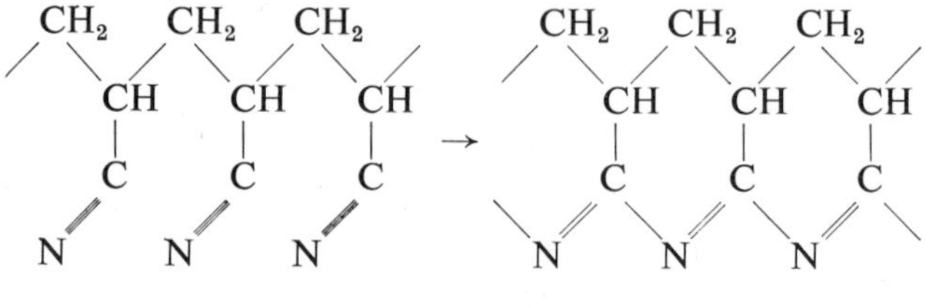

On strong heating, even to 800°, in vacuum only about 10–12% of the poly-mer is volatilized. Of this fraction some 2.9 wt % (or 0.29% of the total polymer weight) is HCN, 5.2% is acrylonitrile monomer, and 3.7% is vinylacetonitrile [15]. (Of course, in air the volatiles would burn to nitrogen oxides, CO_2, and H_2O.) There is little unzipping to monomer in the thermal degradation.

Self-extinguishing acrylonitrile fabric has been produced by treating the surface with phytic acid and urea and heating 8 min at 138° [142]. Phytic acid is not commercially available in quantity but the idea is interesting:

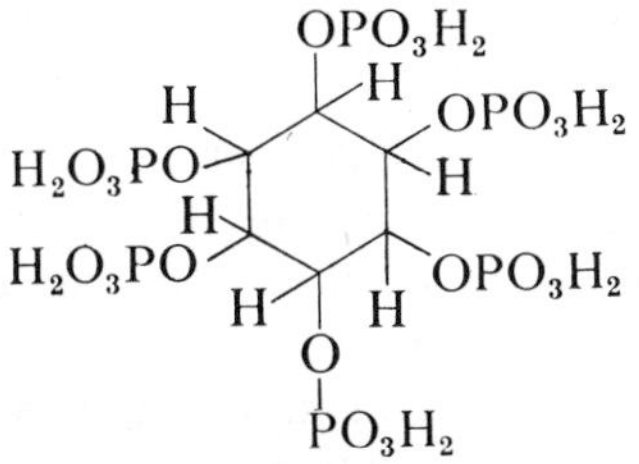

Phytic acid

Improved resistance to flame is claimed for fibers containing trialkyl phosphates which also serve to plasticize the fiber and improve its handling characteristics [143]. As little as 5% of trihexyl phosphate is claimed to be effective. This is the additive approach. A Russian group has suggested phosphonylation of an acrylonitrile modification containing aldehyde groups as follows:

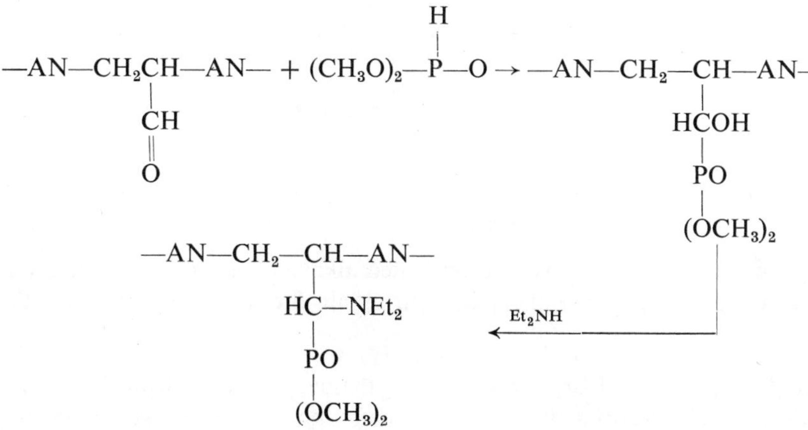

The above reactions are carried out as one step in xylene. A flame-resistant fiber containing 11% CHO groups yielded a phosphorus level of about 9% after boiling in water. (Without the Et_2N treatment the phosphorus moiety was removed by boiling.) The tensile strength was reduced by about 14% and the elongation increased by about 75% over the control fiber [144]. Another way of introducing phosphorus is:

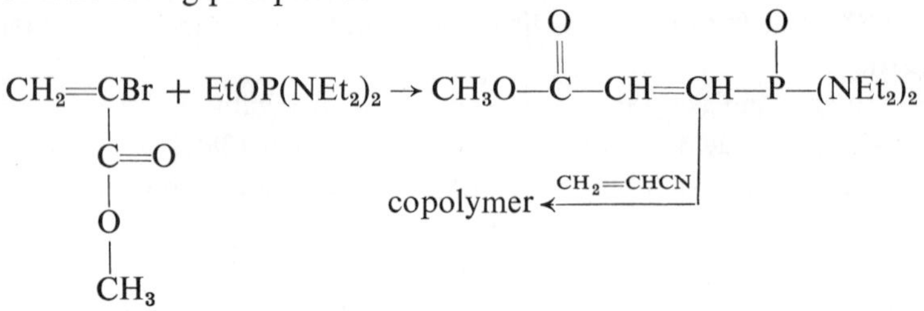

$$
\begin{array}{c}
CH_2{=}CBr \\
\;\;\;| \\
\;\;C{=}O \\
\;\;\;| \\
\;\;\;O \\
\;\;\;| \\
\;CH_3
\end{array}
+ EtOP(NEt_2)_2 \rightarrow
\begin{array}{cc}
O & O \\
\| & | \\
CH_3O{-}C{-}CH{=}CH{-}P{-}(NEt_2)_2 \\
\end{array}
$$

$$\text{copolymer} \xleftarrow{\;CH_2{=}CHCN\;}$$

The phosphonate is used at a level of 20% as a comonomer with acrylonitrile. The result is stated to be white, lustrous, flame-resistant fibers containing about 2.5% phosphorus [145]. Another technique uses a variation of the brominated allyl phosphate concept discussed for cellulosics in Chapter 5. Triallyl phosphate is reacted with CBr_4 to produce an adduct containing 72% bromine and useful especially for adding to acrylonitrile as a flame retarder [146]. α- or β-dialkylphosphonatostyrene may be copolymerized with AN to give fibers of decreased flammability. In one instance 7 parts of AN were copolymerized with 3 parts of the phosphonate at 35° to produce a polymer softening at 195° [146a].

Table 7-17

% $Ca_3(PO_4)_2$	% Phosphonate	% P	% Br	Burning Rate, sec/in.
0	0	0	0	4
15	0	2.9	0	9
0	15	0.7	10.5	17
2.5	12.5	1.0	8.75	SE
5	10	1.4	7	SE
10	5	2.2	3.5	SE

Brominated hydrocarbons or brominated alkane phosphonates have been added along with calcium phosphates to obtain fire retardance. Thus, with

$$
\begin{array}{c}
\;\;\;\;\;\;O \\
\;\;\;\;\;\;\| \\
(Br_2C_3H_5O)_2{-}P{-}C_3H_5Br_2
\end{array}
$$
and $Ca_3(PO_4)_2$ the data shown in Table 7-17 were obtained [147]. The difficulty here lies in handling a two-phase system in the spinning operation: this has serious limitations.

Halogens by themselves appear now to be the fire retardants of choice where reduced flammability but not necessarily nonburning characteristics are required. Chlorine can readily be introduced from vinyl chloride or vinylidene chloride as comonomers [148]. This was suggested in 1960, for example, when an 80:20 copolymer of acrylonitrile and one of these two chlorides was used in blends to produce carpets that burned only slowly and in dense piles were self-extinguishing. A flameproof fiber has been prepared from the following comonomers [149]:

Acrylonitrile	73
Vinylidene chloride	15
Vinyl acetate	6
Acrylamide	6

The tensile strength of these fibers was 1.6 g/denier and the elongation was 28%.

Bromine is, in all cases we have reviewed in this book thus far, consistently more effective than chlorine. So it is not surprising to see in recent references the use of vinyl bromide as a comonomer for acrylic fibers. Self-extinguishing fibers have been made as follows [150]:

	A	B	C	D
Acrylonitrile	60	63.9	66.9	73.1
Vinylidene chloride	40	32.5	23.8	15.7
Vinyl bromide	$\cdots$	3.6	9.3	11.2

The improved performance of bromine over chlorine is evident. Also, the fibers containing vinyl bromide have less shrinkage (ca. 50%) than the copolymer of acrylonitrile and vinylidene chloride. Other ways of incorporating bromine include brominated acrylates [151] which give self-extinguishing products at 15% Br. Use of a haloacrylonitrile has also been suggested [152]. For example, flame resistance results from 10% chlorine obtained by addition of 29% α-chloromethylacrylonitrile and from just 2% bromine as α-bromomethylacrylonitrile. No flammability numbers are presented; rather the fiber is said "not (to) propagate flames."

Mixtures of antimony oxide and halogens are effective. Flame-resistant fibers can be spun from a warm (90°) dimethylacetamide solution in which has been dispersed 2% Sb_4O_6 and 7–8% chlorine derived from, for example, chlorinated methyl stearate [153]. Perhaps a little less bromine is required than chlorine. When 2% Sb_4O_6 was dispersed in the above solvent and 6–7% bromine added as a brominated dialkylamide, self-extinguishing fibers were spun [154]. Bromine has also been added as hexabromobenzene along with Sb_4O_6 and phosphorus as aromatic phosphines or phosphine oxides or sulfides [155]. A comparison between chlorine and bromine is given in Table 7-18. The halogen is added in a polymeric form of the epihalohydrin to achieve greater resistance to extraction of the additive in the coagulating bath [156]. Whereas with 6.5% Sb_4O_6 3.8–4.0% bromine is sufficient to produce self-extinguishing fibers, this amount of chlorine is not.

Table 7-18 Chlorine and Bromine with Sb_4O_6 in Acrylonitrile Fibers [156]

Polyhydrin	Sb_4O_6, wt %[a]	Epihalo-hydrin, wt %	Halogen, wt %	Burning Rate, sec/1 in.
	0	0	0	4
H⎡OCH—CH₂⎤OH	9	0	0	6.5
│	0	20	11.7	7.7
CH₂Br	10.9	2.2	1.3	10
	4.5	4.5	2.6	13
	2.3	6.8	4.0	21
	6.5	6.5	3.8	SE
	4.3	18.7	5.1	SE
	2.2	10.9	6.4	SE
	0	20	7.7	7.7
H⎡OCH—CH₂⎤OH	4.3	8.7	3.3	7.8
│	6.5	6.5	2.5	8.5
CH₂Cl	6.2	10.4	4.0	8.5

[a] Percentages based on final polymer weight. Original reference gives percentages based on acrylic polymer only.

There are some other methods. One such claims nitrogen compounds such as hydroxylamine salts with or without urea or a melamine resin [157]. Another cites ammonium polysulfide [158]. Two others suggest fluoroboric acid and its salts. The fluoroboric acid solution serves as a solvent for the polyacrylonitrile and fibers spun from such a solvent are impregnated with boron and fluorine and have increased flame resistance—probably because of the boron present, not the fluorine [159].

In sum and by interpolating between the references, it appears that polyacrylonitrile fibers can be made flame-resistant by any of the following:

ca. 5% P
1–2% P + 10–12% Cl
1–2% P + 5–10% Br
10–12% Br
10–15% Cl
2% Sb_4O_6 + 8% Cl
2% Sb_4O_6 + 6% Br

The problems with spinning dispersions containing Sb_4O_6 and its increasing cost may minimize its use. The most attractive combination would appear to be 1% phosphorus and 5–10% bromine from the standpoint of the least amount to be added and thus perhaps the least effect on certain key properties. However, the ease of handling a vinyl comonomer and the still high cost of comonomers like allyl phosphorus derivatives suggests that for the present the vinyl halogen compounds will be the materials of choice

Carbonized Acrylonitrile Fibers

It was pointed out on p. 312 that strong heating in vacuum, even to 800°, produced but little in the way of volatiles and that most of the polymer appeared simply to cyclize to a stable ladderlike product. In recent years this phenomenon has been carefully studied for the purpose of producing useful fibers that would be inherently fire-resistant because all the flammable volatiles would have been removed in processing. The uses will be as nearly fireproof items for aircraft interiors and the like and as fiber reinforcements for high-performance polymers such as compressor blades for jet engines [160]. Various conditions have been proposed for the treatment: stepwise heating to 400–475° [161], heating in air to temperatures of 250–275° for 5 min [162], to 345° for up to 30 min, and so on. The time/temperature relations are critical to success. The fiber may then be treated to alter further its surface or to add carding and weaving aids [163]. The product retains some of the original fiber qualities because the carbonizing is a surface phenomenon and many of the important properties are only reduced, not removed. Thus, a tensile strength of 1 g/denier and an elongation of 13.9% [162] can be produced. These are, of course, substantially less than the original fiber but are still reasonable values. Fabrics made from these fibers resist burner flames for short periods very well and are but slowly consumed at high temperatures. Some 3 hr are required at 900° to oxidize one such product [164].

Miyamichi and his colleagues [165] believe that the heating produces the following changes:

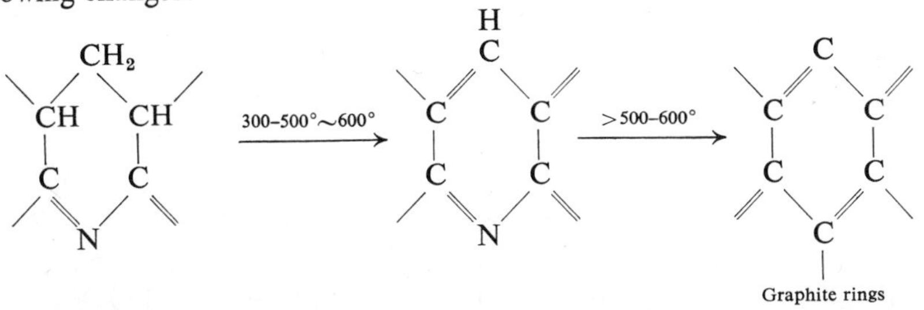

Graphite rings

These workers have studied the carbonizing process by X-rays, electron spin resonance, electric resistance, and many kinds of physical measurements. Their papers should be consulted for further details.

STYRENE

Polystyrene is used in molded shapes and sheets and in foams, but not in fibers. The polymer has long been a subject of research on flammability, especially the foams. Polystyrene ignites easily and burns horizontally at 0.5–2.5 in./min. The products of combustion are primarily CO and CO_2 [2].

Table 7-19 Halogen-Phosphorus Compositions for Polystyrene

Fire Retardant	Amount, wt %	P, wt %	X, wt %	Flame Rating	Reference
Chlorine compositions					
$(ClCH_2CH_2O)_3PO$	13	1.4	4.9	"Flame-resistant"	167
$(Cl_2C_3H_5O)_3PO$	10	0.7	4.9	"Nonflammable"	168
Chlorocarbon (66 % Cl)} $(CH_3\phi O)_3PO$}	31}8}	0.6	11	"Flame-resistant"	169
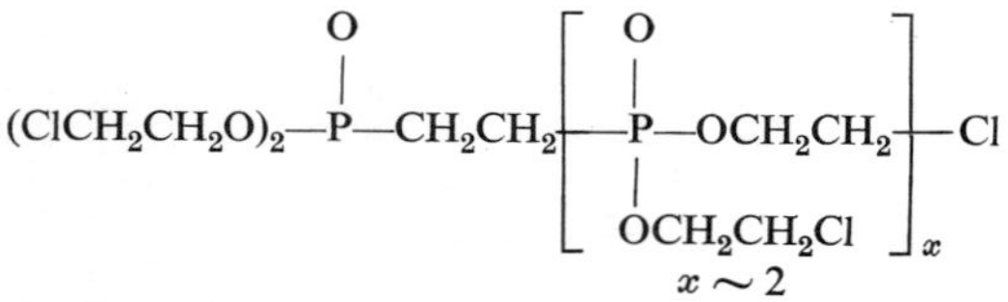 $(ClCH_2CH_2O)_2$—P(=O)—CH_2CH_2—[P(=O)—OCH_2CH_2—Cl, OCH_2CH_2Cl]$_x$ $x\sim2$	5 (F)[a]	0.75	1.5	"Fire-resistant"	170
$Cl_2\phi CH\!=\!CH_2$} $(Br_2C_3H_5O)_3PO$}	25}2}(F)	0.1	10% C. 1.4% Br	"Flame-resistant"	171
Bromophosphates					
$(Br_2C_3H_5)_3PO$	5	0.22	3.4	NB	172
$(Br_2C_3H_5O)_3PO$	2 (F)	0.1	1.4	Coated with chlorine compound containing arsenic, then foamed; foam is flame-retardant	173
$(Br_2C_3H_5O)_3PO$	4	0.2	2.8	SE by ASTM D-536	174
$(Br_2C_3H_5O)_3PO$	3	0.15	2.1	"Burns with difficulty"	175
$(Br_2C_3H_5O)_3PO$	6 (F)	0.3	4.2	"Flame-resistant"	175

Compound					Ref.
$(Br_2C_3H_5O)_3PO$	3–5 (F)	0.03–0.4	0.5–6.0	SE (0.5–3 sec)	175
$(Br_2C_3H_5O)_3PO$	5	0.22	3.4	SE (0–1 sec)	176
$(Br_2C_3H_5O)_3PO$	4 (F)	0.18	2.8	"Fire-retardant foam"	177
$CH_2{=}CHCH_2O{-}\overset{O}{P}{-}(OC_3H_5Br_2)_2$	5 (F)	0.28	2.9	SE	178
Bromophosphonates					
$Br_2C_3H_5{-}\overset{O}{P}{-}(OC_3H_5Br_2)_2$	2	0.1	1.4	NB	179
$BrClC_3H_5{-}\overset{O}{P}{-}(OC_3H_5BrCl)_2$	2	0.1	0.8 Br / 0.4 Cl	NB	179
$BrCH_2CH_2{-}\overset{O}{P}{-}(OCH_2CH_2Br)_2$	4	0.2	2.8	SE	179
Phosphine oxides and phosphines					
Chloroparaffin (70% Cl) $\}$ ϕ_3PO	7.5 0.1	0.01	5.2 Cl	SE < 2 sec	180
Bromoparaffin (75% Br) $\}$ ϕ_3PO	3 0.1	0.01	2.2 Br	"Fireproof"	181
$CH_2{=}CHBr$ $\}$ ϕ_3P	1.5 1	0.12	1.2	SE	182
$(BrCH_2)_3PO$	2	0.2	1.5	NB	182a

[a] F = foamed polymer

On pyrolysis in a vacuum or in an inert atmosphere a large fraction volatilizes—40–75% or so, depending on the temperature—and of this fraction much is styrene monomer—40–50% or 16–35% of the total polymer weight. The monomer may form through unzipping [15]:

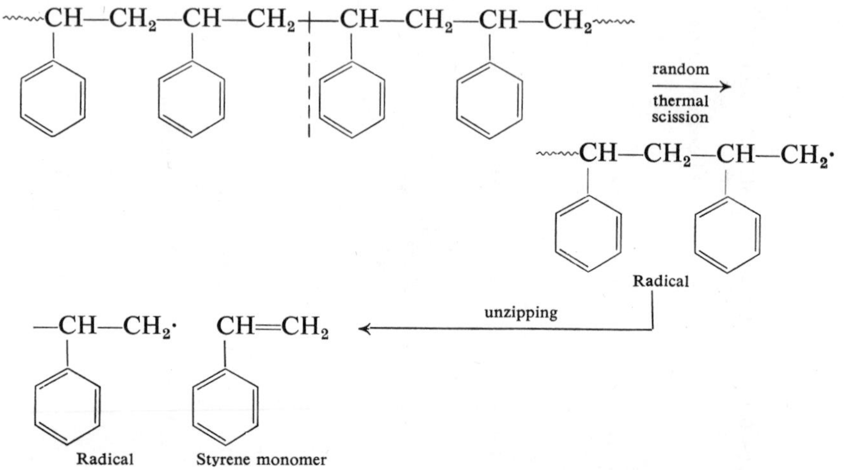

As noted earlier, when the α-carbon is fully substituted, as in poly(α-methylstyrene), unzipping to monomer takes place almost exclusively. Some 95% of this polymer reverts to starting monomer on pyrolysis [2]:

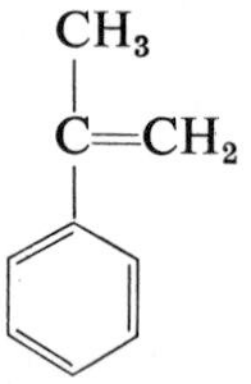

α-Methylstyrene

  Phosphorus by itself has been but little studied as a flame retardant for styrene polymers. There are some good papers dealing with the incorporation of phosphorus in styrene [166] (see Table 2-6) but there are few flammability data on such compounds.

Phosphorus-Halogen Compositions

  Most of the literature on phosphorus-halogen compositions deals with halogenated triesters of phosphoric acid, of phosphorous acid, and the phosphonates. Bromine is the preferred halogen and the following compound is most often used:

$$[BrCH_2CH(Br)CH_2O]_3PO$$

Tris(2,3-dibromopropyl) phosphate

Table 7-19 reviews some of the work in this area. The entries indicate that no more retardant is required for foams than for more dense products. Although foams are the most serious of fire hazards because of their large surface area per unit of weight, the results in the table suggest that the halogen-phosphorus combination is not at all thwarted by this. This may be because the halogen acts in the vapor phase to a large extent and the total surface area of polymer (very high in a foam) exposed has little to do with the volume of vapor space in which combustion of volatiles takes place. The polychlorophosphonate is said to have less deleterious effects on mechanical properties than the monomeric compounds. Other bromoalkyl phosphates have been suggested. For example, tris(brominated) pentaerythritol phosphate with tritolyl phosphate is said to produce a self-extinguishing polymer [183].

Table 7-20 shows another aspect: that of flaming drippings. In this study

Table 7-20   Effect of Fibers on the Flaming Drip Problem in Polystyrene, ASTM D-635 (Modified) (from Gouinlock et al. [176])

Tris(2,3-dibromopropyl) phosphate, wt %	P, wt %	Br, wt %	Glass Fiber, wt %	Drip	Flame Test
0	0	0	0	Yes	0.77 in./min
0	0	0	9.1	No	0.77 in./min
5, 10	0.22, 0.44	3.4, 6.8	0	Yes	SE 0–1 sec
4.5	0.2	3.1	9.1	No	Burning
9.1	0.4	6.2	9.1	No	SE ≥ 118 sec

glass fibers were evaluated as a means of eliminating the problem. The presence of the fibers keeps the fuel at the flame and therefore makes the test more stringent. The addition of 0.5–1.5% phosphorus gives a flame-resistant product when 5% chlorine is present. A lesser amount of phosphorus— 0.1–0.3% by weight—reduces the bromine requirement to 2–4% in phosphate additives. Bromophosphonates appear to be slightly more effective than the analogous phosphates. There is a suggestion at the bottom of Table 7-19 that phosphorus present as a phosphine or phosphine oxide may act synergistically with halogens at only 0.01% phosphorus by weight. This phenomenon must be further explored before we can be sure of its significance. (The phosphines remain expensive at this writing and will find commercial use only at substantially reduced prices.)

Halogen Compositions

Whereas almost all of the phosphorus and phosphorus-halogen compounds are inert additives, many of the halogenated flame retardants are reactive and become part of the finished polymer. The materials may be classified as inert or reactive, bromine or chlorine, cyclic aromatic and aliphatic, and acyclic.

The most direct method of adding halogens in ring compounds is by halogenation of the already polymerized styrene itself. Flame-resistant and self-extinguishing polystyrene results from chlorination to 12–15% Cl using chlorine gas [184] or polymerization in the presence of C_2Cl_6 [185] or Cl_3CCHO [186] in such a way that the chlorine is incorporated in the polymer. A self-extinguishing polystyrene can be produced with only 4–5% bromine, in contrast to chlorine. Bromine, as Br_2, can be reacted with the polystyrene after molding or foaming [187] or right in the polymerization step [188]. Polystyrene can even be brominated by admixing a compound like bromoethane and subjecting the finished product to ionizing radiation [189]. Bromine is reported effective in concentrations as low as 1% by weight, but the 4–7% range seems the most common.

Ammonium bromide in a very finely divided state has been used with some success. From 4 to 7% bromine produces self-extinguishing and non-burning foams and shapes [189]. The nitrogen may assist somewhat with the retardant action. Ammonium bromide should be an economical fire retardant but it presumably has some disadvantages, perhaps in its sensitivity to moisture (the solubility of NH_4Br in H_2O at 0° is 59.8 g/100 g).

Halogenated nonaromatic cyclics have received relatively little attention. A reaction product of styrene or divinylbenzene and "hex" has been suggested as a reactive comonomer [190]. Chlorinated phenoxyethanes have also

been considered as additive retardants [190a]. Brominated saturated rings—pentabromocyclohexane or hexabromocyclododecane, for example—have also been examined as fire retardants [191].

Halogenated styrene can be copolymerized with styrene to put the halogen in the finished polymer itself. Chlorostyrenes can be copolymerized to produce flame-resistant polymers with 10% and up of chlorine [192]. As many as five chlorine atoms may be placed on the ring in such a comonomer. Ring-brominated styrene has not been reported on. A brominated chlorostyrene is the nearest thing to it and here the bromine was added across the ethylenic double bond to produce a nonreactive retardant. This was effective at a 3%

level [193] at which it is claimed the tensile, impact, and elongation properties are unaffected. Another route to polyhalostyrenes is

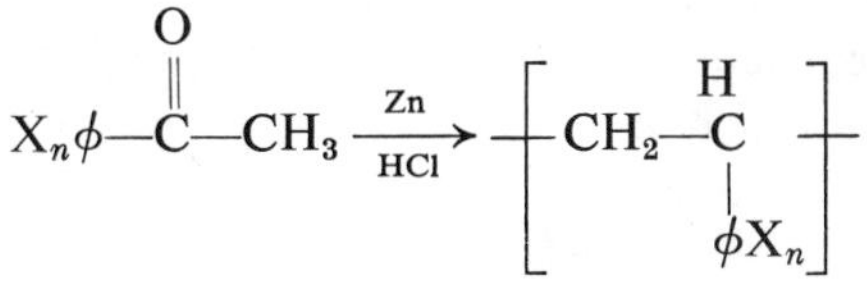

$$X_n\phi-\overset{\overset{\displaystyle O}{\|}}{C}-CH_3 \xrightarrow[\text{HCl}]{\text{Zn}} \left[CH_2-\overset{\overset{\displaystyle H}{|}}{\underset{\underset{\displaystyle \phi X_n}{|}}{C}} \right]$$

where X is chlorine or bromine [194]. This route has not been studied for copolymer possibilities as yet.

About 5% bromine added as a comonomer such as

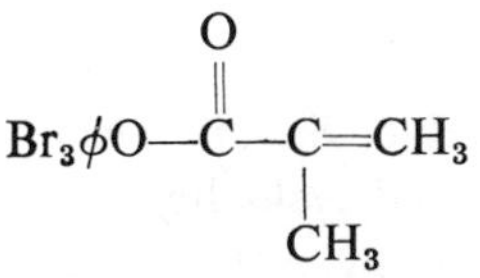

$$Br_3\phi O-\overset{\overset{\displaystyle O}{\|}}{C}-\underset{\underset{\displaystyle CH_3}{|}}{C}=CH_3$$

2,4,6-Tribromophenylmethacrylate

$$Br_2C{=}C(Br)CO_2H$$

2,3,3-Tribromoallylformate [196]

$$Br_2C{=}C(Br)CH_2OH$$

2,3,3-Tribromoallylalcohol [196a]

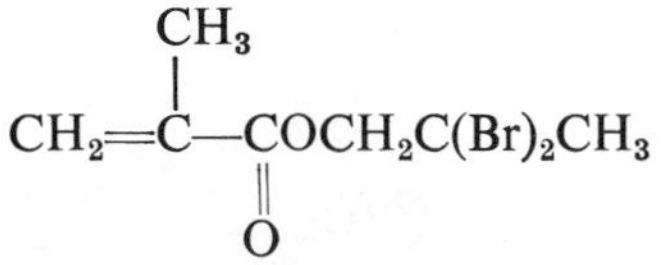

$$CH_2{=}\overset{\overset{\displaystyle CH_3}{|}}{C}-\overset{\overset{\displaystyle }{}}{\underset{\underset{\displaystyle O}{\|}}{C}}OCH_2C(Br)_2CH_3$$

2,2-Dibromopropylmethacrylate [197]

produces polymeric foams or molded products that are rapidly self-extinguishing. The type of brominated comonomer can obviously be varied a great deal and will influence the rate of polymerization, degree of polymerization, cost, and the like. The bromoallyl carboxylates are said to have little effect on physical properties. The bromopropylmethacrylate is claimed not to interfere with polymerization kinetics. And so on. The original literature should be consulted for details. Halogenated diphenyls containing both bromine and chlorine have been used as additives to polystyrene or to ABS resins. A self-extinguishing styrene polymer was produced from

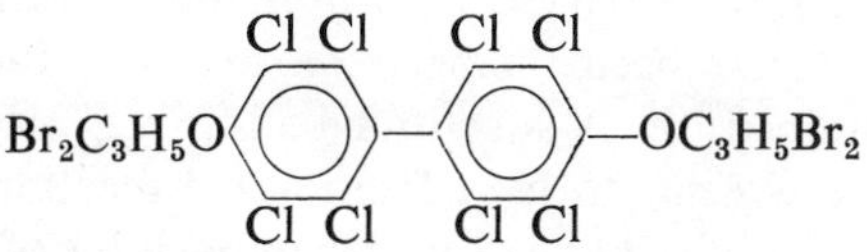

$$Br_2C_3H_5O\overset{\overset{\displaystyle Cl\ Cl}{}}{\underset{\underset{\displaystyle Cl\ Cl}{}}{\bigcirc}}-\overset{\overset{\displaystyle Cl\ Cl}{}}{\underset{\underset{\displaystyle Cl\ Cl}{}}{\bigcirc}}-OC_3H_5Br_2$$

At a 20% level the product contains 7.4% Br and 6.6% Cl and is self-extinguishing by ASTM D-635 [198].

Additive acyclic compounds are almost all bromine derivatives. These linear bromine compounds are effective at slightly lower levels of bromine than are the aromatic materials. As little as 1.5% Br has been found effective when added as dibromomethyldibromocyclohexane [199]; 4% Br is effective as tetrabromobutane [200], brominated polybutadiene [201], or polyvinyl-bromide [202] (this is of special interest as it is the only reference to poly-vinylbromide thus far encountered). Similar bromine levels are effective when added as brominated alcohols [203], esters [204], or fatty acids [205].

Substituted ethanes such as

$$\phi-\underset{\underset{CH_3}{|}}{\overset{\overset{CH_3}{|}}{C}}-\underset{\underset{CH_3}{|}}{\overset{\overset{CH_3}{|}}{C}}-\phi$$

have been used with bromine compounds to reduce the amount of bromine needed. Only 1% bromine (added as a brominated cycloaliphatic) is needed for self-extinguishing properties when 0.1% of the substituted ethane is added [206]. Similar results occur when a fully substituted polymer

$$\left[\begin{array}{c} R_1 \\ | \\ C \\ | \\ R_2 \end{array}\right]$$

is added [207]. Xylyl disulfide reduces the bromine requirement below 2% also [208]. Chelated transition metals [Co(III), Mo(III), Mn(II), Cu(II), Cr(III)] have been shown to reduce the chlorine required to flame-retard polystyrene to 4–8% and the bromine needed to 1.5–3% [209]. N-nitroso secondary amines and N-dichloroamides have an even more dramatic effect, reducing the bromine requirement to 0.4–0.5% by weight of the polymer [210]. Peroxides and hydroperoxides have much the same effect in that the bromine level is reduced to less than 1% [211] and the chlorine level (e.g., from Cl_3C—CCl_3) to 5% or so [212]. Using a brominated phosphonate, addition of 0.2% dicumyl peroxide reduced requirements to 1.6% Br and 0.15% P for a self-extinguishing foam [213].

This phenomenon has been reported on in detail for the two types of materials just mentioned: the N-substituted additives [210] and the peroxides [214, 215]. The effect seems to be a free-radical reaction in which the first step is the production of a radical on decomposition of the additive itself. This radical probably reacts with both the halogen-bearing retardant and the polystyrene. The effect will be a multiple one: the decomposition temperature will be reduced for the polystyrene, the halogen compound may release HX more rapidly [216], or the halogen may be transferred to the α or β carbon atoms in the styrene polymer. (Free-radical initiators can be used to promote halogenation of polystyrene [215].) This is discussed fully in reference 210.

Figure 7-1 shows the dramatic effect of dicumyl peroxide on the bromine requirement. All the substances cited in references 206–215 very likely decompose into free radicals at low temperatures and initiate the same kinds of reactions discussed above for dicumyl peroxide. Eichhorn [214] lists some other substances that achieve the same result.

The free-radical synergist is the latest development in the flameproofing of polystyrene, and by this means additive requirements have been so reduced that the effect on physical properties must be negligible. Whether processing can be adapted to this technique is another matter. And there has yet to appear a report on the effect of such residual sources of radicals on light stability, chemical resistance, and the like. Will flame retardance be permanent? What kinds of radical sources are preferable for long life? One normally would not build a system around an inherently unstable, fugitive source of initiation. Such questions will have to be answered before this very promising concept is commercialized widely.

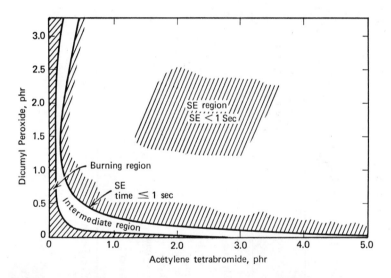

Figure 7-1   Self-extinguishing diagram for polystyrene. (After Eichhorn [212].)

In reference 216 Volans compares the effectiveness of three kinds of brominated organic fire retardants in polystyrene. Table 7-21 summarizes some of the results and shows, in general, a good correlation between the flammability measured in a burn test and pyrolysis rates measured more precisely. That is, the more effective retardants in burn tests also exhibit the lowest rates in pyrolysis. Note that the nuclear brominated styrene comonomer is the least effective (bromine held too strongly in the ring?) and vinyl

Table 7-21 Flammability and Pyrolysis Data as a Function of Type of Bromine Retardant (after Volans [216])

	Burning Lengths in Air at Various Br Levels[a]				(Pyrolysis at 2% Br Rate = K (poly)^n) Exponent (n)	Initial Rate (g/cm^2 hr $\times 10^2$)	
	0% Br	2% Br	4% Br	6% Br		260°	300°
Polystyrene	>9 cm	...	...	...	1	0.94	4.8
Br$_2$φCH=CH$_2$/styrene copolymer	...	>9 cm	8 cm	7.7 cm	1	0.87	3.5
Bis-2,3-dibromopropyl-fumarate/styrene copolymer	...	7	6.4	6	0	0.24	3.0
Brominated cyclodo-decatriene/polystyrene mixture	...	6.3	2	<1	1	0.62	3.2
Vinyl bromide/styrene copolymer	...	3.5	1.2	<1	1	0.47	2.2

[a] Burning lengths are cited by Volans from the literature

bromide is the most effective (vinyl bromide fairly unstable?). The differences pointed up at 2–6% Br seem to be rather marked in this study, more so than one infers from the literature as a whole. There seems little doubt from this work that brominated aliphatics are the materials of choice. Volans presents much useful detail concerning thermal decomposition. He also discusses the effect of adding bromine to flames in various ways; this sheds some light on the mechanism of flame retardation by this halogen.

Antimony-Halogen Compositions

Sb_4O_6 has been used with either bromine or chlorine with phosphorus in polystyrene but the results are not outstanding. At 0.5% P, 6–8% Sb_4O_6, and 7–8% Br, self-extinguishing polystyrene is obtained [217], but this can be achieved without the Sb_4O_6. Where the combinations have proved successful is in coating already formed, low-density polystyrene beads. One of the more common methods of flameproofing these beads is via chlorocarbons and Sb_4O_6. One method calls for 82.5 parts beads, 10 parts chlorocarbon (70% Cl), and 7.5 parts Sb_4O_6 [218]. The beads are coated with the flame retardants by blending and, on expanding, yield a flame-retardant foam. Another coating contains equal parts by weight of PVC and Sb_4O_6 added to the beads by slurrying in water as follows [219]:

25 parts polystyrene beads
5 parts PVC
5 parts Sb_4O_6

There are several patents in this area [218–220]. One [221] even describes immersion of finished foam in a solution of $SbCl_3$ (or $CuCl_2$), but this is an unlikely approach. In general, coating weights run from 15 to 30% of the polymer weight. This is clearly an inefficient means of obtaining fire resistance but it is simple to effect and inexpensive.

Miscellaneous

There is relatively little in this category. Sodium silicate has been used to coat foamed polystyrene [222]. A mixture of alumina, Sb_4O_6, halogens, and phosphorus is used in rubber/styrene-butadiene/polystyrene foamed rubber [223]. Even red phosphorus has been tried in a styrene-acrylonitrile copolymer [224]. Reaction products of halogenated cyclopentadiene are suggested for styrene-butadiene rubber [225]. Expandable polystyrene beads containing brominated hydrocarbons may be coated with vinylidene chloride before expanding the beads into a foam [226].

Ammonium bromide at a 4% level is a possibility for a styrene-butadiene copolymer [227]. A copolymer blend of styrene-acrylonitrile and vinylidene chloride-acrylonitrile has been formulated into a self-extinguishing foam rubber at 22% Cl [228]. Another acrylonitrile-styrene copolymer was made self-extinguishing by adding 25% PVC and 6% Sb_4O_6 [229]. A copolymer of acrylamide and chloromethyl-styrene has been treated with about 1.5% phosphorus added as tributyl phosphine [230]. Finally, a flame-resistant acrylonitrile-butadiene-styrene mixed system has been produced by blending two terpolymers containing bis-2,3-dibromopropylfumarate so that the final polymer contained 13.5% Br [231]. (In a related area a butadiene-acrylonitrile polymer has been flameproofed by treating it with an epoxidizing agent and then condensing with halogenated cyclopentadiene [232].)

Summary

We have noted that the following levels of various combinations are effective:

Phosphorus alone	no good data
Phosphorus-halogens	0.5% P + ~5% Cl, 0.2% P + ~3% Br
Halogens alone	10–15% Cl, 4–5% Br
Halogens plus free-radical initiators	0.1% initiator + 4–8% Cl, 0.1% initiator + 0.5–1.5% Br
Antimony oxide plus halogens or halogen-phosphorus	0.5% P + 6–8% Sb_4O_6 + 7–8% Br, 7.5% Sb_4O_6 + 7% Cl

It is clear that bromine is much superior to chlorine for polystyrene and that a small amount of phosphorus or of a free-radical initiator effectively reduces the bromine level. Antimony oxide does not seem to be very helpful in reducing the need for halogen. This is unlike the situation with most other polymers. The most frequently cited fire-retardant group is the bromine-phosphorus compositions: the phosphates and phosphonates. It seems to be true that bromine in an aliphatic organic molecule is considerably more effective than in an aromatic molecule. For coating expandable beads the combination of halogens and Sb_4O_6 is commonly employed at fairly high add-ons. Finally, the techniques that are effective for polystyrene may also work for many copolymers and terpolymers of styrene, such as those with acrylonitrile and butadiene.

ACRYLONITRILE-BUTADIENE-STYRENE (ABS) RESINS

The terpolymer known as ABS is actually a family of resins whose properties are governed by the ratio of the monomers. Their use in a variety of appliance housings [232a], pipe, and so forth, makes the fire-retardant property particularly important and their fate in combustion has been the subject of considerable controversy in the construction industry. Hilado [2] lists the horizontal burning rate of ABS as 1.0 to 2.0 in./min, a value which is the highest of any in the table he presents. He also indicates that ABS has a rather serious tendency to produce smoke. Madorsky [15] does not discuss the thermal degradation of ABS itself but he briefly mentions both styrene-butadiene and butadiene-acrylonitrile copolymers. For the latter, none of the products characteristic of the degradation of acrylonitrile itself—HCN, for example—are found. Evidently, these materials react very rapidly in the presence of unsaturated fragments arising from the degradation of the butadiene portion.

Oddly, there are very few references in the literature on flame-retarding ABS resins *per se*. Almost all pertinent citations are catalogued under styrene, butadiene, or acrylics rather than under the terpolymer heading. Therefore, the pertinent references are largely under one of those headings; these sections should be consulted (all in this chapter) and the references therein reviewed for application to ABS resins. (See, for example, references 138, 140, 231, and 232 of this chapter.)

ABS resins typically are processed at rather high temperatures (230–260°) and this imposes severe restrictions on the kinds of flame retardants that can be incorporated during processing. (The same difficulty is encountered with nylon and polyester fibers, discussed in Chapter 8.)

Thus, many simple phosphorus esters will decompose to some extent and a salt like NH_4Br [138] might lose a fair amount of ammonia in the process. The degree of instability required in a halogen compound for good flame retardance produces problems in this high-temperature processing. The brominated aliphatics are generally regarded as superior to brominated aromatics. However, the former may not be stable enough to process in molten ABS. At least considerable care would be required. A dimer of hexabromocyclopentadiene has been claimed in an ABS system. It was incorporated in a suspension polymerization step at a bromine level of about 3% [233]. No comment is offered, however, about how the resin is to be formed into shapes and the effect of the processing heat on the color and other properties of the resin.

Bis-2,3-dibromopropylfumarate is a promising flame retardant for ABS resins. It is added at 15–20% by weight (9–12% Br) [233a].

Chlorinated compounds are more stable and are frequently mentioned in the literature. PVC [140] at 35% by weight is effective; at 10% with 5% Sb_4O_6, the same result is obtained. Halogenated diphenyls [234] and derivatives of Diels-Alder adducts of cyclopentadiene ("hex") are effective but very large amounts (~40% Cl) are often needed [234a]. Some flame resistance can be achieved with addition of a comonomer like dibromomethacrylate [235] but stability problems at high temperature would be expected. Similarly, one would predict problems with a brominated phosphate ester [236]. Increased stability in phosphorus compounds is achieved by using lower oxidation states. The phosphinic acids, phosphine oxides, or phosphonium compounds should be more useful and have been tried:

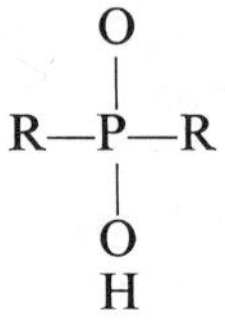

$$R{-}\underset{\underset{H}{\overset{|}{O}}}{\overset{\overset{O}{\parallel}}{P}}{-}R \qquad \text{dicyclohexylphosphinic acid} \qquad \text{Reference 237}$$

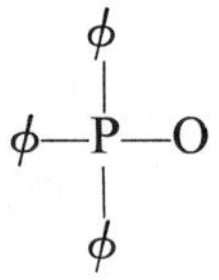

$$\phi{-}\underset{\phi}{\overset{\phi}{P}}{-}O \qquad \text{triphenylphosphine oxide} \qquad \text{Reference 180}$$

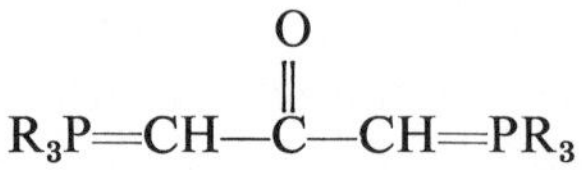

$$R_3P{=}CH{-}\overset{\overset{O}{\parallel}}{C}{-}CH{=}PR_3 \qquad \begin{array}{l}\text{1,3-bis(triphenylphosphor-}\\ \text{anylidine) 2-propanone}\end{array} \qquad \text{Reference 238}$$

However, such materials are still very expensive and therefore of little commercial interest.

Research into flame retardants for ABS materials is very active at this writing. A separate category of literature under this heading will no doubt be very sizable in five years or so. The interested researcher should carefully monitor the literature under each of the three component monomers so as not to miss important reports. He is referred to the sections on "Acrylics," "Polyolefins," and "Styrene" in this chapter.

Addition of a thermally stable polymer like a polysulfone (see p. 423) in large quantity (up to three-fourths of the total weight) makes a composite polymer that may be self-extinguishing [239]. This is a new approach.

RUBBER, ASPHALT, MISCELLANEOUS CO- AND TERPOLYMERS

Rubber

In many ways the flameproofing of rubber [240] is similar to the flameproofing of the polyolefin family (see above). The discussion of polyethylene, polypropylene, and so on should be consulted for comments on mechanisms and general requirements. There may be limitations with rubber owing to vulcanization processes and hot milling conditions, and there are special problems relating to resistance to oxidation for certain uses. These must be kept in mind when screening fire retardants. Thus, ammonium orthophosphates are added only after vulcanization. This is done by immersion and thorough impregnation of the rubber in a highly concentrated solution of, for example, $NH_4H_2PO_4$ [241]. The treatment is said to be useful for both natural rubber and for chlorine-bearing rubber (polychloroprene). A neoprene foam rubber can be made fire-resistant (passes a hot bolt test) by adding 2.8% $(NH_4)_2HPO_4$ and 14% of a urea-formaldehyde resin [242]. This was effected by immersion after the rubber was foamed. (Borates and, in one case, sulfates have been used in much the same way.) Tris(2,3-dibromopropyl) phosphate is used in natural and synthetic rubber formulations [243]. In one synthetic rubber system this phosphate is used at about an 8% level along with 6.6% Sb_4O_6 and 6.6% $Al_2O_3 \cdot 3H_2O$. The rubber also contained some chlorine. The product gave no afterflame and smoldered for less than 1 min [244]. Sb_4O_6 with bromine in rubber is effective; with chlorine, much less so. A variety of conditions in blends of natural rubber and styrene-butadiene copolymers have been reported by Hecker [244]; his paper should be consulted for details.

The halogens have been studied more than other materials for flame-retarding rubber. NH_4Br is one type of compound [245]. Ring-halogenated materials like chlorobenzene, chlorotoluene, chloronaphthalenes, and bromobenzene have been condensed with haloaliphatics to make flame retardants for rubber [246]. About 50% chlorine is needed, in the absence of other flame retardants. To obtain such high levels of chlorine is somewhat difficult. In one case octachloropropane (85–90% Cl) was used [247]. Chlorinated wax (70% Cl) has been tried as an additive in many formulas [248]. PVC and mixtures of vinyl chlorides and vinylidene chlorides have also been evaluated for making fireproof rubber [249]. Bromine and chlorine can be introduced to the rubber molecule by adding, for example, $CBrCl_3$ and either irradiating the mixture or adding free-radical initiators [250]. At 15% $CBrCl_3$ (6% Br, 8% Cl) a self-extinguishing rubber is produced. Contrast these levels with the much higher levels required when chlorine alone is present. The level can be reduced somewhat further (to 12% $CBrCl_3$) by adding 5–10% Sb_4O_6. The $CBrCl_3$ affects tensile strength adversely, however [251].

Combinations of halogens with antimony compounds have been considered for rubber as far back as 1921 when $SbCl_3$ and $Al(OH)_3$ were suggested by a British inventor [252]. At one time (1934) antimony compounds were banned from rubber flooring for the liner "Normandie" by the French government [253], but this now seems to have been an error. A combination of 3.5% Cl (added as a chlorocarbon containing 70% Cl) and 10% Sb_4O_6 in a synthetic butadiene-styrene latex produced a self-extinguishing product with virtually unchanged physical properties [254]. Sb_4O_6 is helpful in lattices made from vinyl chloride or vinylidene chloride [255] (see the section on "Vinyls"). With the chloroparaffin (70% Cl) Sb_4O_6 is used or has been claimed for a variety of rubber products [256]. For example, 4% Sb_4O_6, 4% $Al(OH)_3$, and

2.8% Cl (from the chloroparaffin) produces a foamed ethylene-propylene-dicyclopentadiene terpolymer rubber with reduced flammability [257]. A halophosphate ester may also be included. Thus the following rubber is claimed to be flameproof [258]:

Styrene-butadiene copolymer	100	
ZnO	5	
Carbon black	62.5	4.6% Cl
Chloroparaffin (70% Cl)	10	4.3% Sb_4O_6
Sb_4O_6	10	0.5% P
$Al(OH)_3$	35	
$(ClCH_2CH_2O)_3PO$	10	
	232.5	

Polybutadiene rubber may be flame retarded with hexachlorocyclopentadiene [258a]. The presence of $Al(OH)_3$ seems to help reduce flammability when chlorine is also present [259]. The mechanism of this behavior is not clear but probably $AlCl_3$ is formed, which then participates as a catalyst in the solid phase and as a source of halogen radicals in the flame space. ($AlCl_3$ sublimes at 178°.) One wonders why $Al(OH)_3$ has not received more attention as a substitute for the much more expensive Sb_4O_6 in flame-ratardant applications.

Silica gel is also said to reduce the flammability of rubber, though there are few references to this use [260].

Asphalt

There are many uses of asphalt where reduction in flammability is needed despite the fact that asphalt is not a remarkably flammable material. In roofing, in paper laminations in insulations, and in certain coating and mastic uses a reduced burning tendency is necessary. An intumescing asphaltic product has been prepared by phosphating or sulfating acid petroleum sludges [261]. The phosphated sludge is added at 1–10% of the bituminous product. Asphalt emulsions containing up to 3% of $NH_4H_2PO_4$ or $(NH_4)_2HPO_4$ are said to be fire-resistant on drying [262]. Phosphorus pentasulfide has been tried at a 9% level with additional sulfur at 8% to produce a fire-resistant asphalt mastic for steel and wood [263].

Halogens such as chloronaphthalene [264] have long been known to be effective, but at high levels of addition. Thus, the early citations call for 30% of a chlorinated naphthalene. Chlorinated paraffins [265] (range: 3–20%) have also been suggested. Combinations of halogen compounds and Sb_4O_6 are discussed in a series of references. Chloroparaffin, Sb_4O_6, and $Ca(OH)_2$

were claimed for a variety of end uses in asphalt where the sum of the three fire-retardant components added up to at least 25 % of the asphalt used [266]. Some 7.8 % Cl from a chloroparaffin and 5 % Sb_4O_6 produces fire-resistant asphalt [267]. The chlorine may be added as a chloroparaffin, as a reaction product of chlorocyclopentadiene [268] (e.g., "hex"), or as PVC or other chlorinated vinyl polymers [269]. With all these sources of chlorine, the requirements of antimony and chlorine are similar, that is, 5–10 % of Sb_4O_6 and 5–10 % of Cl. The key point in many applications is cost. Asphalt is an economical polymer but that advantage is lost when expensive additives are used.

Miscellaneous

Gasoline or other hydrocarbon fuels can be rendered safer in storage by means of certain additives. In one patent CCl_3F or $C_2Cl_3F_3$ was added to gasoline to make it fire-resistant in storage [270]. The additive was removed by distillation prior to use. A variety of inorganic salts have been combined to add to gasoline to convert it to a nonflammable liquid [271]. This, of course, makes it useless as a fuel but it retains certain solvent properties of interest. Jet fuel (kerosene) can be made much safer by adding gelling agents [272]. It is possible to adjust the flow properties of a fuel so that under the shear in a pump it becomes free-flowing and can be injected into the combustion chamber. But under less severe stress the fuel becomes a semisolid with little tendency to spread should a tank or line be ruptured. The agent can also reduce vapor pressure dramatically and hence eliminate a serious hazard. Such research studies along with honeycomb-type tanks made of resilient nonflammable polymers will produce much safer aircraft and automotive fuel systems in the immediate future.

Many functional fluids must be fire-resistant, for example, hydraulic fluids for airplanes [273]. Such fluids are outside the scope of this book since they are rendered flameproof by their very nature rather than via additive or reactive fire-retardant chemicals. Many of these fluids are chlorinated aromatics such as the polychloropolyphenyls. Others are phosphate esters [273a] such as tritolyl phosphate. Some brominated aromatic ethers have been mentioned [274]. There are also halogenated aliphatics and esters of various types [275]. Fluorocarbons hold promise because of their inherent stability. Water-base fluids [276] are mentioned frequently in the recent literature because of their potential low cost as well as their fire-safe properties. Functional fluids must possess a series of properties to make them useful, and fire safety is but one of these. Hatton has written a very helpful and valuable book on the subject of functional fluids [277], which the reader is urged to consult.

REFERENCES

[1] H. Vogel, *Flammfestmachen von Kunststoffen*, Huthig Verlag, Heidelberg, 1966, 188 pp.
[2] C. J. Hilado, *Flammability Handbook for Plastics*, Technomic, Stamford, Conn., 1969, 164 pp.
[3] S. Hashimoto and I. Furukawa, *Kyoko Pura such-iikusu*, **4**, 313 (1958); S. Hashimoto, *Kagaku*, **13**, 227 (1958); S. Takagi, Y. Hatogai, H. Hirano, and T. Iwai, *Plast. Age (Osaka)*, **13**, 73 (1967); S. Hoshino, *ibid.*, 35.
[4] P. Klimke, *Kunststoffe*, **56**, 554 (1966).
[5] G. Hazkoto, I. Szalontay, and I. Szondy, *Muanyag Gumi*, **3**, 359 (1966) (in Hungarian).
[6] N. E. Boyer, *Plast. Technol.*, **8**, 33 (1962); J. K. Jacques, *Plast. Inst., Trans. J. Conf. Suppl. No. 2*, 33 (1967).
[7] P. W. Sherwood, *Kunststoffe*, **3**, 281 (1963).
[8] A. L. Baseman, *Plast. Technol.*, **12**, 37 (1966).
[9] D. V. Rosato, *Plast. World*, 28 (February, 1967).
[10] *Mod. Plast.*, 82 (June, 1968).
[10a] M. Mandell, *Electronic Prod.*, **1969**, 18 (February).
[11] W. G. Schmidt, *Plast. Inst. (London), Trans. J.*, **33**, 247 (1965).
[12] J. E. Hauck, *Mater. Design Eng.*, **60**, 83 (1964).
[13] J. E. Hauck, *Mater. Eng.*, **66**, 80 (1967).
[14] J. Schmidt, *Plast. Kautschuk*, **13**, 83 (1966); R. Reichherzer, *Kunstst. Rundsch.*, **13**, 482 (1966).
[14a] A very recent article on the burning of polymers is S. J. Burge, C. F. H. Tipper, Combustion and Flame, **13**, 495 (1969).
[15] S. L. Madorsky, *Thermal Degradation of Organic Polymers*, Wiley-Interscience, New York, 1964, 315 pp.
[15a] Y. Tsuchiya and K. Sumi, *Polym. Lett.*, **6**, 357 (1968); Z. Osawa, T. Shibamiya, and K. Matsuzaki, *Polym. Rept. No. 121*, p. 23 (1968); R. McGuchan and I. C. McNeill, *Eur. Polym. J.*, **4**, 115 (1968).
[16] K. W. Rockey, *Plastics (London)*, 26 (1961).
[16a] A. E. Sherr, H. C. Gillham, H. G. Klein, *Advances in Chemistry Series 85*, Amer. Chem. Soc., Wash., D.C., 1968, pp. 307–326.
[17] C. E. Miles, H. L. Vandersall, and J. W. Lyons, Div. Org. Coatings Plast. Chem., Am. Chem. Soc., Preprints, *28* (1) 237 (1968).
[18] E. V. Kuznetsov and R. S. Devitaeva, *Tr. Kazansk. Khim.-Tekhnol. Inst.*, **30**, 63 (1962).
[19] D. Bellus, Z. Manasek, and M. Lazar, *Vysokomol. Soedin*, **5**, 145 (1963).
[19a] G. J. Listner, *U.S. 3,403,118* (to Johnson & Johnson), Sept. 24, 1968.
[20] B. S. Taylor and M. R. Lutz, *U.S. 3,412,052* (to FMC Corp.), Nov. 19, 1968.
[21] W. L. Mosby, H. C. Gillham, and A. E. Sherr, *U.S. 3,322,860* (to American Cyanamid Co.), May 30, 1967.
[22] J. F. Cannelongo, *U.S. 3,422,047* (to American Cyanamid Co.), Jan. 14, 1969.
[23] I.C.I., Ltd., *Belg. 685,769*, Aug. 19, 1966.
[24] C. Matthews, *Proc. Roy. Aust. Chem. Chem. Inst.*, **34**, 186 (1967).
[25] Dow Chemical Co. (by F. D. Hoerger and H. W. Smeal), *Ger. 1,061,074*, July 9, 1959; Badische Anilin- & Soda-Fabrik A.-G., *Belg. 719,943; 719,944; 719,945; 719,946*, Feb. 26, 1969; *Belg. 720,011; 720,012;* Feb. 27, 1969; *Belg. 720,902*, Mar. 17, 1969; *Belg. 732, 369*, Apr. 30, 1969.
[26] H. M. Anderson, J. D. Gabbert and O. D. S. Deex, *U.S. 3,403,137* (to Monsanto Co.), Sept. 24, 1968.

[27] Dow Chemical Co., *Belg. 684,962*, Aug. 2, 1966.
[28] E. I. duPont de Nemours & Co. (by J. L. Nyce and Roland S.-Y. Ro), *Belg. 622,213*, Mar. 6, 1963; G. E. Waples, Jr., *U.S. 3,316,329* (to Dow Chemical Co.), Apr. 25, 1967.
[29] Allied Chemical Corp., *Brit. 1,024,400*, Mar. 30, 1966.
[30] A. Gumboldt and E. Heitzer, *U.S. 2,913,424* (to Farb. Hoechst A.-G. vorm. Meister Lucius Brüning), Nov. 17, 1959.
[31] J. A. Herbig and I. O. Salyer, *U.S. 3,158,665* (to Monsanto Co.), Nov. 24, 1964.
[31a] Farb. Hoechst A.-G., *Ger. 1,104,693*, Apr. 13, 1961; *Belg. 633,433*, Dec. 10, 1963.
[32] Berk, Ltd. (by D. H. Derbyshire and B. J. Riley), *Brit. 1,068,040*, May 10, 1967.
[33] Farb. Hoechst A.-G. (by E. Grams, H. J. Lenz, and H. Herzberg), *Ger. 1,048,409*, Jan. 8, 1959; W. G. Nelson, IV. *U.S. 3,121,067* (to E. I. DuPont de Nemours & Co.), Feb. 11, 1964.
[34] Anaconda Wire and Cable Co., *Brit. 883,806*, Dec. 6, 1961.
[35] G. B. Feild, *U.S. 2,962,464* (to Hercules Powder Co.), Nov. 29, 1960.
[36] W. G. Nelson, IV, *U.S. 3,121,067* (to E. I. DuPont de Nemours & Co.), Feb. 11, 1964.
[37] R. C. Danison, *U.S. 2,588,362* (to Diamond Alkali Co.), Mar. 11, 1952.
[38] Badische Anilin- & Soda-Fabrik A.-G., *Neth. Appl. 6,413,754*, May 31, 1965.
[39] W. B. Happoldt, Jr., *U.S. 2,480,298* (to E. I. DuPont de Nemours & Co.), Aug. 30, 1949.
[40] J. A. Snyder, *Brit. 845,099* (to Union Carbide Corp.), Aug. 17, 1960.
[41] B. A. Blewis and A. H. Steinberg, *U.S. 3,066,787* (to General Cable Corp.), Appl. Sept. 12, 1958.
[41a] M & T Chemicals, Inc. (by J. R. Leebrick and J. J. Cassidy), *Belg. 627,622*, Jan. 25, 1963.
[42] Abegg & Co. A.-G. Zuerich (by André Schaerer), *Swiss 353,171*, Appl. Apr. 25, 1957.
[43] Hercules Powder Co., *Brit. 991,159*, May 5, 1965.
[44] Farb. Hoechst A.-G., *Belg. 633,433*, Dec. 10, 1963.
[45] Farb. Hoechst A.-G. (by G. Peters, D. Schleede, H. Jochinke, and H. Klug), *Ger. 1,104,693*, Apr. 13, 1961; *Belg. 633,433*, Dec. 10, 1963.
[45a] Farb. Hoechst A.-G. (by H. Peters, J. Kaupp, D. Schleede, and H. Jochinke), *Ger. 1,201,544*, Sept. 23, 1965.
[46] Chemische Werke Huels A.-G. (by H. Sauer, A. Schmidt, and K. Kopetz), *Ger. 1,135,653*, Aug. 30, 1962.
[47] Chemische Werke Huels A.-G. (by K. Kopetz, G. Wick, and F. Merten), *Ger. 1,103,020*, Mar. 23, 1961.
[48] Chemische Werke Huels A.-G., *Brit. 874,006*, Appl. May 24, 1960.
[49] Dynamit Nobel A.-G., *Neth. Appl. 6,613,687*, Mar. 30, 1967.
[50] Chemische Fabrik Kalk G.m.b.H., *Neth. Appl. 6,601,908.* Aug. 19, 1966; H. E. Practzel, H. Jenkner, *U.S. 3,474,067* (to Chemische Fabrik Kalk G.m.b.H.) Oct. 21, 1969.
[51] Chemische Werke Huels A.-G. *Fr. 1,399,195*, May 14, 1965.
[52] Chemische Werke Huels A.-G., *Belg. 661,717*, July 6, 1965.
[53] H. C. Rapp, *U.S. 3,239,482* (to Raychem Corp.), Mar. 8, 1966.
[54] T. H. Ling, *U.S. 3,287,312* (to Anaconda Wire and Cable Co.), Nov. 22, 1966; Rexall Drug and Chemical Co., *Belg. 669,948*, Mar. 22, 1966.
[55] Furukawa Electric Co., Ltd. (by S. Takahashi and T. Kuhara), *Jap. 4471*, Feb. 23, 1967.
[56] Hooker Chemical Corp., *Brit. 1,013,786*, Dec. 22, 1965.

[57] Hooker Chemical Corp. (by A. H. Johnson and S. M. Creighton), *Belg. 612,960*, Feb. 15, 1962.

[58] D. Mahling, H. Tamm, H. Seibt, and E. Scharf, *U.S. 3,419,518* (to Badische Anilin- & Soda-Fabrik A.-G.), Dec. 31, 1968.

[59] Badische Anilin- & Soda-Fabrik A.-G. (by D. Mahling, H. Mueller, H. Seibt, L. Zuern, and E. Scharf), *Ger. 1,235,518*, Mar. 2, 1967; *Fr. 1,360,097*, Apr. 30, 1964.

[60] H. Klug, K. Kuchinka, H. Peters, D. Schleede, and J. Winter, *U.S. 3,275,596* (to Hercules, Inc.), Sept. 27, 1966.

[61] Chemische Werke Huels A.-G., *Neth. Appl. 6,412,657*, Nov. 25, 1965.

[62] Rexall Drug and Chemical Co. (by J. B. Walheim), *Belg. 669,948*, Mar. 22, 1966.

[63] International Standard Electric Corp., *Belg. 668,540*, Feb. 21, 1966.

[64] W. E. Thompson, *U.S. 3,078,250* (to Sun Oil Co.), Feb. 19, 1963.

[65] E. A. Boettner, G. Ball, and B. Weiss, Div. Org. Coatings Plast. Chem., Am. Chem. Soc., Preprints, *28* (1) 311 (1968) T. Morikawa, *Chem. of High Polymers*, **XXV**, 505 (1968); M. V. Neiman (dec.), R. A. Papko, and V. S. Pudon, *Polym. Sci. USSR*, **10**, 975 (1968).

[65a] K. Fischer, Paper presented at the 50th Anniversary Meeting of the Swedish Fire Protection Assoc., Stockholm, Sweden, Apr. 24, 1969.

[66] C. E. Greene and F. J. Maurer, *U.S. 3,283,031* (to General Tire and Rubber Co.), Nov. 1, 1966.

[67] Deutsche Gold- und Silber-Scheideanstalt vorm. Roessler, *Fr. 1,440,823*, June 3, 1966.

[68] R. L. McConnell and H. W. Coover, Jr., *U.S. 3,062,792* (to Eastman Kodak Co.), Nov. 6, 1962.

[69] M. Baer, *U.S. 2,750,351* (to Monsanto Chemical Co.), June 12, 1956.

[70] Japan Chemical Industrial Co., Ltd. (by F. Yamagudi and K. Imamura), *Jap. 9006*, May 1, 1967.

[71] S. Tamlya et al., *Jap. 1395* (to Asahi Chemical Industrial Co.), Feb. 23, 1955; W. Szukiewicz, A. R. Steimle, and P. H. Rhodes, *U.S. 2,898,312* (to Philip H. Rhodes), Aug. 4, 1959.

[72] J. H. Dunn and P. E. Weimer, *U.S. 2,773,046* (to Ethyl Corp.), Dec. 4, 1956; V. A. Voskresenskia, *USSR 110,687*, June 25, 1958.

[73] Dux Chemical Solutions Co., Ltd., and C. A. Redfarn, *Brit. 874,905*, Sept. 12, 1957.

[74] G. G. Zarubin, I. K. Rubstona, M. I. Smirnov, L. D. Pertson, F. F. Dolgov, V. V. Kokorev, and R. D. Zhilind, *Plast. Massy*, **1963**, 7–10.

[75] N. M. Bortnick and M. F. Fegley, *U.S. 2,826,602* (to Rohm and Haas Co.), Mar. 11, 1958.

[76] E. C. Ladd, *U.S. 2,618,899* (to United States Rubber Co.), Nov. 4, 1952; Farb. Bayer A.-G. (by R. Schmitz-Josten, N. Schoen, R. Kubens, and G. Frank), *Belg. 638,799*, Feb. 17, 1964.

[77] G. K. Storm, F. E. Lawlor, and F. Casciani, *U.S. 2,814,602* (to Hooker Electro-chemical Co.), Nov. 26, 1957.

[78] A. H. Lybeck, *U.S. 3,141,850* (to American Enka Corp.), July 21, 1964; N. J. Read, *U.S. 3,014,000* (to Associated Lead Manufacturers, Ltd.), Dec. 19, 1961; N. F. Arone, *U.S. 2,717,216* (to General Electric Co.), Sept. 6, 1955; F. W. Duggan and W. F. Hemperly (to Carbide & Carbon Chemicals Corp.), *Brit. 587,829*, May 7, 1947.

[79] Deutsche Gold- und Silber-Scheideanstalt vorm. Roessler, *Fr. 1,379,121*, Jan. 20, 1964.

[80] G. F. Rugar, *U.S. 2,590,211* (to Diamond Alkali Co.), Mar. 25, 1952.

[80a] U.S. Borax & Chemical Corp., *Belg. 720,570*, Sept. 9, 1968.

[81] P. S. Blatz, *Brit. 1,035,912* (to E. I. DuPont de Nemours & Co.), July 13, 1966.

[82] J. W. Haworth, *Brit. 675,783* (to British Oxygen Co., Ltd.), July 16, 1952.

[83] A. D. F. Toy, *U.S. 2,485,677* (to Victor Chemical Works), Oct. 25, 1949.

[84] Victor Chemical Works (by A. D. F. Toy and K. H. Rattenbury), *Ger. 1,041,251*, Oct. 16, 1958.

[85] Farb. Hoechst A.-G. (by F. Winkler and J. W. Zimmerman), *Ger. 1,116,905*, Nov. 9, 1961; Farb. Hoechst A.-G. (by F. Rochlitz, H. Vilesek, and G. Koch), *Ger. 1,135,176*, Aug. 23, 1956.

[86] Farb. Hoechst A.-G. (by F. Rochlitz and H. Vilesek), *Ger. 1,130,177*, May 24, 9962; G. M. Kosolapoff, *U.S. 2,495,108* (to Monsanto Chemical Co.), Jan, 17. 1950; A. Kuhlkamp and R. Nowack, *Can. 783,641* (to Farb. Hoechst A.-G.), Apr. 23, 1968.

[87] J. H. Johnson and J. E. Fields, *U.S. 3,068,061* (to Monsanto Chemical Co.), Dec. 11, 1962; N. F. Orlov and M. V. Androsova, *Izv. Vyssh. Ucheb. Zaved., Enst. Tekst. Leg. Prom.*, **1967**, 113 (in Russian); N. F. Orlov, L. A. Vol'f, M. V. Androsova, and Yu. K. Kirilenko, *USSR 189,515*, Nov. 30, 1966; N. M. Volgina. A. I. Meos, L. A. Vol'f, and E. E. Nifant'ev, *Zh. Prikl. Khim*, **40**, 209 (1967); K. Tanabe and Y. Ono (to Kurashiki Rayon Co.), *Jap. 10,794*, Dec. 24, 1956.

[88] J. H. Johnson and J. E. Fields, *U.S. 3,210,147* (to Monsanto Co.), Oct. 5, 1965.

[88a] L. A. Tseitlina, N. B. Vanovskaya, L. A. Vol'f, and A. I. Meos, *Khim. Volokna*, **1965** [4], 16.

[89] L. A. Vol'f, A. I. Meos, G. A. Kiselev, and M. A. Sokolovskii, *USSR 166,921*, Dec. 12, 1964; G. A. Kiselev, L. A. Vol'f, A. I. Meos. *J. Appl. Chem. USSR*, **39**, 353 (Jan. 1966).

[90] T. M. Melton, R. A. Matthews, and H. F. O'Connor, *U.S. 2,990,421* (to Virginia-Carolina Chemical Corp.), June 27, 1961.

[91] T. Tsujimura (to Osaka Dye Industries Co.), *Jap. 6650*, Oct. 16, 1954.

[91a] J. R. Jones, *U.S. 3,488,337*, (to Armstrong Cork Co.) Jan. 6, 1970.

[92] Farb. Hoechst A.-G. vorm. Meister Lucius Brüning (by H. Krämer, W. Denk, and G. Messwarb), *Ger. 1,029,156*, Apr. 30, 1958.

[93] Farb. Hoechst A.-G. (by W. Herbst, F. Rochlitz, and H. Vilesek), *Ger. 1,106,963*, May 18, 1961.

[94] D. I. Mendeleev Chemical-Technological Institute (by A. S. Tevlina, S. V. Kotylarova, B. B. Levin, and I. N. Fetin), *USSR 173,407*, July 21, 1965.

[95] C. S. Marvel and J. C. Wright. *J. Polym. Sci.*, **8**, 495 (1952).

[96] S. C. Temin, *U.S. 3,037,950* (to Koppers Co., Inc.), June 5, 1962.

[97] A. Y. Garner, *U.S. 3,010,946* (to Monsanto Chemical Co.), Nov. 28, 1961.

[98] B. S. Taylor, J. L. Thomas, and C. A. Heiberger, *U.S. 3,093,619* (to FMC Corp.), June 11, 1963.

[99] E. D. Holly and W. R. Nammy, *U.S. 2,947,730* (to Dow Chemical Co.), Aug. 2, 1960.

[100] I. J. Gruntlest and E. M. Young, Jr., *Amer. Chem. Soc., Div. Org. Coatings, Plast. Chem., Preprints 21*, No. 1, 113–24 (1962).

[101] M. C. McGaugh and S. Kottle, *J. Polym. Sci., Part A-1*, VC, 1243 (1968)

[102] Peirrefitte-Kalaa Djerda société générale d'engrais et produits chimiques (by C. T. Quesnel), *Fr 1,245,687*, Feb. 1, 1961; Rohm and Haas G.m.b.H. (by Ludwig Hosch), *Ger. 1,032,540*, June 19, 1958.

[103] Imperial Chemical Industries, Ltd., *Fr. 1,364,384*, June 19, 1964; Rohm and Haas Co. (by J. L. O'Brien), *Ger. 1,100,287*, Appl. Mar. 8, 1958.

[104] E. I. duPont de Nemours & Co. (by K. L. Howe), *Belg. 640,389*, May 25, 1964.

[105] P. R. Graham, *U.S. 2,881,147* (to Monsanto Chemical Co.), Apr. 7, 1959; J. W. Baker, *U.S. 2,957,018* (to Monsanto Chemical Co.), Oct. 18, 1960.

[105a] P. A. Blatz, *U.S. 3,449,308* (to E. I. duPont de Nemours & Co.), June 10, 1969.

[106] B. B. Levin, G. S. Kolesnikov, E. F. Rodinova, and I. N. Fetin, *USSR 176,682*, Nov. 17, 1965.

[107] J. L. O'Brien and C. A. Lane, *U.S. 2,934,555* (to Rohm and Haas Co.), Apr. 26, 1960.

[108] J. L. O'Brien and C. A. Lane, *U.S. 3,030,347* (to Rohm and Haas Co.), Apr. 17, 1962.

[109] C. A. Lane, *U.S. 2,934,554* (to Rohm and Haas Co.), Apr. 26, 1960.

[110] A. N. Pudovik, N. G. Khusainova, and E. I. Kashevarova, *Vysokomol. Soedin*, **5**, 1376 (1963).

[111] Deutsche Gold- und Silber-Scheideanstalt vorm. Roessler (by H. Bredereck, W. Ege, and M. I. Iliopulos), *Brit. 1,028,620*, May 4, 1966.

[112] G. Kamai and V. A. Kukhtin, *Zh. Obshch. Khim*, **24**, 1855 (1954).

[113] Teijin, Ltd. (by Toshio Fukado), *Jap. 24,282*, Oct. 29, 1964.

[114] J. B. Dickey and H. W. Coover, Jr., *U.S. 2,780,616* (to Eastman Kodak Co.), Feb. 5, 1957.

[115] J. B. Dickey and H. W. Coover, Jr., *U.S. 2,559,854* (to Eastman Kodak Co.), July 10, 1956.

[116] Rohm and Haas Co., *Neth. Appl. 6,603,629*, Oct. 3, 1966; Rohm and Haas Co., *Belg. 692,146*, Jan. 3, 1967.

[117] H. G. Klein and H. C. Gillham, *U.S. 3,322,716* (to American Cyanamid Co.), May 30, 1967.

[118] F. Fekete, *U.S. 2,831,838* (to Pittsburgh Plate Glass Co.), Apr. 22, 1958.

[119] American Cyanamid Co. (by H. C. Gillham and A. E. Sherr), *Fr. 1,403,292*, June 18, 1965.

[120] W. J. Bailey, *U.S. 3,334,064* (to American Cyanamid Co.), Sept. 6, 1966.

[121] F. J. Lowes, Jr., *U.S. 3,271,343* (to Dow Chemical Co.), Sept. 6, 1966.

[122] Deutsche Gold- und Silber-Scheideanstalt vorm. Roessler, *Neth. Appl. 302,773*, Aug. 25, 1964; Montecatini Società Generale per l'Industria Mineraria e Chimica (by F. Dukli, T. P. Nigra, and R. Casiraghi), *Belg. 645,883*, July 16, 1964; Imperial Chemical Industries, Ltd., *Neth. Appl. 6,410,345*, Mar. 17, 1965.

[123] Chemische Fabrik Kalk G.m.b.H., *Neth. Appl. 6,516,491*, June 20, 1966.

[124] Resart-Gesellschaft Kalkhof & Rose (by H. Schmalz), *Ger. 1,148,751*, May 16, 1963.

[125] Rohm and Haas G.m.b.H. (by W. Koehler), *Ger. 1,134,836*, Aug. 16, 1962.

[126] J. L. O'Brien, *U.S. 2,993,033* (to Rohm and Haas Co.), July 18, 1961.

[127] R. A. Cass and L. O. Raether, *Amer. Chem. Soc., Div. Org. Coatings, Plast. Chem., Preprints 23* (1), 82 (1963).

[128] L. P. Marinelli, T. M. Andrews, and J. C. Montermoso, *U.S. 3,012,018* (to U.S. Dept. of the Army), Dec. 5, 1961; B. F. Goodrich Co., *Belg. 720,954*, Mar. 3, 1969.

[129] A. Boryniec and B. Laszkiewicz, *J. Polym. Sci., Part A-1*, **1**, 1963 (1963).

[130] W. L. Mosby, H. C. Gillham, and A. E. Sherr, *U.S. 3,322,860* (to American Cyanamid Co.), May 30, 1967.

[131] Lonza Elektrizitätswerke und Chemische Fabriken A.-G., *Swiss 245,075*, June 16, 1947.

[132] Badische Anilin- & Soda-Fabrik A.-G. (by B. Vollmert, K. J. Fust, and H. Willersinn), *Ger. 1,066,743*, Oct. 8, 1959.

[133] C. W. Roberts and D. H. Haigh, *U.S. 2,908,712* (to Dow Chemical Co.), July 19, 1960.

[134] A. Armen and R. E. Gentry, Jr., *U.S. 2,945,843* (to Dow Chemical Co.), July 19, 1960.

[135] L. H. Les, *U.S. 3,219,640* (to Dow Chemical Co.), Nov. 23, 1965.

[136] United States Rubber Co., *Neth. Appl. 301,136*, Sept. 27, 1965.

[137] E. H. Hill, J. R. Caldwell, and W. J. Jackson, Jr., *U.S. 3,277,053* (to Eastman Kodak Co.), Oct. 4, 1966.

[138] E. L. McMaster, J. Eichhorn, and F. B. Nagle, *U.S. 3,058,927* (to Dow Chemical Co.), Oct. 16, 1962.

[139] F. J. Lowes, Jr., *U.S. 3,271,344* (to Dow Chemical Co.), Sept. 6, 1966.

[140] Dynamit Nobel A.-G., *Neth. Appl. 6,411,769*, Apr. 13, 1965.

[141] A. F. Balacco, *Ital. 474,135*, Sept. 3, 1952; J. N. Hay, *J. Polym. Sci., Part A-1*, **6**, 2127 (1968).

[141a] J. N. Hay, *J. Polym. Sci., Part A-1*, **6**, 2127 (1968).

[142] Monsanto Co. (by J. J. Hirschfeld), *Belg. 631,087*, Oct. 16, 1963.

[143] T. W. Tarkington and N. T. Anderson, *U.S. 2,949,432* (to Chemstrand Corp.), Aug. 16, 1960.

[144] Z. A. Rogovin, M. A. Tyuganova, G. A. Gabriclyan, and N. F. Konnova, *Khim. Volokna*, **1966**, 27–30.

[145] H. W. Coover, Jr., and N. H. Shearer, Jr., *U.S. 2,790,823* (to Eastman Kodak Co.), Apr. 30, 1957.

[146] G. Palethorpe, *U.S. 3,318,978* (to Monsanto Co.), May 9, 1967.

[146a] H. W. Coover, Jr., and J. B. Dickey, *U.S. 2,743,261* (to Eastman Kodak Co.), Apr. 24, 1956.

[147] Dow Chemical Co., *Neth. Appl. 6,509,569*, Jan. 24, 1967; F. J. Lowes, Jr., *U.S. 3,213,052* (to Dow Chemical Co.), Oct. 19, 1965; F. J. Lowes, Jr., *U.S. 3,242,124* (to Dow Chemical Co.), Mar. 22, 1966.

[148] P. H. Hobson, *U.S. 2,949,437* (to Chemstrand Corp.), Aug. 16, 1960; Monsanto Co., *Neth. Appl. 6,517,189, 6,517,190*, July 1, 1966.

[149] Monsanto Co., *Neth. Appl. 6,515,911*, June 8, 1966.

[150] Monsanto Co., *Belg. 674,638; 674,639*, Dec. 31, 1965; Monsanto Co. (by G. Palethorpe), *Belg. 674,639*, Dec. 31, 1965; G. Palethorpe, *U.S. 3,487,058* (to Monsanto Co.) Dec. 30, 1969.

[151] W. J. Jackson, Jr., J. R. Caldwell, and E. H. Hill, *U.S. 3,143,535* (to Eastman Kodak Co.), Aug. 4, 1964.

[152] E. F. Stroh, *U.S. 3,379,699* (to Monsanto Co.), Apr. 23, 1968.

[153] Monsanto Co., *Neth. Appl. 6,401,139*, Aug. 12, 1964.

[154] W. A. Blackburn and C. H. Apperson, *U.S. 3,313,867* (to Monsanto Co.), Apr. 11, 1967.

[155] American Cyanamid Co., *Belg. 694,133*, Feb. 16, 1967.

[156] F. J. Lowes, Jr., *U.S. 3,271,344* (to Dow Chemical Co.), Sept. 6, 1966.

[157] J. J. Hirshfeld, *U.S. 3,383,240* (to Monsanto Co.), May 14, 1968.

[158] Farb. Bayer A.-G., *Neth. Appl. 6,514,477*, May 10, 1966.

[159] L. L. Crooks and J. J. Hirshfeld, *U.S. 3,177,169* (to Monsanto Co.), Apr. 6, 1965; E. V. Burnthall and J. J. Hirshfeld, *U.S. 3,376,253* (to Monsanto Co.), Apr. 2, 1968.

[160] *Chem. Week*, **1969**, 43 (July); *Oil, Paint Drug Reptr.*, **1969**, 7 (Feb. 10); W. Watt and W. A. Johnson, *New Scientist*, **1969**, 398 (Feb. 20); J. W. Johnson, J. R. Mayorani, and D. G. Rose, *Nature*, **222**, 357 (Jan. 25, 1969); W. T. Gunston, *Sci. J.*, **5** [2], 39 (1969).

[161] V. E. Kotina, A. A. Konkin, and R. M. Kosova, *USSR 138,324*, Nov. 17, 1966; W. N. Turner, F. C. Johnson, *J. Appl. Polym. Sci.*, **13**, 2073 (1969).

[162] W. K. Wilkinson, *U.S. 3,027,222* (to E. I. DuPont de Nemours & Co.), Mar. 27, 1962; Rolls Royce, Ltd., *Belg. 717,817*, July 9, 1968; W. G. D. Carpenter, W. Johnson, T. Lloyd, P. McMullen, and W. Watt, *Belg. 700,655*, June 28, 1967; W. Johnson, L. N. Phillips, W. Watt, *U.S. 3,412,062* (to National Research Development Corp.), Nov. 19, 1968.

[163] I. Barnett, *U.S. 2,913,802* (to Johns-Mansville Corp.), Nov. 24, 1959; Rolls-Royce, Ltd., *Belg. 708,651*, Dec. 28, 1967.

[164] W. J. Vosburgh, *Text. Res. J.*, **30**, 882 (1960).

[165] K. Miyamichi, M. Okamoto, O. Tshizuka, and M. Katayama, *Sen-i Gakkaishi*, **22**, 548 (1966) (in Japanese); K. Miyamichi et al., *ibid.*, **23**, 239 (1967).

[165a] R. H. Still, P. B. Jones, *J. Appl. Polym. Sci.*, **13**, 1555 (1969).

[166] A. B. Davankov, M. I. Kabachnik, V. V. Korshak, Yu. A. Leikin, R. F. Okhovetsker, and E. N. Tsvetkov, *J. Gen. Chem. USSR*, **37**, 1522 (1967).

[167] British Celanese, Ltd. (by W. C. Paterson and A. L. Kennedy), *Brit. 902,903*, Aug. 9, 1962.

[168] A. A. Samuel, R. Bonvet, S. Hittner, and M. de Beaulieu, *Fr.1, 157,174*, May 27, 1958.

[169] H. S. Olson and R. C. Danison, *U.S. 2,582,452* (to Diamond Alkali Co.), Jan. 15, 1952.

[170] U. Bahr, K. Andres, and G. Braun, *U.S. 3,027,349* (to Farb. Bayer A.-G.), Mar. 27, 1962.

[171] Badische Anilin- & Soda-Fabrik A.-G. (by H. Mueller-Tamm, L. Zuern, and E. Stahnecker), *Ger. 1,182,812*, Dec. 3, 1964.

[172] K. H. Krause, *Kunstst. Rundsch.*, **1959**, 337 (August); Badische Anilin- & Soda-Fabrik A.-G. (by H. Mueller-Tamm, F. Statsny, and K. Buchholz), *Ger. 1,046,313*, Dec. 11, 1958; *Ger. 1,050,545*, Feb. 12, 1959.

[173] Shell Research, Ltd. (by G. A. Pogany and P. W. Croft), *Brit. 927,872*, June 6, 1963.

[174] K. Hattori, *U.S. 3,368,916* (to Richardson Co.), Feb. 13, 1968.

[175] Badische Anilin- & Soda-Fabrik A.-G., *Brit. 825,611*, Dec. 16, 1959; Badische Anilin- and Soda-Fabrik A.-G. (by E. Stahnecker), *Fr. 1,388,239*, Feb. 5, 1965.

[176] E. V. Gouinlock, J. F. Porter, and R. R. Hindersinn, Div. Org. Coatings Plast. Chem., Am. Chem. Soc., Preprints, *28* (1) 225 (1968); Badische Anilin- & Soda-Fabrik A.-G., *Neth. Appl. 6,604,209*, Oct. 3, 1966.

[177] Monsanto Chemicals, Ltd. (by N. E. Williams), *Brit. 1,059,777*, Feb. 22, 1967.

[178] A. K. Jahn, *U.S. 3,046,236* (to Dow Chemical Co.), July 24, 1962; Manufacturers de Produits Chimiques du Nord, Etablissements Kuhlmann (by G. Aliotti, D. James and M. Mouly), *Fr. 1,168,661*, Dec. 12, 1958.

[179] B. S. Taylor and M. R. Lutz, *U.S. 3,412,052* (to FMC Corp.), Nov. 19, 1968.

[180] Badische Anilin- & Soda-Fabrik A.-G. (by R. Ilgemann, R. D. Rauschenboch, and G. Foerster), *Fr. 1,411,368*, Sept. 17, 1965.

[181] Badische Anilin- & Soda-Fabrik A.-G. (by R. Ilgemann, R. D. Rauschenboch, and G. Foerster), *Fr. 1,410,556*, Sept. 10, 1965.

[182] Monsanto Chemicals, Ltd. (by P. Volans), *Brit. 975,970*, Nov. 25, 1964; Badische Anilin- & Soda-Fabrik A.-G. (by H. Laib and H. Burger), *Ger. 1,244,396*, July 13, 1967.

[182a] R. B. Clampitt, G. H. Birum, R. M. Anderson, *U.S. 3,468,678* (to Monsanto Co.), Sept. 23, 1969.

[183] C. Moore, *Can. 802,242* (to Dow Chemical Co.), Dec. 24, 1968.

[184] Badische Anilin- & Soda-Fabrik A.-G. (by K. Wintersberger, F. Meyer, and H. E. Knobloch), *Ger. 1,169,122*, Apr. 30, 1964; Badische Anilin- & Soda-Fabrik A.-G.

(by H. Krzikalla, H. Linge, F. Stastny, and F. van Taack-Trakranen), *Ger.* *1,054,236*, Apr. 2, 1959; Styrene Products, Ltd. (by H. M. E. Steiner and A. N. Roper), *Brit.* *841,946*, July 20, 1960.

[185] André Samuel, *Ger.* *1,152,264*, Aug. 1, 1963.

[186] Société Auxiliare de l'Institut Français du Caoutchouc (by P. Branlard, R. Pautrat, and R. Cheritat), *Fr. Addn.* *85,261*, July 9, 1965 (Addn. to *Fr.* *1,377,280*).

[187] Monsanto Chemicals, Ltd. (by D. G. Hare), *Brit.* *1,016,904*, Jan. 12, 1966; Richard Kühn Rohstoffverwertung G.m.b.H. (by H. J. Remmert), *Ger.* *1,050,999*, Feb. 19, 1959; Badische Anilin- & Soda-Fabrik A.-G. (by H. Linge, H. Krzikalla, and K. Buchholz), *Ger.* *962,650*, Apr. 25, 1957.

[188] E. V. Galizia, *U.S.* *3,063,954* (to Koppers Co., Inc.), Nov. 13, 1962; Badische Anilin- & Soda-Fabrik A.-G., *Ger.* *1,002,125*, Feb. 7, 1957.

[189] Dow Chemical Co., *Brit.* *927,118*, May 29, 1963; F. B. Nagle and E. L. McMaster, *U.S.* *3,133,037* (to Dow Chemical Co.), May 12, 1964; R. L. Hill, *U.S.* *3,132,045* (to Dow Chemical Co.), May 5, 1964; M. A. Longstreth, E. L. McMaster, and F. B. Nagle, *U.S.* *3,108,016* (to Dow Chemical Co.), Oct. 22, 1963; Monsanto Co., *Brit.* *1,013,378*, Dec. 15, 1965.

[190] C. W. Roberts, *U.S.* *2,952,711; 2,952,712*, Sept. 13, 1960; *U.S.* *2,956,842* (to Dow Chemical Co.), Jan, 10, 1961.

[190a] Farb. Hoechst A.-G. (by H. Peters, D. Schleede, J. K. Kaupp, and H. Jochinke), *Ger.* *1,139,636*, Nov. 15, 1962.

[191] D. G. Hare and R. J. Stephenson, *U.S.* *3,374,178* (to Monsanto Chemicals, Ltd.), Mar. 17, 1968; M. E. Elder, R. T. Dickerson, and W. F. Tousignant, *U.S.* *3,324,076* (to Dow Chemical Co.), June 6, 1967; Badische Anilin- & Soda-Fabrik A.-G. (by I. Priebe, H. Weber, and H. Willersinn), *Ger.* *1,245,593*, July 27, 1967; Badische Anilin- & Soda-Fabrik A.-G. (by R. Ilgemann and R. D. Rauschenboch), *Fr.* *1,425,563*, Jan. 21, 1966; H. Burger, G. Daumiller, J. Grohmann, E. G. Kastning, H. Mohr, L. Reuter, H. Weber, H. Willersinn, *U.S.* *3,472,799* (to Badische Anilin- & Soda-Fabrik A.-G.), Oct. 14, 1969; H. Jenkner, *U.S.* *3,455,873* (to Chemische Fabrik Kalk G.m.b.H.), Jul. 15, 1969.

[192] W. F. Tousignant, *U.S.* *3,044,999* (to Dow Chemical Co.), July 17, 1962; T. S. Carswell and R. F. Hayes, *U.S.* *2,483,753* (to Monsanto Chemical Co.), Oct. 4, 1949; S. D. Ross, *U.S.* *2,643,270* (to Sprague Electric Co.), June 23, 1953; Badische Anilin- & Soda-Fabrik A.-G. (by K. Wintersberger, F. Meyer, and H. Knobloch), *Belg.* *621,125*, Feb. 6, 1963.

[193] R. M. Price and A. F. Roche, *U.S.* *2,723,963* (to Dow Chemical Co.), Nov. 15, 1955.

[194] Badische Anilin- & Soda-Fabrik A.-G. (by F. Becke and E. Jagla), *Ger.* *1,029,154*, Apr. 30, 1958.

[195] Dow Chemical Co. (by W. F. Tousignant, C. Moore, and M. E. Elder), *Belg.* *624,340*, Apr. 30, 1963.

[196] Dow Chemical Co., *Fr.* *1,477,270*, Apr. 14, 1967.

[196a] Dow Chemical Co., *Belg.* *680,309*, Apr. 4, 1966.

[197] Badische Anilin- & Soda-Fabrik A.-G., *Neth. Appl.* *6,511,535*, Mar. 7, 1966.

[198] Société d'Electro-Chimie, d'Electro-Metallurgie et des Aciéries Electriques d'Ugine, *Fr.* *1,344,284*, Nov. 29, 1963.

[199] Chemische Fabrik Kalk G.m.b.H., *Neth. Appl.* *6,515,049*, May 23, 1966.

[200] R. R. Dreisboch and G. B. Sterling, *U.S.* *2,658,877* (to Dow Chemical Co.), Nov. 10, 1953.

[201] Badische Anilin- & Soda-Fabrik A.-G. (by H. Mueller-Tamm, K. Buchholz, and F. Stastny), *Brit.* *895,609*, May 2, 1962.

[202] Chemische Fabrik Kalk G.m.b.H., *Neth. Appl. 6,601,278*, Aug. 2, 1966; H. E. Practzel, E. F. Wirth, H. Jenkner, *U.S. 3,470,116* (to Chemische Fabrik Kalk G.m.b.H.), Sept. 30, 1969,

[203] Badische Anilin- & Soda-Fabrik A.-G., *Belg. 691,252*, Dec. 15, 1966; W. F. Tousignant and R. T. Dickerson, *U.S. 3,267,070* (to Dow Chemical Co.), Aug. 16, 1966.

[204] R. J. Stephenson, *U.S. 3,401,127* (to Monsanto Chemicals, Ltd.), Sept. 10, 1968.

[205] Sekisui Chemical Co., Ltd. (by Y. Maeda and K. Kuivatsuri), *Jap. 14,217*, Sept. 17, 1962.

[206] Badische Anilin- & Soda-Fabrik A.-G., *Belg. 685,920*, Aug. 24, 1966; Badische Anilin- & Soda-Fabrik A.-G., *Neth. Appl. 6,611,863*, Feb. 27, 1967; Badische Anilin- & Soda-Fabrik A.-G., *Belg. 699,721*, June 9, 1967; H. Burger, H. Mohn, E. Priebe, L. Reuter, H. Weber, *U.S. 3,441,524* (to Badische Anilin- & Soda-Fabrik A.-G.), Apr. 29, 1969.

[207] Badische Anilin- & Soda-Fabrik A.-G., *Belg. 684,493*, July 22, 1966.

[208] J. Eichhorn, *U.S. 3,284,544* (to Dow Chemical Co.), Nov. 8, 1966.

[209] Badische Anilin- & Soda-Fabrik A.-G., *Neth. Appl. 6,400,412*, July 29, 1964.

[210] A. R. Ingram, *J. Appl. Polym. Sci.*, **8**, 2486 (1964); *U.S. 3,274,133* (to Koppers Co., Inc.), Sept. 20, 1966.

[211] Badische Anilin- & Soda-Fabrik A.-G. (by L. Zuern and H. Mueller-Tamm), *Ger. 1,222,670*, Aug. 11, 1966; Farbwerke Hoechst, *Belg. 732,747*, Apr. 8, 1969.

[212] J. Eichhorn, *U.S. 3,124,557* (to Dow Chemical Co.), Mar. 10, 1964.

[213] Dow Chemical Co., *Fr. 1,332,588*, July 19, 1963; Dow Chemical Co. (by A. K. Jahn), *Brit. 959,419*, June 3, 1964.

[214] J. Eichhorn, *J. Appl. Polym. Sci.*, **8**, 2497 (1964).

[215] A. K. Jahn and J. W. Vanderhoff, *Amer. Chem. Soc., Div. Org. Coatings, Plast. Chem., Preprints 23* (1), 61 (1963).

[216] P. Volans, *Plast. Inst., Trans. J., Conf. Suppl. No. 2*, 47 (1967).

[217] Farb. Hoechst A.-G., *Neth. Appl. 6,405,455*, Nov. 23, 1964; *Belg. 648,213*, Nov. 23, 1964; A. Cooper, *U.S. 3,039,991* (to Expanded Rubber Co., Ltd.), June 19, 1962.

[218] J. J. Killoran and G. F. d'Alelio, *U.S. 2,894,918* (to Koppers Co., Inc.), July 14, 1959.

[219] Vereinigte Korkindustrie A.-G. (by H. Fink), *Ger. 1,173,646*, July 9, 1964.

[220] Shell Internationale Research Maatschappij N.V. (by P. J. A. Baersma), *Neth. 102,233*, Aug. 1962; Shell Research, Ltd. (by P. W. Croft, R. H. Lee, and G. A. Pogany), *Brit. 929,652*, June 26, 1963; W. M. B. International A/B (by T. E. Bramstang and E. L. Erlandson), *Brit. 839,862*, June 29, 1960; Rheinhold & Mahla G.m.b.H., *Ger. 1,173,244*, July 2, 1964; Turner Brothers Asbestos Co., Ltd., *Fr. Addn. 88,078*, Dec. 2, 1966 (Addn. to *Fr. 1,339,173*).

[221] V. D. Valgin and V. A. Ushakov, *USSR 132,818*, Oct. 20, 1960.

[222] Badische Anilin- & Soda-Fabrik A.-G. (by F. Stastny, R. Gaeth, B. Schmitt, and V. Haardt), *Belg. 657,782*, June 30, 1965; Rockwool A/B, *Belg. 690,988*, Dec. 9, 1966.

[223] *Chem. Eng. News*, **1968**, 16 (Oct. 28).

[224] Knapsack A.-G., *Belg. 716,175*, June 6, 1968.

[225] Hooker Chemical Corp., *Belg. 687,066*, Sept. 19, 1966.

[226] Badische Anilin- & Soda-Fabrik A.-G. (by F. Stastny and H. Mueller-Tamm), *Fr. Addn. 86,010*, Nov. 26, 1965 (Addn. to *Fr. 1,270,783*).

[227] Dow Chemical Co., *Brit. 890,426*, Feb. 28, 1962.

[228] Badische Anilin- & Soda-Fabrik A.-G. (by G. Daumiller, K. Herrle, and R. D. Rauschenboch), *Belg. 629,640*, Oct. 21, 1963.

[229] Badische Anilin- & Soda-Fabrik A.-G. (by H. Hintz, R. D. Rauschenboch, G. Daumiller, K. Herrle, and H. Willersinn), *Fr. 1,478,925*, Apr. 28, 1967.

[230] J. H. Rossweiler and D. R. Sexsmith, *U.S. 3,068,214* (to American Cyanamid Co.), Dec. 11, 1962.

[231] United States Rubber Co. (by W. Cummings), *Brit. 1,024,778*, Apr. 6, 1966.

[232] Chemische Werke Huels A.-G., *Fr. 1,401,795*, June 4, 1965.

[232a] "Requirements for Polymeric Enclosures of Portable Electrical Appliances," letter from Underwriters' Laboratories, Inc., to Electrical Council of Underwriters' Laboratories and Manufacturers of Listed Equipment, Mar. 3, 1969.

[233] Dow Chemical Co., *Belg. 698,808* June 15, 1967.

[233a] *Mod. Plast.*, **1968**, 86 (June); W. Cummings, *U.S. 3,260,772* (to United States Rubber Co.), July 12, 1966.

[234] Hooker Chemical Corp., *Neth. Appl. 6,515,354*, May 26, 1966; see also reference 195.

[234a] T. S. Grabowski, *U.S. 3,442,980* (to Borg-Warner Corp.), May 6, 1969.

[235] United States Rubber Co., *Neth. Appl. 301,136*, Sept. 27, 1965.

[236] Farb. Hoechst A.-G., *Belg. 648,213*, Nov. 23, 1964.

[237] American Cyanamid Co. (by H. C. Gillham and H. G. Klein), *Fr. 1,407,334*, July 30, 1965; *Chem. Eng. News*, **1968**, 16 (Oct. 28).

[238] A. E. Sherr and H. C. Gillham, *U.S. 3,218,290* (to American Cyanamid Co.), Nov. 16, 1965.

[239] Uniroyal, Inc., *Belg. 721,310*, Sept. 24, 1968.

[240] H. J. Lanning, *Rubber & Plast. Age*, **37**, 227 (1956).

[241] Société Anonyme des Pneumatiques Dunlop, *Fr. 760,712*, Mar. 1, 1934; H. Iwai (to Oriental Rubber Chemical Industries Co.), *Jap. 3839*, Aug. 10, 1953; K. Yasutake, F. Hamazaki, and N. Koi (to Bridgestone Tire Co.), *Jap. 784*, Feb. 6, 1957; F. C. Weissert, *U.S. 2,880,182* (to Firestone Tire & Rubber Co.), Mar. 31, 1959; Firestone Tire & Rubber Co., *Ger. 1,120,127*, Dec. 21, 1961; Dunlop Rubber Co., Ltd. (by D. F. Twiss, I. Kemp, and F. W. Warren), *Brit. 520,428*, Apr. 24, 1940.

[242] E. J. Bethe and T. I. Haggerty, *U.S. 3,033,804* (to United States Rubber Co.), May 8, 1962.

[243] L. Peters, *U.S. 3,279,929* (to R. M. Hollingshead Corp.), Oct. 18, 1966.

[244] K. C. Hecker, *Rubber World*, **1968**, 59 (December).

[245] Société Franco-Belge du Caoutchouc Mousse, *Fr. 856,250*, June 4, 1940.

[246] N. Bennett, *Brit. 478,664* (to Imperial Chemical Industries, Ltd.), Jan. 24, 1938; W. Becker and A. Koch, *U.S. 2,143,470* (to I. G. Farbenind. A.-G.), Jan. 10, 1939.

[247] W. C. Smith, *U.S. 2,545,977* (to Standard Oil Development Co.), Mar. 20, 1951.

[248] T. Kubota and K. Koji (to Yokohama Rubber Co.), *Jap. 4697*, June 18, 1956; The Craigpark Electric Cable Co., Ltd. (by J. R. Macfarlane), *Brit. 521,764*, May 30, 1940.

[249] R. C. W. Moakes and S. H. Morrell, *Proc. Inst. Rubber Ind.*, **2**, 179 (1955).

[250] E. G. Cockbain, T. D. Pendle, and D. T. Turner, *Chem. & Ind.*, **1960**, 318.

[251] National Rubber Producers' Research Assoc. (by T. D. Pendle, D. T. Turner, and E. G. Cockbain), *Brit. 871,320*, June 28, 1961; E. G. Cockbain, T. D. Pendle, E. G. Dole, and D. T. Turner, *Proc. Rubber Technol. Conf. 4th London*, **1962**, 22 (May).

[252] H. Frood and H. P. Alger, *Brit. 183,922*, Apr. 30, 1921.

[253] F. Jacobs, *Rev. Gén. Caoutchouc*, **12**, 3, 30 (1935).

[254] W. A. Rupar, *U.S. 3,393,166* (to Polymer Corp., Ltd.), July 16, 1968.

[255] Polymer Corp., Ltd., *Neth. Appl. 6,512,748*, Apr. 4, 1966.

[256] M. V. Potemkina and L. N. Kireenkova, *Kauchuk i Rezina*, **25**, 25 (1966) (in Russian); D. Lurie, *Fr. 1,138,514*, June 14, 1957.

[257] Uniroyal, Inc., *Neth. Appl. 6,610,966*, Feb. 13, 1967.

[258] Texas-U.S. Chemical Co. (by J. J. Bayerl), *Belg. 645,879*, July 16, 1964.

[258a] G. Witschard, C. T. Bean, *U.S. 3,462,407* (to Hooker Chemical Corp.) Aug. 19, 1969.

[259] T. R. Dawson, *Trans. Inst. Rubber Ind.*, **11**, 391 (1935); I. T. Gridunov, A. S. Sergeev, F. F. Koshelev, A. M. Potopov, and B. S. Puzrin, *Izv. Vyssh. Ucheb. Zaved., Khim. Khim. Tekhnol.*, **9**, 322 (1966).

[260] C. Shaw and W. E. Smith, *Brit. 574,646*, Jan. 15, 1946; I. Gridunov and N. I. Astrakhautseva, *Izv. Vyssh. Ucheb. Zaved., Khim. Khim. Tekhnol.*, **6**, 142 (1963).

[261] J. W. Olsen and C. W. Bechle, *U.S. 2,442,706* (to Anaconda Wire and Cable Co.), June 1, 1948.

[262] W. E. Skelton and C. E. Wilkinson, *U.S. 3,224,890* (to Texaco, Inc.), Dec. 21, 1965.

[263] F. I. L. Lawrence and M. J. Pohorilla, *U.S. 3,279,928* (to Kendall Refining Co.), Oct. 18, 1966.

[264] J. H. Young, *U.S. 1,398,991*, Dec. 6, 1920.

[265] J. H. Young and H. H. Robertson Co., *Brit. 186,861*, Dec. 6, 1921; R. E. Koons, *U.S. 3,342,614* (to Monsanto Co.), Sept. 19, 1967.

[266] L. A. Bierly, *U.S. 2,667,425* (to Presque Isle Laboratories & Manufacturing, Inc.), Jan. 26, 1954.

[267] Hooker Chemical Corp., *Belg. 612,685*, Feb. 15, 1962; Hooker Chemical Corp., *Neth. Appl. 6,402,520*, Sept. 14, 1964; Ruberoid Co., Ltd. (by H. A. F. L. Kremer), *Brit. 1,056,667*, Jan. 25, 1967.

[268] Hooker Chemical Corp., *Belg. 687,067*, Sept. 19, 1966; Hooker Chemical Corp., *Neth. Appl. 6,413,993*, June 3, 1965; Hooker Chemical Corp., *Neth. Appl. 6,612,480* Mar. 21, 1967.

[269] Ruberoid Co., Ltd. (by H. A. F. L. Kremer and E. G. Norman), *Brit. 1,058,451*, Feb. 8, 1967.

[270] A. L. Nugey, *U.S. 3,056,741*, Oct. 2, 1962.

[271] E. Kocchlin, P. Doittau, and E. Valentin, *Fr. 958,447*, Mar. 8, 1950.

[272] J. C. Harris and E. A. Steinmetz, Paper presented at the Nat. Aeronautic Meeting of the Soc. of Auto. Eng., No. 670,365, New York, Apr. 24–27, 1967; also see *USAAVLABS Tech. Rept. 68–79*.

[273] C. M. Murphy and W. A. Zisman, *Prod. Eng.*, **1950**, 109 (September); J. S. Harris, *ibid.*, **25**, 163 (1954); Esso Research & Engineering Co. (by R. Beardeu, Jr., and C. L. Aldridge), *Fr. 1,463,212*, Dec. 23, 1966.

[273a] Q. E. Thompson, *U.S. 3,436,441* (to Monsanto Co.), Apr. 1, 1969; K. L. McHugh and K. A. Nowotny, *U.S. 3,383,318* (to Monsanto Co.), May 14, 1968.

[274] Castrol, Ltd. (by R. A. C. Ker and B. T. Scoltock), *Fr. 1,430,951*, May 27, 1966.

[275] T. W. Campbell and D. J. Lyman, *J. Polym. Sci.*, **55**, 169 (1961).

[276] P. Rakoff, G. J. Colucci, and R. K. Smith, *AD 613368*, available CFSTI, 1965, 20 pp.

[277] R. E. Hatton, *Introduction to Hydraulic Fluids*, Reinhold, New York, 1962, 363 pp.

*[6a] A. D. Delman, J. Macromol, Sci.—Revs. Macromol. Chem. *C3* (2) 281 (1969).

Chapter 8

SYNTHETIC POLYMERS WITH HETEROGENEOUS BACKBONES

In Chapter 7 polymers with all-carbon backbones were reviewed in terms of flame-retardant technology. In this chapter we consider polymers with backbones composed of carbon with other elements, especially oxygen and nitrogen. The principal members of this group are polyurethanes, polyesters, epoxies, polyamides, and phenolic resins. Also included are polycarbonates, aminoplasts, and, briefly, special polymers tailor-made for high-temperature service. This is a rather large and unwieldy class of polymers made more cumbersome by the great number of papers on fire-retardant polyurethanes and polyesters.

POLYURETHANES

Polyurethanes are made by reacting polyisocyanates with polyols:

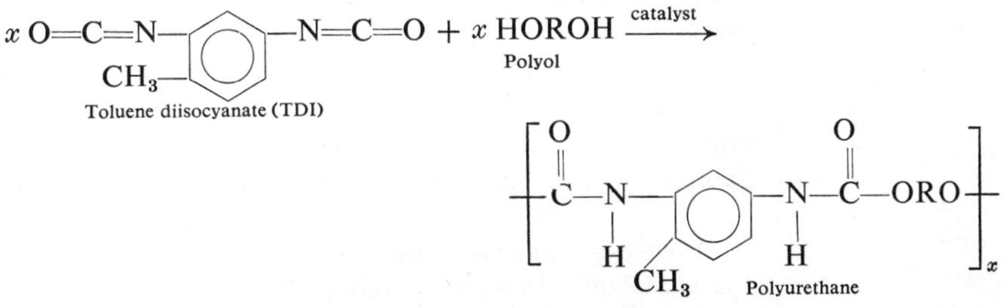

$$x \; O\!=\!C\!=\!N\!-\!\!\bigcirc\!\!-\!N\!=\!C\!=\!O + x \; HOROH \xrightarrow{\text{catalyst}}$$

Toluene diisocyanate (TDI)   Polyol

$$\left[-\overset{O}{\underset{}{\overset{\|}{C}}}\!-\!\underset{H}{N}\!-\!\!\bigcirc\!\!-\!\underset{H}{N}\!-\!\overset{O}{\underset{}{\overset{\|}{C}}}\!-\!ORO\!- \right]_{x}$$

Polyurethane

The catalyst may be an amine (for rigid foams) or a tin salt (for flexible foams). The isocyanate may be as simple as the TDI shown or it may be a monomer of higher molecular weight such as a diphenylmethane derivative

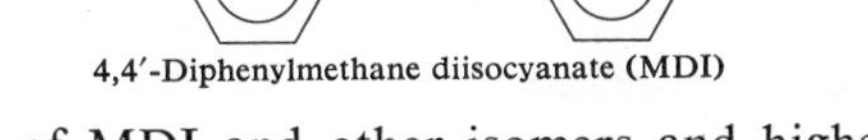

$$OCN\!-\!\!\bigcirc\!\!-\!CH_2\!-\!\!\bigcirc\!\!-\!NCO$$

4,4'-Diphenylmethane diisocyanate (MDI)

or a crude mixture of MDI and other isomers and higher oligomers, for example, polymethylene polyarylisocyanates.

The polyol may be a naturally occurring substance—sucrose, starch (in part), and derivatives thereof, such as propoxylated sucrose—or a synthetic alcohol, for example, polyethers with two or more hydroxyl groups: $HO\ (\ CH_2CH_2O\ )\ _xH$. Derivatives of glycerine and condensed polyglycol-ethers are also used. Cross-linking to produce rigid structures is achieved by using one component with a functionality greater than 2.

Although polyurethanes find use in coatings and other nonfoam areas, the principal outlet is for rigid and flexible foams. The rigid foams [1] are mainly used for insulation and much of this market is sensitive to fire-safety regulations. (See Chap. 5, ref. [2a].) Similarly, the flexible foams, used extensively in mattresses and cushions, are often under fairly strict control as to flammability. The following discussion is devoted almost wholly to fire-retardant foams. It turns out that most of the literature is on rigid foams. This is not to say that there is less need for fire-retardant flexibles but rather to indicate that research has been more fruitful in the rigid area.

Hilado [2, 3] and coworkers and Robitschek [3a] have reviewed some of the factors in the burning of a urethane foam. Frisch [4] has studied flammability of rigids as a function of structure of the monomers. Tilley et al. [5] have evaluated the rate of thermal degradation in terms of structure and they offer detailed speculative paths for the decomposition. Einhorn and coworkers [6, 7] have studied char formation and smoke generation in a series of urethanes with and without fire-retardant additives. Dickert and Toone [8] have reported on the performance of rigids in various fire tests as a function of structure. And Backus and his colleagues [9, 10] have carried out a large number of thermal analyses (using both the thermogravimetric and differential thermal methods) on a variety of systems.

It is clear from several of the foregoing references that the structure of the polyhydroxy compound and of the polyisocyanate has a great effect on flame rating, char formation, and smoke density. For the polyisocyanates, TDI gives the poorest flame ratings and the polymethylene polyarylisocyanates give the best (see Table 8-1). TDI urethanes give the poorest chars—often

Table 8-1   Effect of Isocyanate on
Flammability (Dickert and Toone [8])

Isocyanate	Aromatic Functionality (ring/NCO)	Distance Burned (ASTM D-1692-59T), in.
Modified TDI	0.5	2.0
Crude MDI	1	<0.5
Polymethylene polyphenyl-isocyanate	1	0

Table 8-2   Model Polymer Flammability
Measurements (Backus et al. [10])

Polymer		Flammability ($2\frac{1}{2} \times \frac{1}{2} \times \frac{1}{8}$ in. Strip)			
		In. Burned	Min to Extinguish	Obsd	Rating
Propylene glycol	MDI	0.5	0.5	Melt	SE[a]
	MDI-2.7	0.25	1.0	Char	SE
Dipropylene glycol	MDI	0.5	0.5	Melt	SE
	MDI-2.7	0.25	0.7	Char	SE
Triol 660	MDI	2.5	0.8	Melt	Burned
	MDI-2.7	0.25	1.0	Char	SE
$Ph(OC_2H_5OH)_2$	MDI	2.5	1.0	Melt	Burned
MOCA[c]	MDI	2.5	1.2	Char	Burned
	MDI-2.7	2.5	0.5	Char	Burned
Hydroquinone	MDI	2.5	1.8	Char	Burned
p-Xylenediol	MDI	2.5	1.8	Melt	Burned
p-Phenylenediamine	MDI	2.5	2.3	Char	Burned
m-Xylenediamine	MDI	0.75	0.85	Char	SE
Terephthalic acid	MDI	0.16	2.0	Char	SE
Pyromellitic acid	MDI	0.5	0.7	Char	SE
Cl_4 p-xylenediol	MDI	0.16	0.5	Char	NB[b]
Cl_4 m-xylenediamine	MDI	0	0.5	Char	NB

[a] SE = self-extinguishing.
[b] NB = non-burning.
[c] MOCA = methylene bis-o-chloroaniline.

none at all—and exhibit the lowest melting points and highest vapor pressures at a given temperature. They also produce the least smoke. Polyester linkages give better flame ratings than polyethers [8] and low molecular weight polyethers are superior to higher oligomers [8, 10] (see Tables 8-2 and 8-3). Apparently the higher oligomers provide more opportunity for intramolecular nucleophilic attack, producing volatile (and flammable) fragments. Foams

Table 8-3   Flammability Variation with Polyol Structure
and Equivalent Weight (Backus et al.[10])

Alcohol	Equivalent Weight	Ether groups per OH	Flammability of MDI-Based Polymer
Propylene glycol	38	0	SE
Dipropylene glycol	67	0.5	SE
Triol 660	85	1.0	Burned
$Ph(OC_2H_5OH)_2$	99	1.0	Burned

for which the density of cross-links is low will melt readily and yield no char [6]. In terms of the weight per cross-link, a foam with a value greater than 450 will not produce a char.

Degradation is complex and obviously varies with the starting materials. Unzipping to isocyanate and polyol may occur, especially at lower temperatures. Loss of carbon dioxide and formation of olefins and primary amines occur and in some cases further reaction among fragments produces secondary amines. These reactions are [5]:

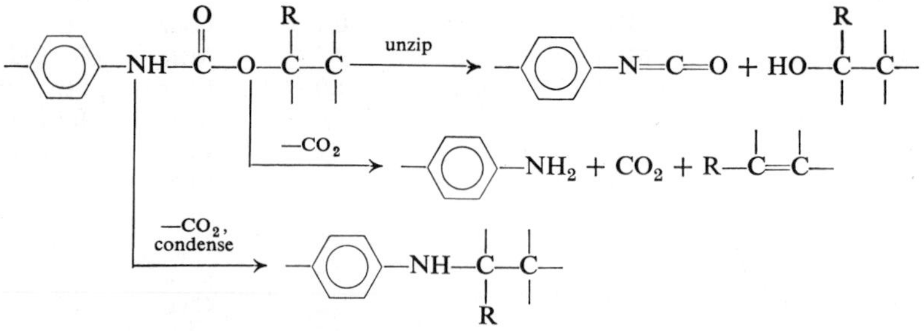

In the presence of oxygen further degradation occurs. The papers by Backus et al. [9, 10] contain the most detail on these points. The volatiles from a typical foam are listed in Table 8-4.

There has been some effort to correlate basic thermodynamic properties to performance in flame tests [4, 8]. Clearly, those parameters that relate to high softening or melting points correlate with flame resistance: the more hydrogen bonds possible, the greater the stability.

Table 8-4 Volatiles on Thermal Degradation of a Polymethylene-Polyarylisocyanate-Propoxylated Sucrose Urethane (Rigid) Foam (Backus et al. [9])

Temperature, °C	Volatiles
80–100	
150–180	
Below 200	CO_2, $CFCl_3$, volatile alkene
Below 240	
Below 300	H_2O, CO_2, CO, alkene. Mixture characterized by —NH, OH, COC, monosubstituted phenyl IR bands; ester, aldehyde, and/or COOH carbonyl IR bands

Table 8-5   Char and Smoke from Rigid Urethane Foams
Containing about 1.5% Phosphorus (Einhorn and Mickelson [6])

Sample Description	Density pcf	% Light Obscuration	Obscuration Time, sec to 50%	100%	% Weight Loss	Char Structure
Propoxylated sucrose TDI, phosphonate foam	2.0	90	0.5	· · ·	100, 40 sec	Melts, no char
Propoxylated sucrose MDI	2.1	96	3.0	· · ·	40	Weak char
Propoxylated sucrose polymethylene polyphenylisocyanate, phosphonate foam	2.1	100	1.0	4.0	26	Strong char

Some of the results from the studies of Einhorn's group on char formation and smoke generation are shown in Table 8-5 and Figures 8-1 and 8-2. The greatest smoke is generated from the least combustible sample—(PAPI ≡ polyarylmethane polyisocyanate)—but the sample produces the best char.

Although most urethanes are made fire-resistant via incorporation of elements specifically for this purpose (P, Cl, Br, Sb, etc.), it is possible by

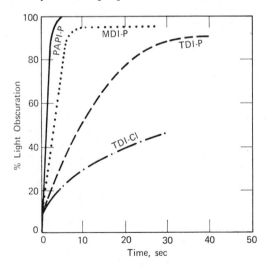

Figure 8-1 Relationship between isocyanate structure and light obscuration in rigid urethanes. (-P means 1.5% phosphorus as reactive fire retardant; -Cl means that chlorinated diphenyl was added.) (From Einhorn et al. [7].)

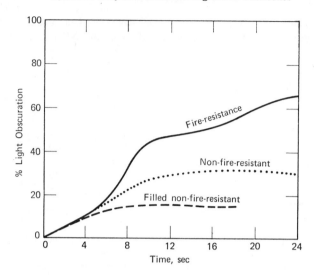

Figure 8-2   Smoke development in flexible urethane foams based on TDI and propoxylated glycerine. (From Einhorn et al. [7].)

structural modification to produce foams with acceptable fire properties. Thus, the information on degradation reviewed above suggests that higher functionality and more aromaticity should raise the decomposition temperature and improve fire resistance. Recently, a foam system has been introduced in which the isocyanates are trimerized to produce an isocyanurate ring:

Very little polyol is used—only enough to provide some resilience. Burnthrough times are increased by a factor of 10 or more. Ball and coworkers have reported on this type of rigid foam [11].

Ring structures in the polyol component are also useful. Trimellitic acid (or its anhydride) provides high functionality and the required stable ring:

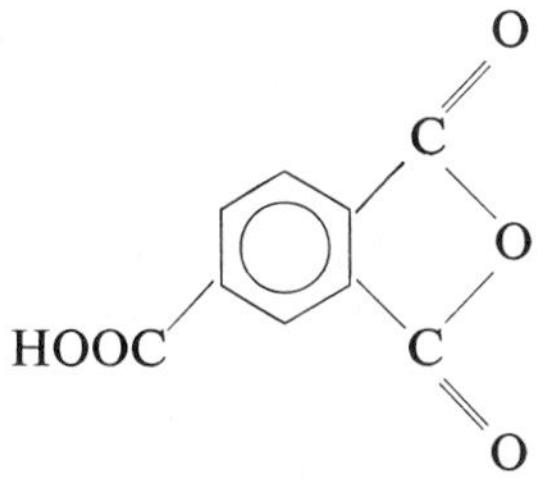

Trimellitic anhydride

It is used with polyetherpolyols and preferably a polymethylene polyarylisocyanate. Foams with weight losses after a flame test as low as 16% (cf. 85% for "ordinary" polyurethane foam) are obtained [12]. Similarly, a phenol hydroxyether has been found to produce flame-resistant urethane foam with MDI [13].

The rings need not be aromatic. The heterocyclic anhydroglucose units in starch are effective in the proper system. Bennett et al. [14] find, for example, that starch used to extend N,N,N',N'-tetrakis(2-hydroxypropyl) ethylenediamine as the polyol in a polymethylene polyarylisocyanate foam yields self-extinguishing foam by ASTM D-1692. Some 30% of starch in the formulation gives good fire resistance even after humid aging.

These are examples, really, of new polymers tailored to meet fire-safety requirements rather than polymers for which fire resistance has been achieved through specific additions of fire-retardant chemicals. As such they represent a significant new approach to the problem—an approach that has been fruitful for urethanes and may well prove to be so for other polymers. (See, for example, the last section of this chapter on "Thermally Stable Polymers.") However, fire retardance even in urethanes is still largely obtained via fire-retardant chemicals and we now examine a rather lengthy list of these.

Phosphates

Systems containing phosphorus in the +5 oxidation state as the sole flame retardant are especially significant because they represent potentially the least expensive solution. Nearly all the systems covered in the literature contain phosphorus in the polyol component. This is achieved in two main ways: reaction of acidic phosphorus compounds with alkylene oxides and esterification via condensed phosphoric acid followed by alkoxylation to remove residual acid groups. The simplest polyol one can obtain readily is perhaps $(HOCH_2CH_2O)_3PO$ [15]. This can be made either from ethylene oxide and H_3PO_4 or from $POCl_3$ and ethylene glycol [16]; the latter combination uses a somewhat more expensive form of phosphorus. Alkylene oxides will react with 100% liquid H_3PO_4 at 50–60°C or so over a period of several hours [17]. The degree of alkoxylation in acid medium is limited. Thus, it is difficult to obtain more than two alkoxy groups per OH group on the phosphorus. Use of a BF_3 catalyst allows more groups to be added [18]. This produces a polyol that is useful in flexible foams.

Higher molecular weights in the polyol are generally desired to reduce volatility, increase water insolubility, and so on. To achieve this, the reaction is run with condensed phosphoric acids. In this case the alkylene oxide reacts with POH groups and the hydroxyl compound so formed reacts with POP groups [19, 76]:

$$
\begin{array}{c}
\overset{\displaystyle O}{\underset{\displaystyle \underset{H}{O}}{-P}}-O-\overset{\displaystyle O}{\underset{\displaystyle \underset{H}{O}}{P}}-OH \;+\; RCH_2\!\!-\!\!CH_2\overset{O}{\diagdown\diagup} \;\rightarrow\; \overset{\displaystyle O}{\underset{\displaystyle O}{-P}}-O-\overset{\displaystyle O}{\underset{\displaystyle O}{P}}-O-\overset{\displaystyle R}{\underset{\displaystyle H}{C}}-CH_2OH
\end{array}
$$

$$
\begin{array}{c}
\overset{\displaystyle O}{\underset{\displaystyle}{-P}}-O-\overset{\displaystyle R}{\underset{\displaystyle H}{C}}-CH_2OH \;+\; -\overset{\displaystyle O}{\underset{\displaystyle \underset{H}{O}}{P}}-O-\overset{\displaystyle O}{\underset{\displaystyle \underset{H}{O}}{P}}- \;\rightarrow
\end{array}
$$

$$
\begin{array}{c}
\overset{\displaystyle O}{\underset{\displaystyle}{-P}}-O-\overset{\displaystyle R}{\underset{\displaystyle H}{C}}-CH_2O-\overset{\displaystyle O}{\underset{\displaystyle \underset{H}{O}}{P}}-OH \;+\; HO-\overset{\displaystyle O}{\underset{\displaystyle}{P}}- \qquad\text{etc.}
\end{array}
$$

The result is a polyphosphate ester that may be highly cross-linked or that may have a high hydroxyl number depending on the stoichiometry. The phosphorylation is difficult to run in polyphosphoric acid because of the high viscosity of the acid. This may be averted or minimized by carrying out the reaction in a large excess of product [19] at 80–130° [20]. By holding the product at 150° for 6 hr self-transesterification (or reorganization) occurs to produce an improved but unidentified product [21]. Certain epoxides are reported to stabilize the product, for example, limonene dioxide [21a].

On occasion the reaction product, a polyol, may be further reacted with an α,β-unsaturated carboxylic acid such as maleic anhydride. The resulting phosphoryl alkenyl ester may then be used in polyester polyols [19].

The starting acid may already be an ester. Thus, P_4O_{10} and various alcohols may first be reacted to give a mixture of mono- and dialkyl phosphoric acids and these in turn may be ethoxylated or propoxylated [22]. The choice of the alcohol permits wide flexibility in controlling the properties of the end products. One may use ethylene glycol [23] or propylene glycol [24] with poly-

$$\overset{\text{O}}{\underset{|}{}}$$

phosphoric acid or with P_4O_{10} or even with RO—P—Cl$_2$ [25] followed by propoxylation. In a similar fashion sorbitol, pentaerythritol, trimethylol propane, glycerol, α-methylglucoside, sucrose, dextrose, and glucose have been used with polyphosphoric acid or with P_4O_{10} followed by reaction with an alkylene oxide to remove acidity [26]. These reactions all yield phosphorus-containing polyols that are easy to make from inexpensive raw materials. The processes are relatively simple to operate. Whether one uses polyphosphoric acid or phosphorus pentoxide is a matter of individual choice. Some producers use one; some, the other. Low cost is the advantage. Some sacrifice in durability is made when these phosphate esters are used because they are somewhat subject to hydrolysis on prolonged exposure to water. This problem is apparently not overly acute and this class of flame retardants is used widely in rigid urethane foams at this writing. About 1.5% phosphorus is required. (See comparative discussion later in this chapter for the effects of levels of phosphorus.)

There are almost no references to nonreactive phosphates that contain no halogen or nitrogen. (For an exception, see the cyclic pyrophosphate ester in reference 26a [toxicity ?].)

Phosphorus(V)-Nitrogen Compositions

Ammonium orthophosphates are little mentioned for urethane foams. They are, of course, very water-soluble and do not dissolve in the urethane components. Nonetheless, diammonium phosphate has been studied by Miles and Lyons [27] in a rigid system and found to perform very well. Foams containing it survived humid aging with good fire resistance. However, on total immersion in running water, the phosphate was leached out and fire resistance was lost. Combinations with antimony oxide have also been suggested [28]. The water solubility of the orthophosphate is largely overcome by use of high molecular weight ammonium polyphosphate [29]. Foams containing this phosphate are self-extinguishing after 2 weeks at 70° and 100% relative humidity. They also survive intact after immersion in running water [27]. Physical properties of the foams are good and fire retardance is excellent [27]. The ammonium polyphosphate suffers from being insoluble in the polyol or isoycanate component, thereby causing potential settling problems in storage of premixed systems. This has thus far proved a barrier to commercial use despite good functional properties.

Ways of getting around this compatibility problem are many. A few that have been tried include melamine phosphate [30], reaction products of

$POCl_3$, and various amines [31] or amides [32] or amine/alcohol combinations [33]. Many of these contain the $-\overset{|}{P}-\overset{|}{N}-$ linkage which is subject to hydrolysis. Simple ethanolamine salts of phosphoric acid [34] are compatible with polyol but are not water-resistant.

Putting phosphorus in the isocyanate is also a possibility. Various derivatives of $P(NCO)_3$ or $OP(NCO)_3$ have been reported [35] but these are not yet available commercially at a reasonable price.

Phosphites

Most phosphites encountered in the urethane field are ultimately converted to phosphonates. There are, however, some patents relating to the use of phosphite esters for foams. The simplest case is hydroxyethylphosphorous acid used with TDI and a polyester polyol to achieve a fire-resistant foam [36]. Dimethyl phosphite can be made to react with polyols such as pentaerythritol or trimethylol propane to obtain polyols with as much as 20% P [37]. But most are triesters, for example, tris(polypropyleneglycol) phosphites [38] and diphosphites [39]. Such multifunctional materials can be isomerized to higher molecular weights by adding base and heating. For example, tris(dipropyleneglycol) phosphite rearranges under vacuum at 130–160° in the presence of a small amount of K_2CO_3 with loss of dipropyleneglycol to give a polymer with glycol bridges [40]. Transesterification may then be effected to produce the desired polyol [41]. Thus, pentaerythritol and triphenyl phosphite afford a tetramer in phosphorus [42]:

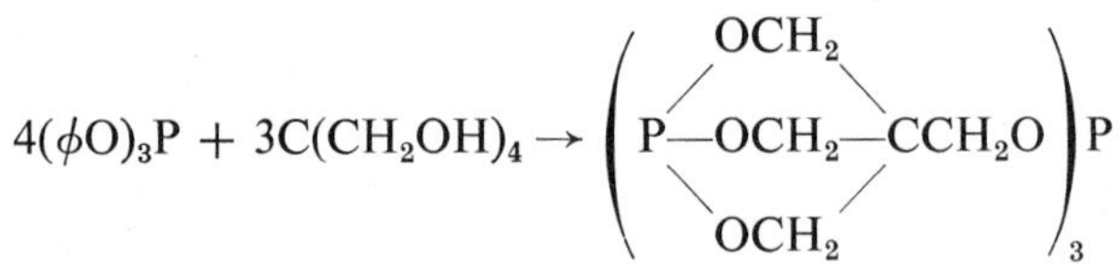

$$4(\phi O)_3 P + 3C(CH_2OH)_4 \rightarrow \left(P \underset{\diagdown}{\overset{\diagup}{\underset{OCH_2}{\overset{OCH_2}{}}}} OCH_2 - CCH_2O \right)_3 P$$

A phosphorus-containing polyester has been reported from phosphorous acid, maleic anhydride, and propylene oxide [43]; another from dimethyl phosphite, phthalic anhydride, and hexanetriol [44].

Phosphonates

Of the phosphonates not containing halogens, perhaps the most widely studied and used is

$$(C_2H_5O)_2 - \overset{\overset{\textstyle O}{\|}}{P} - CH_2N(C_2H_4OH)_2$$

an aminomethylphosphonate [45] (see Chapter 2 for synthesis). The compound is a reactive polyol (two hydroxyls per molecule) and is substituted in polyurethanes for a portion of the polyol normally used. The phosphorus becomes part of the molecule but not part of the polymer backbone. Rather it is in a pendant position so that thermal degradation liberating an acidic phosphorus fragment does not automatically produce scission of the backbone:

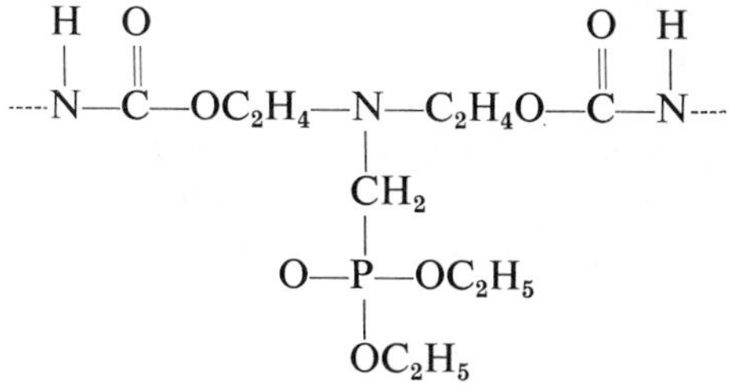

$$\cdots\text{N}\text{—}\overset{\displaystyle \overset{\text{H}}{|}}{}\text{—}\overset{\displaystyle \overset{\text{O}}{\|}}{\text{C}}\text{—OC}_2\text{H}_4\text{—N—C}_2\text{H}_4\text{O—}\overset{\displaystyle \overset{\text{O}}{\|}}{\text{C}}\text{—}\overset{\displaystyle \overset{\text{H}}{|}}{\text{N}}\cdots$$

$$\begin{array}{c}
\text{H} \quad \text{O} \qquad\qquad\qquad\qquad \text{O} \quad \text{H} \\
| \quad\ \| \qquad\qquad\qquad\qquad\ \| \quad | \\
\cdots\text{N—C—OC}_2\text{H}_4\text{—N—C}_2\text{H}_4\text{O—C—N}\cdots \\
| \\
\text{CH}_2 \\
| \\
\text{O—P—OC}_2\text{H}_5 \\
| \\
\text{OC}_2\text{H}_5
\end{array}$$

An entire family of such compounds was studied by Walsh, Uhig, Beck, and their colleagues [46]. A typical formula gives a burning length in ASTM D-1692 of less than $\frac{1}{4}$ in. and self-extinguishing times of less than 15 sec. About 1.5–2.0% phosphorus is desired in the finished foam. About 15% of this retardant is therefore added to the formula and an appropriate amount of ordinary polyol is removed. An example with results is shown in Table 8-6.

Table 8-6 [46]

Formula		Wt %
Component A		
$(\text{C}_2\text{H}_5\text{O})_2\text{P(O)CH}_2\text{N(C}_2\text{H}_4\text{OH})_2$ Hydroxyl No. 419		14.24
Polyol	Hydroxyl No. 456	25.31
Silicone		0.49
Trichlorofluoromethane		12.98
Catalyst		0.28
Component B		
Isocyanate (29.9% NCO)		46.69
		100.00
NCO/OH ratio	1.05	
Cream time, sec	8	
Foam rise time, sec	120	
Tack free time, sec	120	
Density, lb/ft^3	2.3	
% closed cells	91	
10% compression, psi	27	
K factor (after humid exposure)	0.15	
Flammability (ASTM D-1692):		

	Burn Time	In. Consumed
Initial	8.9	$\frac{1}{4}$
1 week at 52°C, 80% relative humidity	10.1	$\frac{3}{16}$

Mickelson and Einhorn [47] report that at this level of the agent the thermal decomposition temperature of a rigid foam is substantially reduced (ca. 75°C in their example—from 390° to 315°C). This is true of all phosphorus-containing urethanes, as it is for cellulosics and many other substrates. The phosphorus, acting via acidic fragments, catalyzes dehydration to char at temperatures below that at which normal decomposition occurs. The lower decomposition temperature therefore is a necessary accompaniment. Smoke levels from urethanes are increased by phosphorus compounds [47]. In one case the time required for a given density of smoke to develop was reduced threefold when the aminomethylphosphonate was present versus the control. (See later paragraphs for comparative data on this compound versus other types of agents.)

A phosphinate analog has recently been patented but it promises to be more expensive because of the scarcity of starting material [47a] (P in the +1 oxidation state).

Phosphonates containing hydroxyl groups on the phosphorus have been studied but these put the phosphorus in the polymer backbone. This is believed to be less effective. Some examples are $(HORO)_2—\overset{\displaystyle O}{\underset{\displaystyle |}{P}}—CH_2NR_2$ [48] and $(HORO)_2—\overset{\displaystyle O}{\underset{\displaystyle |}{P}}—CH_2N(ROH)_2$ [49]. A recent patent claims polyols made by ethoxylating nitrilotris (methylenephosphonic acid) [49a]:

$$N(CH_2PO_3H_2)_3 + CH_2\!\!-\!\!CH_2 \rightarrow N[CH_2PO_3(CH_2CH_2OH)_2]_3, \quad \text{etc.}$$

The starting acid is available commercially (see Chapter 2, p. 57). Molecules containing phosphorus and some aromaticity are

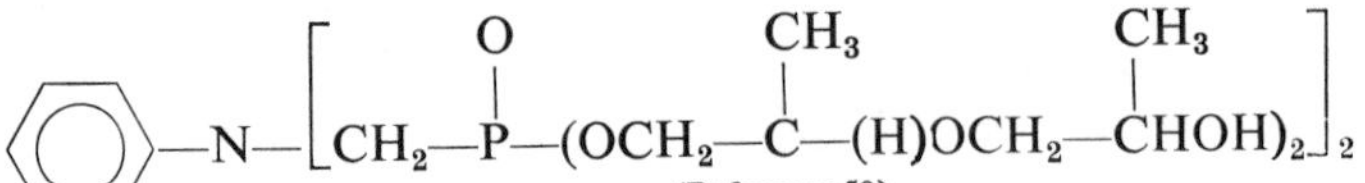

(Reference 50)

and

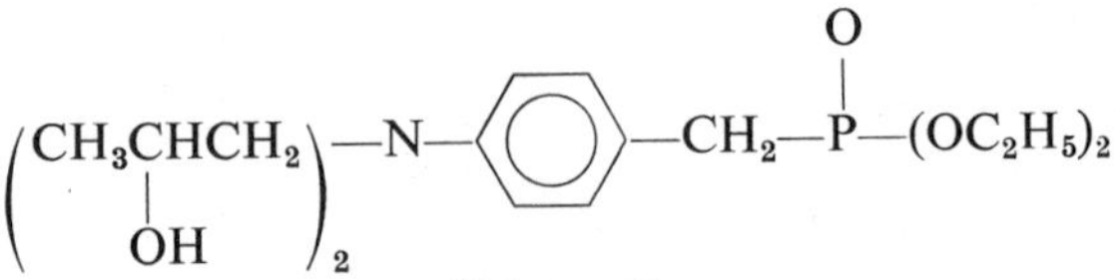

(Reference 51)

The latter structure would be expected to be superior in that the phosphorus will be in a pendant group in the polymer. There has been some work done on reacting an isocyanate with a phosphonate to produce an additive flame retardant of improved compatibility with foam components in storage before foaming. Improved hydrolytic stability is also claimed. Thus, the following is suggested [51]:

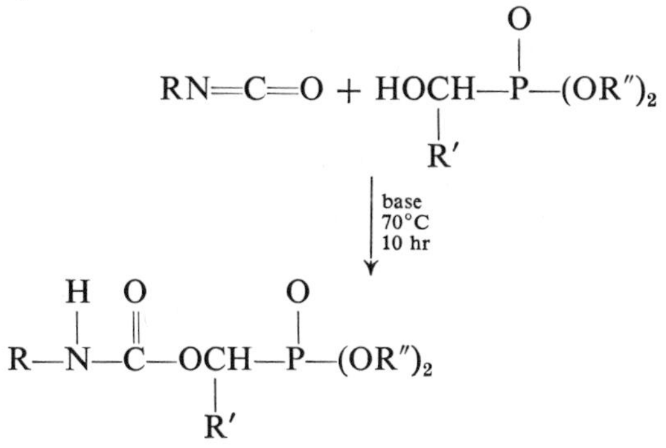

$$
\underset{\substack{|\\ R'}}{\overset{O}{\underset{|}{RN{=}C{=}O + HOCH{-}P{-}(OR'')_2}}}
$$

$$
\Big|\substack{\text{base}\\ 70°C\\ 10\ hr}
$$

$$
\underset{\substack{|\\ R'}}{R{-}\overset{H}{\underset{|}{N}}{-}\overset{O}{\overset{\|}{C}}{-}OCH{-}\overset{O}{\underset{|}{P}}{-}(OR'')_2}
$$

The result is an additive or nonreactive flame retardant. A phosphorus containing diisocyanate such as [52]

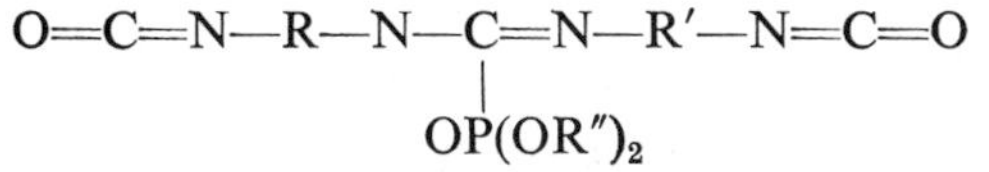

$$
\underset{OP(OR'')_2}{O{=}C{=}N{-}R{-}N{-}C{=}N{-}R'{-}N{=}C{=}O}
$$

is another way to introduce a degree of flame retardance but these compounds have not yet become commercial possibilities. Table 8-7 shows some other phosphonates of some interest. It is significant that of those that are reactive polyols nearly all are constructed so that the phosphorus will not be in the polymer backbone.

Phosphorus-Halogen Compositions

Most materials in this class are additive rather than reactive. They range from red phosphorus plus halogens [63] to halophosphates. Table 8-8 contains a full list of patent references in this category. Tris(chloroethyl) phosphate, made from ethylene oxide and $POCl_3$, is claimed in a number of patents. In two of them [65, 66] addition of lithium salts is said to improve resistance to humidity and heat. A major point in the table is that the synergism, if there is a true one, between phosphorus and halogens is a good bit less than we have seen in Chapter 7. Thus, about 1.5% phosphorus is needed to fire-retard polyurethanes in the absence of halogens. This level is lowered to perhaps 1% by adding 10–15% chlorine. Addition of 4–7% bromine lowers the need for phosphorus to 0.5% or so. Again, we see that 1 part of bromine is equal to about 2 parts of chlorine (by weight) and that the reduction in phosphorus requirement is greater with bromine than chlorine. The halogenated products often present stability problems because they tend to

Table 8-7   Phosphonates Used or Suggested for Polyurethane

Structure	Pendant	In the Backbone	Reference
$\begin{array}{cc} O & R' \\ \| & \| \\ (RO)_2\!-\!P\!-\!CHOH \end{array}$	x		53
$\begin{array}{ccc} O & & O \\ \| & & \| \\ (RO)_2\!-\!P\!-\!R\!-\!C\!-\!N(R'OH)_2 \end{array}$	x		54
$\begin{array}{c} O \\ \| \\ CH_2\!-\!CH\!-\!CH_2\!-\!P\!-\!(OR)_2 \\ \| \quad\ \| \\ OH \ \ OH \end{array}$	x		55
$\begin{array}{c} O \ \ O\!-\!CH_2 \quad CH_2OH \\ \|\diagup \qquad \diagup \\ R\!-\!P \qquad\quad C \\ \ \diagdown \qquad \diagdown \\ O\!-\!CH_2 \qquad R \end{array}$	x		56
$\begin{array}{c} O \\ \| \\ R\!-\!P\!-\!OH + \triangledown CH_3 \\ \| \qquad\quad O \\ R' \end{array}$ (R' = alkyl or alkoxy)	x		57
$\begin{array}{c} O \\ \| \\ CH_2\!-\!\!-\!\!-CH\!-\!CH_2\!-\!P\!-\!(OR)_2,\ \triangledown CH_3 \\ \diagdown\ \diagup \qquad\qquad\qquad\ O \\ O \end{array}$	x		58 58
$CH_2\!=\!CH\langle\bigcirc\rangle PO_3H_2,\ \triangledown CH_3$ $\qquad\qquad\qquad\qquad O$		x	59
$\begin{array}{c} O \\ \| \\ \langle\bigcirc\rangle\!-\!N\!-\!CH_2CH_2\!-\!P\!-\!(OROH)_2 \\ \| \\ H \end{array}$			59a
$\begin{array}{c} CH_3 \\ \| \\ H_2O_3P\!-\!C\!-\!PO_3H_2,\ \triangledown CH_3 \\ \| \qquad\qquad\qquad O \\ OH \end{array}$		x	60
$\begin{array}{c} \ \ C_2H_5 \qquad\quad O \qquad C_2H_5 \\ \ \ \ \| \qquad\qquad\ \| \qquad\quad\ \| \\ \left[(HOCH_2)_2\!-\!C\!-\!CH_2O\right]_2\!-\!P\!-\!CH_2\!-\!C\!-\!(CH_2OH)_2 \end{array}$		x	61
$\begin{array}{ccc} O & & O \\ \| & & \| \\ (R'O)_2\!-\!P\!-\!R\!-\!P\!-\!(OR')_2 \end{array}$	(Additive)		62
$-C\!=\!C\!-\!COOH,\ H_3PO_3,\ RCH_2\!\!-\!\!-\!\!-CH_2$ $\qquad\qquad\qquad\qquad\qquad\quad \diagdown\ \diagup$ $\qquad\qquad\qquad\qquad\qquad\qquad O$		x	62a

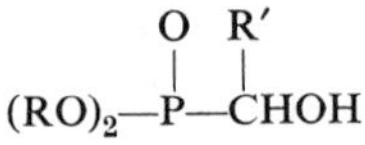

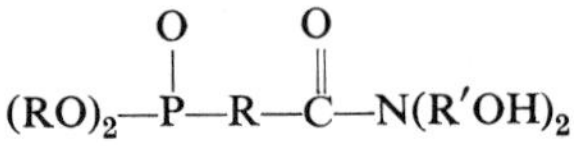

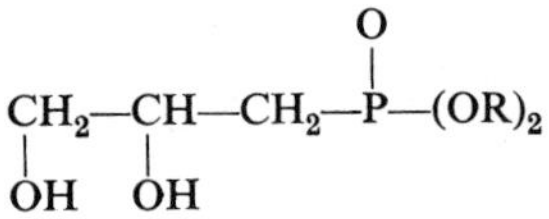

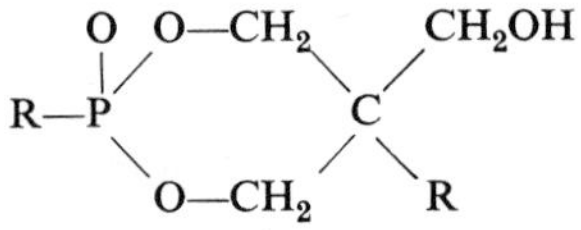

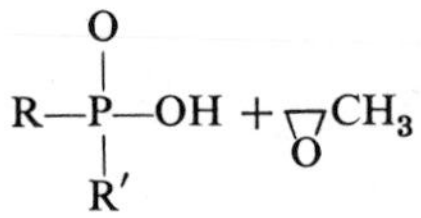

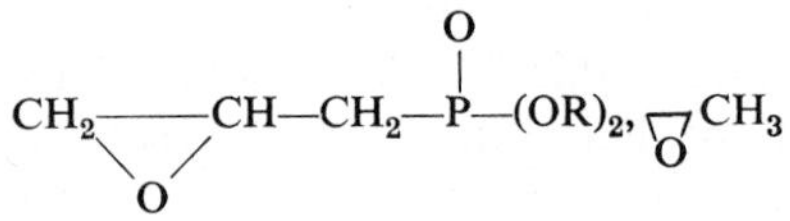

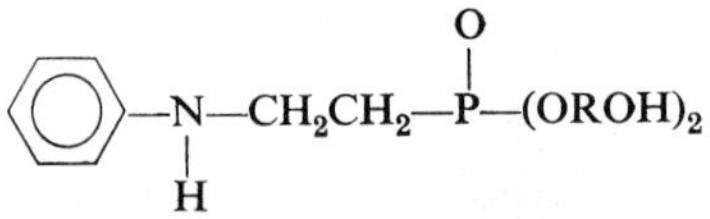

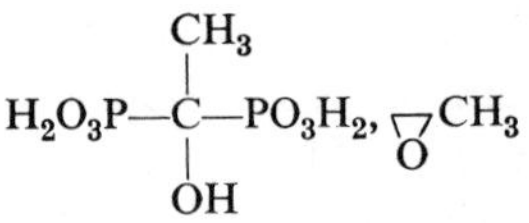

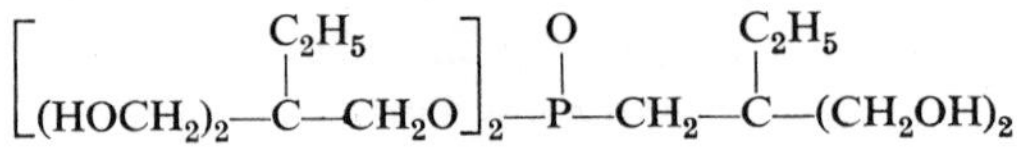

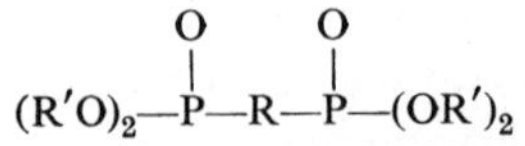

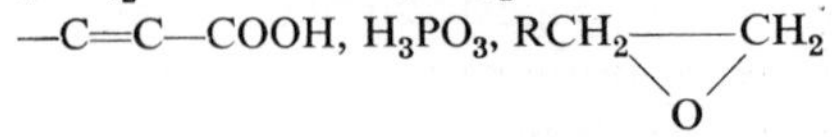

Table 8-8   Halogenated Phosphorus Compounds for Polyurethanes

Compounds	Type	% P	% X	Reference
Chloro Compounds				
Phosphates				
$(ClCH_2CH_2O)_3PO$	Additive	1	3.85	64
$((ClCH_2)_2-CHO)_3PO$	Additive	1	3–4	66
$(ClCH_2-CH-CH_2O)_3PO$ $\quad\;\; \mid$ $\quad\;\; OH$	Reactive	$\cdots$	$\cdots$	67
$ClCH_2-CH_2OH, H_3PO_4$	Reactive	1	10	68
$\left(\begin{array}{c} CH_3CH-CHO \\ \mid \quad\;\; \mid \\ Cl \quad OH \end{array}\right)_3 PO$	Reactive	$\cdots$	$\cdots$	15
$ClCH_2CH_2O-\overset{\overset{\displaystyle O}{\|}}{P}-(OH)_{2'}$ or $(ClCH_2CH_2O)_2-\overset{\overset{\displaystyle O}{\|}}{P}-OH$	Additive	$\cdots$	$\cdots$	69
$Polyol-O-\overset{\overset{\displaystyle O}{\|}}{P}-(OCH_2CH_2Cl)_2$	Reactive	0.5	34	70
$[(HORO)_2POCH_2]_2-C(CH_2Cl)_2$	Reactive	2.7	3.1	71
$\begin{array}{c} H \\ \mid \\ CH_2-O-P-OCH_2CH_2Cl \\ \mid \\ CH_2-O-P-OCH_2CH_2Cl \\ \mid \\ H \end{array}$	Additive	1.0	1.1	72
$CH_3-\!\!\bigcirc\!\!-N-\overset{\overset{\displaystyle O}{\|}}{C}-P(OCH_2CH_2Cl)_2$ with H, OCN	Reactive	3.9	9.0	73
Alkyl phosphates plus polyesters from hexachlorocyclopentadiene derivatives	Additive	1	~15	74
Phosphate polyols plus chlorinated isocyanates	Reactive	2–3	~15	17, 75, 76
Phosphites				
$(ClCH_2CH_2O)_3P$, brominated polyester	Additive	$\cdots$	$\cdots$	77
$(ClCH_2CH_2CH_2O)_3P$	Additive	$\cdots$	$\cdots$	78
$Polyol\ O-P-(OCH_2CH_2Cl)_2$	Reactive	0.43	34	79
Cyanuric chlorides, $(RO)_3P$	Additive	$\cdots$	$\cdots$	80

(continued overleaf)

Table 8-8   (Continued)

Compounds	Type	% P	% X	Reference
Phosphonates				
$[H(OCH_2CH_2CH_2)_2O]_2$—$\overset{\overset{\text{O}}{\|}}{P}$—$CCl_3$	Reactive	· · ·	· · ·	81
$(H(OCH_2CH_2)_2$—$O)_2$—$\overset{\overset{\text{O}}{\|}}{\underset{\underset{\text{OH}}{\|}}{P}}$—$CHCCl_3$	Reactive	0.9	3.1	82
PCl_3, glycol, alkylene oxide, alkane-carboxaldehyde reaction, products	Additive	· · ·	· · ·	83
H_3PO_3, propylene oxide or dipropylene glycol, epichlorohydrin reaction products	Reactive	1.4	15	84
R—$\overset{\overset{\text{O}}{\|}}{\underset{\underset{\text{O—dipropylene glycol}}{\|}}{P}}$—$OCH_2\overset{\overset{\text{Cl}}{\|}}{C}HCH_3$	Reactive	· · ·	· · ·	85
HORO—$\overset{\overset{\text{O}}{\|}}{\underset{\underset{\underset{\underset{\underset{CH_2CH_2Cl}{\|}}{O}}{\|}}{\|}}{P}}$—$CH_2CH_2Cl$	Reactive	· · ·	· · ·	86
$(ClCH_2CH_2O)_2$—$\overset{\overset{\text{O}}{\|}}{P}$—$CH_2$—(ring with NCO, NCO)	Reactive	· · ·	· · ·	87

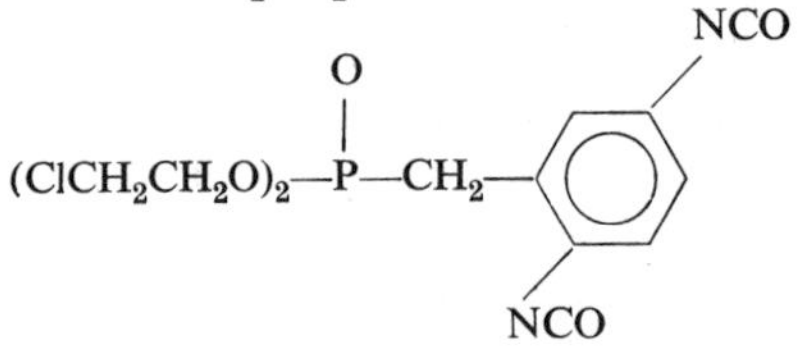

Bromo Compounds

Compounds	Type	% P	% X	Reference
Phosphates				
$(BrCH_2\overset{\overset{}{}}{\underset{\underset{Br}{\|}}{C}}HCH_2O)_3PO$	Additive	0.5	7	47, 88
$(HORO)_2$—$\overset{\overset{\text{O}}{\|}}{P}$—$CH_2CH_2Br$	Reactive	2.9	7.5	89
Brominated phosphate polyol	Reactive	· · ·	· · ·	90
$(Br_3\phi OCH_2\overset{\overset{}{}}{C}HCH_2O)_2$—$\overset{\overset{\text{O}}{\|}}{\underset{\underset{\text{OH}}{\|}}{P}}$—$OROH$	Reactive	· · ·	· · ·	91

Table 8-8 (Continued)

Compounds	Type	% P	% X	Reference
Brominated phosphate ester of prop- oxylated tall oil acid	Reactive	0.6	4	92
Phosphites and phosphonates				
Pentabromodiphenylether plus $(HOCH_2CH_2O)_2PHO$	Reactive	$\cdots$	$\cdots$	93, 94

$(BrCH_2CHCH_2O)_3P$
     |
     Br
 or
          O
          |
 $(BrCH_2CHCH_2O)_2$—P—$CH_2CHCH_2Br)$
     | |
     Br   Br
	Additive	0.12	2.1	95

Polyether from glycidyl phosphonate and glycerol, poly(tribromostyrene)		2	8.5	96

generate HX on prolonged exposure to moisture. Thus, the halogenated phosphonate esters are stored with the isocyanate component in two-part premix systems to avoid inactivating the amine catalyst normally packaged with the polyol.

The bromine-phosphorus additive is an efficient means of adding fire retardance, but the cost is relatively high. In today's market cost is a sensitive issue, with many users seeking zero upcharge for fire safety. As pressure for fire retardants increases in the legislative arena, this factor should gradually change.

Other Phosphorus Compositions

There has been some laboratory study of phosphonitrilics [97]—$(PNX_2)_n$ derivatives—where X is partially replaced by various groups. In some cases a polyol is substituted, rendering the retardant reactive. THPC—$(HOCH_2)_4P^+Cl^-$—has been evaluated either as an additive or prereacted with a polyol (e.g., 6 hr with trimethylolpropane at 175°) and used in place of ordinary polyols [98]. The stability of phosphine oxides can be conferred on the polyol component of a urethane by using some well-known chemistry [99]:

$$\text{THPC} \xrightarrow{\hspace{2cm}} (HOCH_2)_3PO \xrightarrow[\text{(2) hydrolysis}]{\text{(1) epichlorohydrin}}$$

$$(HOCH_2CHCH_2OCH_2)_3PO$$
$$\hspace{2cm} |$$
$$\hspace{2cm} OH$$

Such a compound should be a good deal more hydrolytically stable than structures with —P—O—C— linkages. The penalty in cost for either the phosphine oxides and phosphonium compounds or the phosphonitrilics is excessive and these, even if commercially available sometime in the future, will find only limited use.

Even P_4S_{10}, P_4S_7, and P_4S_3 have been suggested for polyurethanes [99a].

Halogen Compositions

Halogens, principally chlorine or bromine, may be introduced as part of the polyol (most common), as part of the isocyanate, or as an inert additive. Probably the most familiar means of adding chlorine is via the Diels-Alder adducts of hexachlorocyclopentadiene. A commercial product of considerable interest is based on the maleic acid adduct—chlorendic acid:

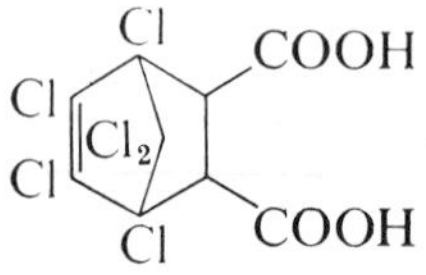

+ triol + alkanedicarboxylic acid − polyester polyol

Chlorendic or HET acid

The polyester polyols are useful in rigid, semirigid, and flexible urethane foams [100] and can be readily adapted to use in any of the three types by selection of the alcohol and the ratio of ingredients in the final foam. Alcohols that have been used include glycerol, trimethylolpropane, and pentaerythritol. Propylene oxide can be used in place of the alcohol and then a linear form will result suitable for flexible foam. The dicarboxylic acid is not essential; when used it is usually adipic acid: $HOOC(CH_2)_4COOH$. Enough halogen-containing compound is used to give about 18–20% Cl in the finished polyurethane.

Related compounds with the halogen on a ring are based on hydronaphthalene dicarboxylic acid [101], a saturated version of chlorendic acid [102] (no double bond at position 5), tetrachlorophthalic anhydride (103], and derivatives of chlorinated bisphenol A [104]. The amount of chlorine needed with the aromatic tetrachlorophthalic anhydride is about the same as with chlorendic acid: ∼18%.

Chlorine can also be obtained in a polyol directly by using

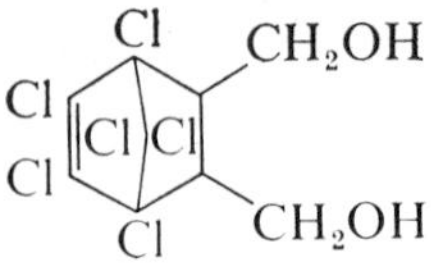

instead of chlorendic acid. This can be reacted as is with the isocyanate [105]. The isocyanate may also be chlorinated directly or adducts with, for example, Cl_3BrC may be formed [106]. Nonreactive chlorinated aromatics that have been studied include chlorinated biphenyl [107] and polyvinylchloride [108]—the latter for polyurethane rubber.

In a series of patents the reaction of chlorinated epoxy compounds with glycols is discussed [109]. The reaction is, for example:

$$Cl_3CCH_2CH{-}CH_2 + HOCH_2CH_2OH \xrightarrow[\substack{ether \\ N_2}]{BF_3}$$
$$\underset{O}{\diagdown \diagup}$$

$$Cl_3CCH_2CH{-}CH_2OCH_2CH_2OH$$
$$\underset{OH}{|}$$

Epichlorohydrin and glycerol may be used with HBF_4. Other polyols that have been employed include erythritols and sorbitol. About 20% chlorine is required to flameproof polyurethane in this manner.

Chloral and polyols react to give polyols with high chlorine content [110]:

$$Cl_3CHO + HOROH \rightarrow Cl_3CHOROH$$
$$\underset{OH}{|}$$

Other chlorinated products tried in urethanes include maleic acid [111], resinous diols [112], mixtures of polyols treated with Cl_2 [113], hexachlorometaxylene [114], pentachlorophenol with epichlorohydrin [115], and chlorinated polyisobutylene [116]. The list could easily be lengthened, but as far as is known none of these is used in any significant way by the urethane foam industry.

There are far fewer references to bromine compounds in urethane systems. Bromine may be added in glycols [117], in a polyester made from succinic acid [118], in the isocyanate [119], or as nonreactive additives, such as bromotoluene or brominated toll oil [120]. Levels of bromine required are considerably lower than for chlorine. 8–10% Br greatly retards burning and 12–14% Br gives a nonburning foam. In a study of tetrabromophthalic anhydride polyester polyols for urethane, Pape, Sanger, and Nametz obtained the results shown in Table 8-9 [121]. In this work as little as 4% Br gave a self-extinguishing foam. The factor of 2 as between bromine and chlorine is thus demonstrated once again. The price of bromine is apparently still too high compared to chlorine to encourage much of a switch to bromine in these low-cost polymers. (But this may change soon.)

Polyvinylidene fluoride has been claimed as a fire retardant for polyurethanes [122]. It must be used with a catalyst—an aluminum compound; otherwise it is too stable to provide the necessary HF.

Table 8-9   Flammability of Urethane Foams Containing Bromine from Tetrabromophthalic Anhydride (after Pape et al. [121])

Recipe	1	2	3	4	5	6	7
Component (eq. wt.)							
Polymeric polyisocyanate (145), eq.	1.05	1.05	1.05	1.05	1.05	1.05	1.05
TBPA diester (324), eq.	0.10	0.15	0.21	0.28	0.30	0.34	0.41
Sorbitol-based polyol (87), eq.	0.55	0.50	0.44	0.42	0.40	0.36	0.34
Polypropylene glycol triol (206), eq.	0.15	0.15	0.15	0.10	0.10	0.10	0.10
Amine-based polyol (73), eq.	0.20	0.20	0.20	0.20	0.20	0.20	0.15
Refrigerant 11, %	16	16	16	16	16	16	16
Surfactant, %	1	1	1	1	1	1	1
Amine catalyst, %	0.3	0.3	0.3	0.3	0.3	0.3	0.3
Properties							
Density, lb/ft^3	1.8	1.9	1.8	2.0	2.1	2.1	2.2
% Br	4.1	6.0	8.0	10.1	11.0	12.1	14.0
Flammability, in. burned (ASTM D-1692)	1.2	0.6	0.3	0.1	0.05	⋯	⋯
Rating	SE	SE	SE	SE	SE	NB	NB

Antimony-Halogen Compositions

Whereas 20% or so of chlorine or 12% or so of bromine is required alone, the presence of antimony reduces these levels considerably [123, 124]. The following mixtures have been suggested:

10% Sb_4O_6 + 10% F [125] (as polyvinylidene fluoride)
4.4% Sb_4O_6 + 3.8% Cl [126] (chlorinated aliphatic)
1.3% Sb_4O_6 + 7.0% Cl [127] (chlorinated aliphatic)
5.9% Sb_4O_6 + 2.4% Cl [128] (chlorinated polyetherpolyol)
2.5% Sb_4O_6 + 2.4% Br [119] (brominated isocyanate)
3.5% Sb_4O_6 + 7.5% Br [129] (brominated aromatic additive)

The most efficient system is clearly the Sb_4O_6-Br mixture. There is one patent on the use of Sb_4O_6 with dibromopropanol [129a]. Pape has obtained a series of results with a tetrabromophthalic anhydride polyester polyol. Some of these are shown in Table 8-10. Despite this good performance, the higher cost of bromine has reduced interest in it. The insolubility of the Sb_4O_6 inhibits its use in prepackaged foam systems because of the settling out of the solid.

Table 8-10   Results with Sb_4O_6 and a Polyester
Polyol Based on Tetrabromophthalic
Anhydride (Pape et al. [121])

% Sb_4O_6	% Br	Rating	In. Burned (ASTM D-1692)
1	2	SE	0.1
1	4	SE	0.1
1	6	SE	0.1
3	2	SE	0.1
3	4	SE, NB	0
3	6	NB	· · ·

Miscellaneous

There has been a little work on Sb(III) esters [130], Sb_4O_6-organoboron-halogen mixtures [131], a boric acid-glycerol ester [132], a brominated boric acid ester [133], fluoborate [108], and some nitrogen compounds [134]. Recently a complex between Sb(III) and hydroxyalkyl esters of phosphoric acid has been synthesized which permits solubility of the Sb(III) in the polyol [135]. This avoids the settling problem.

Comparative Data

There are about a dozen papers containing comparisons among various series of fire retardants for polyurethanes. Without belaboring the question, some of these results are presented here. For data on formulations and details of test variations, consult the original papers.

Piechota [136] found that it mattered little what form phosphorus is in so long as the compound can decompose to acids on heating (probably a poly-acid). He further pointed out that performance maximizes at about 1.5% P and at higher levels fire retardancy actually diminishes (see Figure 8-3). Hilado, Burgess, and Proops [4] have compared the effect of various flame retardants as a function of the type of polyurethane. Figure 8-4 is a sample of some 31 such graphs in the original paper. Obviously there are differences in the various polyol systems. Dickert and Toone [8] list the requirements for various combinations of elements from their studies (Table 8-11). Note that this requirement is to be *nonburning* rather than self-extinguishing. The values differ from those of other investigators, for example, 2% P by several groups, and so on. But the summary is useful. Jolles [137] has given two useful comparisons; one between various halogen compounds (Table 8-12) and one between bromine and chlorine in the presence of antimony oxide. Notice the marked difference in these results between halogenated aromatics and aliphatics. Bromoaliphatics are best, chloroaliphatics are about equal to

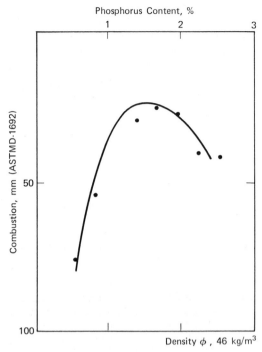

Figure 8-3 Flame-retardant properties of self-extinguishing foams made with increasing amounts of phosphoric acid polyalkylene glycol ester. (Phosphorus content of total foam versus combustion length of foam samples by ASTM D-1692.) (From Piechota [136].)

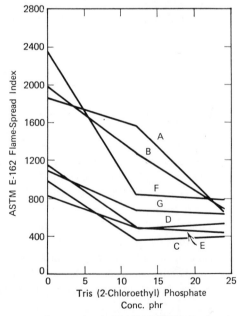

Figure 8-4 Flame spread versus level of additive P-Cl retardant in seven different formulas: A, B, C—sucrose based 1.54, 1.5, 2.0% P in the polyol; D, E—aromatic based 1.0, 1.2% P in polyol; F, G—aromatic-amino polyol 0, 1.0% P in polyol. (After Hilado et al. [4].)

Table 8-11 Amount of Flame-Retarding Elements for Nonburning
Foam by ASTM D-1692-59T (Dickert and Toone [8])

Foam	A	B	C	D	E	F	G	H
Chlorine, %	13	···	···	···	6.5	6.5	···	···
Bromine, %	···	10	···	···	···	···	5	5
Antimony, %	···	···	6	···	···	3	···	3
Phosphorus, %	···	···	···	2	1	···	1	···

Table 8-12 Halogen Required to Produce a
2–3 Sec Burning Time in Rigid Polyurethane
(Jolles [137])

Compound	Halogen, %
Hexachlorobenzene	15.8
Hexachlorocyclohexane	5.23
Hexabromobenzene	6.2
ar-Pentabromoethylbenzene	6.0
1,2-Dibromoethane	3.9
1,2,3,4-Tetrabromobutane	3.1

Table 8-13 Comparison of Fire-Retardant Effectiveness of Chlorine (as Trichloro-
Phenol) and Bromine (as Tribromophenol) in a Rigid Polyurethane Foam (Polyester
Polyol/TDI by ASTM D-1692-59T) (Jolles [137])

	Fire Retardant in Foam, % (w/w)	Halogen Content of Treated Foam, %	Antimony Oxide in Treated Foam, %	Time of Burning, sec	Extent of Burning, in.	Rate of Burning, in./min	ASTM Class
Trichlorophenol	11.7	6.3 Cl	5.0	66.2	1.50	1.39	SE
	10.8	5.8 Cl	2.8	85.8	2.47	1.62	SE
	8.0	5.8 Br	2.5	13.1	0.31	1.10	SE
Tribromophenol	7.8	5.7 Br	5.1	0	0	···	NB
	5.6	4.0 Br	5.0	54.6	1.43	1.44	SE

Table 8-14   Results of Vertical Bar Flammability
Tests—All Systems at 1.8% P[27]

Type of Fire Retardant	Average[a] Flame Spread	Average[a] Weight Loss, %
Control (no retardant)	100	28.9
Ammonium polyphosphate	32.9	10.0
Commercial "reactive" organo- phosphorus compound	38.7	15.8
Diammonium phosphate	36.3	9.2
"Nonreactive" halogenated organophosphorus compound	33.3	14.1
"Nonreactive" nonhalogenated organophosphorus compound	38.6	9.6
Triammonium pyrophosphate	37.2	9.4
Phosphorylated polyol	42.8	15.8

[a] Values for all foams from polyol premixes aged various
ways but prior to any environmental conditioning

bromoaromatics, and chloroaromatics are the least effective. The bromine-antimony combination is clearly superior on a weight basis to the chlorine combination.

Hilado [138] has reported dozens of results on phosphorus-containing foams and has compared behavior in various geometries and with various heat fluxes. Lyon and Applewhite [139] have given comparative data on chlorinated and brominated castor oil versus some commercial phosphorus-bearing retardants. The effect of Sb_4O_6 on the performance is quite striking in this work.

Table 8-15   Immersion Tests on Foams From
Table 8-14

Type of Fire Retardant	Flame Spread	
	Before Immersion	After 7 Days immersion in H_2O
Control (no retardant)	100	100
Ammonium polyphosphate	32.9	35.5
Commercial "reactive" organo- phosphorus compound	38.7	42.2
Diammonium phosphate	36.3	90.0
"Nonreactive" halogenated organophosphorus compound	33.3	30.0
"Nonreactive" nonhalogenated organophosphorus compound	38.6	83.3
Triammonium pyrophosphate	37.2	83.3
Phosphorylated polyol	42.8	36.7

Table 8-16   Aging at 200°C versus
Burning Rate (Anderson [140])

Fire-Resistant Agent	Init. Dist. Burned, in.	Burning Rate, in./min		Crit. Time, min
		Init.	Max.	
P-Polyol A	1.0	1.7	2.6	· · ·
P-Cl-Polyol	1.2	1.7	2.6	· · ·
$CH_3PO(OCH_3)_2$	0.9	1.3	7.5	5
$(CH_3O)_3PO$	0.9	1.4	7.5	5
$Cl_3CPO(OC_2H_5)_2$	1.9	2.4	6.0	5
$(ClC_2H_4O)_3PO$	1.8	1.9	7.5	5
$(ClCH_2CHClCH_2O)_3PO$	1.4	1.7	8.3	5
$(NH_4)_2HPO_4$	1.5	2.0	3.0	· · ·
None	Non-SE	5.9	9.2	· · ·

Miles and Lyons [27] compared a group of commercially available materials in the phosphorus group. Results of vertical bar tests are shown in Table 8-14. Several of the systems appear to be effective after various kinds of exposures but total immersion in water separated the systems cleanly. (See Table 8-15.) This leached out the $(NH_4)_2HPO_4$ and several of the other retardants and only the organic halophosphorus compound, the ammonium polyphosphate, and the two phosphorus polyols resisted leaching. Anderson [140] has addressed himself to the effect of aging at elevated temperatures and at high relative humidities. Some of his findings are shown in Tables 8-16 and 8-17. It is noteworthy again that $(NH_4)_2HPO_4$ does very well in such tests. It would seem that more use might be made of this low-cost flame retardant

Table 8-17   Humid Aging versus Burning Rate
(Anderson [140])

FR Agent	Water Solubility	Burning Rate, in./min		Crit. Time, days
		Init.	Max.	
P-Polyol A	Insol	1.7	2.6	· · ·
P-Polyol B	Sol	1.5	2.3	· · ·
P-Cl-Polyol	Insol	1.7	2.7	· · ·
$CH_3PO(OCH_3)_2$	Sol	1.6	4.9	14
$(CH_3O)_3PO$	Sol	1.3	4.4	14
$Cl_3CPO(OC_2H_5)_2$	Insol	2.5	3.8	· · ·
$(ClC_2H_4O)_3PO$	Insol	1.9	2.7	· · ·
$(ClCH_2CHClCH_2O)_3PO$	Insol	1.9	3.3	· · ·
$(NH_4)_2HPO_4$	Sol	1.6	2.2	· · ·
None	· · ·	5.9	5.9	· · ·

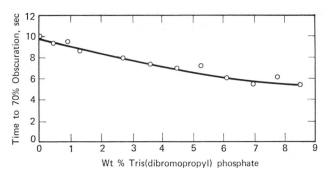

Figure 8-5   Effect of a nonreactive fire retardant on smoke development.

except when immersion in water is anticipated. It does, however, cause a settling problem for prepackaged foam systems.

Finally, Mickelson and Einhorn [47] have studied smoke generation. Figure 8-5 shows a typical curve. Flame-retarded foams often have a tendency to generate excessive smoke. This can be a severe problem and more work needs to be done on many aspects of the smoke hazard.

(A paper, received too late to include excerpts, intercompares the same types of retardants and includes heat-aging data [140a].)

POLYESTERS

This group includes both linear and cross-linked polymers. The simplest are linear polymers made, typically, from one of the phthalic acids and a diol:

$$\text{HOOC}-\underset{}{\bigcirc}-\text{COOH} \quad + \quad \text{HOROH} \longrightarrow$$

$$----\text{ORO}\!\left(\!\!\underset{O}{\overset{}{\underset{\|}{C}}}\!-\!\bigcirc\!-\!\underset{O}{\overset{}{\underset{\|}{C}}}\text{ORO}\!\right)\!\!\underset{O}{\overset{}{\underset{\|}{C}}}$$

Isophthalic and terephthalic acids are used as well as phthalic acid (or their anhydrides). Polyester fibers are spun from the reaction product of terephthalic acid and ethylene glycol. The high temperature of this melt-spinning process ($\sim$300°C) presents special problems of stability for fire retardants. More complex polyesters are used for reinforced shapes and for polyester coatings. In these, unsaturated diacids are incorporated in the polymer and cross-linking is achieved via the points of unsaturation. For example, if some of the phthalic acid is replaced by maleic acid, cross-linking can subsequently be achieved with an agent like divinylbenzene:

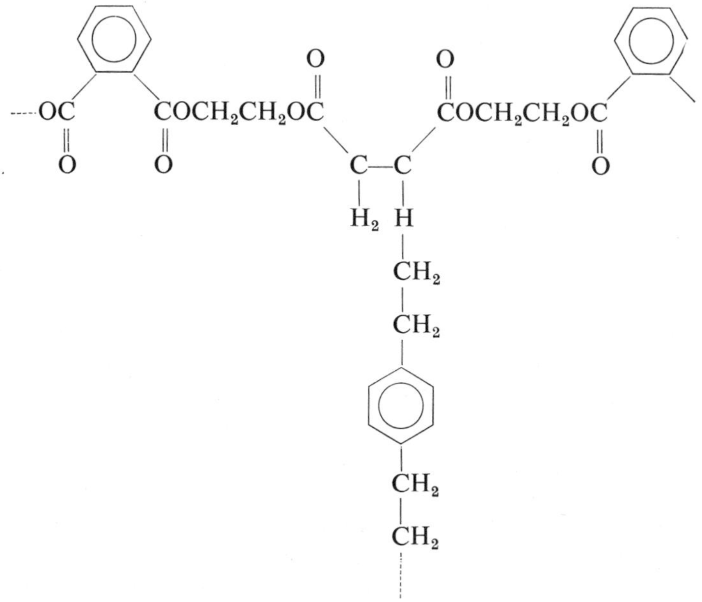

These polyesters are called alkyd resins in the coatings industry. The vinyl polymerization is irreversible and the resin so formed is a thermosetting species whereas the simple polyesters are thermoplastic. The myriad possible combinations of polyacids, polyols, and cross-linking agents lead to almost endless variation in end products [141]. Since there is no one material that can be defined as a polyester, broad statements as to flammability are not possible. Hilado [3] refers to the tendency of polyesters to produce dense smoke and indicates that as a class they are fairly flammable. The amount of research that has been directed to improving their flame resistance attests to the existence of a considerable problem. That problem takes a different form in fibers than in glass-fiber reinforced laminates. These laminates and the filled polyesters have received most of the attention. The fiber problem at this writing remains largely unsolved and is a pressing one in view of forthcoming federal regulations on the flammability of fabrics. (See Chap. 5, ref. [2, 2a].)

Madorsky [142] has reviewed the available information on the thermal degradation of the simple polyester, poly(ethylene terephthalate)—see above. The polymer does not unzip but at about 300°C slowly decomposes. The gaseous products of pyrolysis under nitrogen are shown in Table 8-18. The

Table 8-18   Gaseous Products of Pyrolysis of
Poly(Ethylene Terephthalate)
Under N_2 at 288°C [142]

Product	Mole % of Gaseous Species
CO	8
CO_2	8.7
H_2O	0.8
CH_3CHO	80.0
C_2H_4	2.0
2-Methyldioxolan	0.4
CH_4	0.4
C_6H_6	0.4

principal gaseous product is acetaldehyde. Apparently the degradation proceeds by thermal scission at the —C—O— linkages. The rate at 288°C is
sufficiently slow that melt spinning is readily accomplished without serious
loss.

There have been few good reviews of fire retardants for polyesters.
Roberts [143] has discussed the subject primarily in terms of halogens;
Nametz [144] also discusses the halogens and presents some very useful data
on the effectiveness of bromine compounds (see below). Both Hilado's book
[3] and one of his papers [145] are helpful. A commercial view of the situation has been prepared by Naturman [146].

Phosphorus Compounds

There are few published references to inorganic phosphates used in polyesters. Red phosphorus at 5–10% has been claimed despite the obvious
difficulties [147]. However, ammonium polyphosphate (see p. 35, Chapter 2)
has been used successfully in filled polyesters. The polyphosphate serves both
as filler and as fire retardant. At least 4% P (with about 2% N) is required to
achieve a degree of fire resistance [147a]. Organic phosphates are also
infrequently cited. Triphenyl phosphate is used as a stabilizer for polyesters
[148] and flame-resistant polymers have been made from allyl diphenyl
phosphate [149].

Phosphate ester polyols can be made by simply reacting glycols with

$$RO—\overset{\displaystyle O}{\underset{\displaystyle |}{P}}—Cl_2$$

$POCl_3$ [149a] or with RO—P—Cl₂ [150]. This introduces the possibility of
hydrolysis with rupture of the polymer backbone and subsequent loss
of integrity of the resin. A similar problem may arise with derivatives of

phosphorus isocyanate: $P(NCO)_3$ or $OP(NCO)_3$ [150a]. Phosphorus in a pendant position has been studied in a P—N compound of the formula

$$(RO)_2\overset{\overset{\displaystyle O}{\|}}{P}—N(R'OH)_2$$ [150b]. The problem again will be hydrolytic stability at the P—N bond. In any case, the bulk of the work has been on phosphites and phosphonates.

About 5% phosphorus incorporated as an organic phosphite or phosphonate is effective. Hindersinn and Boyer [151] discuss the incorporation of trialkyl phosphites into polyesters. They place the phosphorus in a pendant position on the chain to avoid chain scission in the event of hydrolysis. Beginning with unsaturated groups from maleic acid, the phosphite is added without difficulty to produce a phosphonate resin. At 5.3% phosphorus a burning rate of 0.25 in./min by ASTM D-757-49 is obtained.

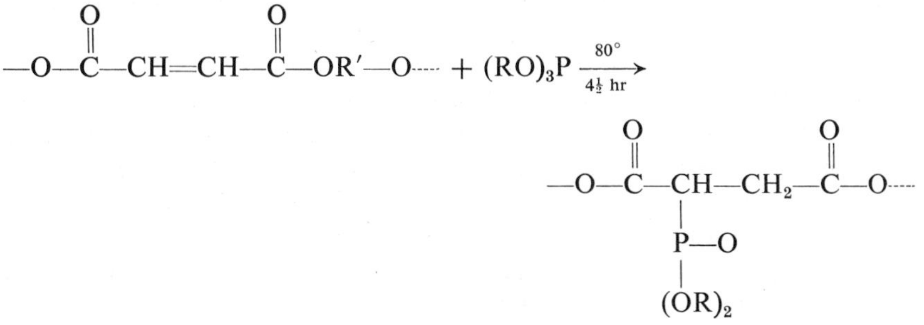

$$—O—\overset{\overset{\displaystyle O}{\|}}{C}—CH{=}CH—\overset{\overset{\displaystyle O}{\|}}{C}—OR'—O{\cdots} + (RO)_3P \xrightarrow[4\frac{1}{2}\ hr]{80°}$$

$$—O—\overset{\overset{\displaystyle O}{\|}}{C}—\underset{\underset{\displaystyle (OR)_2}{\underset{\displaystyle |}{P—O}}}{\underset{\displaystyle |}{CH}}—CH_2—\overset{\overset{\displaystyle O}{\|}}{C}—O{\cdots}$$

Successful use in acrylate- or styrene-modified polyester-glass laminates is discussed in this paper. Bahr et al. have added dialkyl phosphite to castor oil to form a pendant phosphonate and then prepared a polyester for use in urethanes [151a]. Trialkyl phosphites will react with maleic anhydride units at the anhydride portion in, for example, styrene-maleic anhydride copolymers [152]. Cyclic phosphites react with carboxylic acids to form adducts useful as flame retardants [153]. One supposes that a mixed anhydride is formed of questionable hydrolytic stability. (For example, acetyl phosphate is almost as unstable as acetic anhydride.) Dialkyl phosphites react with the double bond in the maleate group, as do trialkyl phosphites. A catalyst is required along with higher temperatures. Thus

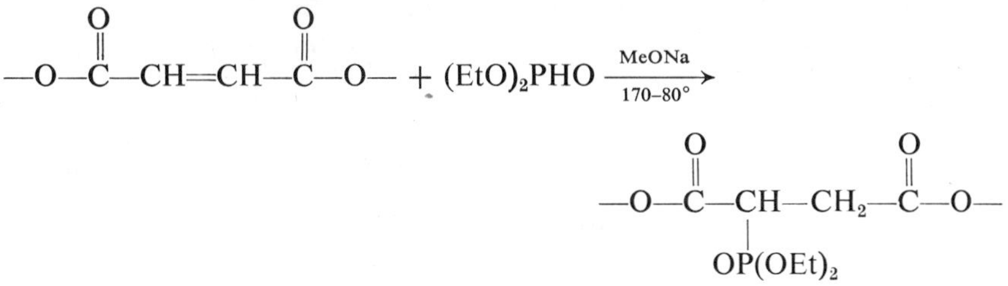

$$—O—\overset{\overset{\displaystyle O}{\|}}{C}—CH{=}CH—\overset{\overset{\displaystyle O}{\|}}{C}—O— + (EtO)_2PHO \xrightarrow[170-80°]{MeONa}$$

$$—O—\overset{\overset{\displaystyle O}{\|}}{C}—\underset{\underset{\displaystyle OP(OEt)_2}{\displaystyle |}}{CH}—CH_2—\overset{\overset{\displaystyle O}{\|}}{C}—O—$$

The product is a hygroscopic solid containing 11% phosphorus and is self-extinguishing [154]. The dialkyl phosphites are readily transesterified to produce polymers of polyols and phosphite with phosphorus in the backbone [155]:

$$n\text{RO}\!-\!\underset{\underset{\text{O}}{|}}{\overset{\overset{\text{H}}{|}}{\text{P}}}\!-\!\text{OR} + n\text{HOR}'\text{OH} \rightarrow \text{RO}\!\left(\!\underset{\underset{\text{O}}{|}}{\overset{\overset{\text{H}}{|}}{\text{P}}}\text{OR}'\text{O}\!\right)_{\!n}\!\!\text{H} + (2n-1)\text{ROH}$$

Basic catalysts enhance the reaction rate.

Phosphonates may be used as reactants. Phenyl phosphonate has been incorporated in polyesters by transesterification to yield phosphorus-terminated resins: $\text{RO}\!-\!\underset{\underset{\phi}{|}}{\overset{\overset{\text{O}}{|}}{\text{P}}}\!-\!\text{ORO}\!-\!\overset{\overset{\text{O}}{\|}}{\text{C}}\!-\!\text{R}'\!-\!\overset{\overset{\text{O}}{\|}}{\text{C}}\text{-----}$, where R is alkyl or H.

Some acidity seems unavoidable. Phosphorus levels from 4.2 to 5.6% have been obtained and the polymers were rated as self-extinguishing in a fire tube test, losing less than 10% in weight [156]. When vinylphosphonic acid is used, in ester form, as a comonomer for unsaturated polyesters, at 5% phosphorus the resin ignites with difficulty but is self-extinguishing [157]. Allyl phosphonic acid dichloride may be used to provide cross-linked polymers with diols (a diene may be used also) [158]:

$$\begin{array}{ccc}
\text{CH}\!=\!\text{CH}_2 & & \text{---CH}\!-\!\text{CH}_2\text{---} \\
| & & | \\
\text{CH}_2 & & \text{CH}_2 \\
| & & | \\
\text{Cl}\!-\!\text{P}\!-\!\text{Cl} + \text{HOROH} \rightarrow & \left(\text{ORO}\!-\!\text{P}\!-\!\text{OR---}\right)_x \\
| & & | \\
\text{O} & & \text{O}
\end{array}$$

Extensive cross-linking can be achieved by means of dialkenyl alkenylphosphonates [159]. At 4% phosphorus in a resin made with bisphenol A and phenylphosphonic acid dichloride, self-extinguishing properties are obtained [160] (other polyols, aromatics, and aliphatics may be used):

$$\text{----O}\!-\!\underset{\underset{\text{O}}{|}}{\overset{\overset{\phi}{|}}{\text{P}}}\!-\!\text{O}\!-\!\langle\bigcirc\rangle\!-\!\underset{\underset{\text{CH}_3}{|}}{\overset{\overset{\text{CH}_3}{|}}{\text{C}}}\!-\!\langle\bigcirc\rangle\!-\!\text{O}\!-\!\underset{\underset{\text{O}}{|}}{\overset{\overset{\phi}{|}}{\text{P}}}\!-\!\text{O----}$$

Transesterification reactions have been employed with, for example, $(\phi\text{O})_2\!-\!\underset{}{\overset{\overset{\text{O}}{|}}{\text{P}}}\!-\!\text{CH}_2\text{Cl}$ [161] and $(\phi\text{O})_2\!-\!\overset{\overset{\text{O}}{|}}{\text{P}}\!-\!\text{CH}\!=\!\text{CH}_2$ [162].

A fairly new polyphosphonate has been claimed for use in polyesters [163]:

$$
\text{H}_2\text{O}_3\text{P}-\overset{\overset{\displaystyle \text{CH}_3}{|}}{\underset{\underset{\displaystyle \text{H}}{|}}{\underset{\displaystyle \text{O}}{\text{C}}}}-\text{PO}_3\text{H}_2 \;+\; \text{CH}_3-\text{CH}\underset{\displaystyle \text{O}}{\diagdown\!\diagup}\text{CH}_2 \;\rightarrow
$$

$$
\left[\left(\underset{\displaystyle \overset{|}{\text{OH}}}{\text{CH}_3\text{CHCH}_2\text{O}}\right)_{\!2}-\overset{\overset{\displaystyle \text{O}}{\|}}{\text{P}}-\right]_{2}\overset{\overset{\displaystyle \text{CH}_3}{|}}{\underset{\underset{\displaystyle \text{CH}_2\text{CHCH}_3}{|}}{\underset{\displaystyle \text{O}\quad\text{OH}}{\text{C}}}}
$$

Lower oxidation states of phosphorus have been but little studied in polyesters. Triallylphosphine has been evaluated at 30% in a polyester system [164] (5.5% phosphorus). Tetrakis(hydroxymethyl) phosphonium chloride reacts with polycarboxylic acids in a melt to produce a polyester useful for fabric finishing [165].

Phosphorus-Halogen Compositions

Although most work has been directed to halogens in polyesters (see below), the amount of halogen can be reduced by adding phosphorus. Table 8-19 lists the several ways that phosphorus and halogens have been combined for this purpose. From the entries it appears that 1% P and 15–20% Cl combine to produce self-extinguishing polyesters and that 1.5% P and 20% Cl provide excellent results. By switching to bromine, similar results may be obtained at a much lower halogen level. Again, the factor of 2 favoring bromine over chlorine is confirmed. Bis(β-chloroethyl) vinylphosphonate (called a phosphinate in several of the Russian articles—improperly translated into English) is reported on by several groups. When heated, this molecule condenses with evolution of dichloroethane [184]:

$$
(\text{ClCH}_2\text{CH}_2\text{O})_2-\overset{\overset{\displaystyle \text{O}}{\|}}{\text{P}}-\text{CH}\!=\!\text{CH}_2 \xrightarrow{210\text{–}240^\circ}
$$

$$
\text{ClCH}_2\text{CH}_2\!\left[\!\text{O}-\overset{\overset{\displaystyle \text{O}}{\|}}{\underset{\underset{\displaystyle \text{CH}\!=\!\text{CH}_2}{|}}{\text{P}}}-\text{OCH}_2\text{CH}_2\!\right]_{n}\!\!\text{Cl} + (n-1)\text{CH}_2\text{ClCH}_2\text{Cl}
$$

Therefore, much of its chlorine should be lost in processing molten polyesters. Many of the phosphonates cited in Table 8-19 are relatively expensive, as are the hexachlorocyclopentadiene adducts and the bromine compounds

Table 8-19   Phosphorus-Halogen Combinations for Fire-Resistant Polyesters

Phosphorus Compound	Halogen Compound	Polyester Type	Remarks	Reference
H_3PO_4 reacted with glycidyl ether or alkyleneoxide	Tetrachlorophthalic anhydride	Unsat., [a]maleic	1% P, 18% Cl, SE 3 sec	166
H_3PO_4, $ClCH_2CH\!-\!CH_2$ (epoxide, O)	$ClCH_2CH\!-\!CH_2$ (epoxide, O)	Unsat., maleic	2% P, 23% Cl	167
...	$(ClCH_2CH_2O)_3PO$	Unsat., for glass mat; styrene, methacrylics	1.3% P, 4.7% Cl	168
$(ClCH_2CH_2O)_3P$	Tetrachlorophthalic anhydride	Unsat., fumaric, styrene	0.6% P, 35% Cl	169
$(ClCH_2CH_2O)_3P$	...	Unsat., maleic	Reacted with maleate to give phosphonate in polyester	170
Et_3PO_4	HET acid (or anhydride)	Unsat., maleic	Transesterification of phosphate ester?	171
...	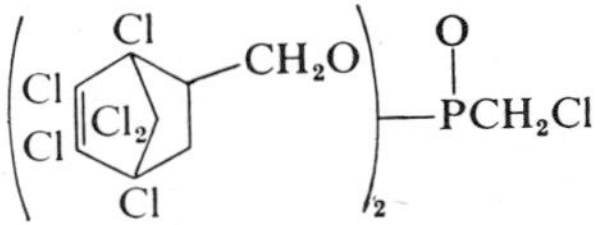 (tetrachloro aromatic structure) $\left(CH_2O\!-\!\overset{O}{\underset{}{P}}\!-\!PCH_2Cl\right)_2$	Unsat., fumaric	1% P, 15% Cl, SE	172
$POCl_3$, (tetrachloro aromatic CH_2OH structure)	... 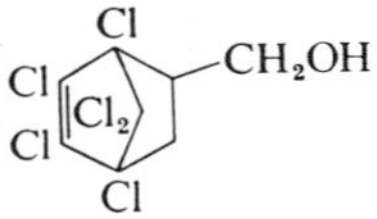	Sat, propoxylated	...	173

[structure: chlorinated bicyclic imide with S, NOP—$(OMe)_2$, Cl, Cl_2, Cl, Cl]	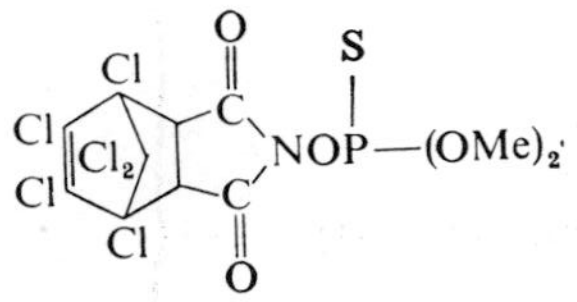 $\cdots$	$\cdots$	Unsat., maleic	174
$(RO)_2PHO$	Cl_3CHO	$\cdots$	$\cdots$	175
$(CH_3O)_2PHO$	HET acid	$\cdots$	Unsat., maleic, styrene	176
$(\phi O)_3P$	Tetrachlorophthalic anhydride or "hex" adducts	$\cdots$	Unsat., maleic	177
$(CH_3O)_3P$	HET acid	0.8% P, 10% Cl, SE but just barely	Unsat., fumaric, styrene	178
$(EtO)_3P$	Tetrachlorophthalic acid	$\cdots$	Unsat., maleic	179
$\cdots$	$(ClCH_2\text{—}CH\text{—}CH_2O)_2\text{—}P\overset{O}{=}$ $CH_2\text{=}CHCH_2O\overset{O}{=}$	2.1% P, 9.8% Cl	Unsat., maleic glass fibers	180
$\cdots$	$(ClCH_2CH_2O)_2\text{—}P\overset{O}{=}\text{—}CH\text{=}CH_2$	5% P, 12% Cl	Unsat.	181
$(ClCH_2CH_2O)_2\text{—}P\overset{O}{=}\text{—}CH_2\text{=}CH_2$	Tetrachlorophthalic anhydride	0.6–0.9% P, 17–21% Cl	Unsat.	182
$(ClCH_2CH_2O)_2\text{—}P\text{—}CH\text{=}CH_2$	HET acid	$\cdots$	Unsat., fumaric	183

(a) Unsat. = unsaturated type.

(continued overleaf

Table 8-19 (Continued)

Phosphorus Compound	Halogen Compound	Polyester Type	Remarks	Reference
Homopolymer of $(ClCH_2CH_2O)_2P(=O)-CH=CH_2$		. . .	Cross-links and losses CH_2ClCH_2Cl	184
$(CH_2=CHCH_2O)_2P(=O)-CH_2Cl$	Chlorinated polyvinyl compound	Unsat., maleic	. . .	185
. . . $(C_3H_7O)_2P(=O)-CH_2Cl$	. . .	Unsat., maleic glass fibers	. . .	186
. . . $ClCH_2CH_2PCl_2$, $HOCH_2CH_2OH$	. . .	Unsat., maleic	. . .	187
$CH_3-CH(OH)-P(=O)-(OMe)_2$	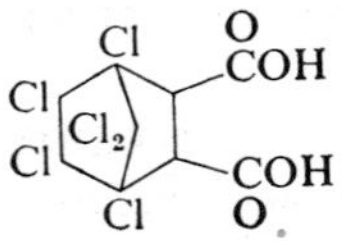	Unsat., maleic	. . .	188
$CH_3-C(OH)-(P(=O)-OH)_2)_2$ and $ClCH-CH_2$ (epoxide, O) condensate	HET anhydride	Unsat., maleic	1.5% P, 20% Cl, SE 7 sec	189

Compound / Structure		Type	Properties	Ref.
...	(aromatic phosphonic structure) $\overset{OH}{\underset{}{}}$ O $CH-P-OH$ OR, ring X_n	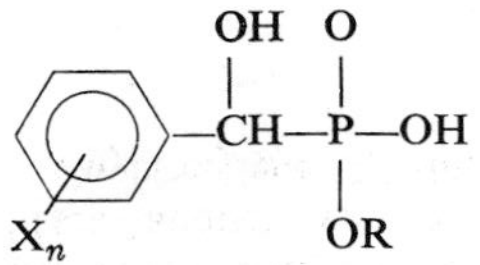 Sat., foam	...	189a
$\phi-\overset{O}{\underset{}{P}}-(OH)_2$		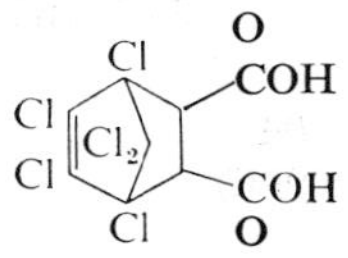 Unsat., maleic	...	190
Phosphorines	$ClCH-CH_2$ (O epoxide)	...	...	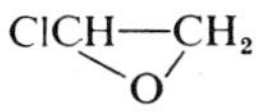 191
$(PNCl_2)_x$	Tetrachlorophthalic anhydride	Unsat., maleic	...	192
Bromine compositions				
...	$Br_2CH_2CH_2-\overset{O}{\underset{}{P}}-Cl_2$, glycols	Unsat., maleic	2.2% P, 5.7% Br light stable, SE 1.5 sec	193
...	Organo P—Br compounds	Unsat., maleic	2.2% P, 5.7% Br	194
$BrCH_2CH_2O-\overset{O}{\underset{ClCH_2CH_2O}{P}}-OCH_2-\overset{CH_2Br}{\underset{CH_2Br}{C}}-CH_2O-\overset{O}{\underset{OCH_2CH_2Cl}{P}}-OCH_2CH_2Br$		Unsat., maleic		195
$(CH_2=CHCH_2O)_3PO$	$Br_3\phi OCH=CH_2$	Unsat., fumaric	...	196
H_3PO_4	Dibromosuccinic acid	Unsat., maleic	...	197
...	$(BrCH_2CHCH_2O)_3PO$ and analogs, Br	...	...	198

Table 8-20   Halogenated Materials Used or Suggested as Fire Retardants for Polyesters

Class	Compound	Remarks	Reference
Alkylene oxides	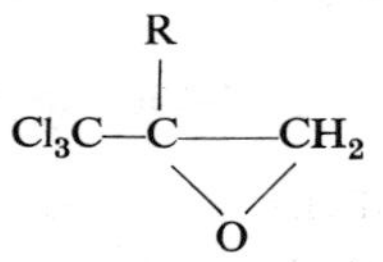 Cl_3C—C(R)—CH_2 epoxide	SE at 25% Cl	205
Chlorohydrins	$ClCH_2CH_2OH$	Improved fire resistance at even 12% Cl	206
	$(ClCH_2)_2C(CH_2OH)_2$	19% Cl	207
	$HOCH_2$—CCl_2—CO—CCl_2—CH_2OH	...	209
Unsaturated cyclics	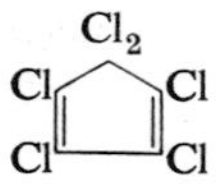 "Hex" (hexachlorocyclopentadiene structure)	Up to 50% Cl when used as cross-linker	210, 211
	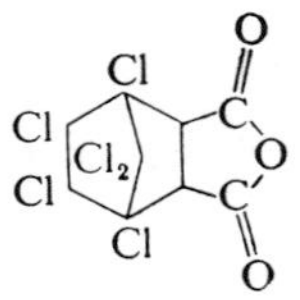 Chlorendic anhydride, HET anhydride (structure)	Used at 25–30% Cl, many references	212, 213
	Chlorendic acid, HET acid; "Hex" adducts with: divinyl ether, cyclobutene, vinylacetylene, 2,5-dihydrofuran, Cl_2		214, 215, 216, 217, 218

Saturated cyclics	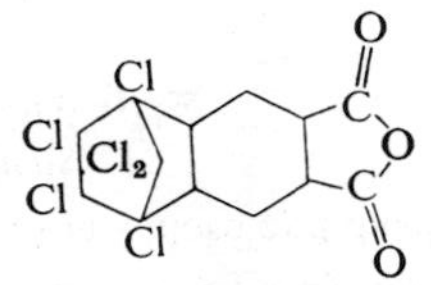 Hexachloro-endo-methylene octahydronaphthalene dicarboxylic anhydride		219
	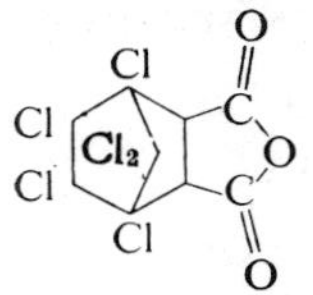 Hexachloro-endo-methylene tetrahydrophthalic anhydride	Can get 35–45% Cl in polyester	220
Aromatics	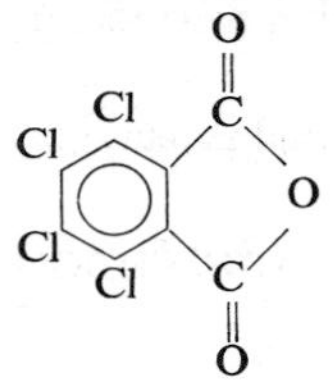 Tetrachlorophthalic anhydride	Up to 27% Cl possible in unsat. polyester; adding 2–3% Et_3PO_4 or other phosphorus compound is helpful	221 222 223

(continued overleaf)

Table 8-20 (Continued)

Class	Compound	Remarks	Reference
	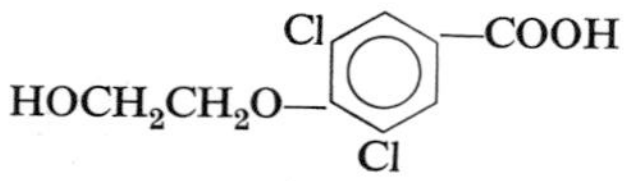 HOCH$_2$CH$_2$O—⬡(Cl)(Cl)—COOH	Self-polymerized, contains 30% Cl	224
	Chlorinated phenol derivatives	Up to 34% Cl from pentachlorophenol compounds	225
	Halogenated bisphenol A and analogs	...	226
	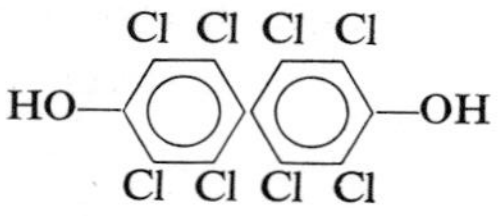 HO—⬡(Cl)(Cl)(Cl)(Cl)—⬡(Cl)(Cl)(Cl)(Cl)—OH	Up to 50% Cl in polyester with chloroacetic acid	227
	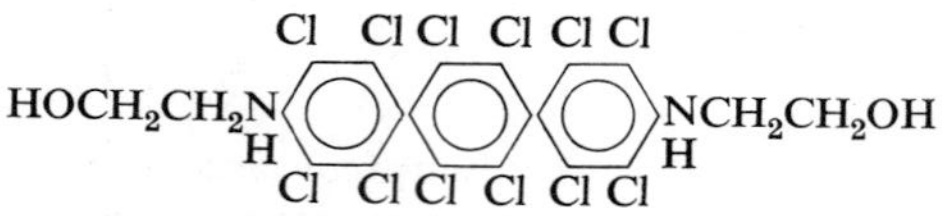 HOCH$_2$CH$_2$N(H)—⬡(Cl)(Cl)(Cl)(Cl)—⬡(Cl)(Cl)(Cl)(Cl)—N(H)CH$_2$CH$_2$OH (or the biphenyl homolog)	40–50% Cl in polyester	228
	Monochloro and dichlorostyrene	...	229
Bromine compositions Aliphatic	Br$_4$C$_{12}$H$_{22}$	11% Br in polyester (additive, nonreactive)	230

Structure	Notes	Ref.
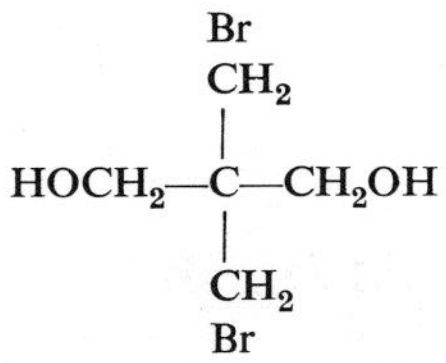 $\text{HOCH}_2-\overset{\overset{\displaystyle \text{CH}_2-\text{Br}}{\vert}}{\underset{\underset{\displaystyle \text{CH}_2-\text{Br}}{\vert}}{\text{C}}}-\text{CH}_2\text{OH}$	Up to 28% Br is possible	231, 232
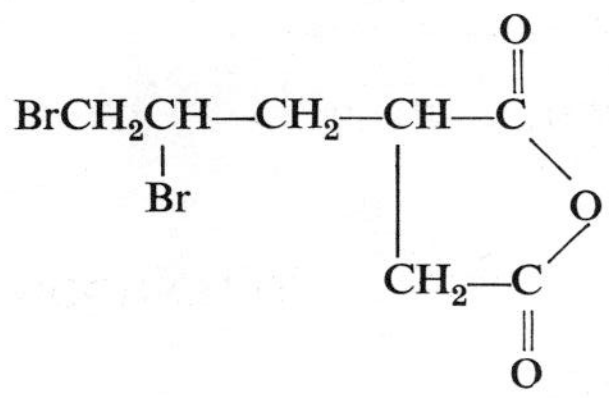 $\text{BrCH}_2\text{CH}-\text{CH}_2-\text{CH}$ (cyclic anhydride), Br, CH$_2$–C=O	24% Br in styrenated polyester	233
HO–(aromatic ring)–CH–CH$_2$Br, Br		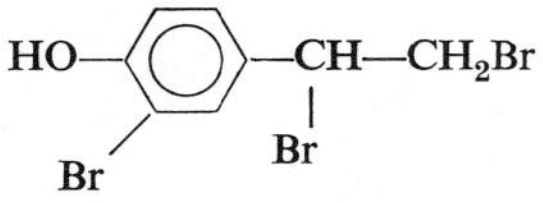 234
Saturated cyclics (bicyclic anhydride, Br Br)	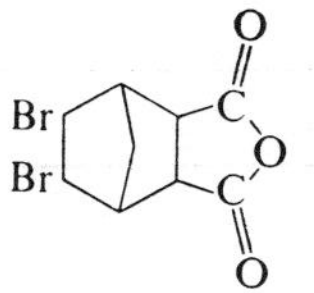 21% Br in polyester, SE	235

(continued overleaf)

Table 8-20 (Continued)

Class	Compound	Remarks	Reference
Aromatics	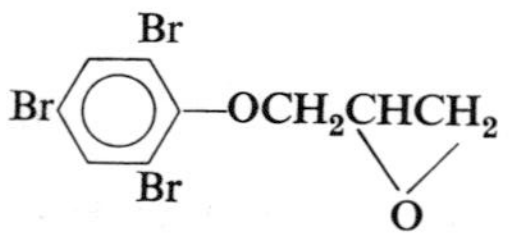 Br-substituted phenyl OCH_2CHCH_2O (glycidyl ether)	12% Br in polyester, SE	236
	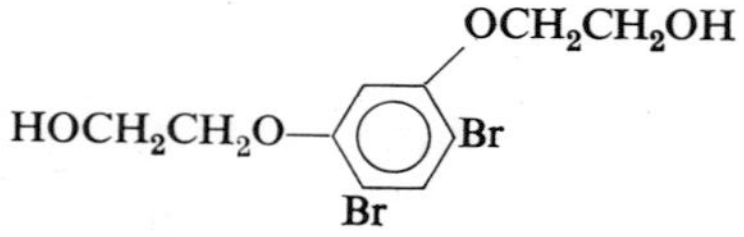 $HOCH_2CH_2O$-dibromophenyl-OCH_2CH_2OH	17.3% Br in polyester, SE	237
	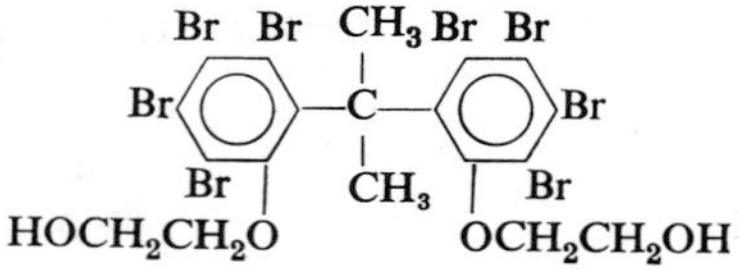 $HOCH_2CH_2O$-(Br,Br)phenyl-$C(CH_3)(CH_3)$-(Br,Br)phenyl-OCH_2CH_2OH	23% Br in polyester, NB	238

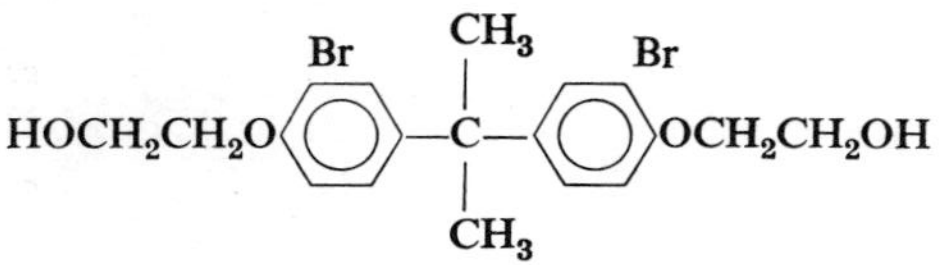

SE polyester at 17.5% Br — 239

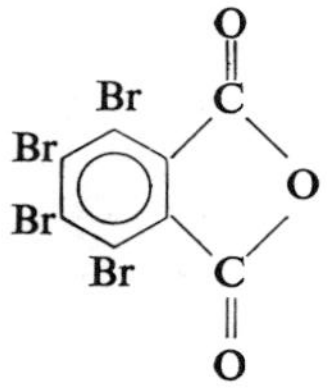

240, 224
241

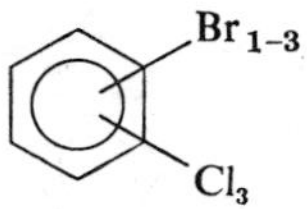

242

Table 8-21   Combinations of Antimony Compounds with Halogen in Polyesters

Class	Compounds	Remarks	Reference
Sb_4O_6 alone	Sb_4O_6	Highly filled resin	243
$SbCl_3$	$CH_2{=}CHCH_2O{-}\overset{\overset{\displaystyle O}{\|\|}}{C}{-}CCl_3$, $SbCl_3$	12% Cl, 2.7% Sb_4O_6 equiv.	244
Chlorinated aliphatics	Chloroparaffin, $Sn(OH)_3$	$\cdots$	245
	C_2Cl_6, Sb_4O_6	$\cdots$	246
Saturated cyclics	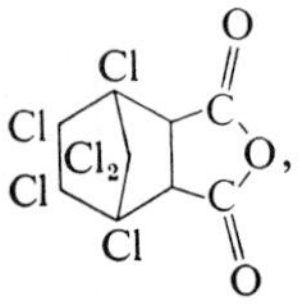 , Sb_4O_6	$\cdots$	247
Unsaturated cyclics	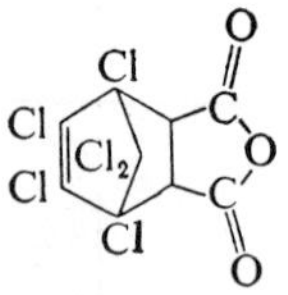 (or di acid), Sb_4O_6	$\cdots$	248, 249
	Chlorinated cycloaliphatics and triazines Sb_4O_6	$\cdots$	249a
Aromatics	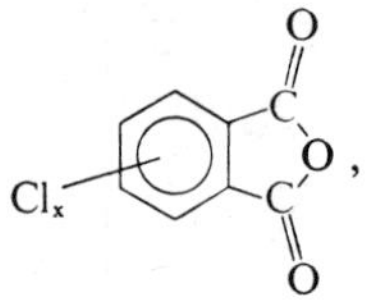 , Sb_4O_6	$\cdots$	250

Bromine compositions

Class	Compounds	Remarks	Reference
Saturated cyclics	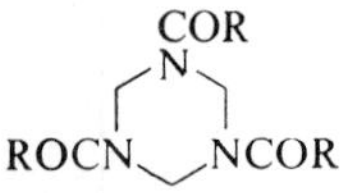 $R = Br_2C_3H_5O$, Sb_4O_6	$\cdots$	251
	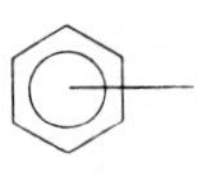 $-Br_6$, Sb_4O_6	9.7% Br 4.5% Sb_4O_6, SE <1 sec	252
Aromatics	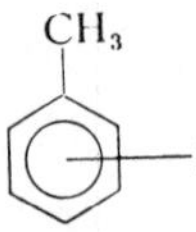 $-Br_5$, Sb_4O_6	8% Br, 8% Sb_4O_6, SE 10 sec	253

Table 8-21   (Continued)

Class	Compounds	Remarks	Reference
Aromatics	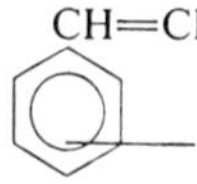 CH=CH$_2$ benzene ring —Br$_2$, Sb$_4$O$_6$	10% Br, 5% Sb$_4$O$_6$	254
	Br, Br, Br, Br tetrabromophthalic anhydride, Sb$_4$O$_6$	6% Br, 5% Sb$_4$O$_6$, 100 in Hooker HLT-15 test	224, 255

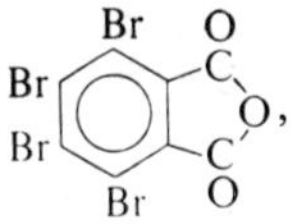

Organic antimony compounds

	KSb tartrate, chloroparaffin		256
	Sb Caproate, HET acid	22% Cl, 1% Sb$_4$O$_6$ (equiv.)	257
	Sb(OEt)$_3$, tetrachlorophthalic acid	19.6% Cl, 4.5% Sb$_4$O$_6$ (equiv.) SE at once	258
	Sb polyethylene glycol, 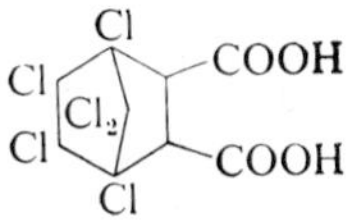 Cl, Cl, Cl, Cl, Cl$_2$ —COOH, —COOH		259
	Sb alkylene glycoxides, Cl compounds		260
	Sb(OCH$_2$CH=CH$_2$)$_3$, 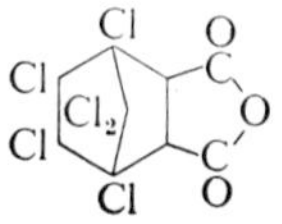 Cl, Cl, Cl, Cl, Cl$_2$ (anhydride structure)	25% Cl, 2.3% Sb$_4$O$_6$ (equiv.)	261
	ϕ_3SbCl$_2$, HET acid	18% Cl, 1.7% Sb$_4$O$_6$ (equiv.)	262
	ϕ_3Sb, tetrachlorophthalic anhydride		263
	(CH$_2$=CHCH$_2$O—Sb(OCH$_2$CHCH$_2$Br)$_2$ | Br		264

Since polyester resins are inherently inexpensive, flame retardance in most cases must be achieved by use of similarly low-cost flame-retardant materials. For the large volume uses the retardants of choice will be simple chlorinated materials and the lowest-cost phosphorus compounds that are compatible. This favors chloroethylphosphates and phosphites, for example, and militates against (PNCl$_2$)$_x$, phosphorines, and so on. Indeed, despite the

Table 8-22   Flame-Resistant Polyesters Based on TCPA [222]

Exp. No.	Mole Ratio TCPA/Ma	Equiv. Ratio OH/COOH	Mole Ratio Styrene/Ma	% Styrene	% Cl_2	HDT, °C No TEP	HDT, °C 3% TEP	Flame Retardancy D-635 SE time, sec No TEP	D-635 SE time, sec 3% TEP	D-757 mm/sec to SE in sec No TEP	D-757 mm/sec to SE in sec 3% TEP
1	0.5	1.02	0.99	25	17.2	112	105	88	5	25/147 (0.17)	15/122 (0.12)
2	0.5	1.02	1.33	30	16.1	120	106	>180	...	43/320 (0.14)	...
			1.97	40	13.8	130	...	>180	71	62/413 (0.15)	32/231 (0.14)
3	1.0	1.03	1.52	25	22.3	104	97	25	7.5	22/125 (0.18)	14/117 (0.12)
4	1.25	1.03	1.78	25	23.8	88.5	62	63	10	35/230 (0.15)	19/164 (0.12)
5	1.25	1.00	1.78	25	24.0	86.5	77	46	12	30/184 (0.16)	18/122 (0.15)
6	1.50	1.02	2.05	25	24.4	83	73	...	1.5	...	13/150
7	0.5	1.10	1.17	25	16.5	97	...	68	28	34/220 (0.16)	23/137 (0.17)

No.											
8	1.0	1.10	1.55	25	22.0	100	90	...	15	...	...
				28			95		21		
9	1.25	1.07	1.81	25	23.5	97	82	78	7	33/190 (0.17)	16/108 (0.15)
10	1.25	1.10	1.82	25	23.4	94	83	...	2	...	15/128 (0.12)
11	1.25	1.10	1.82	25	23.4	90	80	...	3	...	14/136 (0.10)
12	1.25	1.10	1.82	25	23.4	91	77.5	64	9	37/226 (0.17)	16/98 (0.16)
13	1.25	1.10	1.82	25	23.4	90	78.5	47	14	28/152 (0.18)	23/152 (0.15)
14	1.25	1.10	1.82	25	23.4	92.5	80	41	10	28/143 (0.19)	16/105 (0.15)
15	1.25	1.10	1.82	25	23.4	...	82	...	2	...	14/125 (0.11)
16	1.25	1.10	1.82	25	23.4	89	74	45	5	25/132 (0.19)	...
17	1.25	1.10	1.82	25	23.4	94	...	92	...	41/271 (0.15)	...
18	1.48	1.10	2.43	25	24.2	79	68	52	10	27/175 (0.15)	15/100 (0.15)

Ma = maleic anhydride
TEP = triethyl phosphate

Table 8-23 Properties of TCPA Polyester Resins [223]

| | Molar Composition | | | | | | | | | Viscosity | | | |
Resin	TCPA	MA	HBPA	NPG	EG	Benzoic Acid	% Cl	Acid No.	% Styrene	No TEP, cks	3% TEP, cks	Excess OH	HDT 3% TEP, °C
A	2.50	2.20	···	···	5.02	0.10	22.1	20.7	27	2485	1525	3.0	87
B	2.00	2.41	···	4.59	···	0.10	17.1	26.8	27	9800	4630	3.0	95
C	A 50/50 blend of A and B						19.6	···	27	4800	2930	3.0	91
D	2.00	1.74	···	2.60	1.30	0.10	19.2	26.5	30	1400	980	3.0	91
E	2.10	1.40	1.18	···	2.39	0.10	18.7	29.0	30	1800	1000	2.0	102

MA = maleic anhydride
HBPA = hydrogenated bisphenol A
NPG = neopentyl glycol
EG = ethylene glycol
TEP = triethyl phosphate

synergistic effects of phosphorus compounds, a considerable amount of the fire-retardant polyester sold today is made from halogenated products without phosphorus. So we now turn to the category of halogen-containing polyesters.

Halogen Compositions

A great many halogen compounds have been evaluated in polyester compositions. Some of these have been reviewed by Schulze [199], Roberts [200], and Nametz [145]. Approaches vary widely and range from adding NH$_4$Cl [201] (up to 50% by weight) to chlorinating the polyester after it is made [202]. In some cases highly chlorinated inerts such as chlorocarbons [203] are added but these are exceptions. One of the comonomers normally carries the halogen: the polyol, the polyacid, the unsaturated species, or the cross-linking agent. Or a substance that enters into the reactions and becomes a pendant group or a terminating group may also be used; one such would be vinyl chloroacetate [204]. In Table 8-20 the classes of materials covered in the literature are listed by categories. In Table 8-21 are references to halogen-antimony compositions.

Vogt et al. [205] have presented considerable information on the properties of polyesters containing 20.5% Cl made from Cl$_3$C—CH—CH$_2$ and indicate

$$Cl_3C-CH-CH_2$$
$$\diagdown\diagup$$
$$O$$

the burning rate by ASTM D-757 in casting form is 0.18 in./min unchanged by boiling in water for 48 hr. By ASTM D-635, the samples are nonburning.

There are many references to reaction products of hexachlorocyclopentadiene ("hex") with various unsaturates. This chemistry was reviewed early by Robitschek and Bean [210]; the interested reader might start with their article. Similarly, synthesis of a variety of chlorinated polyesters has been explored by two Russian groups in terms of chemistry and the resultant physical properties of the resins [208, 213]. Polyesters based on "hex" reacted with maleic anhydride are widely used and may be formulated so as to give resins with Class A or noncombustible ratings in the ASTM E-84 tunnel test. Such resins are often very resistant chemically in addition to their fire resistance [265].

In recent years tetrachlorophthalic anhydride (TCPA) has been much discussed and is finding acceptance as an economically attractive flame-retardant component. There are two papers on this retardant with a wealth of data on viscosity, heat distortion temperature (HDT), results of various strength tests, and flammability ratings [222, 223]. Increasing HDT improves flame resistance. Flame resistance appears to level off above 22% Cl or so in these systems. Increasing styrene content decreases flame resistance. Tables 8-22–8-26 summarize some of the results. Note the

Table 8-24   Flame Retardancy of Laminates [223]
All Experimental Samples Contain 3% TEP. 72 Hr Exposure
at 90°C

Resin[a]	D-757, SE Time, sec			D-757, Burning Rate, in./min		
		After Exposure to			After Exposure to	
	Original	Water	10% NaOH	Original	Water	10% NaOH
A	111	123	113	0.17	0.13	0.13
B	203	241	245	0.26	0.22	0.28
C	169	175	151	0.26	0.22	0.28
D	179	183	192	0.21	0.24	0.25
E	231	230	242	0.26	0.29	0.31
Commercial control	379	308	282	0.36	0.31	0.37

[a] See Table 8-23 for the composition of the resins.

excellent self-extinguishing properties achieved when phosphorus is also present; for example, as triethyl phosphate in Table 8-22. It seems clear that good results may be obtained either when the chlorine level is of the order of 25% or when some 1% P is present with 16–18% Cl. When Sb_4O_6 is used at 2%, the same reduction in chlorine appears possible.

Bromine is, once again, more effective on a weight basis than chlorine. Figure 8-6 shows comparative data for polyesters based on TCPA and tetrabromophthalic anhydride (TBPA). More of Nametz's results are

Table 8-25   Value of Some Phosphorus
Compounds in TCPA Resins as Measured
by ASTM D-635 [223]

Synergists	%	SE Time, sec
Antimony oxide	2	0.8
PHOS-CHECK P-30[a]	3	1.3
PHOSGARD C-22-R[b]	3.3	6.1
Triallyl phosphate	3.5	16.5
Triethyl phosphate	3	23.8

[a] An ammonium polyphosphate
[b] Chlorinated polyphosphonate ester

Table 8-26 Some Comparisons of Fire-Resistant Behavior in
TCPA-Based Resins Containing Phosphorus as
Hydroxyethyl Phosphates [166]

% Cl	% P	Simple Burning Test	ASTM Spec. 757–49, in./min
17	0.7	SE, 3 sec	0.36
17.1	1	SE, 3 sec	0.22
16.5	0.7	SE, 3 sec	0.35
18.1	0.75	SE immediately	0.17
19.8 (0% Sb$_4$O$_6$)	0	Burns completely	0.52
19.8 (5% Sb$_4$O$_6$)	0	SE 5–10 sec	0.16
25	0	SE 7 sec or longer	0.21

illustrated for TBPA in Figures 8-7 and 8-8 which show that phosphorus is
again helpful in reducing the halogen requirement and that very respectable
tunnel ratings can be achieved. In Figure 8-7 the HLT-15 test data need a
note of explanation. In this test the resin is repeatedly exposed to an ignition
source. If it passes, in a prescribed manner, five such exposures, a rating of
100 is awarded. Figure 8-9 compares bromine and chlorine in this test with
phosphorus present; the difference between the two halogens is striking.

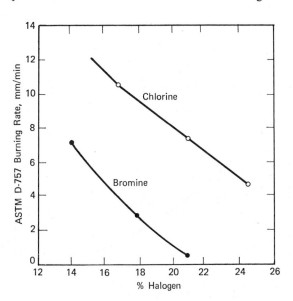

Figure 8-6 Comparative performance of polyesters based on chlorine (TCPA) and bromine
(TBPA). (After Nametz [144].)

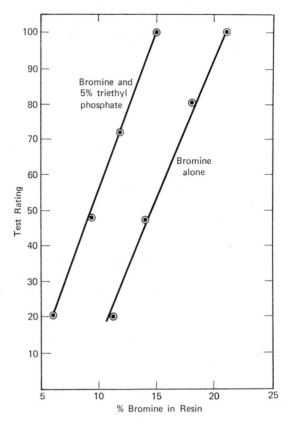

Figure 8-7   Results in the Hooker HLT-15 test for TBPA resins [255].

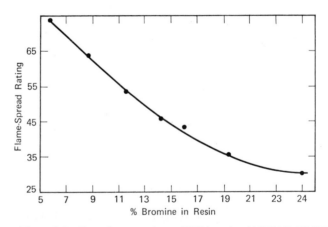

Figure 8-8   Tunnel test results on TBPA resins (ASTM E-84) [255].

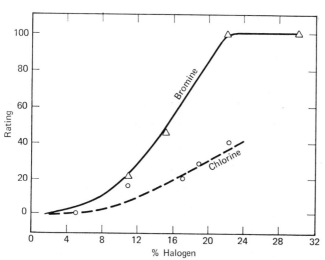

Figure 8-9 Performance of TBPA and TCPA polyesters in the Hooker HLT-15 test with 3% TEP present [224].

Antimony trioxide—Sb_4O_6—is effective with TBPA as shown in Figure 8-10 and Table 8-27. At 2% Sb_4O_6 between 8 and 9% Br is sufficient as against 16–18% Cl. (The factor of 2 is seen again between Br and Cl.) Similar results have been noted with brominated aliphatics [231]. However, brominated toluene is somewhat less effective and requirements for both Br

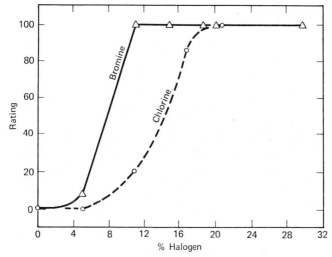

Figure 8-10 Effect of 2% Sb_4O_6 on TBPA and TCPA polyesters in the Hooker HLT-15 test [224].

Table 8-27   Results of Flame Tests on TBPA
Polyester Containing Sb_4O_6 [145]

Br, %	Sb_4O_6, phr	HLT-15	D-757, mm/min	D-635
			Burning Rate or Score	
6	0	0	15.4	Burning
6	1	0	10.1	SE
6	2	4	5.3	NB
6	5	100	2.8	NB
9	0	0	10.9	SE
9	1	20	4.3	NB
9	2	100	2.3	NB
9	5	100	1.5	NB
12	0	16	3.3	NB
12	1	100	1.5	NB
12	2	100	1.5	NB
12	5	100	1.5	NB
15	0	20	1.4	NB
15	1	100	1.2	NB
15	2	100	1.5'	NB
15	5	100	1.4	NB

and Sb_4O_6 are higher with this aromatic additive than with the reactive TBPA. Pentabromotoluene was found in this work to be better than the corresponding phenols, ethers, or less brominated toluenes [253]. When dibromostyrene is used good flame retardance is obtained at 5% Sb_4O_6 and 10% Br—requirements higher than with TBPA [254].

An interesting basic paper on fire-resistant polyesters from Czechoslovakia discusses the mechanisms involved in flame retardation with halogens and carefully compares various organic chlorine compounds [249]. The author concludes that most materials act similarly at a given chlorine level in the presence or absence of Sb_4O_6. However, he finds that resins based on chlorendic acid or its anhydride (HET) are superior by a measurable amount (see Table 8-28). He rationalizes this by observing that HET acid decomposes on heating to the starting materials and the hexachlorocyclopentadiene is more effective as a free-radical inhibitor than the HCl produced by the decomposition of the other chlorinated fire retardants. This explanation is, however, largely speculative. It is of interest to find that fundamental thermodynamic calculations (e.g., of heats of combustion) are not helpful in selecting the preferred material. However, they do show that ΔH combustion is lower when chlorine is present (down from ~6 kcal/g to ~3 kcal/g). Obviously, kinetic factors are critical in the combustion process.

Table 8-28   Comparisons of Various Types of Halogen Compounds on the Flammability of Polyester Resins (Zvonar [249])

Modifying Component	ChTHF	TChF	HET	ChPG	DChEF
Burning of resins containing 7.8% Cl modified by various raw materials					
Reduction of weight at various temperatures, g/cm					
500°C	0.61	0.65	0.10	0.64	0.51
650°C	1.69	1.23	0.20	1.63	1.45
800°C	0.80	1.99	1.79	1.61	1.66
950°C	1.36	1.99	1.51	1.79	1.69
Length of flame, cm	8.6	9.0	5.0	9.0	9.6
Average burning rate, cm/min	4.3	4.3	2.6	4.0	4.0
Burning of resins containing 7.8% Cl and 5% Sb_4O_6 modified by various raw materials					
Reduction of weight at various temperatures, g/cm					
500°C	0.63	0.57	0.07	0.69	0.67
650°C	1.00	0.99	0.23	0.57	0.62
800°C	1.01	0.89	0.59	0.76	0.82
950°C	1.13	1.33	0.78	0.96	1.11
Length of flame, cm	4.6	5.3	3.0	5.3	7.0
Average burning rate, cm/min	3.0	2.6	1.3	2.0	2.6

ChTHF = monochlorotetrahydrophthalic acid
TChF = tetrachlorophthalic anhydride
HET = hexachloro-endo-methylene tetra-
         hydrophthalic anhydride
ChPG = 3-chloropropyleneglycol-1, 2
DChEF = dichloro-endo-methylene hexa-
         hydrophthalic anhydride

Miscellaneous

There are scattered references to other flame retardants, for example, $CaSO_4 \cdot 2H_2O$ [266], and to borate esters formed *in situ* by addition of H_3BO_3 [267] or by direct addition of, for example, $[BrCH_2CH(Br)CH_2O]_3B$ [268]. Some 5% of H_3BO_3 confers fire retardance. Filled polyesters containing $Al_2O_3 \cdot 3H_2O$ at 38% or clay-filled resins with Sb_4O_6 and $Al_2O_3 \cdot 3H_2O$ are rated self-extinguishing by ASTM D-635 [269].

Summary

After a great deal of research there are now a variety of methods for achieving flame resistance in polyesters. The requisite chemical levels are:

$\sim$5% P (organophosphorus compounds)
1% P and 15–20% Cl
2% P and 6% Br
25% Cl
12–15% Br
2% Sb_4O_6 and 16–18% Cl
2% Sb_4O_6 and 8–9% Br

Which system to use will be based on cost, importance of degree of plasticization caused by the additive, weather and uv resistance required, and so on. For filled opaque resins the Sb_4O_6 systems or inorganic phosphates can be used. Yellowing is less of a problem with the bromine systems than with the chlorine ones; but bromine costs more for the same performance, and so on. The phosphorus-halogen mixtures are receiving much attention at present because of their good cost/performance.

Polyester Fibers

Only four references specific to fibers as opposed to general resins were discovered. Fibers of polyester today are largely polyethylene terephthalate

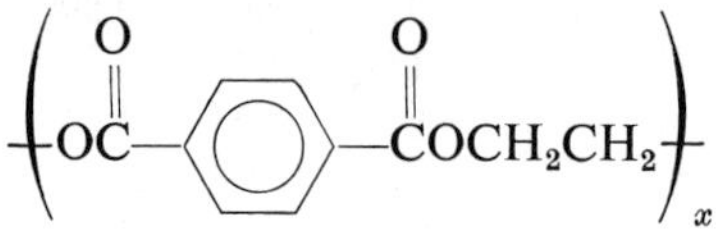

made from ethylene glycol and terephthalic acid. These are spun at nearly 300°C from a melt and a flame retardant must be able to withstand such a temperature without decomposing or reacting with the other ingredients in an undesired way. One patent discusses the use of $(CH_3O)_2P(H)O$ at such a level as to give 1.7% P in the finished fiber. Presumably transesterification occurs and some of the terephthalate is replaced with phosphite. The fibers are pale white after processing at 290° and 12 min Hg in the presence of Na and Mg [270]. In an analogous case, ethylene glycol phenylphosphonate

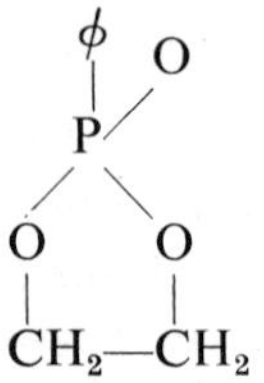

Ethylene glycol phenylphosphonate

was used [271]. $(ClCH_2CH_2O)_2P(O)CH_2CH_2Cl$ has been added to polyester melts for fibers to give not more than 0.5% P for improved dyeability [272]. Considerable self-polymerization, along with transesterification, will occur.

Flame-retardant polyester fibers are sorely needed now. Federal regulations will affect many items of apparel and perhaps sheeting in certain end uses. The cotton formerly used in such items is rapidly being replaced by durable-press polyester/cotton blends. The fire-resistant treatments that work on cotton by and large are not applicable to these blends. It is expected that the solution ultimately will come from adding fire-retardant comonomers to the polyester melts before they are spun. The blended fibers may then be treated—as fabric—with a finish that will protect the cotton fraction. This is not being practiced today because the technology is lacking. However, research is under way in many laboratories. The compounds must be stable, must not initiate charring reactions at 300°, and must not degrade the fabric in terms of strength and other important properties. They should be either reactive soluble inerts or thoroughly dispersible inerts that are not abrasive to the spinnerets. Some of these requirements may be found in the various compounds listed in foregoing sections for polyester resins.

Some basic data that are important to the understanding of this problem have been gathered very recently by Tesoro et al. [273]. They have determined the limiting oxygen indices (LOI) for a variety of fibers (see Table 8-29) and

Table 8-29   Limiting Oxygen Indices
for Various Fibers [273]

Fabric, 100%	Weight, oz/yd^2	LOI
Acrilan	6.5	0.182
Arnel triacetate	6.5	0.184
Acetate[a]	6.5	0.186
Polypropylene[a]	6.5	0.186
Vinylon (PVA)	6.5	0.197
Rayon	6.5	0.197
Cotton	6.5	0.201
Nylon[a]	6.5	0.201
Polyester	6.5	0.206
Wool (dry-cleaned)	7.0	0.252
Dynel	6.5	0.267
Nomex N-4274	4.8	0.282
Rhovyl (PVC) "55"	6.5	0.371

[a] Cotton thread stitched through sample

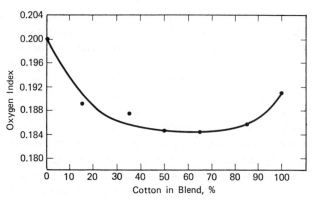

Figure 8-11   Relationship of LOI to blend composition for polyester/cotton felts [273].

for polyester/cotton blends (see Figure 8-11). The blends are more flammable (lower LOI) than either component. Finishes on blends in which the cotton was treated with diammonium orthophosphate (DAP) and the polyester with tris(2,3-dibromopropyl) phosphate (TBPP) were then studied. These treatments were selected so that only one component fiber or the other could be selectively flameproofed. Subsequent analysis showed this did indeed occur. The results are shown in Table 8-30. An LOI of about 0.28 corresponds to a fabric that would pass the vertical flame test (with a char length <5 in.). The combined effect of both flame retardants is to improve markedly the flame resistance but not to an LOI of 0.28. Very large amounts of TBPP are

**Table 8-30   Polyester/Cotton Fabric (50/50)
Test Fabrics, S/9503, 5.0 Oz/Yd² [274]**

% DAP (OWF)	% TBPP, (OWF)	% Phosphorus Calcd	% Phosphorus Found	LOI
1.6	0	0.34	0.45	0.215
5.0	0	1.06	1.12	0.234
0	2.7	0.12	0.18	0.188
0	6.4	0.28	0.23	0.204
0	7.0	0.31	0.32	0.205
0	17.0	0.75	0.81	0.248
0.8	2.4	0.29	0.41	0.215
0.7	4.6	0.35	0.32	0.230
0	0	· · ·	· · ·	0.176

effective in raising the LOI from the untreated value of 0.176 to about 0.25—but even this may not be enough. A recent patent refers to use of bis(hydroxymethyl) phosphinic acid, as the ammonium salt, to treat polyester/cotton fabric. This is cured at elevated temperatures and gives a good hand. No data on resistance to laundering are given in the abstract [273a].

Kruse [273b] has made a study of a variety of blended fabrics and comes to much the same conclusions as Tesoro et al. He too finds that both components must be treated unless one is less than 15% of the total.

These are early results, necessary for making any sort of progress on the problem. It is clear that blends of fibers require that attention be directed to all components and that the technical challenge is substantial. Nonetheless, the approach to take is clearly shown by the knowledge already at hand for the components. It is now necessary to work within the limiting confines presented by the fiber spinning operation or the exigencies of developing a finish for more than one fiber in a fabric.

EPOXY RESINS

Epoxy resins are made by reacting alkylene oxide groups (a) with themselves in the presence of a catalyst, (b) with amines, or (c) with carboxylic acid anhydrides [274]:

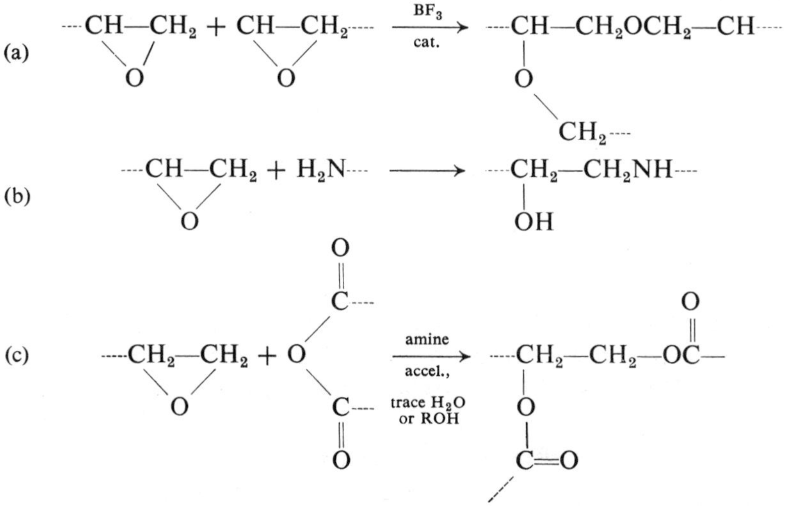

The epoxide is usually in the form known as a glycidyl ether. These are made from epihalohydrins and diols (or polyols):

$$2CH_2\!\!-\!\!CHCH_2Cl + HOCH_2\cdots \rightarrow CH_2CHCH_2OCH_2\cdots$$

Glycidyl ether

The diol is often bisphenol A or one of its derivatives:

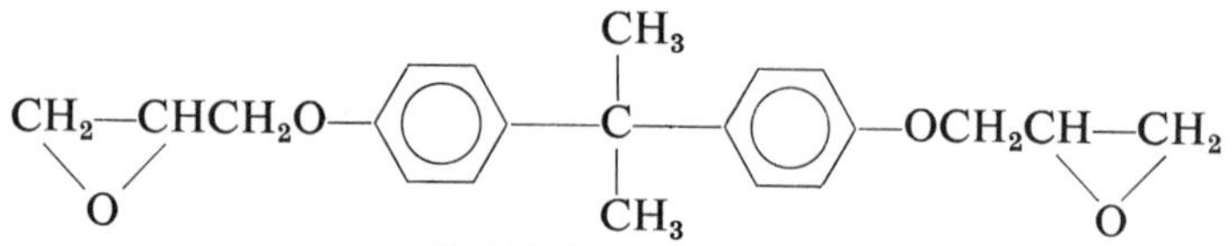

$$CH_2-CHCH_2O-\underset{\diagup}{\overset{O}{\diagdown}}-\bigcirc-\overset{\overset{\displaystyle CH_3}{|}}{\underset{\underset{\displaystyle CH_3}{|}}{C}}-\bigcirc-OCH_2CH-CH_2$$

Glycidyl ether of bisphenol A

Pentaerythritol [275], glycerine, and other polyols may be used. Epoxy resins are flammable and ignite readily; filled epoxies may be very flame-resistant. The LOI for several epoxies are:

	LOI
Epoxy, conventional	0.198
Epoxy, cycloaliphatic	0.198
Epoxy, filled 50% Al_2O_3	0.250
Epoxy, filled 60%	
$Al_2O_3 \cdot 3H_2O$	0.408

(A LOI of 0.28 or so is comparable to a rapidly self-extinguishing system.) The combustion products are principally CO, CO_2, propylene, and ethylene [142].

A detailed study has been published of the thermal decomposition of one epoxy resin: one cross-linked with nadic methyl anhydride (NMA) [275a,

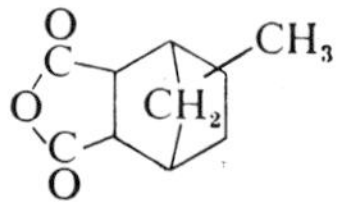

NMA

330]. The decomposition occurs in several steps and is complex. Degradation begins at 240° or so and continues up to 450° or higher. The volatile products, kinetics of the process, and some thermal parameters were defined.

Phosphorus Compounds

In filled resins an inorganic salt like ammonium polyphosphate (relatively insoluble in water) has been successfully employed. Some 2% phosphorus is required [276] (see Table 8-33). Some patents in this area are listed in Table 8-31. From the relatively little that has been done in this category it appears that 5–6% P is required from such compounds, not unlike the requirements for polyesters which these resins closely resemble.

Table 8-31 Phosphorus Compounds Claimed as Fire Retardants for Epoxies

Compound	Epoxy Type	Remarks	Reference
$(NH_3PO_3)_x$ (insol in H_2O)	Bisphenol	1.8% P (flame spread 35)	276, 277
$P—(N(CH_3)_2)_3$	Bisphenol	2.4% P	
[cyclic phosphonitrilic structure]	Conventional	Use 50% of P compound	278
$R—P—(OH)$, $RO—P—(OH)_2$	Conventional		279
$CH_2—CH—CH_2$ (epoxide)	Homopolymer, BF_3 cat.	16% P	280
$(EtO)_2—P—CH—C$ ... $OP(OEt)_2$	Cyclic aliphatic epoxy	5.8%	281
$(HOCH_2)_4P^+Cl^-$	Conventional	5.8%	282
$(HOCH_2)_4P^+Cl^-$	$HOCH_2CH_2Cl$, $(HOCH_2)_3(C_3H_3N_6)$	6%	283
$(HOCH_2)_3PO$	$CH_2—CHCH_2Cl$ (epoxide)		285
[phenyl phosphine oxide diglycidyl structure]	...	...	284

Table 8-32   Phosphorus-Halogen Compositions in Epoxy Resins

Phosphorus Compound	Halogen Compound	Resin Type	Remarks	Reference
P_4 (red)	Brominated bisphenol A	SiO_2 filled bisphenol A	2% P, 4.8% Br; SE <10 sec by ASTM D-635	286
PCl_3, cyclic C_{12} diene, epoxide	Br_2	...	<1% P, 2.4% Br	287
—Rxn product of PCl_3, glycidyl compounds		Bisphenol A	1.6% P, 5.6% Cl; NB, ASTM D-635	288
Bisphenol A		Bisphenol A	~3% P, 6012% Cl	289
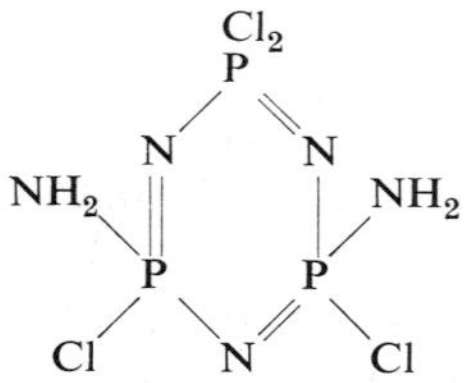 (phosphonitrilic ring structure; PCl_3)		Alkylene oxide or conventional epoxy	...	290
$(NH_4PO_3)_x$, dichlorane or brominated epoxies		Bisphenol	1.8% P, 17% Br, flame spread of 14!	276
$(ClCH_2CH_2O)_3PO$		Conventional	1.5% × 5% Cl; SE 0–10 sec	291
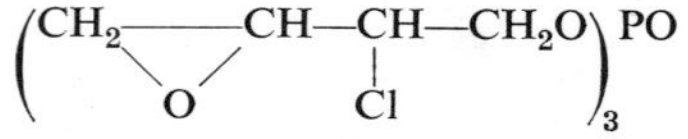 $\left(CH_2\!-\!\underset{O}{CH}\!-\!CH\!-\!\underset{Cl}{CH_2O} \right)_3 PO$		Bisphenol A	...	292
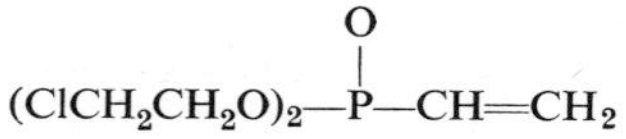 $(ClCH_2CH_2O)_2\!-\!\underset{O}{P}\!-\!CH\!=\!CH_2$		...	...	293

Phosphorus compound for reference 290 (ring structure):

$$\begin{array}{c} Cl_2 \\ P \\ N \diagdown \quad \diagup N \quad NH_2 \\ \| \qquad \diagdown P \diagdown \\ N \diagup \quad N \quad Cl \\ \diagdown P \diagup \\ NH_2 \quad Cl \end{array} \qquad PCl_3$$

—Ran product of RPCl₂, diglycidyl ethers:

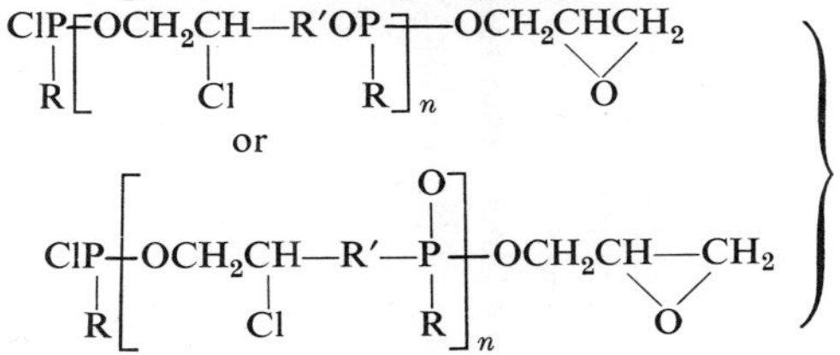

Structure	Description	Note	Ref.
$ClP\!-\!OCH_2CH\!-\!R'OP\!-\!OCH_2CHCH_2$ with $\overset{|}{R}\ \overset{|}{Cl}\ [\ \overset{|}{R}\ ]_n$ $\overset{O}{\triangle}$ or $ClP\!-\!OCH_2CH\!-\!R'\!-\!P\!-\!OCH_2CH\!-\!CH_2$		3% P, 12% Cl	294
$(CH_3O)_2\!-\!P\!-\!O$, $\overset{H}{\underset{|}{P}}$	Glycidyl ether of butenediol reacted with "hex"	2.3% P, 40% Cl	275
$R\!-\!O\!-\!P\!-\!OR'$	...	...	296
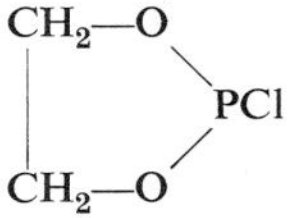 $CH_2\!-\!O$ PCl $CH_2\!-\!O$	Bisphenol A	2.9% P, 3.3% Cl	297
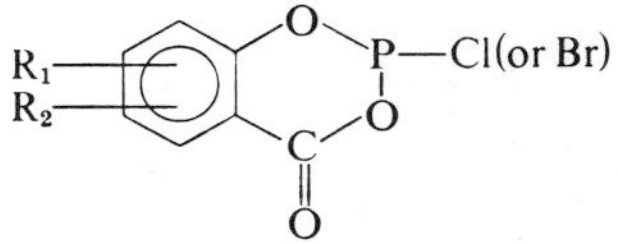 $P\!-\!Cl(or\ Br)$ ring structure, $R_1, R_2 = Cl, Br, alkyl, H$	Bisphenol A	...	298

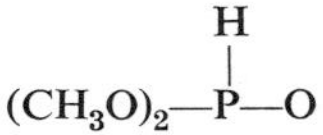

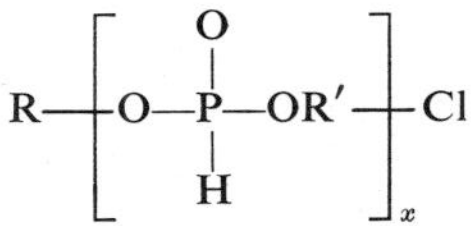

Phosphorus-Halogen Compositions

A good deal more has been done with phosphorus-halogen mixtures than with phosphorus by itself [285]. Table 8-32 summarizes the more significant work. From 1.5% to 2% P and 5–6% Cl provides self-extinguishing epoxies. It makes some difference whether the chlorine is on an aromatic nucleus or in an aliphatic portion of the resin, the latter being preferred [288]. Noncombustible resins (flame spreads <25) have been made with ammonium polyphosphate (of low water solubility) and brominated epoxies or with a highly chlorinated additive. Table 8-33 summarizes the work on this new

Table 8-33   Ammonium Polyphosphate with Halogenated Compounds in Epoxy Resins [276]

	Parts by Weight			
Resin				
Regular	60	100	60	60
Brominated (48% Br)	40	· · ·	40	40
Curing agent	13	15	15	30
Fire resistance				
APP (30% P, 14% N)	7	7	7	7
Dichlorane (78% Cl)	· · ·	20	· · ·	· · ·
$Al_2O_3 \cdot xH_2O$	· · ·	· · ·	125	· · ·
% P	1.8	1.5	1.8 (resin basis)	1.6
% Cl	· · ·	11	· · ·	· · ·
% Br	16	· · ·	16 (resin basis)	· · ·
Flame spread	14	28.0	25	28
Ht. of intumescence, in.	1.5	1.0	1.0	1.5
Smoke density	Low	Mod	Low	Mod
Afterburn time, sec	0	0	5	0

inorganic polyphosphate [276]. Tris(β-chloroethyl) phosphate and bromo analogs are commonly encountered commercial flame retardants. The results of one series of tests on the chloro compound are shown in Table 8-34 in terms of physical properties of the resins as well as their flame resistance. From these findings it appears that the flame-resistant resins are tougher and have higher softening points than the control.

Halogen Compositions

Without phosphorus, more halogen is required. Table 8-35 reviews the situation with respect to chlorine and bromine compounds. (There is one reference in the literature to fluorine [299] but this element is known to be relatively ineffective.) About 26–30% Cl or 13–15% Br is effective [317];

Table 8-34 Results with Tris(β-Chloroethyl) Phosphate in an Epoxy System [291]

Formulation	Parts	Room Temp. Flex. Strength, psi		Rockwell Hardness M scale	HDT, °C	Elevated Temp Flexural Strength, psi	Flame Retardance ASTM D-635
		Precure	Postcure				
Epoxidized polybutadiene	100						
Di-Phenylenediamine	21						
Tris(β-chloroethyl) phosphate	2.5	17,800	15,100	110	128	2,900	SE, 1 in.
Resorcinol	5						
Epoxidized polybutadiene	100						
m-Phenylenediamine	24						
Tris(β-chloroethyl) phosphate	5	19,500	14,000	115	140	5,400	SE, 1 in.
Resorcinol	5						
Epoxidized polybutadiene	100						
m-Phenylenediamine	24						
Tris(β-chloroethyl) phosphate	10	20,200	15,300	119	175	6,400	NB
Resorcinol	5						
Epoxidized polybutadiene	100						
m-Phenylenediamine	24						
Tris(β-chloroethyl) phosphate	20	20,700	20,300	120	180	8,200	SE, in.
Resorcinol	5						
Epoxidized polybutadiene	100						
m-Phenylenediamine	24	17,100	15,100	107	113	785	Burning, 0.50 in./min
Resorcinol	5						

Table 8-35   Halogen Compounds in Epoxies

Compound	Resin Type	Remarks	Reference
Chlorine compounds			
Hexachlorocyclopentadiene	Bisphenol A + maleic anhydride	38% Cl nonflammable	300
$HOCH_2-\overset{\displaystyle Cl}{\underset{\displaystyle Cl}{C}}-\overset{\displaystyle Cl}{\underset{\displaystyle Cl}{C}}-CH_2OH$	Epichlorohydrin	32–45% Cl	301
$CH_2=CHCH_2Cl$	Allyl glycidyl ether	30% Cl	302
...	...	18% Cl	303
Chloroepoxycyclohexane derivatives	Variety of conventional epoxies		
Chlorendic anhydride	Bisphenol A or Novolak type	28–30% Cl	304
		~20% Cl	305
$Cl_5\phi-NH-CH_2CH_2-N(CH_2CH-CH_2)_2$ with epoxide (O)		(~16 Br)	
(or Br analog)			
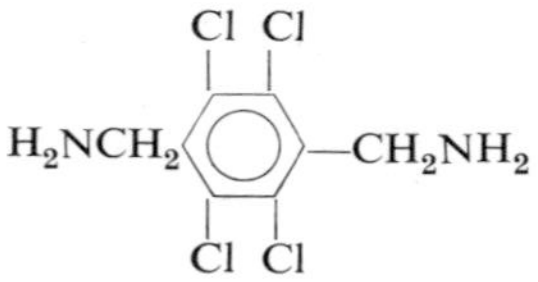 H_2NCH_2—(ring with Cl Cl / Cl Cl)—CH_2NH_2	Bisphenol A	...	306
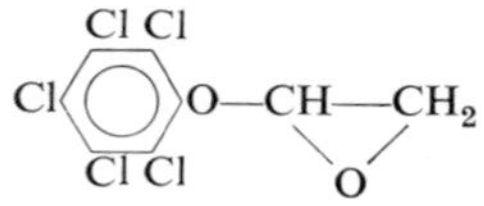 (ring with Cl Cl / Cl Cl)—$O-CH-CH_2$ with epoxide (O)	Bisphenol A	~28% Cl	307

Structure			
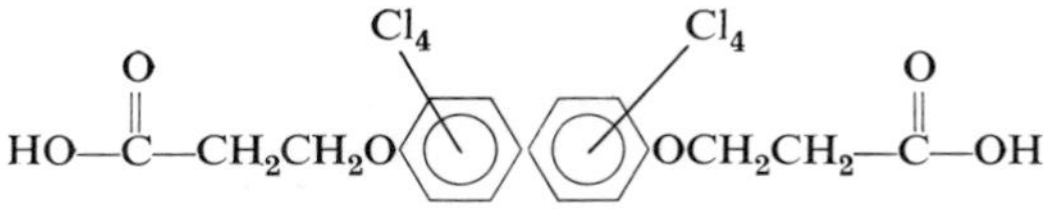 $HO-C(=O)-CH_2CH_2O-$ (with Cl_4, Cl_4 aromatic rings) $-OCH_2CH_2-C(=O)-OH$	Butanediol epoxide	$\sim$31 % Cl	308
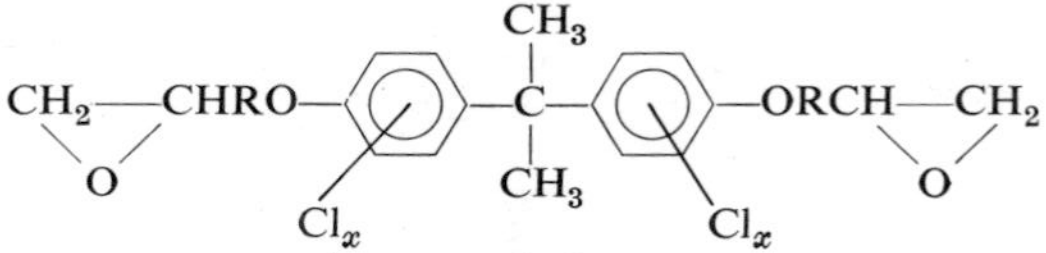 $CH_2-CHRO-$ (epoxide) aromatic ring with Cl_x, CH_3-C-CH_3, aromatic ring Cl_x, $-ORCH-CH_2$ (epoxide) where $x = 1-2$	Conventional	18–27 % Cl	309
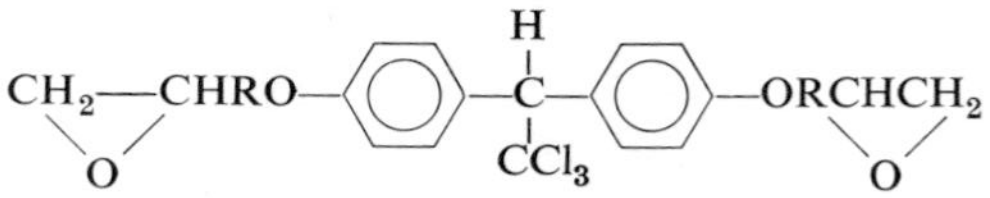 $CH_2-CHRO-$ (epoxide) aromatic ring, $H-C-CCl_3$, aromatic ring $-ORCHCH_2$ (epoxide)	Conventional	$\sim$25 % Cl SE 2–3 sec	310
Bromine compounds			
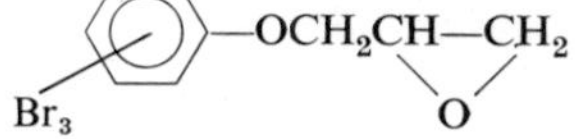 OCH_2CH-CH_2 (epoxide) aromatic ring with Br_3	Bisphenol A	$\ldots$	311
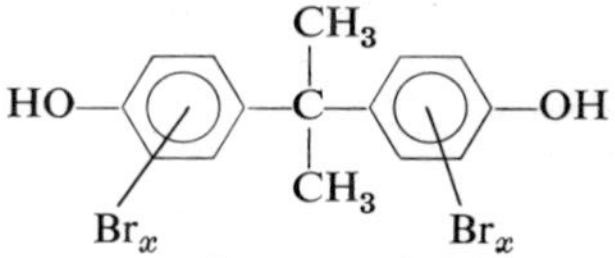 $HO-$ aromatic ring Br_x, CH_3-C-CH_3, aromatic ring Br_x $-OH$ where $x = 1-2$	Bisphenol A	13–20 % Br	312

(continued overleaf)

Table 8-35 (Continued)

Compound	Resin Type	Remarks	Reference
Sb$_4$O$_6$-halogen combinations			
Chlorendic anhydride	Novolak epoxy	1.4% Cl + 25% Sb$_4$O$_6$	313
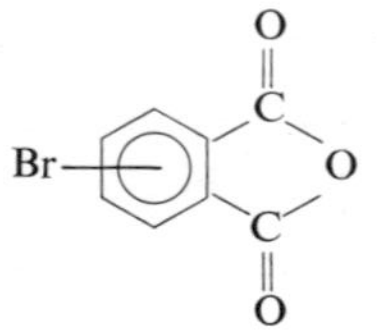	Conventional	5% Br + 3% Sb$_4$O$_6$	314
	Bisphenol A	8% Br + 2.4% Sb$_4$O$_6$ (equiv.)	315
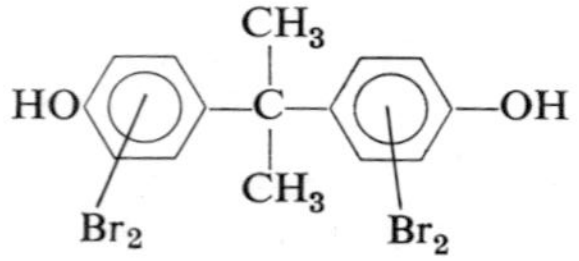	Bisphenol A	...	316

Table 8-36   Bromine and Chlorine in Two Different
Epoxy Resins [318]

Halogen added as: $\langle\bigcirc\rangle$NHCH$_2$CH$_2$N$\left(\text{CH}_2\!-\!\text{CH}\overset{\displaystyle}{\underset{\displaystyle \text{O}}{\diagdown\!\!\diagup}}\text{CH}_2 \right)_2$

X$_5$

where X = Cl or Br

Resin Type	HDT, °C	Halogen, % wt	SE Time, sec	Extent of Burn, in.
Novolak	141	20% Cl	4	9/32
Bisphenol A	126	20% Cl	3	18/32
Novolak	162	16% Br	2	18/32
Bisphenol A	133	16% Br	3	13/32

Sb_4O_6 reduces this requirement dramatically. Some data comparing chlorine and bromine when added in the amine curing agent are shown in Table 8-36 [318]. Again bromine is superior to chlorine. The Novolak resin is more stable and gives a better self-extinguishing time with bromine than does the conventional bisphenol A type. Nametz has claimed the following burning characteristics for brominated epoxies based on tetrabromobisphenol A [312]:

% Br	Burning Rate, in./min	SE Time, sec
2–3	0.2	45
5	0	11
10	0	8

Epoxy resins and their decomposition products may be studied by various forms of spectroscopy, especially ir and mass, and by gas chromatography. Resins based on tetrabromobisphenol A have been carefully studied by ir [319]. The decomposition products from a series of halogenated resins exposed to conditions simulating a fire were shown to be, in the halogen-bearing gaseous fraction, largely HX. No phosgene-like structures were detected. Some low molecular weight alkylhalides were also present and some elemental halogen—Cl_2, Br_2 [320].

Boron Compositions

About 20% of B_2O_3 renders an epoxy self-extinguishing. The acid shortens the curing time or pot life by a catalytic effect. This must be counteracted in the formula [321]. Borates such as calcium borate may also be employed [322]. Esters of boron acids such as trimethoxyboroxine [323, 324], a borolane, or a borinane [325] may be used. These compounds will dissolve in the resin and be thoroughly dispersed therein. When 23% of trimethoxyboroxine—13% B_2O_3—is added, fair self-extinguishing properties (~30 sec) are obtained. This can be boosted further with boric acid or sodium borate [324].

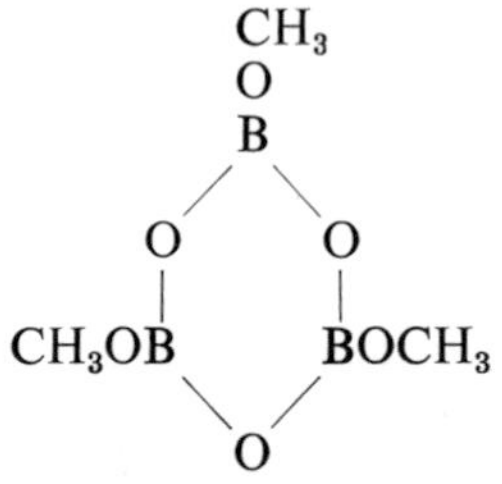
Trimethoxyboroxine

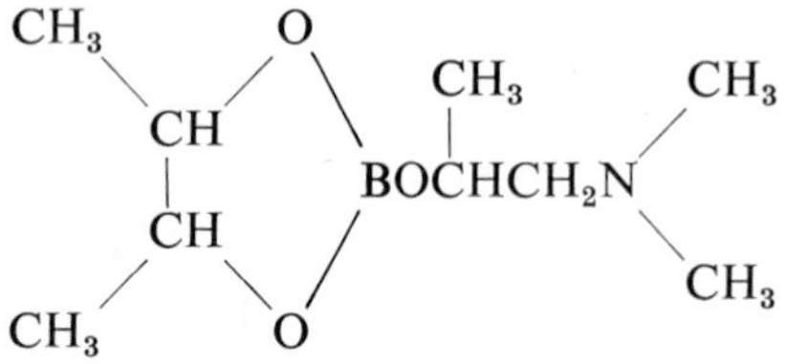
Borolane

POLYAMIDES

The principal polyamides of interest are wool, silk, and synthetics, especially Nylon 6 and 66. Wool and silk are natural fibers comprised of proteinaceous matter characterized by the polypeptide link, $-\overset{\text{H}}{\overset{|}{\text{N}}}-\overset{\text{O}}{\overset{\|}{\text{C}}}-$. Wool is inherently flame-resistant. It has a LOI of 0.25 [273] and generates only moderate amounts of smoke [142]. Indeed, blends of wool with asbestos fiber have been suggested for clothing in high-risk occupations [326]. And wool rugs are considered adequately flame-resistant by most workers. On combustion, wool tends to generate a lot of char and relatively little flammable gas. Silk, another natural protein, is also not very flammable. Both fibers contain a high proportion of nitrogen. Two Japanese references deal with organic nitrogen compounds [327] or a variety of phosphates and borates [328] to improve further silk's flame resistance. Most efforts have been bent to making synthetic polyamides at least as flame-resistant as their natural counterparts.

Nylon

The two principal commercial polyamides are Nylon 6 and Nylon 66. The former is made by polymerizing either ϵ-aminocaproic acid or its lactam. Nylon 66 is made from adipic acid and hexamethylenediamine:

$$1/x\,\mathrm{H_2N(CH_2)_5COOH} \rightarrow \left(\overset{\text{O}}{\overset{\|}{\text{C}}}-\overset{\text{H}}{\overset{|}{\text{N}}}-(\text{CH}_2)_5 \right)_x$$

Nylon 6

$$1/x\,\mathrm{HOOC(CH_2)_4COOH} + 1/x\,\mathrm{H_2N(CH_2)_6NH_2} \rightarrow$$

$$-\overset{\text{H}}{\overset{|}{\text{N}}}\left(\overset{\text{O}}{\overset{\|}{\text{C}}}(\text{CH}_2)_4\overset{\text{O}}{\overset{\|}{\text{C}}}-\overset{\text{H}}{\overset{|}{\text{N}}}(\text{CH}_2)_6-\overset{\text{H}}{\overset{|}{\text{N}}} \right)_x\overset{\text{O}}{\overset{\|}{\text{C}}}\cdots$$

Nylon 66

The two have similar properties:

	Glass Trans. Temp, °C	Melting Point, °C
Nylon 6	75	220
Nylon 66	80	265

Fibers are spun from a melt at temperatures near 300°. This imposes a limitation on the types of additives or comonomers one can add, as it also does for polyester fibers.

Madorsky [142] has reviewed some of the work on the thermal decomposition of nylon. The significant finding is that, on pyrolysis, as much as 95% of the products is nonvolatile. Of the 5% of Nylon 6 which was volatile but noncondensable at room temperature, no nitrogen was found; 50% of the gas was CO_2, 35% H_2O, and 6% benzene. The nitrogen remained in the high-boiling fraction. Hasselstrom et al. [329] studied the decomposition of Nylon 66 and found some ammonia in the volatiles. The evidence points to breaking of both N—C bonds and CH_2—CO bonds. Recent work on pyrolysis kinetics has been reported by Goldstein [330].

Of course, these two nylons melt in a flame and this causes some difficulty. One hears of burns inflicted not by flaming nylon but by molten droplets. Thus, the low melting points may be a key part of the fire-safety problem.

Some fire retardants actually aggravate the flame problem in nylon [329]. Addition of nitrogen compounds is helpful and thiourea in the form of aminoplast resin coatings has been found useful [331] with 4% add-on of $(NH_2)_2C{=}S$ being claimed. Melamine-formaldehyde resins have also been studied [332] but as often as not they have been found ineffective [333].

The effective amounts of various ammonium salts are shown in Table 8-37 (after Douglas from Dorset [333]). Note that the ammonium phosphate is not especially useful. Douglas has correlated these results with the ability of the additives to *lower* the melting point of nylon, as shown in Table 8-38 [333]. To prevent leaching of the thiourea, application must be done as part of a resin finish. It is not added to the melt for spinning (mp, thiourea: 182°; decomposes thereafter).

Table 8-37 [333]

Flameproofing Agent	Minimum Effective Amount	
Thiourea	3.05%	
Ammonium thiocyanate	5.91%	
Ammonium bromide	9.14%	
Ammonium sulphate	11.33	Not sufficient
Ammonium dihydrogen phosphate	17.7	to flameproof

Table 8-38 [333]

Flameproofing Agent	Melting Point of Nylon Net after Having Dried into it a 10% Solution of the Flameproofing Agent	
	Dry, °C	Wet, °C
None	250–255	240–245
Ammonium dihydrogen phosphate	250–255	240–245
Ammonium chloride	235–240	195–205
Ammonium bromide	235–240	195–205
Ammonium sulphate	245–250	195–205
Ammonium sulphamate	245–250	195–205
Ammonium nitrate	230–235	205–215
Ammonium thiocyanate	240–245	195–205
Thiourea	235–240	195–205

Low flammability in Nylon 66 has been achieved by copolymerizing a phosphorus monomer of the following structure

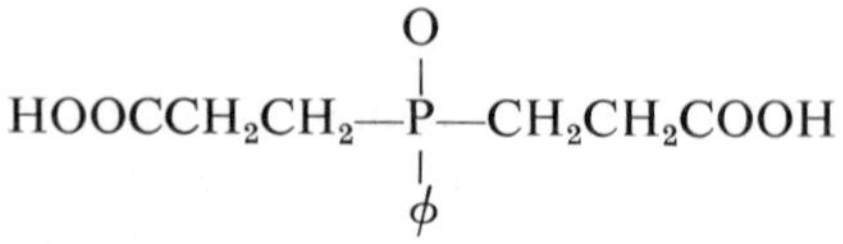

$$HOOCCH_2CH_2 - \overset{\overset{\displaystyle O}{\displaystyle \|}}{\underset{\displaystyle \phi}{P}} - CH_2CH_2COOH$$

so as to yield 3.5% P in the fiber. This structure has inherently high stability (three P—C bonds—a phosphine oxide) and can survive the melt-spinning process [334]. A carboxyphenyl phosphonate has also been added to molten nylon for modifying dyeing properties rather than for improving flame resistance [335].

Chlorinated compounds added to nylon improve its flame resistance (and likely reduce still further its resistance to yellowing from uv light). Derivatives of hexachlorocyclopentadiene may be used; for example,

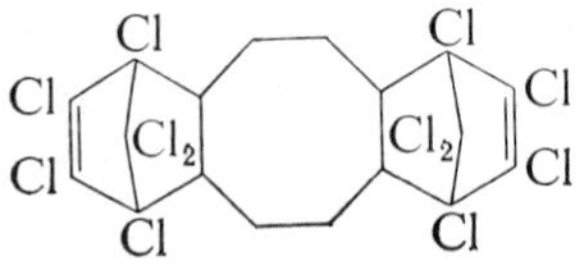

is stable in molten Nylon 66. (Adding ZnO to this provides a boost in flame resistance [336].) Chlorinated diphenyl alone or with antimony oxide is helpful. With 3.5–7% chlorine from diphenyl, self-extinguishing behavior is obtained [337]. With 6% Cl and 10% of a metal oxide (Sb_4O_6, SnO, SnO_2, PbO, etc.) self-extinguishing nylon is produced which does not drip in a flame. Dripping is apparently prevented by rapid charring reactions. The chlorine may be added as chlorinated polyphenyls, polynuclear compounds (e.g., naphthalene), and so on [338]. Finally, a coating for nylon composed of

PVC, Sb_4O_6, and a lead oxide is said to render nylon nondripping in a flame and self-extinguishing [339].

Molding-grade nylon has been fire retarded in the patent literature by metal halides (5+%) [339a], Sb_4O_6 with chlorocarbons [339b], and with melamine derivatives [339c].

Higher Nylons and Aromatic Analogs

There are a variety of other polyamides known as the higher nylons [340]. These aliphatic polymers have not made significant commercial inroads. The aromatic polyamides are currently receiving a great deal of attention as high-performance materials. Polyimides [341] and polyimidazoles are also in the forefront. There are commercial products in each category. The aromatic character of the polymers raises the melting point so that they char before they melt; they are essentially nonburning. Tradenames like Nomex, Kapton, and acronyms like PBI are becoming common terms in the polymer language. Some of these are shown below:

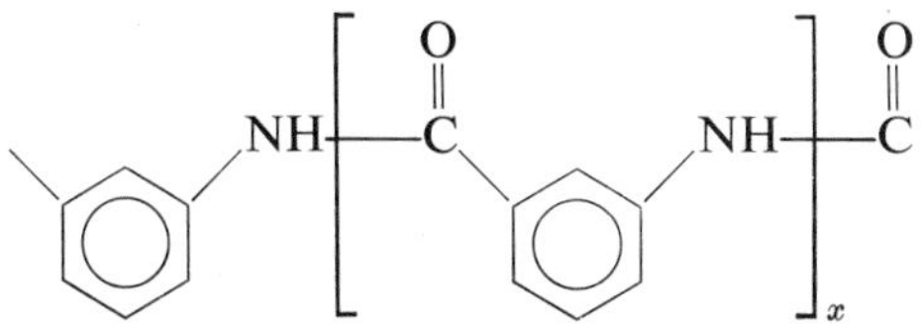

poly-m-phenylene isophthalamide   mp 424°C

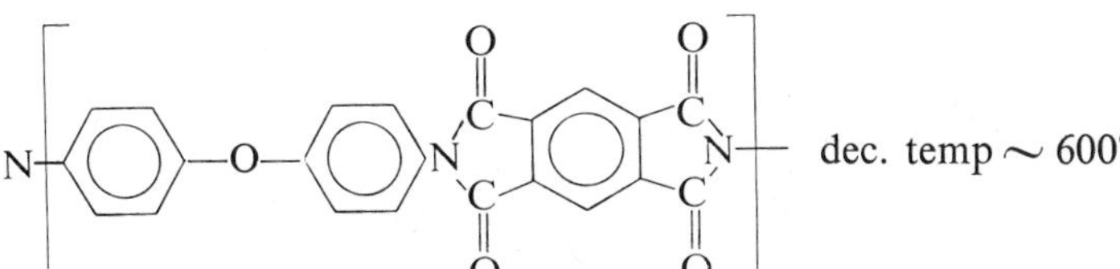

Nomex?

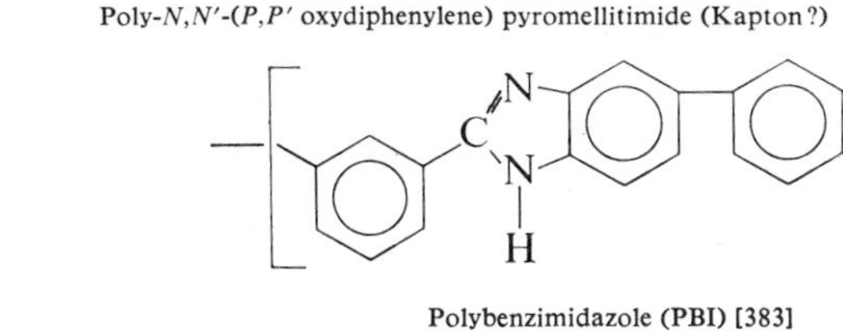

dec. temp ∼ 600°C

Poly-*N,N'*-(*P,P'* oxydiphenylene) pyromellitimide (Kapton?)

Polybenzimidazole (PBI) [383]

These materials are not only nonburning in the usual tests but retain good textile properties even after prolonged exposure to temperatures unheard of for ordinary fibers. They are of obvious interest for space suits and in aircraft interiors and the like. It is not unreasonable to look to this kind of fiber structure for applications in draperies and other consumer items in the future.

The subject of inherently incombustible fibers of this kind is really outside the scope of this book. The interested reader should consult the references in reference 342 for more discussion.

Two additional items of significance are the use of chlorine or bromine to halogenate finished polyamides to enhance their properties and such treatment as a precursor to carbonization to produce truly flameproof fibers [343]. Carbonized fibers may be based on aromatic polyamides treated carefully by controlled heating and in the presence of various gases, for example, heating a polyamide, in tension, through an oxidizing stage to pyrolysis all the way to 2500° to form graphite fibers [344]. These fibers are finding use in a variety of high-performance composites for uses such as turbine blades.

Surface treatment of some of these aromatic polymers with sulfur improves their flame resistance [344a] and polyimides based on benzothiazole are very thermally stable [344b].

PHENOLICS

Phenolic resins include the acid-catalyzed Novolak noncurable resins and the thermosetting alkaline-cured Bakelite types. Novolaks are mixtures of materials of the structural type:

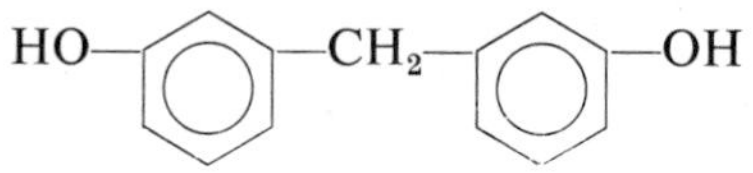

$$HO-\underset{\text{}}{\bigcirc}-CH_2-\underset{\text{}}{\bigcirc}-OH$$

Bakelites are cross-linked:

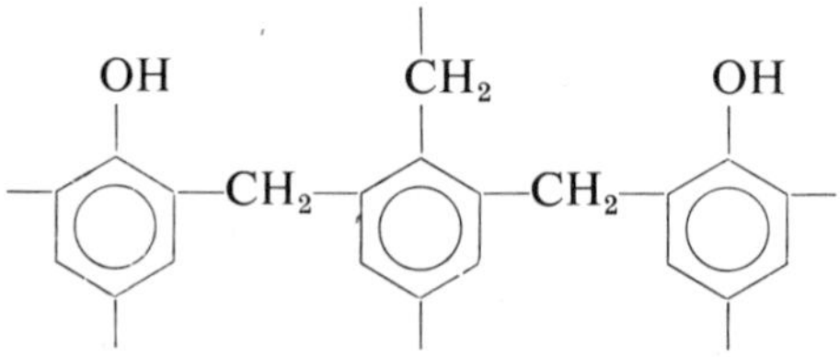

$$
\begin{array}{ccc}
OH & CH_2 & OH \\
| & | & | \\
-\bigcirc-CH_2-\bigcirc-CH_2-\bigcirc- \\
| & | & |
\end{array}
$$

They are made from formaldehyde and phenol. Phenolics are not especially flammable and often give self-extinguishing ratings when tested in the horizontal position [3]. The smoke generated is not copious. The aromaticity of these materials makes them stable and inherently resistant to oxidation. Madorsky indicates that degradation at moderate temperature ($\approx 360°$) proceeds by thermal scissions at the $CH_2-\phi$ bond, yielding fragments which give rise to acetone, propylene, and propanol. The aromatic nucleus remains in the high-boiling residue which comprises some 75% of the starting weight [142]. The low volatility at elevated temperatures further confirms the low flammability of the resins. Phenolics are used as sacrificial coating (ablative materials) on space craft. Studies of their behavior under simulated degradation have recently been published [344c].

Improvements in flame resistance have nonetheless been sought. Amines or amides have been added to provide some NH_3 in the volatiles in the hope of reducing combustibility [345]. About 2% P and 2% N from a product of $POCl_3$ and NH_3 in a phenolic is effective [346].

A nonflammable molding powder has been prepared from phenol, CH_2O, and $P(NCO)_3$ or $OP(NCO)_3$. Phenol reacts with these:

$$OP(NCO)_3 + 3HO—\phi \rightarrow (\phi O—\overset{\overset{O}{\|}}{C}—\overset{\overset{H}{|}}{N}—)_3PO$$

This may then be added to the resin at a level so as to give about 1% P and 1.3% N [347].

Adducts of polyols or aminoalcohols with a condensate of NH_3 and P_4O_{10} are liquids with a degree of compatibility with the phenolic resins. At $\sim$5% P and $\sim$5-6% N excellent flame resistance is obtained [348].

$\phi O—\overset{\overset{O}{|}}{P}—Cl_2$ can be reacted with an early intermediate in phenolic processing to incorporate the P in the resin:

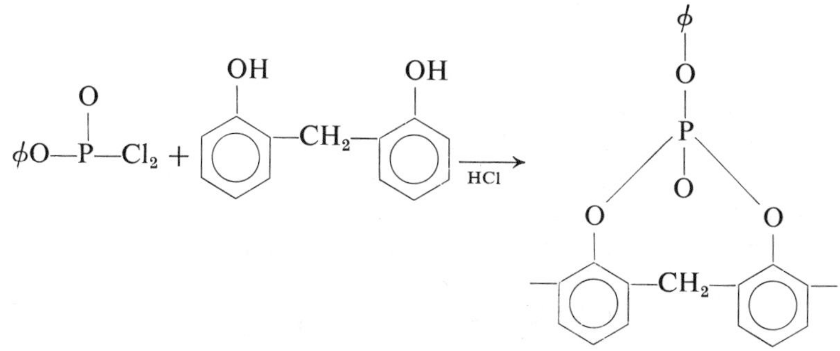

About 6% P can be incorporated readily in this way and, if desired, halogens can be added on the phenyl ring in the phosphorus compound.* A similar sequence with a reactive phosphine oxide has just been reported. Other members of Group VA can be similarly made part of the resin, that is, arsenic and antimony compounds [349, 350]. These same researchers have also placed Si and B in phenolics in like manner. They, too, improve flame resistance [350].

A resin containing 13% Cl, 3.7% Br, and $\frac{3}{4}$% P from additives—bis-(bromoethyl) chloroethyl phosphate and a chlorinated polyphenyl—was self-extinguishing in 23 sec [351]. Mixtures of dichlorophenols can be copolymerized in phenolics and, at a level of about 16% Cl, thereby achieve fire resistance (self-extinguishing in <4 sec) [352]. The resin itself can be chlorinated after polymerization by introducing chlorine gas [353]. (This can only be done with Novolaks.) Chlorinated polyphenyls can be physically admixed [354] as can chlorinated saturated ring compounds [355]. In a three-step process, hexachlorocyclopentadiene is reacted with an alkenone, this product reacted with phenol to give the substituted bisphenol

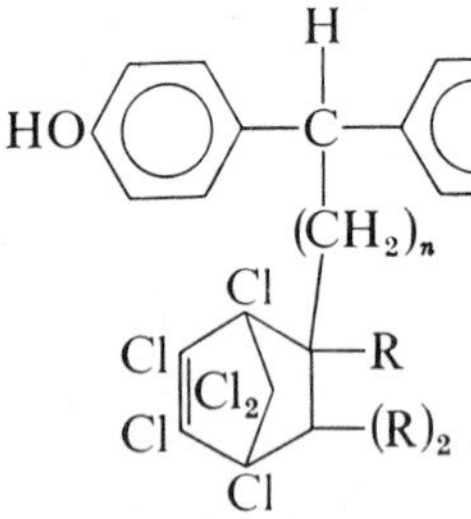

and then in turn is finally reacted with CH_2O. The resin contains as much as 38% Cl and is self-extinguishing [356].

Boron as boric acid or boron trioxide added at levels of from 5 to 11% B_2O_3 equivalent provides nonburning phenolics [357]. From 3.5 to 5% sodium borate has also been used [358].

It is important to repeat that phenolic resins are not difficult to render flame-resistant because they are not especially flammable in the first place. Indeed, by stepping up the cross-linking of phenolics a new flame-resistant fiber (Kynol[TM]) has been produced. The increased cross-linking improves the resin's resistance to chemical oxidizers and provides serviceability at 150°C in oxygen. This is yet another new fiber based on new structural arrangements for its flame resistance rather than the addition of the traditional elements [359].

MISCELLANEOUS

In the following sections are discussed polymers for which there are relatively few literature references as to fire-safe properties. This is not to downgrade the importance of the polymers or to say that they require no improvement in fire resistance. Also included are a few remarks on the rapidly expanding class of thermally stable materials, although this is truly outside the scope of this book.

Aminoplasts

This category of resins is usually made from formaldehyde and an amide such as urea or an amine, for example, melamine. The high nitrogen content decreases flammability in the manner of the polypeptides. Both urea and melamine-formaldehyde resins are self-extinguishing in the horizontal position [3]. Often these resins are cured with an acid solution. When this is H_3PO_4, additional flame resistance results [360]. Tris(2,3-dibromopropyl) phosphate can be used as an additive (7–20 wt %) [361]. The remaining references are to placing phosphorus in the polymer by using a phosphorus-bearing monomer. Resins containing 15% P and 14% N are prepared from

$$\phi—\overset{\displaystyle O}{\overset{\|}{P}}—(NH_2)_2 \text{ or } \phi O—\overset{\displaystyle O}{\overset{\|}{P}}—(NH_2)_2 \text{ and } CH_2O \text{ [362]. These produce non-}$$

burning surfaces resistant to, for example, cigarette burns. Resins made from a melamine-formaldehyde system and a phosphate ester polyol are described as intumescent coatings in a recent patent [362a]. Guanylurea phosphate has been used in place of urea with formaldehyde [363]. Phosphorus can be added to a melamine resin by using the following phosphine oxide:

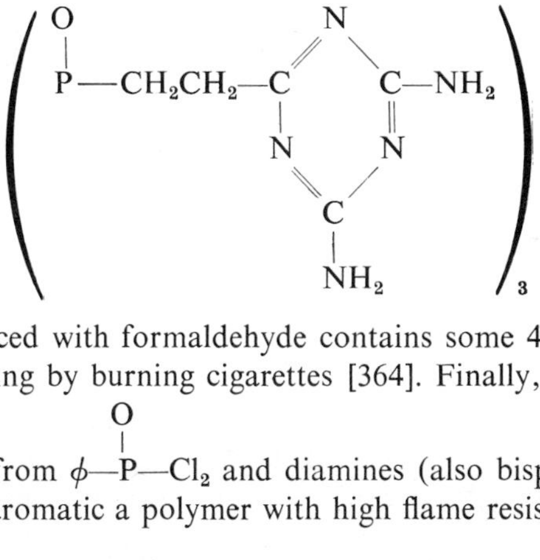

The resin produced with formaldehyde contains some 4–5% P and 30% N and resists marring by burning cigarettes [364]. Finally, there is a series of

$$\text{polymers made from } \phi—\overset{\displaystyle O}{\overset{\|}{P}}—Cl_2 \text{ and diamines (also bisphenols). When the}$$

diamine is also aromatic a polymer with high flame resistance should result [365], such as

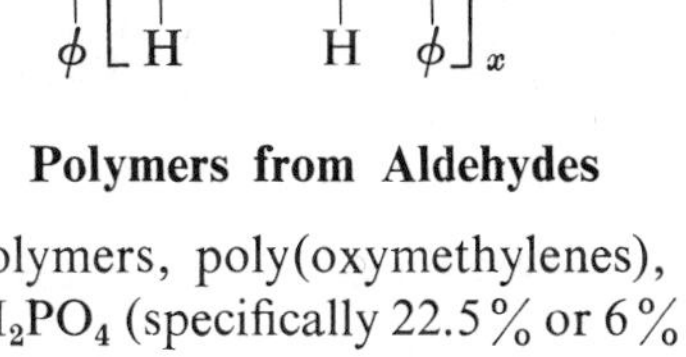

Polymers from Aldehydes

The formaldehyde polymers, poly(oxymethylenes), can be flameproofed by adding >20% $NH_4H_2PO_4$ (specifically 22.5% or 6% P and 2.7% N) [366].

Polymers made from chlorinated acetaldehydes are flame-resistant from the high chlorine level contained. Thus, chloral—Cl_3CCHO—and dichloro-acetaldehyde are copolymerized and stabilized by treatment of the residual OH groups with acetic anhydride. Similar polymers can be made from chloral and CH_2O [367].

A flame-resistant polyacetal is prepared from chlorinated xylylene glycol and dichlorobenzaldehyde to give a final polymer with benzyl alcohol that contains 9% Cl [368].

Allylic Polymers

Resins prepared from allylphthalate, allyl acetate, and the like are readily rendered more flame-resistant by copolymerizing a monomer rich in halogen. Thus, allyl chlorendate may be used. When Sb_4O_6 is also added (e.g.,

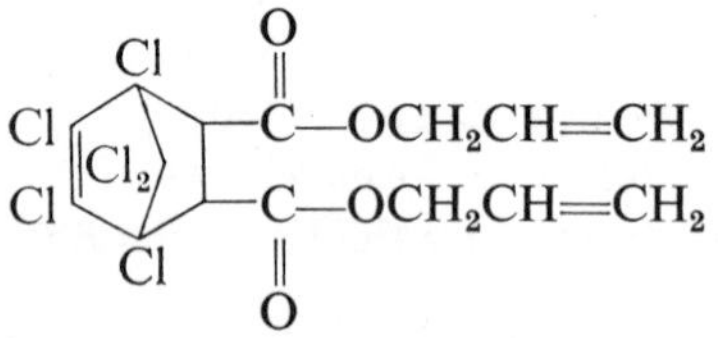

$$
\begin{array}{c}
\text{Cl} \quad\quad \text{O} \\
\text{Cl} \quad\quad \| \\
\text{Cl} \overbrace{\;[Cl_2]\;}^{}-\!\!-\text{C}\!-\!\text{OCH}_2\text{CH}=\text{CH}_2 \\
\text{Cl} \quad\quad\quad -\text{C}\!-\!\text{OCH}_2\text{CH}=\text{CH}_2 \\
\text{Cl} \quad\quad \| \\
\quad\quad\quad \text{O}
\end{array}
$$

10% Sb_4O_6 plus 4.5% Cl) the resin is self-extinguishing by ASTM D-635 and has a burning rate of 0.15 in./min by ASTM D-757. Thomas and Wright have reported on such systems in some detail [369]. Another tack in filled allylic resins is the use of additive halogens, such as perchlorocyclic aliphatics at 5–10% Cl with 8–12% hydrated alumina [370]. MgO has been found to reduce the smoking tendency of such resins at levels of from 0.4 to 1.0% by weight [371].

Polycarbonates

These resins are polyesters of CO_2 and are made by a variety of processes. Typical LOI's range from 0.26 to 0.28. They are self-extinguishing in the horizontal position [142]. Phosphorus has been incorporated by adding species of the type $R\!-\!\overset{\text{O}}{\underset{}{P}}\!-\!Cl_2$ so as to form linkages such as [372]

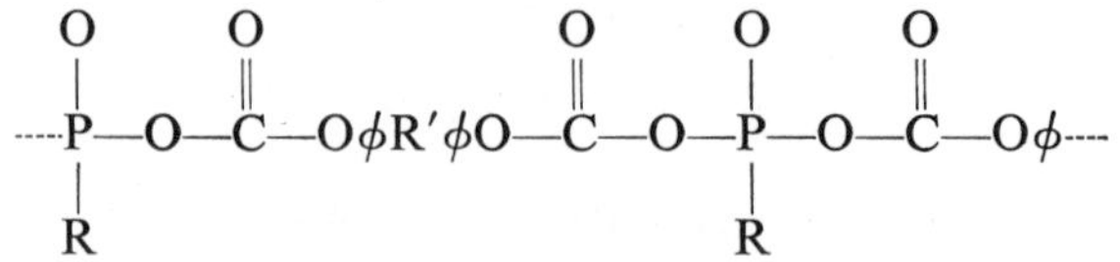

$$
\cdots\!P\!-\!O\!-\!C\!-\!O\phi R'\phi O\!-\!C\!-\!O\!-\!P\!-\!O\!-\!C\!-\!O\phi\!\cdots
$$

Flame resistance most often is attained with halogens, either as additives or as part of the monomer. Chlorinated cycloaliphatics [373] and brominated aromatics [374] have both been suggested, the latter being itself a carbonate ester:

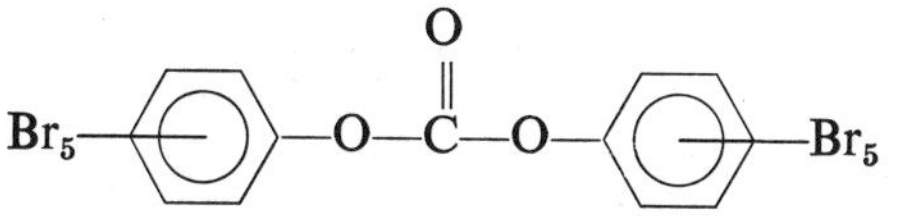

Halogenated bisphenol A's are perhaps the easiest means of incorporating the flame retardant. Even fluorine has been added in this way [375]. As little as 5% Br is effective. The resins so obtained are of the following structure [376, 377]:

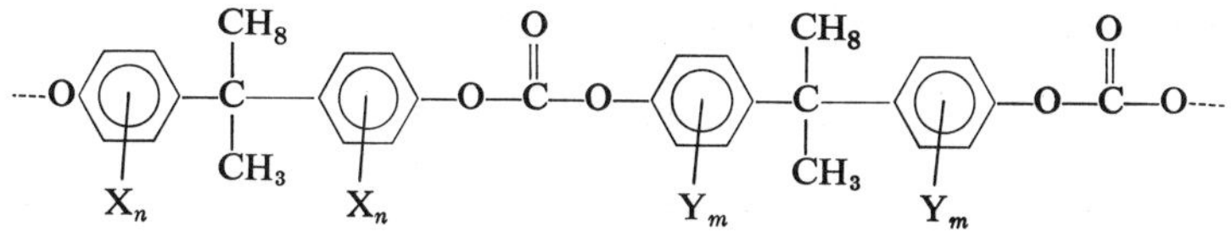

where X = halogen, $n = 1-2$, and Y = halogen or hydrogen, $m = 1-4$. Resins containing 5% Br are self-extinguishing in <2 sec [378]. The resin from tetrachlorobisphenol A will contain some 30% Cl and should be very flame-resistant [376]. The aromatic nature of the basic resin, of course, provides considerable stability for the molecule. Recently a commercial application of one of these systems has appeared for which a nonburning rating is claimed by ASTM D-635 and a LOI of 0.435 is given [379].

Truly Miscellaneous

Linoleum can be made "incombustible" by adding phenyl phosphates according to an old patent (1913) [380]. Leather may be made flameproof by using tetrakis(hydroxymethyl) phosphonium chloride in the "fat liquoring" process as given in a new patent [381]. And rosin by-products known as Vinsol resins may be made fire-resistant by adding H_3BO_3, borax, or phenyl phosphates [382]. And so on.

THERMALLY STABLE POLYMERS

There is an entirely new field of polymer chemistry in the area of thermally stable polymers. This has been prompted by massive government efforts—the military and space programs when the need for light-weight extra-stable materials has pushed back frontiers in several directions. High-temperature fibers have been alluded to earlier. Frazer's volume of collected papers is useful as a reference [342]. Useful, too, are several chapters in the Mark, Cernia, and Atlas book [341], especially the chapters by Black and Preston and by Chambion. Even more recently a lengthy review has appeared by J. I. Jones [383]. In this review the structural chemistry is laid out so that one gets a complete picture of the synthesis effort that has been undertaken.

There are few data on properties, however, and none of these works gives results of conventional fire testing. This is probably because the materials in almost all cases would be nonburning by the usual tests.

The most recent compilation is that edited by Preston on the proceedings of a symposium held by the American Chemical Society on thermally stable fibers [383a]. This covers the fiber aspect through late 1968 and is recommended highly.

Most of the polymers that have found practical utility are highly aromatic; many have essentially no oxidizable substituents. Thus, hydrogens are often found only on aromatic nuclei and nitrogen is linked to three other stable carbon groups; for example, in the polyimides:

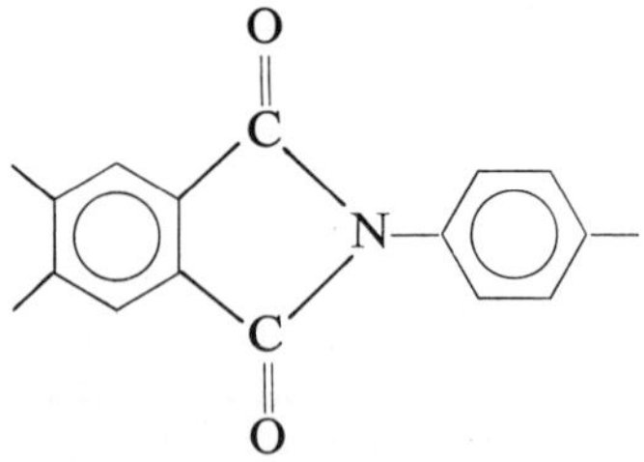

Analogs of flammable substances become very flame-resistant when the aromatic ring is interposed. The aromatic polyesters are a case in point. (These are called polyarylates by the Russians [384].) An example is

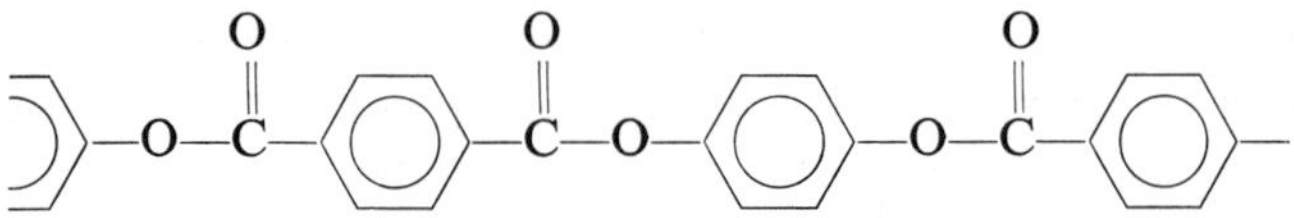

from terephthalate and p-dihydroxybenzene. Aromatic polyethers are another example. PPO is an acronym that is becoming familiar. It refers to one of the polyphenylene oxides [383]:

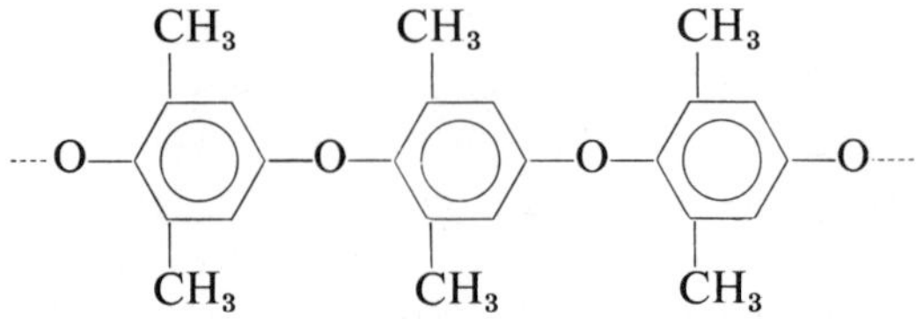

PPO

Halogenated polyphenylene oxides have also been reported for use when flame-retardant requirements are severe [385, 386]. The limiting case in this family is polyphenylene. Synthesis of branched polyphenylenes has recently been described [387]. Thermal degradation studies on p-polyphenylene and chlorinated derivatives (up to 8 % Cl) have been reported with emphasis on ablation uses. The partially chlorinated material shows real promise [388]. The aromatic polyamides and imides have already been discussed (see the section on "Nylon"). Polyaromatic sulfones, for example, are also very

stable; the above structure has been commercialized as "Polysulfone" [383]. It is clear that in the future many of the flame-resistant materials that will be required will come from new structures rather than from old structures with flame retardants either grafted on or incorporated as additives. A book on flame-resistant materials written in 1989 may very well be largely devoted to a discussion of such new structures—with P, Sb, Cl, Br, and so on, assuming a secondary role. Such is the evolutionary nature of applied chemistry.

REFERENCES

[1] *Chem. Eng. News*, **1966**, 24 (Sept. 5).

[2] C. J. Hilado, *J. Cell. Plast.*, **4**, 339 (1968).

[3] C. J. Hilado, *Flammability Handbook for Plastics*, Technomic, Stamford, Conn. 1969, 164 pp.

[3a] P. Robitschek, *J. Cell. Plast.*, **1**, 395 (1965).

[4] C. J. Hilado, P. E. Burgess, Jr., and W. R. Proops, *ibid.*, **4**, 67 (1968); K. C. Frisch, *ibid.*, **1**, 321 (1965).

[5] J. N. Tilley, H. G. Nodeau, H. E. Reymore, P. H. Waszeciak, and A. A. R. Sayigh, *ibid.*, **4**, 56 (1968).

[6] I. N. Einhorn and R. W. Mickelson, Div. Org. Coatings Plast. Chem., Am. Chem. Soc., Preprints, *28*(1) 291 (1968).

[7] I. N. Einhorn, R. W. Mickelson, B. Shah, and R. Craig, *J. Cell. Plast.*, **4**, 188 (1968).

[8] E. A. Dickert and G. C. Toone, *Mod. Plast.*, **42**, 197 (1965).

[9] J. K. Backus, W. C. Darr, P. G. Gemeinhardt, and J. H. Saunders, *J. Cell. Plast.*, **1**, 178 (1965).

[10] J. K. Backus, D. L. Bernard, W. C. Darr, and J. H. Saunders, *J. Appl. Polym. Sci.*, **12**, 1053 (1968).

[11] G. W. Ball, G. A. Haggis, R. Hurd, and J. F. Wood, *J. Cell. Plast.*, **4**, 248 (1968); I.C.I., Ltd., *Belg. 697,411*; *680,380*, Apr. 21, 1967; Apr. 29, 1966.

[12] Standard Oil Co., *Neth. Appl. 6,604,906*, Oct. 17, 1966; Standard Oil Co., *Neth. Appl. 6,516,709*, June 24, 1966.

[13] Chemische Fabrik Kalk G.m.b.H. (by A. Heslinga and P. J. Napjus, *Ger. 1,197,614*, July 29, 1965.

[14] F. L. Bennett, F. H. Otay, and C. L. Mehltretter, *J. Cell. Plast.*, **3**, 369 (1967).

[15] H. Bernstein, D. S. Greidinger, and Chemicals & Phosphates, Ltd., *Israeli 17,744*, Aug. 27, 1964.

[16] Imperial Chemical Industries, Ltd. (by J. H. Taylor), *Brit. 935,926*, Sept. 4, 1963.

[17] B. R. Franko-Filipasic, *U.S. 3,324,202* (to FMC Corp.), June 6, 1967; A. L. Austin, R. J. Hartman, and J. T. Patton, Jr., *U.S. 3,439,067* (to Wyandotte Chemicals Corp.), Apr. 15, 1969; R. Merten, O. Bayer, G. Braun, and H. Kaiser, *U.S. 3,428,577* (to Farb. Bayer A.-G.), Feb. 18, 1969.

[18] R. J. Hartman and L. C. Pizzini, *U.S. 3,317,638* (to Wyandotte Chemicals Corp.), May 2, 1967.

[19] Farb. Bayer A.-G., *Ger. 1,106,489* (by V. Trescher, G. Baren, and H. Nordt), May 10, 1961; P. E. Pelletier and F. Pelletier, *U.S. 3,369,060* (to Wyandotte Chemicals Corp.), Feb. 13, 1968; Merten, Bayer, Braun, and Kaiser, reference 17; T. E. Ronay and R. D. Dexheimer, *U.S. 3,441,544* (to Richardson Co.), Apr. 29, 1969.

[20] Upjohn Co., *Belg. 696,571*, Apr. 4, 1967.

[21] J. T. Patton, Jr., and R. J. Hartman, *U.S. 3,417,164* (to Wyandotte Chemicals Corp.), Dec. 17, 1968.

[21a] H. E. Hill, D. W. Kaiser, *U.S. 3,439,068* (to Olin Mathieson Chemical Corp.), Apr. 15, 1969.

[22] M. Wismer, R. Township, H. P. Doerge, P. R. Mosso, and J. F. Foote, *U.S. 3,407,150* (to Pittsburgh Plate Glass Co.), Oct. 22, 1968; Hooker Chemical Corp. (by R. R. Hindersinn, M. Worsley, and B. O. Schoepfle), *Brit. 919,067*, Feb. 20, 1963.

[23] M. Wismer, H. P. Doerge, P. R. Mosso, and J. F. Foote, *U.S. 3,314,902* (to Pittsburgh Plate Glass Co.), Apr. 18, 1967.

[24] Atlas Chemical Industries, Inc. (by J. D. Zech and E. C. Ford, Jr.), *Belg. 642,446*, July 13, 1964; Atlas Chemical Industries, Inc., *Neth. Appl. 6,400,198*, July 16, 1964.

[25] Thiokol Chemical Corp. (by E. R. Bertozzi, H. L. Elkin, R. F. Hoffman, B. J. Sutker, and S. Schreiber), *Fr. 1,365,331*, July 3, 1964; H. W. Coover, Jr., and R. L. McConnell, *U.S. 2,952,666* (to Eastman Kodak Co.), Sept. 13, 1960.

[26] Farb. Bayer A.-G. (by H. Holtschmidt and G. Braun), *Ger. 1,143,638*, Feb. 19, 1963; N. V. W. A. Scholtens Chemische Fabriken, *Neth. Appl. 6,407,577*, Jan. 6, 1965; J. J. Anderson, *U.S. 3,251,785* (to Socony Mobil Oil Co.), May 17, 1966; M. R. Lutz, *U.S. 3,251,828* (to FMC Corp.), May 17, 1966; Lanko Chemicals, Ltd., *Neth. Appl. 6,504,036*, Oct. 1, 1965; FMC Corp., *Fr. 1,410,280*, Sept. 10, 1965; Merck & Co., Inc. (by D. Shew and B. W. Greenwald), *Fr. 1,374,250*, Oct. 2, 1964; *Can. 796,745*, Oct. 15, 1968.

[26a] W. M. Lanham, *U.S. 3,159,591* (to Union Carbide Corp.), Dec. 1, 1964.

[27] C. E. Miles and J. W. Lyons, *J. Cell. Plast.*, **3**, 539 (1967).

[28] W. Abbotson and R. Hurd (to Imperial Chemical Industries, Ltd.), *Brit. 792,016*, Mar. 19, 1958.

[29] Monsanto Co. (by J. C. Barnett), *Fr. 1,462,604*, Dec. 16, 1966; Monsanto Co., *Neth. Appl. 6,516,724*, June 23, 1966; J. C. Barnett, *U.S. 3,423,343* (to Monsanto Co.), Jan. 21, 1969.

[30] Pittsburgh Plate Glass Co., *Brit. 942,616*, Nov. 27, 1963.

[31] Farb. Bayer A.-G. (by G. Oertel, O. Bayer, and G. Braun), *Ger. 1,143,021*, Jan. 31, 1963; E. N. Walsh and E. H. Uhig, *U.S. 3,440,222* (to Stauffer Chemical Co.), Apr. 22, 1969.

[32] Pittsburgh Plate Glass Co., *Brit. 918,636*, Feb. 13, 1963.

[33] Pittsburgh Plate Glass Co., *Brit. 918,636*, Feb. 13, 1963; Stauffer Chemical Co., *Neth. Appl. 6,407,376*, Jan. 4, 1965; Albright & Wilson, Ltd. (by P. A. T. Haye and D. Eastwood), *Ger. 1,163,018*, Feb. 13, 1964.

[34] C. C. L. Hwa and P. Robitschek, *U.S. 3,247,134* (to Hooker Chemical Corp.), Apr. 19, 1966; W. Von Bonin, H. Holtschmidt, and H. Piechota, *U.S. 3,312,638* (to Farb. Bayer A.-G.), Apr. 4, 1967.

[35] Imperial Chemical Industries, Ltd., *Belg. 689,523*, Nov. 9, 1966; I.C.I., Ltd. (by H. C. Fielding), *Brit. 938,367*, Oct. 2, 1963; *Brit. 982,931*, Feb. 10, 1965.

[36] M. Kaplan and M. Koral, *U.S. 3,007,884* (to Allied Chemical Corp.), Nov. 7, 1961.

[37] V. I. Kirilovich, I. K. Rubtsova, L. I. Pokrovskii, R. V. Khinich, and A. A. Federov, *Plast. Massy*, **1966**, 10 (in Russian).

[38] L. Friedman, *U.S. 3,142,650* (to Union Carbide Corp.), July 28, 1964.

[39] L. Friedman, *U.S. 3,142,651* (to Union Carbide Corp.), July 28, 1964; L. Friedman, *U.S. 3,220,961* (to Weston Chemical Corp.), Nov. 30, 1965; *U.S. 3,246,051* (to Weston Chemical Corp.), Apr. 12, 1966.

[40] K. Yamamoto, M. Iwasa, K. Ema, and N. Arai, *U.S. 3,422,037* (to Mitsui Kagaku Kogyo Kabuskiki Kaisha Matsui Chemical Industries Co., Ltd.), Jan. 14, 1969.

[41] A. Guttag, *U.S. 3,144,419* (to Union Carbide Corp.), Aug. 11, 1964.

[42] J. Langrish and P. D. Perry, *U.S. 3,429,837* (to Imperial Chemical Industries, Ltd.), Feb. 25, 1969; C. F. Baranauckas and I. Gordon, *U.S. 3,412,051* (to Hooker Chemical Corp.), Nov. 19, 1968.

[43] Farb. Bayer A.-G. (by P. G. Boiron), *Brit. 1,043,832*, Sept. 28, 1966.

[44] Farb. Bayer A.-G. (by H. Jonas, P. Hoppe, and E. Walaschewski), *Ger. 1,138,219*, Oct. 18, 1962.

[45] T. M. Beck and E. N. Walsh, *U.S. 3,235,517* (to Stauffer Chemical Co.), Feb. 15, 1966; Stauffer Chemical Co. (by R. H. Rosenberg and R. S. Cooper), *Fr. 1,388,667*, Feb. 12, 1965; Mobay Chemical Co. (by P. G. Gemeinhardt), *Fr. 1,438,582*, May 13, 1966; Farb. Bayer A.-G., *Belg. 714,132*, Apr. 24, 1968.

[46] E. N. Walsh, E. N. Uhig, and T. M. Beck, *Amer. Chem. Soc., Div. Org. Coatings, Plast. Chem., Preprints 23* (1), 1 (1963).

[47] R. W. Mickelson and I. N. Einhorn, Div. Org. Coatings Plast. Chem., Am. Chem. Soc., Preprints *28*(1) 272 (1968).

[47a] Stauffer Chemical Co. (by G. R. Price), *Can. 800,044*, Nov. 26, 1968.

[48] L. Friedman, *U.S. 3,309,342* (to Union Carbide Corp.), Mar. 14, 1967; R. Merten, O. Bayer, G. Braun, and H. Kaiser, *U.S. 3,428,578* (to Farb. Bayer A.-G.), Feb. 18, 1969.

[49] Stauffer Chemical Co., *Neth. Appl. 6,613,257*, Apr. 24, 1967.

[49a] Stauffer Chemical Co., *Belg. 723,248*, Oct. 31, 1968.

[50] Farb. Bayer A.-G. (by R. Merten, O. Bayer, G. Braun, and H. Kaiser), *Ger. 1,206,152*, Dec. 2, 1965.

[51] Upjohn Co., *Brit. 1,062,869*, Mar. 22, 1967; A.-M. Nashu, H. E. Reymore, Jr., and A. A. R. Sayigh, *U.S. 3,437,607* (to Upjohn Co.), Apr. 8, 1969; U. Bahr and H. Holtschmidt, *U.S. 3,365,530* (to Farb. Bayer A.-G.), Jan. 23, 1968 (see also *Can. 761,948*, June 27, 1967); Farb. Bayer A.-G., *Belg. 714,132*, Appl. Apr. 24, 1968.

[52] P. Fischer, H. Holtschmidt, G. Oertel, and G. Braun, *U.S. 3,334,056* (to Farb. Bayer A.-G.), Aug. 1, 1967.

[53] G. H. Birum, R. M. Anderson, and R. B. Clampitt, *U.S. 3,385,801* (to Monsanto Co.), May 28, 1968; U. Bahr, G. Oertel, G. Nishk, and M. Dahm, *U.S. 3,410,812* (to Farb. Bayer A.-G.), Nov. 12, 1968 (see also *Can. 802,285*, Dec. 24, 1968).

[54] Stauffer Chemical Co., *Belg. 716,291*, June 7, 1968.

[55] Chemische Fabrik Kalk G.m.b.H., *Neth. Appl. 6,510,388*, Feb. 15, 1966.

[56] Hooker Chemical Corp., *Neth. Appl. 6,502,541*, Aug. 30, 1965.

[57] B. O. Schoepfle, R. R. Hindersinn, and M. Worsley, *U.S. 3,214,396* (to Hooker Chemical Corp.), Oct. 26, 1965; *U.S. 3,249,562*, May 3, 1966.

[58] Shell Internationale Research Maatschappij N.V., *Neth. Appl. 6,600,681*, July 21, 1966.

[59] Canadian Industries, Ltd. (by J. M. Turner), *Can. 694,407*, Sept. 15, 1964.

[59a] Farb. Bayer A.-G. (by G. Oertel, G. Braun, and O. Bayer), *Can. 780,507*, Mar. 12, 1968.

[60] Albright & Wilson, Ltd. (by J. K. Jacques), *Fr. 1,455,978*, Oct. 21, 1966; J. K. Jacques, *U.S. 3,458,457* (to Albright & Wilson (Mfg.) Ltd.), Jul. 29, 1969.

[61] C. F. Baranaukas and I. Gordon, *Can. 792,785* (to Hooker Chemical Corp.), Aug. 20, 1968.

[62] L. Friedman, *U.S. 3,225,010* (to Weston Chemical Corp.), Dec. 21, 1965.

[62a] Merten, Bayer, Braun, and Kaiser, reference 17.

[63] Farb. Bayer A.-G. (by H. Piechota and H. Wirtz), *Ger. 1,173,641*, July 9, 1964.

[64] C. C. Clark, H. M. Schroeder, and L. R. Garrison, *U.S. 3,375,220* (to Textron, Inc.), Mar. 26, 1968; N. C. W. Judd, *Materials Protection*, **1968**, 30 (August); Hooker Chemical Corp., *Neth. Appl. 6,506,721*, Nov. 29, 1965; Imperial Chemical Industries, Ltd., *Neth. Appl. 6,611,484*, Feb. 17, 1964; Dunlop Rubber Co., Ltd. (by J. Fisbein and S. D. Taylor), *Ger. 1,169,654*, May 6, 1964; A. Khawam, *U.S. 2,953,533* (to Allied Chemical Corp.), Sept. 20, 1960; T. M. Powanda, *U.S. 3,171,819* (to Celanese Corp. of America), Mar. 2, 1965; I.C.I., Ltd., *Belg. 685,586*, Aug. 16, 1966; I.C.I., Ltd. (by J. F. Chapman and R. Hurd), *Brit. 857,876*, Jan. 4, 1961; I.C.I., Ltd., *Neth. Appl. 6,612,405*, Mar. 6, 1967; F. M. Kujawa, *U.S. 3,424,470* (to Hooker Chemical Corp.), Jan. 23, 1969.

[65] W. Lindlow and E. J. Hensch, *U.S. 3,041,296* (to Celanese Corp. of America), June 26, 1962; G. H. Wiech and R. A. Dragon, *U.S. 3,041,295* (to Celanese Corp. of America), June 26, 1962.

[66] R. J. Polachek, *U.S. 3,041,293* (to Celanese Corp. of America), June 26, 1962.

[67] W. M. Lanham (to Union Carbide Corp.), *Brit. 922,198*, Mar. 27, 1963.

[68] A. V. Mercer, H. Green, and J. E. Jones, *U.S. 3,412,048* (to Shell Oil Co.), Nov. 19, 1968.

[69] Daihachi Chemical Industry Co., Ltd. (by I. Umemura and A. Hatano), *Jap. 17,158*, Sept. 29, 1966.

[70] FMC Corp. (by M. R. Lutz, G. Nowlin, and H. Stange), *Belg. 622,006*, Dec. 14, 1962.

[71] R. F. W. Ratz and A. D. Bliss, *U.S. 3,423,486* (to Olin Mathieson Chemical Corp.), Jan. 21, 1969.

[72] G. A. Haggis, D. Williams, and A. M. Wooler (to Imperial Chemical Industries, Ltd.), *Brit. 927,175*, May 29, 1963.

[73] Farb. Bayer A.-G. (by G. Oertel, H. Holtschmidt, G. Nishk, and G. Braun), *Ger. 1,127,583*, Apr. 12, 1962.

[74] Hooker Chemical Corp., *Fr. 1,297,489*, June 29, 1962; R. R. Hindersinn and M. Worsley, *U.S. 3,208,960* (to Hooker Chemical Corp.), Sept. 28, 1965; Hooker Chemical Corp., *Neth. Appl. 6,400,468*, July 23, 1964; F. M. Kujawa and M. Worsley, *U.S. 3,297,596* (to Hooker Chemical Corp.), Jan. 10, 1967; Hooker Chemical Corp., *Neth. Appl. 302,931*, Oct. 25, 1965.

[75] G. Nowlin and D. Warren, *U.S. 3,218,272* (to FMC Corp.), Nov. 16, 1965.

[76] FMC Corp. (by M. R. Lutz), *Brit. 1,029,465*, May 11, 1966.

[77] Farb. Bayer A.-G., *Belg. 616,916*, May 15, 1962.

[78] Union Carbide Corp., *Fr. 1,298,363*, July 13, 1962; Farb. Bayer A.-G. G. (by Nishk and G. Braun) *Ger. 1,106,067*, May 4, 1961; *Ger. 1,129,686*, May 17, 1962.

[79] M. R. Lutz, G. Nowlin, and H. Stange, *U.S. 3,321,555* (to FMC Corp.), May 23, 1967.

[80] Mitsubishi Chemical Industries Co., Ltd. (by M. Tokisava), *Jap. 5436*, Mar. 6, 1967.

[81] L. Friedman, *U.S. 3,159,605* (to Union Carbide Corp.), Dec. 1, 1964; Pure Chemicals, Ltd., *Fr. 1,395,595*, Apr. 16, 1965.

[82] M. S. Larrison, *U.S. 3,328,493* (to Union Carbide Corp.), June 27, 1967; Pure Chemicals, Ltd. (by L. Friedman), *Brit. 1,011,120*, Nov. 24, 1965.

[83] G. H. Birum, R. B. Clampitt, and R. M. Anderson, *U.S. 3,332,893* (to Monsanto Co.), July 25, 1967; G. H. Birum, *U.S. 3,317,510* (to Monsanto Co.), May 2, 1967; G. H. Birum and R. B. Clampitt, *U.S. 3,391,226* (to Monsanto Co), July 2, 1968.

[84] Shell Internationale Research Maatschappij N.V., *Belg. 660,828*, Sept. 9, 1965.

[85] Weston Chemical Corp. (by L. Friedman), *Can. 786,369*, May 28, 1968.

[86] Farb. Bayer A.-G. (by U. Bahr, G. Braun, and G. Nishk), *Ger. 1,170,636*, May 21, 1964.

[87] Farb. Bayer A.-G. (by H. Holtschmidt, E. Degener, and G. Braun), *Ger. 1,154,105*, Sept. 12, 1963.

[88] Dow Chemical Co., *Brit. 924,945*, May 1, 1963; Union Carbide Corp., *Belg. 611,748*, Jan. 15, 1962; J. Eichhorn, *U.S. 3,379,656* (to Dow Chemical Co.), Apr. 23, 1968.

[89] Nederlandse Organisatie voor Toegepast-Natuurwetenschappelijk Onderzoek ten Behoeve van Nijverheid, Handel en Verkeer, *Belg. 623,822*, Feb. 14, 1963.

[90] Farb. Bayer A.-G. (by V. Trescher, G. Braun, and M. Dahm), *Ger. 1,160,172*, Dec. 27, 1963.

[91] Distillers Co., Ltd., *Belg. 670,728, 670,729*, Apr. 8, 1966.

[92] Farb. Bayer A.-G. (by R. Merten, O. Bayer, G. Braun, and H. Kaiser), *Ger. 1,206,580*, Dec. 9, 1965; (by H. Piechota and H. Wirtz), *Belg. 617,305*, Aug. 31, 1962.

[93] Farb. Bayer A.-G. (by M. Dahm), *Ger. 1,112,822*, Appl. June 10, 1960.

[94] Chemische Fabrik Kalk G.m.b.H. (by W. Lampe and H. E. Proetzel), *Ger. 1,169,119*, Apr. 30, 1964.

[95] Albright & Wilson, Ltd. (by H. Coates and M. V. Cooksley), *Brit. 1,069,524*, May 17, 1967.

[96] Chemische Fabrik Kalk G.m.b.H., *Neth. Appl. 6,610,686*, Mar. 1, 1967.

[97] Shell Internationale Research Maatschappij N.V. (by R. W. Ashworth and A. V. Mercer), *Brit. 1,020,254*, Feb. 16, 1966; *Neth. Appl. 6,408,119; 6,414,233; Neth. Appl. 6,402,984*, Jan. 19, 1965; June 10, 1965; Sept. 22, 1964; Albright & Wilson, Ltd. (by L. G. Lund), *Ger. 1,776,841*, Aug. 27, 1964.

[98] C. F. Baranauckas and I. Gordon, *U.S. 3,248,429* (to Hooker Chemical Corp.), Apr. 26, 1966; Lanitro Chemical, Ltd., *Belg. 705,441*, Oct. 20, 1967; Imperial Chemical Industries, Ltd. (by J. H. Taylor), *Brit. 925,570*, May 8, 1963.

[99] Albright & Wilson, Ltd. (by P. A. T. Hoye and H. Coates), *Brit. 974,033*, Nov. 4, 1964; E. H. Kober and G. F. Ottmann, *U.S. 3,404,187* (to Olin Mathieson Chemical Corp.), Oct. 1, 1968; Olin Mathieson Chemical Corp. (by E. H. Kober and G. F. Ottmann) *Can. 803,566*, Jan. 7, 1969.

[99a] N. J. Clark and F. McCollough, Jr., *U.S. 3,294,712* (to Stauffer Chemical Co.), Dec. 27, 1966.

[100] Hooker Chemical Corp. (by P. Robitschek), *Ger. 1,142,234*, Jan. 10, 1963; Hooker Chemical Corp., *Belg. 611,246*, Dec. 29, 1961; (by M. Worsley) *Belg. 621,572*, Dec. 14, 1962; *Brit. 873,974*, Appl. Nov. 22, 1957; *Brit. 891,995*, Mar. 21, 1962; P. Robitschek, *U.S. 3,156,659*, Nov. 10, 1964, and *U.S. 3,058,925*, Oct. 16, 1962 (both to Hooker Chemical Corp.); M. Worsley and R. Hindersinn, *U.S. 3,391,092* (to Hooker Chemical Corp.), July 2, 1968; F. M. Kujawa and M. Worsley, *U.S. 3,275,606* (to Hooker Chemical Corp.), Sept. 27, 1966; J. C. Tapas and I. J. Dissen, *U.S. 3,098,047* (to Velsicol Chemical Corp.), July 16, 1963.

[101] Universal Oil Products Co. (by R. M. Lusskin, F. Backer, and J. R. Larson) *Fr. 1,411,373*, Sept. 17, 1965.
[102] Imperial Chemical Industries, Ltd. (by R. R. Aitken, J. F. Chapman, and A. M. Wooler), *Brit. 895,967*, May 9, 1962.
[103] Columbian Carbon Co., *Neth. Appl. 6,601,190*, Aug. 19, 1966.
[104] P. W. Morgan, *U.S. 3,391,111* (to E. I. duPont de Nemours & Co.), July 2, 1968.
[105] R. R. Hindersinn and M. Worsley, *U.S. 3,055,849*, Sept. 25, 1962, and *U.S. 3,146,220*, Aug. 25, 1964 (both to Hooker Chemical Corp.).
[106] C. A. Erickson and D. Warren, *U.S. 3,250,749* (to FMC Corp.), May 10, 1966; W. J. Farrissey, F. P. Recchia, and A. A. R. Sayigh, *U.S. 3,437,680* (to Upjohn Co.), May 8, 1969.
[107] H. L. Heiss, *U.S. 3,102,875* (half to Monsanto Chemical Co. and half to Mobay Chemical Co.), Sept. 3, 1963; Monsanto Chemical Co., *Brit. 793,780*, Apr. 23, 1958.
[108] J. A. Parker, S. R. Riccitiello, W. J. Gilwee, and R. Fish, *SAMPE J.*, 41 (April/May, 1969); Michelin & Cie, *Neth. Appl. 6,509,905*, Feb. 1, 1966.
[109] Allied Chemical Corp. (by R. A. Wiles and E. R. Degginger), *Fr. 1,379,825*, Jan. 27, 1964; Olin Mathieson Chemical Corp., *Brit. 1,005,657*, Sept. 22, 1965; H. A. Bruson and J. S. Rose, *U.S. 3,269,961* (to Olin Mathieson Chemical Corp.), Aug. 30, 1966; Wyandotte Chemicals Corp., *Neth. Appl. 6,415,197*, Dec. 29, 1964; Olin Mathieson Chemical Corp. (by S. Fuzesi and F. A. Perico), *Belg. 629,996*, Oct. 21, 1963; Olin Mathieson Chemical Corp. *Fr. 1,371,674*, Sept. 9, 1964; Olin Mathieson Chemical Corp., *Belg. 637,779*, Mar. 29, 1964; Olin Mathieson Chemical Corp. *Neth. Appl. 6,413,843*, May 31, 1965; H. A. Bruson and J. S. Rose, *U.S. 3,244,754* (to Olin Mathieson Chemical Corp.), Apr. 5, 1966.
[110] Courtaulds, Ltd. (by B. J. Wakefield), *Brit. 1,037,323*, July 27, 1966; J. S. Rose, D. R. Shine, and J. V. Karabinos, *U.S. 3,137,661* (to Olin Mathieson Chemical Corp.), June 16, 1964.
[111] Farb. Bayer A.-G. (by R. Merten and O. Bayer), *Ger. 1,178,587*, Sept. 24, 1964.
[112] S. P. Rowland, E. G. Pritchett, and N. L. Hofmann, *U.S. 3,284,511* (to National Distillers and Chemical Corp.), Nov. 8, 1966.
[113] Farb. Bayer A.-.G (by H. Piechota, R. Merten, V. Trescher, and H. Wirtz), *Belg. 628,045*, May 29, 1963.
[114] J. P. Stallings and D. H. Wagner, *U.S. 3,305,497* (to Diamond Alkali Co.), Feb. 21, 1967.
[115] Deutsche Solvay Werke G.m.b.H. (by G. Faerber), *Ger. 1,048,694*, Jan. 15, 1959.
[116] Badische Anilin- & Soda-Fabrik A.-G. (by F. Meyer), *Ger. 1,177,816*, Sept. 10, 1964.
[117] Imperial Chemical Industries, Ltd. (by R. R. Aitken, J. F. Chapman, and E. A. Packer), *Brit. 895,966*, May 9, 1962; Farb. Bayer A.-G. (by E. Degener, H. Holtschmidt, and G. Braun), *Belg. 616,865*, May 15, 1962; Dow Chemical Co., *Neth. Appl. 6,601,266*, Aug. 4, 1966; Shell Internationale Research Maatschappij N.V., *Neth. Appl. 6,408,118*, Jan. 10, 1965.
[118] R. M. Anderson and J. C. Wygant, *U.S. 3,256,506* (to Monsanto Co.), June 14, 1966.
[119] F. W. Berk & Co., Ltd. (by Z. E. Jolles and B. J. Riley), *Brit. 1,079,984*, Aug. 16, 1967, R. C. Nametz, R. D. Deanin, and P. M. Lambert, *SPE Trans.*, 4, 251 (1964).
[120] Mobay Chemical Co. (by J. H. Saunders), *Belg. 618,391*, June 29, 1962.
[121] P. G. Pape, J. E. Sanger, and R. C. Nametz, *J. Cell. Plast.*, 4(11), 438 (1968).
[122] Raychem Corp., *Belg. 677,506*, Mar. 8, 1966.
[123] W. M. Lanham, *U.S. 3,075,927* (to Union Carbide Corp.), Jan. 20, 1963.

[124] F. W. Berk & Co., Ltd. (by Z. E. Jolles), *Brit. 994,087*, June 2, 1965.

[125] Raychem Corp., *Neth. Appl. 6,603,028*, Sept. 9, 1966.

[126] Allied Chemical Corp. (by E. R. Degginger, E. A. Dickert, and R. A. Wiles), *Belg. 657,423*, Apr. 15, 1965; E. A. Bulygina and F. A. Kryuchkov, *Plast. Massy*, **1968**, 33 (January).

[127] Diamond Alkali Co., *Neth. Appl. 6,414,132*, June 8, 1965; *Neth. Appl. 6,602,655*, Sept. 2, 1966.

[128] S. M. Postol, *U.S. 3,260,637* (to Union Carbide Corp.), July 12, 1966.

[129] J. J. Pedjac and C. T. Pumpelly, *U.S. 3,275,578* (to Dow Chemical Co.), Sept. 27, 1966.

[129a] Imperial Chemical Industries, Ltd. (by J. F. Chapman and B. F. Jones), *Brit. 889,720*, Feb. 21, 1962.

[130] Allied Chemical Corp. (by E. A. Dickert), *Belg. 657,422*, Apr. 15, 1965.

[131] J. J. Pedjac and C. T. Pumpelly, *U.S. 3,284,376* (to Dow Chemical Co.), Nov. 8, 1966.

[132] I. S. Bengelsdorf and W. G. Woods, *U.S. 3,250,732* (to United States Borax & Chemical Co.), May 10, 1966.

[133] W. G. Woods, D. Luruccia, and I. S. Bengelsdorf, *U.S. 3,189,565* (to United States Borax & Chemical Co.), June 5, 1965 (see also *U.S. 3,250,797*).

[134] Jefferson Chemical Co., Inc., *Brit. 1,002,272*, Aug. 25, 1965; Allied Chemical Corp. (by E. A. Dickert and M. R. Hallinan), *Fr. 1,375,991*, Oct. 23, 1964.

[135] F. J. Welch and H. J. Paxton, Jr., *U.S. 3,412,125* (to Union Carbide Corp.), Nov. 19, 1968.

[136] H. Piechota, *Kunstst. Rundsch.*, **12**, 191 (1965); H. Piechota, *J. Cell. Plast.*, **1**, 186 (1965).

[137] Z. E. Jolles, *Plast. Inst. Trans. J.*, *Conf. Suppl. No. 2*, 3–8 (1967).

[138] C. J. Hilado, Div. Org. Coatings Plast. Chem., Am. Chem. Soc., Preprints, *28*(1) 265, 317 (1968).

[139] C. K. Lyon and T. H. Applewhite, *J. Cell. Plast.*, *3*, 91 (1967).

[140] J. J. Anderson, *Ind. Eng. Chem.*, *Prod. Res. Develop.*, **2**, 260 (1963).

[140a] D. B. Parrish, R. M. Pruitt, *J. Cell. Plastics*, **5**, 348 (1969).

[141] L. Schwartz, *Paint Varn. Prod.*, **57**, 65 (1967).

[142] S. L. Madorsky, *Thermal Degradation of Organic Polymers*, Wiley-Interscience, New York, 1964, pp. 272–76.

[143] C. W. Roberts, *SPE Trans.*, **3**, 111 (1963).

[144] R. C. Nametz, *Ind. Eng. Chem.*, **59**, 99 (1967).

[145] C. J. Hilado, *Ind. Eng. Chem. Prod. Res. Develop.*, **7**, 81 (1968).

[146] L. Naturman, *Plast. Technol.*, **1968**, 37 (July).

[147] Chemische Werke Albert, *Belg. 715,110*, May 14, 1968 [147a] C. E. Miles, H. L. Vandersall, and J. W. Lyons, Div. Org. Coatings Plast. Chem., Am. Chem. Soc., Preprints, *28*(1) 237 (1968).

[148] H. Zimmermann, *Faserforch. Textiltech.*, **19**, 372 (1968).

[149] K. Hayakawa, K. Kawase, and T. Matsuda, *Kogyo Gijutsu Shikensho Hokoku*, **7**, 538 (1958).

[149a] W. F. Brucksch, Jr., and L. H. Howland, *U.S. 2,583,356* (to United States Rubber Co.), Jan. 22, 1952.

[150] Farb. Hoechst A.-G. (by C. Henck, F. Rochlitz, H. Schmidt, H. Vilczek, and J. Winter), *Ger. 1,117,305*, Appl. May 31, 1958.

[150a] Imperial Chemical Industries, Ltd. (by H. C. Fielding), *Brit. 995,241*, June 16, 1965.

[150b] Walsh and Uhig, reference 31.

[151] R. R. Hindersinn and N. E. Boyer, *Amer. Chem. Soc., Div. Org. Coatings, Plast. Chem., Preprints* **23** (1), 50 (1963); *Ind. Eng. Chem., Prod. Res. Develop,* **3**, 141, (1964); N. E. Boyer, R. R. Hindersinn, and C. T. Bean, Jr., *U.S. 3,278,464* (to Hooker Chemical Corp.), Oct. 11, 1966.

[151a] Bahr, Oertel, Nishk, and Dahm, reference 53.

[152] J. E. Herweh, N. L. Miller, and A. C. Poshkus, *U.S. 3,262,918* (to Armstrong Cork Co.), July 26, 1966.

[153] Chemische Werke Albert, *Belg. 705,280,* Oct. 18, 1967.

[154] A. N. Pudovik, G. I. Evstaf'ev, and R. A. Cherkasov, *Dokl. Akad. Nauk, USSR,* **145,** 344 (1962).

[155] V. I. Kirilovich, I. K. Rubtsova, and E. L. Gefter, *Plast. Massy,* **1963,** 20; Farb. Bayer A.-G. (by G. Nishk, H. Holtschmidt, and I. Ugi), *Ger. 1,190,186,* Apr. 1, 1965.

[156] S. S. Spasskii, V. I. Kodolov, A. I. Kopylov, N. A. Obolonskoya, and A. I. Tarasov, *Plast. Massy,* **1965,** 13.

[157] V. V. Korshak, I. A. Gribova, and M. A. Andreeva, *ibid.,* **1963,** 11; A. Yuldashev and S. Tulyaganov, *Akad. Nauk Uzb. SSR Tashkent Dokl.,* **21,** 38 (1964).

[158] I. K. Rubtsova, E. L. Gefter, A. Yuldashev, and P. A. Moshkin, *Plast. Massy,* **1961,** 22; M. F. Sorokin and I. Manovicin, *Mater. Plast.,* **3,** 136 (1966) (in Romanian).

[159] A. D. F. Toy and L. V. Brown, *U.S. 2,586,884* (to Victor Chemical Works), Feb. 26, 1952; Lensovet Technological Institute (by L. N. Mashlyakovskii, B. I. Ionin, I. S. Okhrimenko, and A. A. Petrov), *USSR 183,385,* June 17, 1966.

[160] Farb. Hoechst A.-G. (by W. Stark, H. Vileseck, J. Winter, and F. Rochlitz), *Ger. 1,128,652,* Apr. 26, 1962; *Ger. 1,110,411; 1,164,093,* July 6, 1961; Feb. 27, 1964; *Brit. 883,754; 893,674,* Dec. 6, 1961; Apr. 11, 1962; J. R. Caldwell and W. J. Jackson, Jr., *U.S. 3,378,523* (to Eastman Kodak Co.), Apr. 16, 1968; R. L. McConnell and H. W. Coover, Jr., *U.S. 2,926,145* (to Eastman Kodak Co.), Feb. 23, 1960; Gelsenberg Benzin A. -G., *Neth. Appl. 6,401,071,* Aug. 17, 1964; polymerization of $\phi P(O)Cl_2$ is discussed by F. Millich, C. E. Carraher, Jr., *J. Polym. Sci. A-1,* **7,** 2669 (1969).

[161] A. D. F. Toy and K. H. Rattenbury, *U.S. 2,918,449* (to Victor Chemical Works), Dec. 22, 1959; H. W. Coover and M. A. McCall, *U.S. 2,682,522* (to Eastman Kodak Co.), June 29, 1954.

[162] I. K. Rubtsova, E. L. Gefter, A. Yuldashev, and P. A. Moshkin, *Plast. Massy,* **1961,** 13.

[163] Albright & Wilson, Ltd., *Fr. 1,455,979,* Oct. 21, 1966; Albright & Wilson (Mfg.) Ltd., (by J. K. Jacques), *Brit. 1,115,043,* May 22, 1968.

[164] R. S. Ludington and W. W. Young, *U.S. 3,009,897* (to Westinghouse Electric Corp.), Nov. 21, 1961.

[165] W. A. Reeves and J. D. Guthrie, *U.S. 2,913,436* (to U.S. Dept. Agr.), Nov. 17, 1959.

[166] P. J. Apice, *U.S. 3,433,854* (to Allied Chemical Corp.), Mar. 18, 1969; Dr. Beck & Co. G.m.b.H., *Brit. 1,031,791,* June 2, 1966; Marles-Kuhlmann-Wyandotte, *Belg. 732,251,* Apr. 28, 1969.

[167] C. W. McGary, Jr., and P. L. Smith, *U.S. 3,419,642* (to Union Carbide Corp.), Dec. 31, 1968.

[168] Montecatini Società Generale per l'Industria Mineraria e Chimica (by I. Dakli and F. Formi), *Belg. 615,925,* Apr. 30, 1962.

[169] Farb. Bayer A.-G. (by U. Bahr, K. H. Andres, and G. Braun), *Ger. 1,098,707*, Feb. 2, 1961.

[170] Farb. Bayer A.-G. (by G. Nishk, U. Bahr, and K. Andres), *Ger. 1,122,252*, Jan. 18, 1962; *Ger. 1,174,075*, July 16, 1964.

[171] J. D. McGovern and G. B. Duhnkrack, *U.S. 2,822,340* (to American Cyanamid Co.). Feb. 4, 1958; L. A. Lundberg, *U.S. 2,819,247* (to American Cyanamid Co.), Jan. 7, 1958.

[172] Chemische Werke Huels A.-G. (by G. Maaks, H. Wieschollek, W. Guinlich, and H. Kraemer), *Ger. 1,188,804*, Mar. 11, 1965.

[173] D. S. Raden, *U.S. 3,406,225* (to Velsicol Chemical Corp.), Oct. 15, 1968.

[174] Hitachi Chemical Industry Co., Ltd. (by W. Koga, T. Watanabe, and H. Kochi), *Jap. 20,234*, Nov. 25, 1966.

[175] Scientific Research Institute of Synthetic Resins (by T. K. Samigulin, I. M. Kafengouz, and A. P. Kafengouz), *USSR 182,330*, May 25, 1966.

[176] W. Cummings, *U.S. 2,824,085* (to United States Rubber Co.), Feb. 18, 1958.

[177] Chemische Werke Witten G.m.b.H., *Brit. 960,556*, June 10, 1964.

[178] N. E. Bayer, R. R. Hindersinn, and C. T. Beary, Jr., *U.S. 3,321,553* (to Hooker Chemical Corp.), May 23, 1967.

[179] Badische Anilin- & Soda-Fabrik A.-G. (by H. Willersinn and H. Distler), *Ger. 1,128,146*, Apr. 19, 1962.

[180] Progil S. A. (by J. Fritz), *Fr. 1,394,776*, Apr. 9, 1963.

[181] J. K. Craver, *U.S. 3,163,627* (to Monsanto Co.), Dec. 29, 1964; A. Yuldashev, I. K. Rubtsova and P. A. Moshkin, *Plast. Massy*, **1962**, 10; Monsanto Chemical Co., *Brit. 866,384*, Apr. 26, 1961; P. Z. Li, Z. V. Mikhailova, L. N. Scdov, E. L. Kaganova, and E. L. Gefter, *Plast. Massy*, **1960**, 9; R. G. Beaman, *U.S. 2,854,434* (to E. I. duPont de Nemours & Co.), Sept. 30, 1958.

[182] P. Z. Li, Z. V. Mikhailova, and L. V. Bykova, *Plast. Massy*, **1964**, 12.

[183] Farb. Bayer A.-G. (by U. Bahr and K. Andres), *Ger. 1,115,015*, Oct. 12, 1961; P. Robitschek and C. T. Bean, *U.S. 2,931,746* (to Hooker Chemical Corp.), Apr. 5, 1960.

[184] E. L. Gefter and A. Yuldashev, *Plast. Massy*, **1962**, 49.

[185] American Cyanamid Co. (by L. A. Lundberg and I. H. Updegraff), *Belg. 621,119*, Feb. 6, 1963.

[186] G. B. Duhnkrack and C. H. Dugliss, *U.S. 2,877,204* (to American Cyanamid Co.), Mar. 10, 1959.

[187] Farb. Hoechst A.-G. (by F. Rochlitz and H. Vilesek), *Ger. 1,165,262*, Mar. 12, 1964.

[188] Farb. Bayer A.-G. (by H. Ruppert, H. Schnell, and L. Goerden), *Ger. 1,074,262*, Jan. 28, 1960.

[189] Albright & Wilson, Ltd., *Fr. Addn. 88,915*, Apr. 14, 1967.

[189a] E. D. Weil, E. Dorfman, J. Linder, *U.S. 3,467,622* (to Hooker Chemical Corp.), Sept. 16, 1969.

[190] Japan Reichhold Chemicals Industries, Inc. (by H. Suzuki, A. Hirayama, H. Maeda, and S. Nomura), *Jap. 3696*, Apr. 21, 1961.

[191] Dynamit Nobel A.-G., *Belg. 671,780*, Nov. 3, 1965.

[192] T. Akita and J. Okazawa (to Riken Synthetic Resin Co.), *Jap. 9588*, Nov. 14, 1957.

[193] Chemische Fabrik Kalk G.m.b.H. (by A. Schors and A. Heslinga), *Ger. 1,203,464*, Oct. 21, 1965; Nederlandse Organisatie voor Toegepast-Natuurwetenschappelijk Onderzoek ten Behoeve van Nijverheid, Handel en Verkeer, *Belg. 619,217*, Oct. 15, 1962.

[194] Chemische Fabrik Kalk G.m.b.H. (by A. Heslinga and P. J. Napjus), *Ger. 1,186,211,* Jan. 28, 1965.
[195] Monsanto Chemical Co. (by R. M. Anderson and G. H. Birum), *Fr. 1,349,088,* Jan. 10, 1964.
[196] Chemische Werke Huels A.-G. and Chemische Fabrik Kalk G.m.b.H. (by K. Heidel, A. Schick, H. Jenkner, and H. E. Praetzel), *Ger. 1,247,015,* Aug. 10, 1967.
[197] Chemische Werke Albert, *Neth. Appl. 6,412,649,* May 3, 1965.
[198] Manufactures de Produits Chimiques du Nord, Etablissements Kuhlmann (by D. James), *Fr. 1,168,662,* Dec. 12, 1958.
[199] H. Schulze, *Plasti Kautschuk,* **11,** 131 (1964).
[200] C. W. Roberts, *SPE Trans.,* **3,** 111 (1963).
[201] F. G. Singleton and P. J. Kennan, *U.S. 3,061,492* (to H. H. Robertson Co.), Oct. 30, 1962.
[202] Eastman Kodak Co. (by J. R. Caldwell and W. J. Jackson, Jr.), *Fr. 1,429,537,* Feb. 25, 1966.
[203] K. M. Bell, *Plast. Inst., Trans. J., Conf. Suppl. No. 2,* 27 (1967).
[204] Solvay & Cie, *Belg. 546,667,* Oct. 1, 1956.
[205] H. C. Vogt, P. Davis, E. J. Fujiwara, and K. C. Frisch, 24th Annual Tech. Conf., 1969, Reinf. Plast./Composites Div., S.P.I., Inc., Section 16-C, 1; A. Blaga and M. J. Skrypa, *U.S. 3,328,485* (to Allied Chemical Corp.), June 27, 1967; Wyandotte Chemicals Corp. (by P. Davis), *Belg. 639,670,* May 8, 1964; G. Elfers and P. Davis, *U.S. 3,274,163* (to Wyandotte Chemicals Corp.), Sept. 20, 1966; Solvay & Cie, *Belg. 545,915,* Sept. 10, 1956.
[206] Société des Usines Chimiques Rhone-Poulenc, *Fr. Addn. 75,304,* Sept. 8, 1961 (Addn. to *Fr. 1,192,272*); Takeda Pharmaceutical Industries, Ltd. (by J. Ogura and K. Matsumoto), *Jap. 7097,* 1960; Chemolimpex Magyar Vegyiaru-Kulkeres Kedelmi Vallalot, *Belg. 704,463,* Sept. 28, 1967.
[207] I. M. Al'shits, G. A. Shtraikhman, R. G. Luchko, and Kh. V. Tsubina, *Zh. Prikl. Khim.,* **34,** 468 (1961); I. M. Al'shits, N. M. Grad, R. G. Luchko, and Kh. V. Tsubina, *Plast. Massy,* **1962,** 12; Badische Anilin- & Soda-Fabrik A.-G. (by H. Willersinn and G. Dietrich), *Ger. 1,074,259,* Jan. 28, 1960.
[208] P. Z. Li, Z. V. Mikhailova, and Yu. S. Makarova, *Plast. Massy,* **1963,** 15.
[209] M. Kororudz and W. K. Langdon, *U.S. 3,387,060* (to Wyandotte Chemicals Corp.), June 4, 1968.
[210] P. Robitschek and C. T. Bean, *Ind. Eng. Chem.,* **46,** 1628 (1954).
[211] Westinghouse Electric Corp. (by J. K. Allen), *Ger. 1,006,154,* Apr. 11, 1957; E. M. Evans (to Distillers Co., Ltd.), *Brit. 782,057,* Aug. 28, 1957.
[212] P. Robitschek and C. T. Bean, Jr., *U.S. 2,783,215,* Feb. 26, 1957; *U.S. 2,863,794; 2,890,144; 2,898,256,* Dec. 9, 1958; June 9, 1959; Aug. 4, 1959 (all to Hooker Chemical Corp.); R. C. Slagel, G. P. Shulman, and F. M. Young, *Ind. Eng. Chem., Prod. Res. Develop.,* **6,** 100 (1967); H. P. Marshall and R. E. Davies, *U.S. 2,880,193* (to Celanese Corp. of America), Mar. 31, 1959; K. W. Rockey, *Intern. Reinforced Conf. 3rd. London,* **1962,** 11.0–11.12; D. Deverell, *Brit. Plast.,* **39,** 650 (1966).
[213] V. D. Valgin, V. N. Demin, and E. B. Petrilenkova, *Plast. Massy,* **1963,** 14.
[214] P. Robitschek and C. T. Bean, Jr., *U.S. 2,779,700; 2,779,701* (to Hooker Electrochemical Corp.), Jan. 29, 1957.
[215] J. C. Wygant and R. M. Anderson, *U.S. 3,270,089* (to Monsanto Co.), Aug. 30, 1966.
[216] Badische Anilin- & Soda-Fabrik A.-G. (by H. Willersinn, G. Dietrich, K. J. Fust, H. Friederich, and H. P. Siebel), *Ger. 1,099,742,* Feb. 16, 1961.
[217] P. E. Hoch, *U.S. 3,154,591* (to Hooker Chemical Corp.), Oct. 27, 1964.

[218] Ruhrchemie A.-G. (by H. Feichtinger), *Ger. 959,229*, Feb. 28, 1957.

[219] Badische Anilin- & Soda-Fabrik A.-G. (H. P. Siebel, H. Willersinn, K. J. Fust, and H. Friederich), *Ger. 1,093,090*, Nov. 17, 1960; C. W. Roberts, D. H. Haigh, and R. J. Rathsack, *J. Polym. Sci.*, **8**, 363 (1964); Dow Chemical Co. (by D. H. Haigh and C. W. Roberts), *Brit. 990,339*, Apr. 28, 1965; Universal Oil Products Co. (by R. M. Lusskin, F. Backer, and J. R. Larson), *Fr. 1,462,179*, Dec. 16, 1966.

[220] Toyo Koatsu Industries, Inc. (by T. Okada, H. Watanabe, N. Tonami, and Y. Miyake), *Jap. 13,697*, Sept. 19, 1960; G. Nishk and E. Müller, *U.S. 2,912,409* (half to Mobay Chemical Co. and half to Farb. Bayer A.-G.), Nov. 10, 1959; Farb. Bayer A.-G. *Brit. 800,215*, Aug. 20, 1958; V. Zvonar, *Sb. Praci Nenasycenych Polyesterovych Prykyricich*, **1960**, 91; Velsicol Chemical Corp. (by J. C. Tapas and I. J. Dissen), *Ger. 1,033,891*, July 10, 1958.

[221] C. S. Shull, E. P, Benzing, R. A. Cass, and R. J. Rohrbacker, *Proc. Annual Tech. Conf.*, *S.P.I.*, *Reinf. Plast. Div.*, *22nd, Washington, D.C. 1967*, Section 6-B, 10 pp.; T. Akita and J. Okazawa, *Reinf. Plast.*, **2**, 43 (1956); Badische Anilin- & Soda-Fabrik A.-G. (by H. Willersinn, H. Wurzler, K. J. Fust, M. Minsinger, and H. Friederich), *Ger. 1,097,674*, Jan. 19, 1961; B.A.S.F. (by I. Ziegler, R. D. Rauschenbach, F. Meyer, R. Allwinn, and K. Wintersbarger), *Belg. 643,139*, July 29, 1964.

[222] Shull, Benzing, Cass, and Rohrbacker, reference 221.

[223] Y. C. Chae, W. M. Ruiehart, C. S. Shull, R. A. Cass, R. J. Rohrbacker, *Proc. Annual Tech Conf. S.P.I. Reinf. Plastics Div.*, *22nd, Washington, D.C., 1967*, Section 6-E, 8 pp.; H. N. Meggos, Y. C. Chae, Annual Tech. Conf. SPI Reinf. Plastics Div., 25th, Washington D.C., 1970, Section 9-A, pl.

[224] R. C. Nametz, J. DiPietro, and I. N. Einhorn, Div. Org. Coatings Plast. Chem., Am. Chem. Soc., Preprints, *28* (1) 204 (1968); Imperial Chemical Industries, Ltd. (by M. K. McReath and I. Goodman), *Brit. 985,611*, Mar. 10, 1965.

[225] Pittsburgh Plate Glass Co., *Brit. 893,341*, Apr. 11, 1962; M. Wismer, *U.S. 3,043,881* (to Pittsburgh Plate Glass Co.), July 10, 1962; Dynamit Nobel A.-G., *Belg. 708,369*, Dec. 21, 1967; Mobay Chemical Co., *Belg. 699,028*, May 25, 1967.

[226] F. B. Cramer, *U.S. 3,388,097* (to E. I. duPont de Nemours & Co.), June 11, 1968; Farb. Bayer A.-G., *Brit. 843,746*, Aug. 10, 1960.

[227] Société d'Electro-Chimie, d'Electro-Metallurgie et des Aciéries Electriques d'Ugine (by L. Szobel, M. Troussier, and J. Vuillement), *Fr. 1,322,878*, Apr. 5, 1953.

[228] Société d'Electro-Chimie, d'Electro-Metallurgie et des Aciéries Electriques d'Ugine (by L. Szobel and L. Parvi), *Fr. Addn. 85,621*, Sept. 17, 1965 (Addn. to *Fr. 1,336,751*); *Fr. 1,336,208*, Aug. 30, 1963.

[229] Badische Anilin- & Soda-Fabrik A.-G. (by H. Willersinn, G. Dietrich, and F. Stolp), *Ger. 1,103,021*, Mar. 23, 1961; L. C. Rubens, C. F. Thompson, and R. M. Nowak, *Kunstst. Rundsch.*, **13**, 405 (1966).

[230] E. M. Barrentine, R. D. Goold, F. Roselli, and V. Toggweiler, 24th Annual Tech. Conf., 1969, Reinf. Plast. Composites Div., S.P.I., Inc., Section 16-B, 8 pp.; P. Herte, R. Kiessig, and H. J. Steinbach, *Ger. (East) 56,098*, May 20, 1967.

[231] W. C. Weaver and E. R. Larsen, Div. Org. Coatings Plast. Chem., Am. Chem. Soc., Preprints, *28*(1) 196 (1968).

[232] Imperial Chemical Industries, Ltd., *Belg. 711,582*, Mar. 1, 1968.

[233] R. M. Anderson and J. C. Wygant, *U.S. 3,243,480* (to Monsanto Co.), Mar. 29, 1966.

[234] Chemische Werke Huels A.-G. (by F. A. Fries and H. Wiesschollek), *Ger. 1,125,174*, Mar. 8, 1962.

[235] Monsanto Co. (by E. J. Prill and J. C. Wygant), *Brit. 1,079,015*, Aug. 9, 1967.

[236] Dow Chemical Co. (by L. F. Sonnabend), *Belg. 628,199*, Aug. 7, 1963.

[237] United States Rubber Co., *Brit. 924,323*, Apr. 24, 1963.

[238] United States Rubber Co. (by M. Borr and K. E. MacPhee), *Ger. 1,149,899*, June 6, 1963.

[239] M. Borr and K. E. MacPhee (to Dominion Rubber Co., Ltd.), *Can. 663,542*, May 21, 1963.

[240] Deutsche Akademie der Wissenschaften zu Berlin, *Fr. 1,294,986*, June 1, 1962; Michigan Chemical Corp., *Brit. 988,304*, Apr. 7, 1965.

[241] R. C. Nametz and R. J. Nulph, *Proc. Annual Tech. Conf., S.P.I. Reinf. Plast. Div., 20th, Chicago, 1965*, Section 11-C, 6 pp.; Michigan Chemical Corp. (by R. C. Nametz and R. J. Nulph), *Belg. 655,793*, Mar. 16, 1965.

[242] Etablissements Kuhlmann (by J. Benoit, M. Hermant, and J. Bryks), *Fr. 1,396,394*, Apr. 23, 1965.

[243] G. G. Roberts and H. Nakagawa, *U.S. 3,284,378*, Nov. 8, 1966.

[244] Farb. Hoechst A.-G. vorm. Meister Lucius & Brüning (by R. Sartorius and F. Schuelde), *Ger. 1,033,409*, July 3, 1958.

[245] Toyo Koatsu Industries, Inc. (by T. Okada, K. Watanabe, N. Tonami, A. Sabashi, and M. Namba) *Jap. 7285*, 1960; I. M. Al'shits, L. A. Gladkaya, N. N. Grad, V. V. Meshcheryakov, and Kh. V. Tsubina, *Plast. Massy*, **1966**, 68.

[246] A. A. Samuel, *Fr. 1,109,057*, Jan. 20, 1956.

[247] V. D. Elarde, *U.S. 2,885,380* (to Western Electric Co., Inc.), May 5, 1959; Licentia Patent-Verwaltungs G.m.b.H. (by P. Nowak), *Ger. 948,192*, Aug. 30, 1956.

[248] P. Robitschek and J. L. Ohnstead, *U.S. 2,909,501* (to Hooker Chemical Corp.), Oct. 20, 1959.

[249] V. Zvonar, *Chem. Prum.*, **12**, 321 (1962).

[249a] R. Zimmermann, W. E. Busch, F. Reiners, *U.S. 3,470,094* (to Chemische Werke Albert), Sept. 30, 1969.

[250] Allied Chemical Corp., *Brit. 842,760*, July 27, 1960; T. C. Baker, *U.S. 2,680,105* (to Allied Chemical & Dye Corp.), June 1, 1954; F. B. Shaw, *U.S. 2,921,867* (to Continental Can Co., Inc.), Jan. 19, 1960.

[251] Chemische Werke Albert, *Belg. 666,847*, Jan. 14, 1966.

[252] Chemische Fabrik Kalk G.m.b.H. (by R. Sckell), *Ger. 1,191,569*, Apr. 22, 1965.

[253] J. A. Rhys, *Reinf. Plast. (London)*, **11**, 357 (1967).

[254] F. E. Bennett, L. Chesner, and R. Preston, *Appl. Plast.*, **10**, 45 (1967).

[255] Michigan Chemical Corp. (by R. C. Nametz and R. J. Nulph), *Belg. 671,425*, Feb. 14, 1966; M. Narkis, M. Grill, G. Leeser, *J. Appl. Polym. Sci.*, **13**, 535 (1969).

[256] Diamond Alkali Co. (by J. E. Dereich), *Ger. 1,026,951*, Mar. 27, 1958; M. M. Gherson, *Peusez Plast.*, **1960** (19697), 55.

[257] B. O. Schoepfle and P. Robitschek, *U.S. 3,031,425* (to Hooker Chemical Corp.), Apr. 24, 1962.

[258] Farb. Bayer A.-G. (by E. Eimers and L. Goerden), *Ger. 1,089,967*, Sept. 29, 1960.

[259] Riken Synthetic Resin Co., Ltd. (by T. Akita and I. Okazawa), *Jap. 4696*, Apr. 25, 1963.

[260] M. Worsley, B. N. Wilson, and B. O. Schoepfle, *U.S. 3,054,760* (to Hooker Chemical Corp.), Sept. 18, 1962; Peter Spence & Sons, Ltd., *Brit. 844,555*, Aug. 17, 1960.

[261] Peter Spence & Sons, Ltd. (by L. Williams and R. Sidlow), *Brit. 837,696*, June 15, 1960.

[262] B. O. Schoepfle, B. S. Marks, and P. Robitschek, *U.S. 2,913,428* (to Hooker Chemical Corp.), Nov. 17, 1959.

[263] Chemische Werke Witten G.m.b.H. (by E. Behnke and H. Wulff), *Ger. 1,109,886*, Appl. Aug. 13, 1958.

[264] Diamond Alkali Co. (by F. F. Roselli), *Fr. 1,388,520*, Feb. 5, 1965.

[265] "Hetron®Advanced Materials for Corrosion Control," *Bull. D-8 and D-800*, Hooker Chemical Corp., No. Tonawanda, N.Y.

[266] United States Rubber Co., *Brit. 870,331*, June 14, 1961.

[267] J. P. Stalego, *U.S. 3,218,279* (to Owens-Corning Fiberglass Corp.), Nov. 16, 1965; G. F. Neucetti and A. Lucchesi, *Ann. Chim. (Rome)*, **52**, 546 (1962).

[268] F. F. Roselli (to Diamond Alkali Co.), *Brit. 1,034,279*, June 29, 1966.

[269] W. J. Connolly and A. M. Thornton, *Mod. Plast.*, **43**, 154 (1965); J. McNally, M. Bawn, and C. Morgan, 24th Annual Tech. Conf., 1969, Reinf. Plast./Composites Div., S.P.I. Inc., Section 16-E, 4 pp.

[270] Albright & Wilson, Ltd. (by H. Coates), *Brit. 862,539*, Mar. 15, 1961.

[271] Dainippon Printing Ink Manufacturing Co., Ltd. (by Y. Iwakura, H. Kakiki, and K. Ichikawa), *Jap. 20,771*, Oct. 30, 1961.

[272] F. Jakob and H. Hoyer, *U.S. 3,412,070* (to Farb. Hoechst A.-G. vorm. Meister Lucius & Brüning), Nov. 19, 1968.

[273] G. C. Tesoro, *Textilveredlung*, **2**, 435 (1967); G. C. Tesoro, S. B. Sello, and J. J. Willard, *Text. Res. J.*, **39**, 180 (1969).

[273a] Farbwerke Hoechst, A.-G., *Belg. 721,643*, Sept. 30, 1968.

[273b] W. Kruse, *Proc. Study Conf. Textile Flam. & Consumer Safety*, Gottlieb-Dutweiler, Inst. Econ. Social Studies, Ruschlikon-Zurich,Switzerland, Jan. 23–24, 1969, p. 137.

[274] P. F. Bruins, ed., *Epoxy Resin Technology*, Wiley-Interscience, New York, 1968; H. Lee and K. Neville, *Handbook of Epoxy Resins*, McGraw-Hill, New York, 1967; R. E. Burge and J. R. Hallstrom, *SPE J.*, **20**, 75 (1964).

[275] J. M. Jordan, F. W. Michelotti, and M. Zief, Div. Org. Coatings Plast. Chem. Am. Chem. Soc., Preprints, **28**(1) 335 (1968).

[275a] E. S. Freeman, A. J. Becker, *J. Polym. Sci. A-1*, **6**, 2829 (1968).

[276] *Tech. Bull. 0667*, Celanese Resins, Div. of Celanese Coatings Co., Louisville, Ky.

[277] Schering A.-G. (by M. Wallis and E. Griebsch), *Ger. 1,154,624*, Sept. 19, 1963.

[278] S. M. Zhivukhin, V. B. Tolstoguzov, and V. V. Kireev, *USSR 176,392*, Nov. 2, 1965; A. F. Nikolaev, E-T. Wang, T. A. Zyryanova, G. A. Balaev, E. V. Lebedva, and K. S. Afanas'eva, *Plast. Massy*, **1966**, 17.

[279] Farb. Hoechst A.-G., *Belg. 719,901*, Aug. 23, 1968.

[280] Chemische Fabrik Kalk G.m.b.H., *Belg. 673,460*, Appl. Nov. 19, 1965.

[281] P. L. Smith and C. W. McGary, Jr., *U.S. 3,236,863* (to Union Carbide Corp.), Feb. 22, 1966.

[282] A. L. Bullock, W. A. Reeves, and J. D. Guthrie, *U.S. 2,916,473* (to U.S. Dept. of Agr.), Dec. 8, 1959.

[283] Albright & Wilson, Ltd., *Brit. 816,069*, July 8, 1959.

[284] *Chem. Eng. News*, **1960**, 36 (Feb. 1).

[285] D. Parrett and E. Leumann, *Chimia*, **16**, 72 (1962).

[286] H. Jenkner and H. Eberhard, *U.S. 3,373,135* (to Chemische Fabrik Kalk G.m.b.H.), Mar. 12, 1968.

[287] David G. Hare and R. J. Stephenson, *U.S. 3,374,187* (to Monsanto Chemicals, Ltd.), Mar. 19, 1968; *Brit. 1,048,840*, Nov. 23, 1966.

[288] R. M. Waters and J. C. Smith, *U.S. 3,372,208* (to Dow Chemical Co.), Mar. 5, 1968.

[289] Farb. Hoechst A.-G. (by O. Manz, F. Rochlitz, and D. Schleede), *Ger. 1,113,826*, Appl. Feb. 13, 1958.

[290] A. Wende and D. Joel, *Plaste Kantschuk*, **13**, 99 (1966).

[291] J. J. Rizzo, *U.S. 3,312,636* (to FMC Corp.), Apr. 4, 1967; P. L. Smith (to Union Carbide Corp.), *Brit. 1,002,185*, Aug. 25, 1965; Richardson Co., *Brit. 980,000*, Jan. 6, 1965; Y. Tanaka, M. Seki, and M. Murata, *Shikizai Kyokaishi*, **40**, 19 (1967).

[292] G. J. Trudel, *U.S. 3,433,809* (to Canadian Industries, Ltd.), Mar. 18, 1969; *Brit. 1,061,616*, Mar. 15, 1967.

[293] V. D. Valgin, E. A. Vasil'eva, V. A. Sergeeva, E. L. Gefter, and A. Yuldadiev, *USSR 168,881*, Feb. 26, 1965.

[294] Union Carbide Corp. (by F. J. Welch and H. J. Paxton, Jr.), *Can. 805,228*, Jan. 28, 1969.

[295] CIBA, Ltd., *Belg. 620,538*, July 23, 1962.

[296] Scientific Research Institute of Plastics (by V. I. Kirilovich, S. M. Shner, I. K. Rubtsova, A. E. Rabkina, and M. A. Tikhonova), *USSR 183,379*, June 17, 1966.

[297] Dynamit Nobel A.-G., *Neth. Appl. 6,602,522*, Aug. 29, 1966.

[298] W. Vogt, D. Janssen, and H. Richtzenhain, *U.S. 3,378,526* (to Dynamit Nobel A.-G.), Apr. 16, 1968; *Neth. Appl. 6,515,176*, May 25, 1966; *Belg. 672,467*, Nov. 17, 1965.

[299] J. Nelson, J. S. Sconce, and P. Robitschek, *U.S. 2,833,681* (to Hooker Electrochemical Co.), May 6, 1958.

[300] CIBA, Ltd., *Belg. 611,277*, June 7, 1962.

[301] M. E. Chiddix and R. W. Wynn, *U.S. 2,951,854*; *2,951,829* (both to General Aniline and Film Corp.), Sept. 6, 1960.

[302] Mitsubishi Chemical Industries, Ltd. (by S. Nakano, D. Nishikawa, and N. Inomata), *Jap. 7063*, Apr. 7, 1965.

[303] C. W. McGary, Jr., and C. T. Patrick, Jr., *U.S. 2,936,292* (to Union Carbide Corp.), May 10, 1960.

[304] P. Robitschek and S. J. Nelson, *Ind. Eng. Chem.*, **48**, 1951 (1956); W. L. Bressler and J. C. Smith, *U.S. 2,834,790* (to Dow Chemical Co.), May 13, 1958; C. S. Ilaido, C. T. Bean, and P. Robitschek, *U.S. 2,992,196* (to Hooker Chemical Corp.), July 11, 1961; W. L. Bressler and J. C. Smith, *U.S. 2,834,496* (to Dow Chemical Co.), June 17, 1958; C. G. Schwarzer, *U.S. 3,355,511* (to Shell Oil Co.), Nov. 28, 1967.

[305] Dow Chemical Co. (by B. J. Bremmer), *Brit. 1,017,270*, Jan. 19, 1966.

[306] W. E. Prescott and W. L. Bressler, *U.S. 2,989,502* (to Dow Chemical Co.), June 20, 1961.

[307] Deutsche Solvay-Werke G.m.b.H. (by G. Faerber), *Ger. 1,022,945*, Feb. 21, 1957; CIBA, Ltd., *Brit. 962,020*, June 24, 1964.

[308] Société d'Electro-Chimie, d'Electro-Metallurgie et des Aciéries Electriques d'Ugine (by J. Vuillemont), *Fr. 1,323,734*, Apr. 12, 1963.

[309] M. Wismer, *U.S. 3,016,362* (to Pittsburgh Plate Glass Co.), Jan. 9, 1962; K. Ting, C. Wang, F. Li, S. Kao, and K. Sheu, *K'o Hsueh T'ung Pao*, **10**, 36 (1962); Deutsche Solvay-Werke G.m.b.H. (by G. Faerber), *Ger. 1,034,356*, July 17, 1958; Pittsburgh Plate Glass Co., *Brit. 897,290*, May 23, 1962; A. R. A. Cass and E. T. Reaville, *Amer. Chem. Soc., Div. Paint, Plast., Printing Ink Chem., Preprints 19*, **2**, 77 (1959).

[310] Politechnika Warszawska (by Z. Brzozowski and S. S. Porejko), *Pol. 48,724*, Dec. 15, 1964, Zjednoczenie Przemyslu Organicznego i Tworzyw Sztucznch "Erg," *Brit. 1,051,335*, Dec. 14, 1966; S. Porejko and Z. Brzozowski, *Polimery*, **8**, 280 (1963); Politechnika Warszawska (by Z. Brzozowski and S. Porejko), *Pol. 47,344*, Nov. 25, 1963.

[311] Farb. Hoechst A.-G., *Belg. 687,317*, Sept. 23, 1966.

[312] B. J. Bremmer, *U.S. 3,294,742* (to Dow Chemical Co.), Dec. 27, 1966; Dow Chemical Co. (by B. J. Bremmer), *Belg. 617,496*, Nov. 12, 1962; Minnesota Mining and Manufacturing Co., *Fr. 1,318,840*, Feb. 22, 1963; Minnesota Mining and Manufacturing Co., *Belg. 614,524*, Aug. 28, 1962; R. C. Nametz, *U.S. 3,268,619* (to Michigan Chemical Corp.), Aug. 23, 1966; R. C. Nametz, *U.S. 3,058,946* (to

Michigan Chemical Corp.), Oct. 16, 1962; Minnesota Mining and Manufacturing Co., *Brit. 1,000,633*, Aug. 11, 1965.

[313] Weyerhaeuser Co. (by J. D. Frichette), *Belg. 629,381*, Oct. 21, 1963.

[314] Peter Spence and Sons, Ltd. (by L. Williams), *Brit. 1,010,204*, Nov. 17, 1965.

[315] Peter Spence and Sons, Ltd. (by R. Sidlow and L. Williams), *Brit. 953,206*, Mar. 25, 1964.

[316] DeBell and Richardson, Inc., *Brit. 967,259*, Aug. 19, 1964; R. H. Snedeker, *U.S. 3,405,199* (to Union Carbide Corp.), Oct. 8, 1968.

[317] See p. 40 of Bruins' book, reference 274; E. J. Morrisey, 24th Annual Tech. Conf., 1969, Reinf. Plast. Composites Div., S.P.I., Inc., Section 16-F, 6 pp.

[318] B. J. Bremmer, *Ind. Eng. Chem., Prod. Res. Develop.* **5**, 340 (1966).

[319] H. W. Mackinney and J. T. Spalik, *SPE Trans.*, **5**, 39 (1965).

[320] R. I. Thrune, *Amer. Chem. Soc., Div. Org. Coatings, Plast. Chem., Preprints 23*, 15 (1963).

[321] General Electric Co. (by R. Alton Skiff), *Ger. 1,089,167*, Sept. 15, 1960; R. A. Skiff, *U.S. 3,073,799* (to General Electric Co.), Jan. 15, 1963.

[322] Chemische Werke Albert, *Belg. 715,110*, Nov. 14, 1968.

[323] L. E. Brown and J. B. Harshman, Jr., *AEC Accession No. 35935, Rept. No. M.L.M.-CF-64-8-1.*

[324] H. H. Chen and A. C. Nixon, *SPE Trans.*, **5**, 90 (1965).

[325] W. G. Woods, W. D. English, and I. S. Bengelsdorf, *U.S. 3,257,347* (to United States Borax and Chemical Corp.), June 21, 1966.

[326] A. Johnson, *J. Text. Inst.*, **39**, 561 (1948).

[327] T. Unishi, *Kogyo Kagaku Zasshi*, **69**, 2343 (1966).

[328] S. Nomura, *Jap. 5500*, July 9, 1956.

[329] T. Hasselstrom, H. Coles, C. Balmer, M. Harrigan, M. Keeler, and R. J. Brown, *J. Text. Res.*, **22**, 742 (1952).

[330] H. E. Goldstein, Div. Org. Coatings Plast. Chem., Am. Chem. Soc., Preprints, *28*(1) 131 (1968).

[331] J. J. Nimes, R. Polansky, and W. F. Herbes, *U.S. 3,308,098* (to American Cyanamid Co.), Mar. 7, 1967; L. J. Moretti and W. N. Nakajima, *U.S. 2,922,726* (to American Cyanamid Co.), Jan. 26, 1960; T. F. Cooke and P. B. Roth, *J. Text. Res.*, **26**, 229 (1956).

[332] M. R. Burnell, *U.S. 2,953,480* (to American Cyanamid Co.), Sept. 20, 1960.

[333] B. C. M. Dorset, *Text. Mfr.*, **87**, 243 (1961).

[334] Institute of Organic Chemistry, Academy of Sciences, USSR (by B. A. Arbuzov, M. A. Sokolovskii, P. M. Zavlin, and G. M. Vinokurova), *Brit. 1,012,009*, Dec. 1, 1965.

[335] J. B. Ballentine and L. W. Crovatt, Jr., *U.S. 3,365,427* (to Monsanto Co.), Jan. 23, 1968.

[336] I.C.I., Ltd., *Belg. 711,582*, Mar. 1, 1968.

[337] R. R. Lunt, Jr., *U.S. 3,440,211* (to E. I. duPont de Nemours & Co.), Apr. 22, 1969.

[338] W. F. Busse, *U.S. 3,418,267* (to E. I. duPont de Nemours & Co.), Dec. 29, 1968.

[339] Farb. Hoechst A.-G. (by E. Hilscher), *Ger. 1,150,044*, June 12, 1963.

[339a] W. F. Busse, *U.S. 3,468,843* (to E. I. duPont de Nemours and Co.), Sept. 23, 1969.

[339b] Monsanto Co., *Belg. 723,394*, Nov. 5, 1968.

[339c] FarBenfabriken Bayer, *Belg. 729,877*, Mar. 14, 1969.

[340] *Mod. Plast.*, pp. 86–90 (August, 1968); Rhone Poulenc, *Belg. 721,205*, Sept. 20, 1968; J. Chambion, in *Man-Made Fibers*, Vol. 2, H. Mark, E. Cernia, and S. M. Atlas, eds., Wiley-Interscience, New York, 1968, pp. 435 ff.

[341] N. Vlasova and A. G. Cherno, *Sov. Plast.*, **19**, 27 (October, 1967).

[342] Teijin, Ltd., *Belg. 716,017*, May 31, 1968; Rhone Poulenc, *Belg. 714,046*, Apr. 23, 1968; *Chem. Eng. News*, **1968**, 40 (Sept. 23); J. H. Ross, *Mod. Text. Mag.*, **1969**, 18 (January); W. B. Black and J. Preston, in *Man-made Fibers*, reference 340, pp. 297–400; A. H. Frazer, "High Temperature Resistant Fibers," *J. Polym. Sci., Part C, Polymer Symposia, No. 19*, Interscience, New York, 1967, 300 pp.; Rhone Poulenc, *Belg. 714,069*, Apr. 23, 1968.

[343] Monsanto Co., *Belg. 718,184*, July 17, 1968.

[344] Rolls-Royce, Ltd., *Belg. 719,961*, Aug. 26, 1968.

[344a] Monsanto Co., *Belg. 721,194*, Sept. 20, 1968.

[344b] S. S. Hirsch, *U.S. 3,448,080* (to Monsanto Co.), June 3, 1969.

[344c] R. M. Ross, Div. Org. Coatings Plast. Chem., Am. Chem. Soc., Preprints, *28*(1) 146 (1968); G. S. Learmonth, D. P. Searle, *J. Appl. Polym. Sci.*, **13**, 437 (1969).

[345] Comp. de St-Gobain, *Belg. 715,792*, May 5, 1968; Comp. de St-Gobain, *Belg. 696,274*, Mar. 3, 1967.

[346] H. K. Nason and M. L. Nielson, *U.S. 2,661,341* (to Monsanto Chemical Co.), Dec. 1, 1953.

[347] Imperial Chemical Industries, Ltd. (by J. M. Pollock and N. H. Ray), *Brit. 920,722*, Mar. 13, 1963.

[348] E. W. Snyder, *U.S. 3,413,380* (to Union Carbide Corp.), Nov. 26, 1968.

[349] A. F. Shepard and B. F. Dannels, *U.S. 3,409,571* (to Hooker Chemical Corp.), Nov. 5, 1968; A. F. Shepard and B. F. Dannels (to Hooker Chemical Corp.), *Can. 801,114*, Dec. 10, 1968.

[350] B. F. Dannels and A. F. Shepard, *Amer. Chem. Soc., Div. Org. Coatings, Plast. Chem., Preprints 27*, **1**, 125 (1967).

[351] Monsanto Co., *Brit. 969,095*, Sept. 9, 1964.

[352] Olin Mathieson Chemical Corp., *Brit. 807,020*, Jan. 7, 1959; G. M. Wagner, *U.S. 2,816,090* (to Olin Mathieson Chemical Corp.), Dec. 10, 1957; G. M. Wagner, *U.S. 2,814,607* (to Olin Mathieson Chemical Corp.), Nov. 26, 1957.

[353] G. Gavlin and W. M. Boyer, *U.S. 3,038,882* (to Richardson Co.), June 12, 1962.

[354] Plastugil, *Belg. 696,720*, Apr. 6, 1967.

[355] Hooker Chemical Corp. (by M. E. Hull, III, and E. W. Simpson), *Fr. 1,364,821*, June 26, 1964.

[356] C. G. Schwarzer, *U.S. 3,369,056* (to Shell Oil Co.), Feb. 13, 1968.

[357] S. Shimizu, *Jap. 13,073*, June 24, 1965; Dynamit Nobel A.-G., *Belg. 616,587*, May 15, 1962; R. W. Quailes and J. A. Baumann, *U.S. 3,298,973* (to Union Carbide Corp.), Jan. 17, 1967; C. Elmer and J. J. Mestdagh, *U.S. 3,352,744* (to Monsanto Co.), Nov. 14, 1967.

[358] Formica International, Ltd. (by D. C. Lowe and A. S. Carello), *Brit. 901,663*, July 25, 1962; Formica International, Ltd., *Fr. 1,354,840*, Mar. 13, 1964.

[359] *Chem. Eng. News*, **1969**, 16 (Apr. 21); *Chem. Week*, **1969**, 75 (Apr. 19).

[360] Allied Chemical Corp. (by P. J. Mason), *Fr. 1,432,889*, Mar. 25, 1966; Allied Chemical Corp., *Brit. 1,021,248*, Mar. 2, 1966; R. L. Wells and G. H. Justice, *U.S. 3,383,338* (to Allied Chemical Corp.), May 14, 1968.

[361] Badische Anilin- & Soda-Fabrik A.-G., *Belg. 699,677*, June 8, 1967.

[362] Omnium de Produits Chimiques pour l'Industrie et l'Agriculture (by G. L. Quesnel), *Fr. 1,145,836*, Oct. 30, 1957; (by G. L. Quesnel and J. J. Franck), *Fr. 1,167,050*, Nov. 19, 1958.

[362a] R. E. Sempert, R. R. Harris, *U.S. 3,440,201* (to the Sherwin Williams Co.), Apr. 22, 1969.

[363] R. A. Pingree and R. C. Ackerman, *U.S. 2,488,034* (to Sun Chemical Corp.), Nov. 15, 1949.

[364] H. P. Wohnsiedler, *U.S. 3,410,750* (to Formica Corp.), Nov. 12, 1968.

[365] T. Ogawa, T. Nishimatsu, and Y. Minoura, *Makromol. Chem.*, **114**, 275 (1968).

[366] Celanese Corp. of America, *Neth. Appl. 6,503,940*, Sept. 28, 1965; Celanese Corp. of America (by J. F. Megee), *Can. 773,159*, Dec. 5, 1967.

[367] E. I. duPont de Nemours & Co. (by C. E. Lorenz), *Ger. 1,181,417*, Nov. 12, 1964; Diamond Alkali Co., *Neth. Appl. 6,408,673*, Feb. 8, 1965; *Neth. Appl. 6,410,272*, Mar. 22, 1965.

[368] Celanese Corp. of America, *Neth. Appl. 6,507,357*, Dec. 13, 1965; M. Slovinsky, *U.S. 3,442,863* (to Celanese Corp.), May 6, 1969.

[369] J. L. Thomas and C. L. Wright, *Proc. Annual Tech. Conf., S.P.I., Reinf. Plast. Div., 1967*, Section 6-D, 8 pp.; FMC Corp., *Belg. 698,639*, May 18, 1967.

[370] Rogers Corp., *Belg. 685,791*, Aug. 22, 1966.

[371] FMC Corp., *Belg. 699,508*, June 5, 1967.

[372] Scientific Research Institute of Plastics, *Fr. 1,402,407*, June 11, 1965.

[373] W. J. Jackson, Jr., and J. R. Caldwell, *U.S. 3,357,942* (to Eastman Kodak Co.), Dec. 12, 1967.

[374] D. B. G. Jaquiss, *U.S. 382,207* (to General Electric Co.), May 7, 1968.

[375] D. I. Mendeleev Chem.-Technol. Inst. (by I. P. Losev, O. V. Smirnova, et al.), *USSR 170,665*, Apr. 23, 1965.

[376] Farb. Bayer A.-G., *Brit. 857,430*, Dec. 29, 1960.

[377] L. Bottenbruch, G. Fritz, and H. Schnell, *U.S. 3,062,731* (to Farb. Bayer A.-G.), Nov. 6, 1962.

[378] General Electric Co. (by J. K. S. Kim), *Fr. 1,386,646*, Jan. 22, 1965.

[379] *Plast. Tech.*, **1968**, 21 (October); *Materials Eng.*, **1968**, 42 (October); *Plast. World*, **1969**, 56 (April).

[380] A. Maschke, *Ger. 286,690*, May 9, 1913.

[381] R. N. Jones, *U.S. 3,419,344* (to Swift & Co.), Dec. 31, 1968.

[382] H. J. Kanth, *U.S. 2,396,575*, Mar. 12, 1946.

[383] J. I. Jones, *J. Macromol. Sci.-Revs. Macromol. Chem.*, C2 (2), 303 (1968); *Chem. Eng. News*, **1969**, 27 (May 19); Plastics Design & Processing, *1969*, 29 (Dec.).

[383a] J. Preston, ed., "High Temperature Resistant Fibers from Organic Polymers," *Appl. Polym. Symposia No. 9*, Interscience Publishers, New York, 1969, 412 pp.

[384] Inst. Elemental-Org. Comp., Acad. Sci. USSR (by V. V. Korshak, S. V. Vinogradova, A. S. Lebedeva, and I. H. Bulgakova) *USSR 183,386*, June 12, 1966; (by V. V. Korshak, S. V. Vinogradova, and S. N. Salazkin), *USSR 183,935*, July 9, 1966.

[385] H. S. Blanchard and H. L. Finkbeiner, *U.S. 3,256,243* (to General Electric Co.). June 14, 1966.

[386] *Chem. Eng.*, **1968**, 120 (Sept. 23).

[387] R. M. Lurie, C. K. Mullen, and S. F. D'Urso, Div. Org. Coatings Plast. Chem., Am. Chem. Soc., Preprints, *28*(1) 73 (1968).

[388] D. N. Vincent, ibid, *28*(1) 38 (1968).

*[348a] E. B. Trostyanskaya, E. S. Venkova, L. V. Aristovskaya, B. M. Kovarskaya, I. Ya. Slonim, *UDC 678.632.01:536.495:543.872*, RAPRA 44C–968. (See Page 417).

AUTHOR INDEX

Only authors of articles are indexed. Patents may be located by reference to the appropriate subject matter via the Subject Index.

SUBJECT INDEX

Specific chemical compounds are only indexed if they are discussed in detail in the text. Compounds in tabulations can be located via the entry "Tables."

451